BIOLOGICAL TECHNIQUES

BIOLOGICAL TECHNIQUES

Collecting, Preserving, and Illustrating Plants and Animals

JENS W. KNUDSEN
Associate Professor of Biology
Pacific Lutheran University

HARPER & ROW, Publishers   New York

BIOLOGICAL TECHNIQUES:

Collecting, Preserving, and Illustrating Plants and Animals

Copyright © 1966 by Jens W. Knudsen

Printed in the United States of America. All rights reserved. No part of this book may be used or reproduced in any manner whatsoever without written permission except in the case of brief quotations embodied in critical articles and reviews. For information address
Harper & Row, Publishers, Incorporated, 49 East 33rd Street, New York, N.Y. 10016.

Library of Congress Catalog Card Number: 66-10839

Contents

Preface   ix

1. General Collecting Sites and Techniques   1
2. The Algae   26
3. The Fungi   38
4. The Mosses and Liverworts   62
5. The Higher Plants (Tracheophyta)   71
6. The Protozoans   97
7. The Sponges   114
8. The Coelenterates and Ctenophorans   119
9. The Lower "Worm" Phyla   130
10. The Higher Worm Phyla   156
11. The Mollusks and Brachiopods   167

12. *The Bryozoans* 193

13. *The Insects* 204

14. *The Crustaceans* 266

15. *Other Arthropods: Spiders, Scorpions and Their Allies* 286

16. *The Echinoderms: Sea Stars, Sea Cucumbers, and Others* 299

17. *The Lower Chordates* 311

18. *The Fishes* 318

19. *The Amphibians* 346

20. *The Reptiles* 359

21. *The Birds* 373

22. *The Mammals* 402

23. *Vertebrate Skeletal Techniques* 432

24. *Scientific Illustration* 446

Appendix A. *Some Display Methods* 493

Appendix B. *Slide-Making* 498

Appendix C. *Reagents and Solutions* 504

Appendix D. *Narcotizing Agents* 510

Index 515

Many individuals have influenced me in the field
of biology, and thus this book is dedicated to . . .

Three outstanding teachers who showed me the worth
of their profession, not only in their instruction,
but in their constant concern for the personal well-being
of their students: Mrs. Irene Creso, Dr. John S. Garth,
and in tribute to the late Norman T. Mattox.

That handful of exceptional students whose interest and ability
have made teaching an enjoyable challenge:
Dave Wake, Loyde and Bert, Sandy and Dixie, Mugs and Ron,
Dave and Jack.

My constant field companions
and two of the "hottest tiger beetle men in the business,"
my wife Winona and little Jim.

Preface

This book has been written as a textbook for courses in biological techniques or museum techniques, and to meet the needs of those who are involved in working with natural history material. Descriptions of techniques for field study, collection, preservation, illustration, and the like, are usually unavailable to the student and often seem to be wanting altogether. All too frequently neither teachers nor students can save or work with specimens that are available to them, because they lack information on preservation, for example. The truth is that a lot of literature on biological techniques is available, but it is scattered through various scientific journals back over a span of 50 to 100 years and locating it often requires a lot of time spent in a well-stocked university library. Therefore, laboratory studies, research, and degree programs may suffer from the lack of up-to-date information in this field.

Biological techniques are usually quite simple but may require specialized equipment. This book not only describes the techniques in detail, but also gives instructions and diagrams for the construction of the equipment needed. Thus amateur naturalists or high school teachers and their students who lack specialized equipment will find this work extremely useful. The techniques presented, however, have not been modified for use by the amateur but are those used by the professional biologist working with his particular specialty.

This textbook also provides general consideration of major habitats and methods for studying and collecting specimens in each. Methods of field preservation, the taking of field notes and color notes, methods of photography, and other information presented can be applied to all natural groups of organisms.

Each major group of plants and animals is discussed separately and in detail from the point of view of how to locate, collect, preserve, and study its species. This format avoids numerous repetitious chapters and emphasizes the proper preservation of specimens. Consequently, however, some phyla are regrouped (thus losing a strict taxonomic order). The worm phyla are divided into two chapters; this arrangement is based on the great similarity of biological techniques for these groups, and the fact that many collectors cannot separate the minor phyla. Also, several invertebrate phyla are taxonomically rearranged.

Since this book is by nature a source book of vast quantities of diverse information, it seemed necessary, at the time of writing, to make each section complete in itself. Thus the sections are cross-referenced, and they may include material that is repeated under other topics the reader would not normally consult. In every case, the reader should use the index to locate items that he needs to complete his particular study.

Certain techniques are credited and cited, whereas others are not. Many techniques recorded in the literature, for example, the pinning of insects, are so well known generally and have been handed down from person to person for so long that their origin is either obscure or universally recognized; thus special reference would seem superfluous. However, those techniques which are new or little known are cited. In addition, a general bibliography is provided for each major group of plants or animals to aid the student in pursuing his particular interest.

A textbook preface is, to some extent, a place where the author can step out of his formal role and explain why he felt compelled to write what he did. In an age when the vast acres of wilderness are falling steadily before the bulldozers of economic expansion, one wonders if it wouldn't be wiser to say nothing of collecting scientific specimens in the interest of prolonging what now exists. Indeed, it would be a nightmare to stimulate the needless collection of specimens. However, aside from recognition of the very great need for information concerning biological techniques, part of my objective in writing this text is in the interest of conservation and the elimination of the fantastic waste of plant and animal material that apparently cannot be kept alive. For by using what is available to us, by collecting wisely, and by preserving specimens so that they will last for years of repeated study, there will ultimately be conservation not now realized. Consider all of the times when students bring specimens to class, when road kills are observed, when hunters abandon their kill, when birds destroy themselves against our windows, or when wilderness areas are cleared for new construction. By possessing easy-to-use instructions, teachers and students alike may conserve what is already available and enhance their teaching and educational experience as well as their own personal enjoyment.

All of the out-of-doors is our great heritage and should be that of the generations to come. Its study, use, and protection become our privilege, while our personal discovery in the field becomes one of our supreme sources of enjoyment. Since we must use our privilege intelligently and ever be mindful of the delicate balance of nature, this book has been written for the amateur, the student, and the professional biologist alike, with the hope that the instructions set down will permit greater enjoyment and conservation of our natural resources.

JENS W. KNUDSEN

Tacoma, Washington

BIOLOGICAL TECHNIQUES

CHAPTER 1

General Collecting Sites and Techniques

Hopefully, one of the most provocative things that can be said in a book that deals primarily with collecting and preservation is "don't collect"—unless there is a need which justifies taking natural history specimens, such as research and classroom or museum work. Obviously, some forms of collecting will not disturb the balance of nature. Insects, pelagic organisms, some fishes and plant forms may well tolerate considerable collecting. However, always be aware of the strength of any population you sample, and be sure that you collect very sparingly where specimens are few and collect not at all if things are rare. There are exceptions to these dicta, of course, governed by need and the use to which the specimens are put. Shooting a rare bird to show to fifth-grade students or even college freshmen could not be justified; whereas, with reason and careful judgment, it may be justified for taxonomic work on a high level.

Bear in mind also that collecting without any facility for preserving and storing specimens is almost sure to be wasteful. Be equipped to handle specimens before you collect. Be aware of the destruction that often occurs in collecting. Marine collectors may destroy the habitat of thousands of species and hundreds of thousands of individual plants and animals while collecting only three or four "pretty shells." They do this simply by turning rocks over in their search without returning them to their original position.

The success of class field trips, on the other hand, must be gauged by what the student learns rather than by the number of specimens captured and killed. Students may actually learn a great deal more—say, in the study of salamanders—by the work involved in hunting, locating, capturing, and observing specimens—only to release them again. Therefore, collect, but always with an eye to conservation and to need.

Plants and animals do not occur at random but are precisely located in their microhabitats. They live within the limits of their physical tolerance—that is to say, they live only in that microhabitat to which they are adapted. Although this factor seems obvious, the physical and biological factors which govern distribution do not at first appear to be obvious. The collector must (1) be cognizant of those physical factors which make up the habitat in which he will collect; (2) know how those physical factors affect plants and animals and thus serve as "permitting" or "limiting" factors in distribution; and (3) also understand how these physical factors inhibit the metabolic or reproductive activity of those organisms he wishes to collect before he can predict where to collect. When the collector begins to recognize the physical factor or factors that dominate a given habitat and couples this with his knowledge of the habitat requirements of plants and animals, he will become successful in his endeavor.

Although each group of plants and animals may require techniques and collecting equipment peculiar to it (and these will be discussed separately), it is the intent of this chapter to explore collecting sites, techniques, and equipment that will be generally applicable to all the groups described in this book. In general the three major "ecozones" or spheres where biological specimens are collected are (1) the marine or oceanic zone, (2) the terrestrial zone, and (3) the aquatic or fresh-water zone. These zones tend to merge and cannot be cleanly separated from one another, yet they will be treated individually for practical purposes.

THE MARINE ENVIRONMENT

Factors Effecting Zonation

The following physical factors are often very complex and subtle in the ways they influence plant and animal distribution. The discussion here, however, will be directed to show the most obvious kind of influence exerted by each factor, and to point out what this should mean to the collector. Even a thorough knowledge of oceanography is no substitute for years of collecting and observation in the marine zone.

Substrate. Every type of marine substrate (mud, sand, gravel, rock rubble, boulders, etc.) will support some kind of plant or animal life. Plants and animals require of their substrate a place to burrow or attach or hide; they often require a supply of food, protection from other physical or biological factors, and a place where reproduction can be completed. For example, although sand is suitable for many burrowing organisms it will not suffice for all. In some habitats wave action will constantly disturb the sand and render it unsuitable for burrowing, while hydrogen sulfide will make life impossible in habitats where waves are absent. Animals which are sessile and must attach to some solid substrate are totally inhibited by sandy substrates. By following this line of reasoning you can quickly determine how each type of substrate is favorable for some organisms but totally unsuitable for others.

Waves. The presence or absence of wave action has a significant influence on plant and animal distribution. From the positive point of view waves oxygenate sea water, prevent stagnation, create local currents, and may transport food to sessile organisms. From the negative point of view, however, the friction or tremendous impact created by translating waves inhibits all organisms except those which are constructed to withstand such pressure.

Currents. Currents are extremely important in the marine habitat. They have the role of transporting food and cool, oxygenated water into the habitat while removing waste, decaying material, and excreta from it. Currents are responsible for the transport of larvae out of and back into the shoreline habitat. Local currents may be produced by waves, the changing of the tide, local winds, or offshore currents. Regardless of the origin, the presence of currents serves as a permitting or limiting factor in distribution.

Tides. The periodic lowering of the water level in the intertidal area has the effect of exposing organisms to the air, to high or low temperatures, to direct sunlight, and to direct wave attack. The vertical difference between the extreme high and low tides from one area to the next is probably not a significant factor. For whether the tidal exchange is only 3 feet or 33 feet the duration of ebb and flow is about the same. Therefore, the number of minutes or hours of exposure during any low tide is the significant factor, rather than the height of the tide itself. In some areas there is a very marked zonation of plants and animals in the intertidal zone; those species which can tolerate exposure the best are most broadly distributed, whereas those with the least tolerance are found at the lower levels of the intertidal zone.

Temperature. The metabolic rate of all marine invertebrates may be influenced by temperature. Therefore, an organism can only exist within the limits of its critical high and low levels of tolerance for life and for reproduction. Usually the temperature of an intertidal area is similar to that of the offshore water. However, in areas which are exposed at low tide or in which the water is seldom exchanged with offshore water, local temperature variation will be significant. Many plants and animals are greatly or narrowly tolerant of temperature fluctuation and will be found only where conditions are tolerable.

Light. Although all photosynthetic plants require light, they are zoned according to the quality and quantity of light available. Various colors (wave lengths) become extinct as light penetrates the ocean. Thus, plants which are very tolerant in their reaction to light quality and quantity may be broadly distributed, whereas other species are found only where the light is suitable. Likewise, light has a profound effect on animal distribution and on behavioral patterns. Furthermore, the changing length of daylight frequently serves as a trigger mechanism to initiate plant or animal reproduction or growth cycles. Therefore, a knowledge of the effect of light on any given organism is usually essential to locating that species within its broad habitat.

Other Factors. The concentration in sea water of dissolved oxygen, dissolved carbon dioxide, nutrients, and the like may all be critical for plant or animal distribution. Likewise, the presence or absence of food organisms, predators, or parasites may also serve as limiting or permitting factors in distribution. The influence of such conditions is difficult to detect and evaluate, though it may be profound.

Intertidal Zones and Their Populations

Unprotected Rocky Coast. Figure 1–1 shows an aerial diagram of an unprotected rocky headland projecting into the ocean. The term "unprotected" refers to wave action and is applicable to areas that receive almost the full force of wave activity. So long as a wave does not interact with its substrate, we may assume that there is an equal amount of energy in any given part of the wave crest. Note how simple wave refraction (bending)

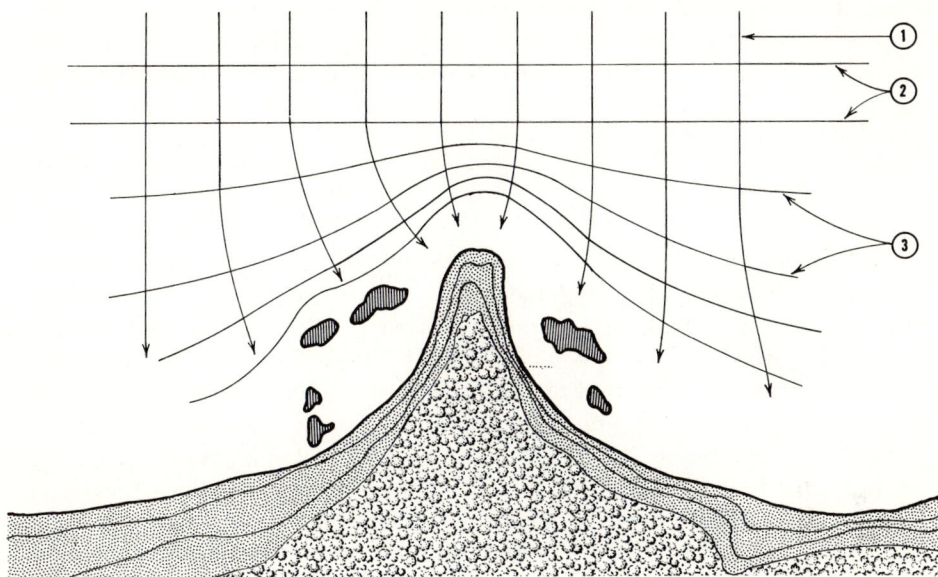

FIG. 1–1. An aerial diagram of a rocky headland projecting out into the ocean, showing how refracting waves distribute more energy on the headland than in the adjacent areas. (1) Lines of equal energy which arbitrarily divide wave fronts into energy units; (2) straight wave fronts moving toward the shore; (3) refracting or bending wave fronts which are reacting with the relatively shallow bottom extending out from the rocky headland.

causes more energy to be delivered on the rocky headland than in the surrounding areas. Therefore, the wave attack is always most severe on the unprotected rocky coast zone, as contrasted to adjacent protected areas. Primary plants and animals living in the unprotected zone must have some means of protection, such as a tough skeleton, in order to withstand the wave impact. Chief among the population in this zone are the starfishes, mussels, barnacles, and certain species of algae. Secondary organisms may hide behind or under such primary species. Interestingly enough, many organisms are restricted to the unprotected zone, since they require the severe wave action either for oxygen or for food. In general, the number of species of plants and animals is noticeably less here than in the protected rocky coast zone.

Protected Rocky Coast. Rocky habitats of different degrees of protection are immediately adjacent to the unprotected areas. Frequently the two sides of a large rock will harbor unprotected- and protected-zone species. As one moves from the outermost, unprotected rocks he will encounter greater and greater degrees of protection and, hence, a changing species composition. The protected rocky coast zone is by far the most heavily populated by both plant and animal species. Sessile and motile species are abundant here and will be subzoned according to the effect of waves, tide, availability of light, and so on.

Exposed Sand. Sand which is more or less exposed to wave action will be found between rocky headlands and may extend along a coastline for hundreds of miles. Sand is constantly shifted by wave action and will not permit the attachment of plants or animals. Therefore, only those species which can live upon the sand or burrow into it are likely to be found in this zone. The presence or absence of offshore rocks or reefs and the slope of the bottom will determine the severity of the wave attack. Many species of mollusks are found only where the waves are most severe, since they have a high oxygen requirement. Along sandy beaches one should not neglect the drift algae that are pushed high up on the beach by wave and storm action. Here many organisms from the offshore waters will be found, as well as a host of local species that make their living by feeding on drift kelp.

Bays and Inlets. Perhaps bays, estuaries, and inlets are the most variable in physical characteristics. Every type of substrate may be found in such a habitat, as well as extremes in range of tide, temperature, salinity, oxygen, pollutants, and so on. In extremely large bodies of water such as Puget Sound wave action may be of a purely local nature. Tidal currents become very important in flushing and providing fresh oxygen and food. The plant and animal populations will change noticeably as one moves away from the open ocean toward the head of the bay, and as the substrate changes from one type to another. Usually the permanent dwellers in this habitat can withstand extreme variation in temperature, salinity, and other physical factors.

Other Zones or Habitats. Unnatural habitats which are very productive for collecting are pilings or floating docks, signal buoys, and the like. Since floating docks are never exposed by the low tide the populations there may be noticeably different from those on pilings which are always exposed in part by the tide. Such habitats will provide a number of plants and animals that are hard to collect in other situations (e.g., sponges, hydroids, and tunicates).

COLLECTING METHODS AND EQUIPMENT

Exposed Intertidal Zones

Equipment. The kinds of equipment and personal clothing used on a collecting trip will depend on the zone, the local weather conditions, and the type of plant or animal material sought. Hip boots are excellent for intertidal collecting, in that they permit you not only to wade into deep water, but to kneel down in shallow water without getting wet. If hip boots are not worn, one should plan on getting thoroughly wet and wear tennis shoes which will afford the best footing. On rocky substrates one should have his legs protected and may also benefit from wearing cotton gloves. A little sidepack or knapsack is handy to hold such things as field notebooks, pencils, and plastic bags.

Use plastic buckets or large plastic bags for collecting plants or animals. Metallic buckets are not as suitable in that the metal may kill delicate specimens. Small vials with stoppers or small plastic bags and rubber bands are handy for isolating small delicate species. A pocketknife is useful for prying sessile plants and animals from rocks, but is dangerous and should be used with care on slippery substrates. Remember to oil the pocketknife well before and after collecting. A geology pick is handy for turning rocks and chipping off samples. One should keep his collecting equipment to a minimum and have it well organized so that his hands are as free as possible for collecting.

Low-tide periods frequently occur at night, thus necessitating the use of lights. No portable electric light can surpass the brilliance of the pressure gasoline lantern. The lantern will burn brilliantly for hours and is preferred by the writer, even though it may be damaged by splashing or dropping. Miner's headlights (flashlights) are preferred by many collectors for intertidal collecting, in that both hands are left free. However, flashlights that appear brilliant at home will seem very dim in the intertidal zone, because light is so quickly absorbed by plant material. A supply of batteries is needed: eight to twelve flashlight batteries are required on a single evening's collecting. A spare "emergency" flashlight should always be carried at night when collecting alone, for if you break your light and become stranded in the intertidal zone you will be in extreme danger. The writer, having been stranded on both rocky and sandy beaches at night in heavy fogs without a

8 Biological Techniques

second light, finally adopted the practice of placing a gasoline lantern on shore above the high-tide mark as a guide to work toward in case of emergency.

Collecting Procedure. Consult tide tables to select the time for collecting. When collecting on rocky substrates, walk slowly and carefully and crouch low in order to avoid slipping and falling. Attempt to walk on rough rocks or barnacles and avoid smooth rock surfaces, as these may be covered with slippery species of algae. When collecting around breaking waves, learn to count the waves so as to know when large waves will approach and to hear these waves before they reach you. Always collect with a companion or two for safety's sake. When working out on rocky reefs away from the high-tide line, be very conscious of the amount of time that you have been collecting and do not get trapped by the incoming tide. In the Northwest many beach collectors have been trapped against rocky cliffs or out on rocky reefs and have been drowned by the incoming tide.

While collecting in rocky areas observe plants and animals on all rock surfaces and collect only those which are needed. Be careful to turn rocks back to their original position for conservation's sake. Some species of non-mucous-secreting seaweed and a little seawater should be placed in the bucket for receiving animal specimens. Filling the bucket with water is unnecessary, burdensome, and often damaging to the specimen. From time to time flood the bucket with water and then drain the water off again. Be aware of the fact that some species of animals will die and disintegrate very rapidly (especially sea cucumbers and the like), or they may expel sperm and eggs (starfishes), or in one way or another pollute the water and kill other animal specimens in the same container. Some species of plants and animals secrete slime or mucus that may damage other delicate forms. The use of small plastic bags to isolate these species may overcome this problem.

For collecting on sand, mud, or gravel substrates similar techniques are used, although one frequently has to dig and screen large quantities of the substrate in order to locate animal specimens. The need for care not to overcollect and to leave the habitat as undisturbed as possible cannot be overemphasized.

Subintertidal Collecting

The methods used in subintertidal collecting will depend entirely upon the nature of the substrate and the kind of organisms sought. In unprotected and protected rocky coast zones skin diving, with or without compressed air equipment, proves to be the best method. In colder or deeper waters insulated diving suits and oxygen equipment are necessary, whereas a snorkle tube, faceplate, and pair of swim fins are enough for warm, shallow waters. In collecting from a boat in intermediate depths over sand or mud bottoms, various devices such as the dredge, orange-peel bucket, and dippers may be used.

Dredges and Dredging. A dredge consists essentially of a strong net attached to a heavy frame which is pulled along the substrate in order to obtain plants and animals. The size of the dredge is determined mostly on the basis of how it will be pulled across the bottom, rather than on the organisms to be obtained. In both fresh-water and marine collecting hand dredging with small gear is very effective; large dredges measuring 8 to 10 feet across may be used only by large fishing vessels. It must be remembered that dredges frequently become caught on rocks and other obstructions and must be somewhat "snag-proof"—or else, expendable. In addition, the net may become torn on rocks or other obstructions and should be protected by a stout outer net made of screen wire, rope, or chain, depending on the substrate and size of the dredge.

In Fig. 1–2, A, B, and C show a small marine dredge with a mouth opening of 8 by 24 inches. The hoop is made of $\frac{1}{4}$- by 3-inch bar metal but may be made of $\frac{3}{4}$-inch pipe or any other material. Holes are drilled in the bar metal so that the net can be wired in place. The net may be made of hardware cloth or fish netting. A tow rope is attached to the dredge by means of two arms extending ahead of the hoop and constructed of $\frac{3}{8}$-inch rod iron. The distal end of the arm is bent into a ring; the proximal end is triangular and passes through a piece of $\frac{1}{2}$-inch pipe welded to the frame of the dredge. The pipe (Fig. 1–2A) serves as a hinge for the arm. Note that the tow rope does not attach to both arms, but rather to one arm alone (Fig. 1–2B). The second arm is tied to the first by means of medium-weight twine which will break and free the dredge should it become hung up on the bottom, as illustrated in Fig. 1–2C. This dredge may be modified by using a rope sling in place of the arms. However, a twine safety device is still important.

There are several ways in which a collector with little equipment may use the hand dredge. The first method is quite simple. Attach one end of a $\frac{1}{2}$-inch rope, 500 feet long, to some object on shore and row out until all of the rope is payed out. Attach the dredge to the rope and drop this to the bottom. Return to shore, pull the dredge in hand over hand, and examine its contents on the beach.

Dredging can be done in deeper water without returning to the beach, by pulling the dredge from a skiff equipped with an 18-horsepower outboard motor. However, the dredge must be pulled in by means of a winch or by

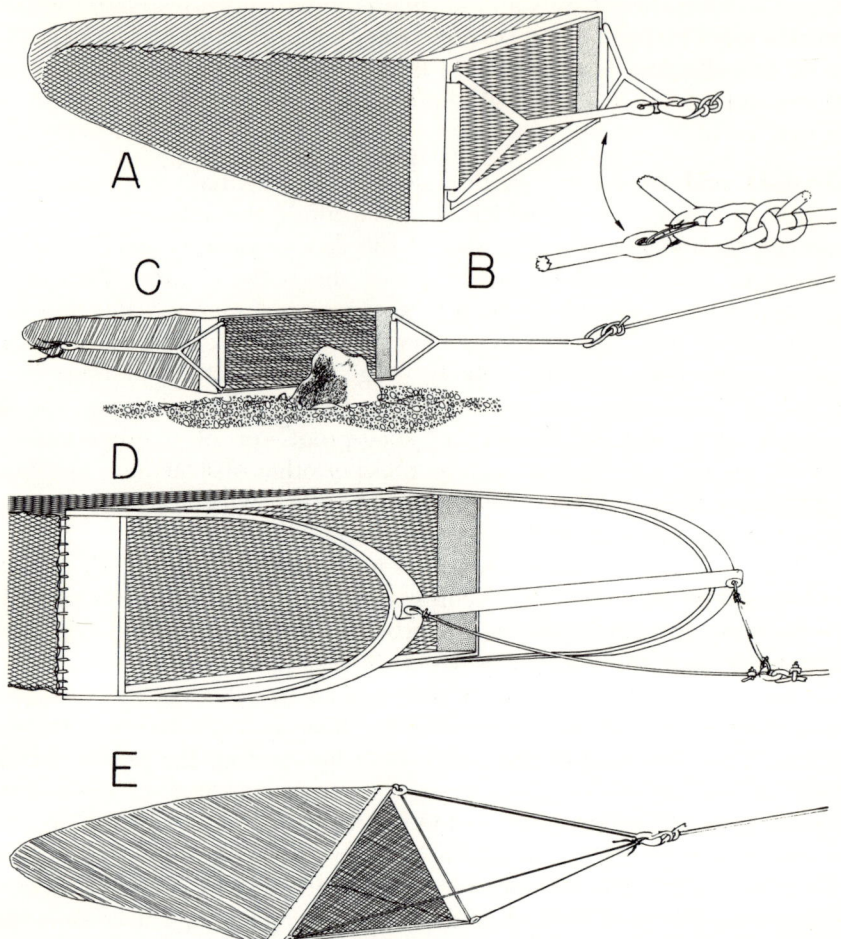

FIG. 1–2. Dredges. A. Biological dredge. B. The detail of securing the two dredge arms by means of string. C. The dredge releases itself from the substratum. D. The frame of a large dredge with a skid. E. A lightweight dredge.

hand. When dredging in open water, one must be prepared to stop quickly if the dredge becomes hung up on the bottom. Some collectors attach a large float to the dredge rope for emergencies. This is thrown overboard when the dredge becomes hung up and thus lessens the danger of the dredge rope's breaking.

The best dredging method, however, is to attach a large pulley to some object on shore before the dredging operation begins. When the dredge rope is pulled in to shore with the outboard motorboat, a companion on shore quickly passes the end of the rope through the pulley and gives it back to the operator of the boat. The boat then pulls the rope seaward until the signal is given that the dredge has come up on the beach.

Figure 1–2D shows the framework of a large dredge which is provided with a sled and a chain attachment for the tow cable. This kind of dredge will readily foul on rocks or other obstructions, but is very useful on sandy or muddy bottoms.

Figure 1–2E shows a light-weight dredge designed to be used in freshwater ponds or lakes or in marine collecting. The frame is made of $\frac{1}{4}$- by 3-inch bar metal built in a triangular pattern. The size of the triangle may be 18 inches or larger on each side. A stout ring is welded to each corner of the triangle. A single 3-foot metal arm made of $\frac{1}{4}$-inch rod and provided with a ring at either end is attached to one corner of the dredge. Heavy twine or light rope is then attached to the other two corners and tied into the ring as illustrated. Should this dredge become fouled on the bottom the string will break and will usually release the dredge. Quarter-inch minnow netting, some other netting, or hardware cloth may be used for the bag.

Bottom Samplers. Thousands of species of minute invertebrates living on or in the surface muds can be obtained only by collecting and processing bottom samples. When a population census is being made any sampler that will bring up part of the substrate will be satisfactory, whereas more elaborate gear is needed for quantitative work. Among the few pieces of satisfactory equipment for quantitative work are the orange-peel bucket and the Petersen grab. The orange-peel bucket (Fig. 1–3, A and B) is a modified tool used in construction work. It has four jaws that plunge into the substrate, picking up a particular quantity of mud or sand. The area covered by the bucket is known and the volume of mud can easily be determined; hence the number and kind of organisms per unit of substrate can be determined. Dwarf orange-peel buckets are available commercially (Hytech Corp., 6803 West Boulevard, Inglewood 3, California, and others). These weigh 35 pounds, have a capacity of 100 cubic inches, cover an area of $11\frac{1}{2}$ inches in diameter, and are very satisfactory when used from a skiff.

Figure 1–3C shows a scoop sampler designed for the U.S. Public Health Service which is made from any convenient container with a U-shaped metal arm and a ring for attachment. This is pulled along the bottom until a sample is obtained. One of the drawbacks of this device is that the surface area covered is unknown and some of the organisms may become lost as a sample is pulled to the surface.

A shallow-water sampler designed by the writer is shown in Figure 1–3D. This consists of a tin can bolted to a short stick. This, in turn, is hinged to

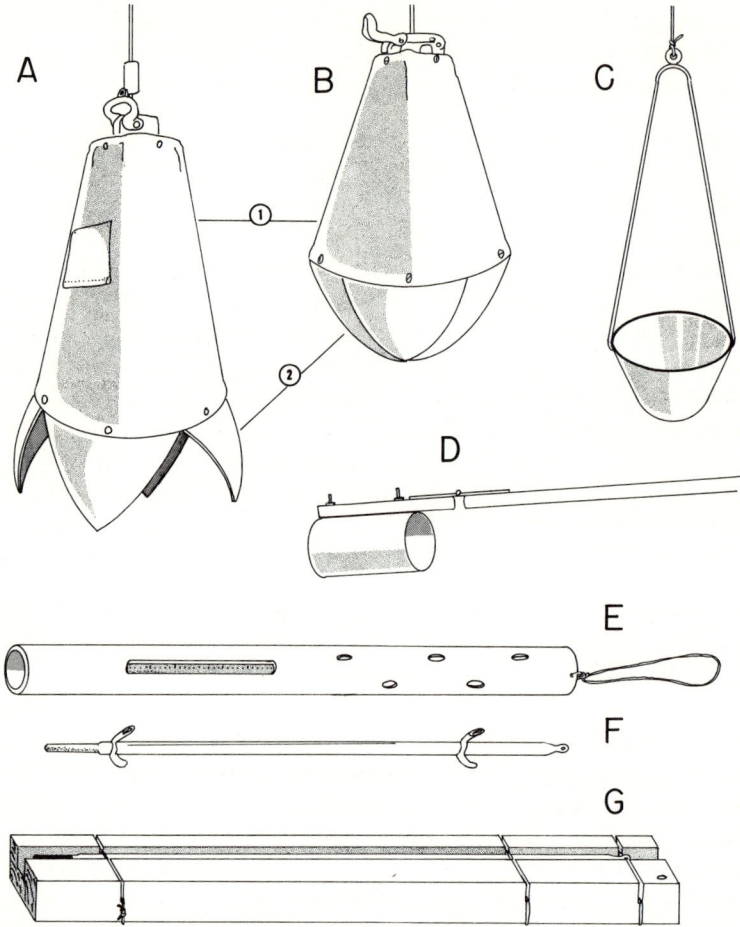

FIG. 1–3. Other collecting equipment. A–B. The orange-peel bucket opened and closed. C. A dipper for bottom sampling. D. A tin can bottom sampler. E. A tubular thermometer holder. F. The thermometer supported by two pieces of rubber tubing. G. A wooden thermometer holder. (1) The canvas cover which prevents the loss of the sample, (2) the jaws.

a long stout handle which can be maneuvered from shore or from a skiff. The hinge permits the scoop to be pulled along the bottom from any angle and thus makes the sampler a versatile collecting device.

Preserve a small quantity of surface mud in 5-percent formalin and another small sample in 75-percent alcohol so that microscopic plants and animals can be removed in the laboratory. These samples should be labeled as to locality and depth of water in which they were taken. Wash the remainder of the sample on a series of screens so that the plants and animals present can be picked out and preserved in appropriate solutions.

Thermometers. In any kind of aquatic or marine collecting thermometers will be useful when complete data are being taken. Conventional laboratory thermometers reading in $2/10$ of 1° C. are quite suitable. Small stick thermometers already enclosed in a metal case are available. Figure 1–3, E and F, show a protective case made out of $1/2$-inch metal or plastic tubing which has a slot cut in it to permit the reading of the thermometer. The thermometer is held by two short pieces of rubber surgical tubing (Fig. 1–3F) and is tied in place by means of a string (Fig. 1–3E). Another kind of thermometer holder can be made by ripping a groove in a small board which is slightly longer than the thermometer (Fig. 1–3G). Make three shallow saw cuts across the board. Tie the thermometer firmly in place, passing the string through two of these cross cuts. Then pass a string through the eye of the thermometer and tie it in the third cut. The groove receiving the thermometer must be deep enough that the thermometer will lie below the surface of the board.

Pelagic Collecting

The pelagic zone of the ocean consists of all of the water mass above the substrate and thus includes the entire surface of the ocean. Organisms living there may be planktonic—that is, organisms which simply drift with the water—or nectonic, that is, organisms which can swim sufficiently fast to be independent of ocean currents. In the latter category fishes and squid are the chief forms while thousands of species of both adult and larval plants and animals are found in the plankton. Large tow nets or night lights are used for collecting necton, and plankton nets are used for the smaller forms.

Plankton Nets. Plankton nets are usually $9\frac{1}{2}$ inches in diameter and 35 inches long, tapering to a point. The meshes vary from very coarse to very fine, ranging from 20, 40, 75, 125, 175, to 200 meshes per inch. Silk is the standard material used in making plankton nets, and muslin is used to reinforce the ring. The tapered tip of the net is either tied shut or tied around a small vial which collects the plankton. Materials for making plankton nets are available, but handmade nets are less suitable than those purchased from biological supply houses. Because of the high cost of such nets (fifteen to fifty dollars) they should be carefully washed in fresh water and air dried to prevent rotting or other damage.

Plankton nets are either towed through the water on the end of a heavy cord or thrown out from a skiff and then retrieved. They can be made to "fish" at various depths by adding weights just ahead of the net and by carefully controlling the length and angle of the line. Another method of using the weighted plankton net is to drop it straight into the water and retrieve it repeatedly until a sample is obtained.

Night Lighting. One rather exciting form of collecting is "night lighting," which consists of hanging a light near the surface of the water from the side of a ship or the edge of a pier. This technique works best where no other lights will distract the marine organisms. After a brief period of time plankton will slowly work its way up toward the light and can be removed with a plankton net or dip net. Small and large fishes, squid, and even swimming crabs will appear under the night light. The composition of the species will change every few hours during the night so that one can profitably collect all night long.

Bulk Field Preservation

Dredges, deep nets, plankton nets, and intertidal collectors all gather large quantities of material indiscriminately. Not all of this material can be preserved in the same way, nor is it desirable to attempt to preserve all of the specimens. An outline of some of the choices that must be made in the field with regard to the treatment of specimens is as follows:

A. Specimens to be kept alive for transportation to the laboratory for subsequent culturing or preservation must be isolated.
B. Plants and animals to be preserved in the field must be sorted as follows:
   1. Animals needing narcotization (those which contract or autotomize) must be sorted for the several probable techniques to be used and subsequently preserved by the appropriate method.
   2. Organisms requiring direct preservation must be sorted as follows:
      a. Specimens to be preserved in alcohol.
      b. Specimens to be temporarily or permanently preserved in formaldehyde solution, either normal or neutralized.
      c. Specimens to receive special fixation for tissue or anatomical studies.

Each group of plants or animals may require some special technique depending on its nature and the intended use of the specimen. Often huge quantities of specimens are preserved in the field en masse with 5-percent to 10-percent formalin (preferably neutralized). Formalin will affect specimens with calcium carbonate structures, but is usually quite suitable for temporary preservation. In the laboratory these specimens are washed, sorted, or soaked out, and reassigned to permanent preservative solutions according to their particular need.

Treatment of Plankton

Because the content of the plankton net will be a conglomerate of various species, it is advisable to preserve small portions in 5-percent formalin, in hot (50–60° C.) FAA, and in alcohol. Formalin and alcohol will cause many of the forms to contract, whereas hot FAA poured over the concentrated plankton will kill numerous specimens in an expanded condition. Because the alcohol may cause precipitation of salts from the sea water, a change of preserving solution should be made soon after the initial preservation. Small vials with stoppers that will not trap and hold the plankton (as will cotton or porous corks) should be used to hold the plankton. These, in turn, are placed in screw-top jars after the proper collecting data have been enclosed.

Slide Preparation. In the laboratory plankton may be stored in vials or portions may be prepared on slides. When the latter technique is used masses of plankton may be stained and generally prepared on slides or individual specimens may be selected from the stained material and isolated on slides. Several stains are generally used: borax carmine, iron-haematoxylin, fast green, and the like. Place plankton to be stained in a small vial and wash several times with oxygen-free water (previously boiled and cooled). Pour off most of the water and add stain. After about an hour carefully pour off stain, add fresh oxygen-free water, and, after the plankton has had time to settle, pour off the water and replace with 50-percent ethyl alcohol. Dehydrate and infuse with xylene, and mount in balsam or permount as directed in Appendix B, or omit dehydration and mount directly in Turtox CMC-10. Plankton may also be mounted directly from water or alcohol in CMC-S which stains and mounts (see Appendix B).

Transporting Live Plants and Animals

Some general problems encountered in transportation are (1) maintaining proper temperatures, (2) reducing the metabolic rate, (3) supplying adequate amounts of oxygen, (4) preventing pollution of sea water or fresh water, (5) preventing overcrowding and (6) allowing for the general incompatibility of different kinds of organisms. The time of day, the air temperature, the distance the organisms must be transported, and the number of organisms, all help to determine the methods of transportation.

Containers. Avoid using metal containers, unless they have a porcelain lining, as they will give off toxic substances. Plastic or glass is best, as neither will contaminate the specimens. (Some notable exceptions for glass containers will be mentioned in the discussion of fresh-water algae.) When delicate animals are transported one should include some nonmucous-secreting algae which will provide hiding places and prevent damage from motion. Small arthropods such as crabs are prone to fight with one another unless they are provided with hiding places. Do not overcrowd the container with either plants or animals. Most marine invertebrates which normally survive exposure at low tide are best transported in wet seaweed with almost no water.

Aeration en Route. If specimens are to be transported in water, place only a few inches of water in each container to maintain a high ratio of surface area to volume; this permits a greater degree of diffusion of oxygen. Car motion will often create small waves in the container, facilitating oxygen diffusion. The presence of some plants will also provide oxygen for short periods of time, but cannot be relied upon for long shipping periods. If water temperatures become higher than those of the habitat additional aeration is essential. Mechanical agitation of the water or pumping of air through the water by means of a hand syringe or tire pump may be essential. The best technique for aeration is as follows: Take a superinflated inner tube, a length of rubber tubing, and an aquarium stone along on the collecting trip. When you are ready for aeration, partially unscrew the valve to release air, place the rubber tube over the valve stem, and stick the aquarium stone in the opposite end of the rubber tube. Place the aquarium stone in the container and, if necessary, readjust the inner-tube valve until the proper flow of air is achieved. Stop at as many gas stations as necessary to reinflate the inner tube en route. Battery-driven aquarium pumps or even cylinders of compressed oxygen may be substituted if they are available.

Temperature Control

Increased temperature increases the metabolic rate and, hence, the rate of respiration and excretion of most plants and animals. Keep containers out of the sun at all times in transport. Inexpensive camp iceboxes will maintain low temperatures in small containers, or smaller containers may be placed in a water bath in a large tub or bucket. Reduce the temperature in the water bath by adding small quantities of ice or by frequently changing the bath water at gas stations.

THE TERRESTRIAL ENVIRONMENT

We are all more familiar with the terrestrial environment than with the marine, and thus less needs to be said about it. One difference between collecting in fields and meadows as compared to marine collecting is that one does not collect all kinds of organisms simultaneously, but usually seeks out some limited group of plants or animals. Biologists on extended field trips in foreign countries may attempt to collect most of the available organisms, but this is an exception to the rule. Before collecting locally study the group of organisms you are interested in as to habitat requirements, seasons, life histories, equipment necessary for collecting, and so on.

Factors Affecting Biotic Distribution and Zonation

To a great degree the latitude of a given collecting site will determine seasonal temperature patterns and light distribution. The length of daylight changes in a significant way from month to month throughout the year, so that each latitude has its own peculiar annual daylight cycle. Increasing or decreasing periods of daylight may have a profound effect on local temperatures. However, temperatures also are greatly modified by the presence or absence of moisture in the air and in the soil, air currents, the cover (trees, shrubs, and so on) available or the lack of such cover, the nature of the substrate, and other factors. Light and temperature so profoundly influence living organisms (see the preceding discussion of "Marine Environment") that various species will be zoned in their distribution within a small area of land, according to their preference or tolerance for light and temperature.

Moisture is another factor that is highly variable both on the latitudinal plane and within any small habitat. The annual rainfall is partially controlled by the latitude of a given habitat, but is highly modified by local topography and wind currents. For example, on the Olympic Peninsula the annual rainfall exceeds 150 inches on the outer coast, but drops to just 11 inches 40 air miles away to the northeast. This remarkable difference is caused by the Olympic Mountains which intercept the rain from ocean-borne winds. Even in small areas moisture differences may be profound. For example, any variation in soil make-up may influence the soil's ability to hold or lose water. With increasing or decreasing amounts of vegetation, shade, organic litter, and the like the moisture content will vary within a given habitat.

We may interpret these few ideas as meaning that any given habitat, such as a meadow or hillside or forest, will be extremely complex in its physical make-up. Many species of plants and animals are broadly tolerant to variations in moisture content, temperature, and light, whereas others are narrowly tolerant to any changes in these factors. The field collector must become familiar with local ecology and learn to recognize changes in such important factors as temperature, light, moisture, and soil composition. He must be thoroughly familiar with the habitat requirements of those species he wishes to collect. With considerable field experience, a knowledge of habitat requirement, and a knowledge of local ecology the collector will soon recognize that plants and animals are distributed in orderly fashion in what at first appeared to be a haphazard conglomerate.

THE FRESH-WATER ENVIRONMENT

The fresh-water environment merges with the terrestrial to provide a variety of habitats that are entirely different from any other. Plants and animals show definite patterns of distribution based on physical environmental factors. With every subtle shift of even a single factor such as temperature or light, the nature of the community population will change. The density of water affords support for many animals and plants that do not exist in the terrestrial environment. Bodies of fresh water which may afford good collecting vary in size from the water trapped in the footprint of a cow or a small mudpuddle or bird bath to large lakes, and from small trickles of water seeping from a roadside bank to rivers the size of the Columbia or Mississippi.

Factors Affecting Distribution and Zonation

Light. Light is important in plant photosynthesis and the creation of food chains for aquatic life. The factor of depth alone is not serious in excluding light from the depths of most bodies of water. Seasonally, however, organic debris and plankton may accumulate in sufficient quantities to restrict light penetration. Thus, the heaviest concentrations of plant and animal life will be found in the shallower waters where light and favorable temperatures permit abundant growth. Because some aquatic animals are diurnal, whereas others are nocturnal in behavior patterns, the collector should sample a habitat under all light conditions.

Temperature. Daily and seasonal variations in temperature may dictate the appearance or disappearance of certain organisms. During unfavorable temperature conditions certain organisms will be present only in resistant stages, thus necessitating a knowledge of the life cycle of the organism sought. Water temperature may be more or less uniform in small bodies of standing water, but decidedly stratified in larger bodies. Temperature controls water density, which may in turn affect distribution. Water becomes most dense at 4° C., but actually becomes lighter in weight between 4° C. and the freezing point. This phenomenon frequently causes a complete seasonal overturn for water in a lake or pond and is responsible for ice forming on the water's surface rather than at the bottom.

Oxygen and Carbon Dioxide. In smaller bodies of water the availability of O_2 and CO_2 is less of a problem than in large bodies of water. The tremendous amount of organic material accumulated, the lack of water overturn, the stratification of temperature, all make the distribution of O_2 difficult in deep water. The collector should note that during the summer period stagnation may occur in deep water, thereby limiting plant and animal life to the shallower regions where O_2 is available.

Substrate. The substrate of any pond, lake, stream, or river will greatly influence the kinds of plants and animals that are present. Organisms that require points of attachment, hiding places, partial protection from light, and so on, naturally seek out rocky substrates rather than soft mud. In collecting, bear in mind that changes in substrate often reflect differences in water currents, available organic material, and other factors, and hence may influence distribution.

Currents. Currents tend to organize the biota into definite patterns of distribution and may serve as a requirement or limiting factor in distribution. Organisms become orientated to either facing the current or being protected from it. The degree of force of currents and differences in stream depth, width, or slope will further affect distribution. The current becomes a source of oxygenated water and of food for plants and animals, and serves to remove their excretion.

Food. Obviously, the availability of food will control the distribution and the presence or absence of aquatic organisms. The availability of food tends to be cyclic in that the amounts of sunlight, heat, inorganic nutrients, and other factors are cyclic. In consequence, plant growth tends to be cyclic or seasonal and, following this, zooplankton become cyclic. In turn, larger aquatic organisms will reflect this chain of events by their presence or absence.

Aquatic Collecting Methods and Equipment

General Equipment. Collecting methods and equipment must be geared to the organisms sought. The aquatic dip net proves invaluable for collecting swimming organisms and for bringing debris to the shore. The dredge net is designed to collect large quantities of debris from shallow water. Such debris is sorted on shore and will yield numerous species of plants and animals. Special dippers, kitchen strainers, or aquarium nets are also useful for gathering isolated specimens. In running water the stream net and dip net are useful for catching specimens that have been dislodged from their hiding places. The construction and use of these nets is described in Chapter 13. Occasionally, a large pipette is useful for getting bottom samples. This apparatus is made by placing a 50-cubic-centimeter rubber bulb on the end of a long length of plastic tubing. The plankton net, biological dredge, and scoop samplers described above are considered among the most important fresh-water collecting devices. See Welch (1948) for a thorough discussion of the techniques used in fresh-water ecology and measurement.

General Collecting Methods. In collecting within the permanent aquatic habitat utilize the plankton net for sampling the pelagic plants and animals. These organisms are highly cyclic; they will change from week to week throughout the season and require a continuing program of sampling. The number of organisms near the surface may also vary daily with changing light conditions or with minute but rhythmic temperature fluctuations.

Aquatic vegetation harbors a tremendous number of invertebrate species. These populations may differ greatly between those associated with plants growing on the surface and those harbored by plants growing in deep water. In shallow water carefully cut vegetation free from the substrate and float it into the dip net without disturbing the organisms. Next, hunt through the vegetation with a field dissecting microscope or hand lens for attached forms. Place similar vegetation in a clean plastic bucket of water and add a small quantity of formalin. This will drive the motile forms from the vegetation and ultimately kill them, permitting them to settle to the bottom. Carefully remove the vegetation and filter the fluid to obtain the specimens.

When working along the bottom substrate, carefully remove rocks and examine both the upper and under surfaces. Large sticks and logs should be checked, the bark removed, and, if necessary, the log broken open to expose the hordes of specimens living there. Small pieces of gravel should be examined underwater with the hand lens. Large quantities of organic bottom debris should be brought in with a dredge net and carefully picked through on the shore, or placed in a screen and washed, or examined in small quantities with the dissecting microscope or hand lens.

When collecting from stream bottoms, place a stout dip net against the bottom and carefully lift out stones so that any animals that are dislodged will float down into the net. Each time the stream widens or narrows, deepens or shallows, has a change in slope or in the nature of the substrate, there will be a change in the animal and plant population. All macroscopic and microscopic crevices should therefore be examined.

A strikingly different population of plants and animals may be found in temporary ponds or puddles. If the pond is dry, collect samples of the uppermost bottom mud with a hand trowel or pocketknife and put them into filtered pond water. After several days or weeks at room temperature many plants and invertebrate organisms which were present in resistant stages will appear. Otherwise, temporary water is sampled with the same techniques as that of permanent lakes or ponds.

SHIPPING PRESERVED SPECIMENS

Dry Specimens

Dried specimens such as sponges, echinoderms, coral, algae, and the like, should be carefully supported by tissue paper in individual cardboard boxes with the data included. These boxes, in turn, are placed in a large, strong cardboard box and completely surrounded by crumpled newspaper or other soft packing. Paradichlorobenzene crystals or naphthalene flakes may well be included as a fumigant, especially if the material must go through customs. The purpose of packing dry material is to prevent shaking and jarring (tissue paper support around the inner box) and crushing (the outer box). See Chapter 13 for methods of shipping dried insects.

Specimens Preserved in Liquid

Specimens preserved in liquid present problems of weight and leakage. Material such as formalin may not be shipped through the regular mail and will have to be sent by freight. The size of the organisms must be considered first, and will be handled in one of three ways: (1) very small specimens must be kept in vials (preferably plastic) with small quantities of preservative, data, and a cotton stopper or screw cap. An alternative is to place the specimens, data and preservative in a small plastic freezer bag. (2) Specimens of small and intermediate size may be grouped according to collecting locality and date, wrapped in cheesecloth (along with the field data), and tied with a piece of string. (3) Large specimens, such as amphibians or fishes, may be individually tagged with field data.

Once the specimens are processed with their data they should be placed in a stout plastic bag, along with some additional paper toweling and a small quantity of the preservative. Carefully tie the top of the plastic bag after expelling most of the air and place the bag inside a screw-cap or snap-cap tin can or metal drum. If necessary, add padding so that the specimens will not bounce around inside the drum. An alternative is to put the bag into a second and third plastic bag, each sealed in turn, and to ship this in a pasteboard box padded with crumpled newspaper.

Over the past fifteen years many field collectors in museums have used regular home-canning devices for the packing and shipping of specimens in tin cans. Preserved specimens are prepared as described above, placed in cans, padded, and sealed. Taylor (1950) thoroughly reviews the techniques required and mentions that many kinds of plants and animals including coral, echinoderms, algae-encrusted rocks, fungi, soil samples, and the like, can easily be shipped in this manner.

FIELD NOTES

Specimens soon become worthless unless they are accompanied by adequate data. Each collector should therefore keep a notebook in which the field data are arranged chronologically, in addition to putting slips of data in the specimen containers. You may choose between a bound, field notebook of good quality, high rag content paper or the looseleaf variety. Collectors using the looseleaf notebook transfer their field notes into a master notebook which always remains in the laboratory, and thus avoids the risk of losing past data. There are many suitable field notebooks on the market, but one of the most satisfactory for work in the Pacific Northwest is the "Rite in the Rain" #311 (Darling Corp., Tacoma, Washington), which is entirely waterproof. The paper offered by this company can be used underwater or in pouring rain with equal facility, will not wrinkle badly, and will always dry out and look very respectable. Notes should always be taken with a medium-grade pencil or with waterproof ink—not with ink that will run if subjected to water.

Three types of data should always be taken for complete notes. First, the locality according to state, county, nearest township, and approximate location in miles distance from the nearest township should be given, with the idea in mind that you or some other collector may wish to locate the spot with little difficulty. Do not use local names that are not established on land maps, as they will soon become valueless. By like token, do not simply give the name of a bay or town or lake; if you do your specimens may eventually have to be discarded for lack of adequate data. For example, there are seven "Clear Lakes" in the state of Washington, three lakes in one county alone bearing this title. Thus "Clear Lake" specimens are valueless unless additional data are given. Second, describe the habitat in which you collect in detail sufficient to help a new collector locate the locality. Kinds of vegetation, whether you are dealing with a meadow or forest or river, and so on, should be noted. Third, data concerning the kinds of specimens and the conditions under which they were collected, including weather conditions, hiding places, kinds of activity, and so on, should be jotted down.

COLOR NOTES

Occasionally it is necessary to make color notes in the field of some plant or animal whose color will be destroyed or altered before it can be examined in the laboratory. Robert Ridgway (1912) prepared an excellent but elaborate color standard as did Maerz and Paul (1950). Both of these books have colors arranged in small squares across and down the page, each keyed alphabetically and numerically so that the worker may match the color square with the actual specimen and code the color in the field notebook. The Maerz and Paul book is available; Ridgway is out of print. Neither, however, is ideal for actual field collecting. A second popular alternative to color standards would be the use of color photographs. However, unless some known color standard, such as the color cards for house paints or water colors, is included in the photograph, the accuracy of the photo cannot be guaranteed. With a color standard the photograph can be compared with the standard in the laboratory and a judgment made of the degree of difference existing between the photograph and the standard.

By far the simplest and most direct method of making field color notations is to use an inexpensive box of watercolor paints. This comes equipped with a small brush and mixing pan and is small enough to pack in a field kit. First make a crude sketch of the specimen, showing the areas where particular colors are located. If, for example, you have a crab such as *Grapsus grapsus*, which is orange-red on the back and light sky blue on the undersurface, mix a small quantity of red and yellow and paint a small area of this on the margin of the paper. When you hold the paper against the specimen itself you will quickly see whether the paint blotch is too red or too orange, whether you should add yellow or blue to make it lighter or darker, and so on. By experimentation and frequent comparison of colors you will eventually get one color blotch that actually matches the specimen. Encircle this color blotch and put an arrow to the place on the specimen that it matches most exactly. Use the same procedures for the light blues or other colors that may be present. Do not, unless you wish, attempt to paint the drawing you have made. In the laboratory the color blotches may be analyzed and more carefully described for publication.

REFERENCES

Allee, W. C., et al., 1949, *Principles of Animal Ecology*, Saunders, Philadelphia.
Allee, W. C., and K. P. Schmidt, 1951, *Ecological Animal Geography*, John Wiley & Sons, N.Y.
Beaufort, L. F. de, 1951, *Zoogeography of the Land and Inland Waters*, Sidgwick and Jackson, London.
Buchsbaum, Ralph, 1948, *Animals Without Backbones*, Univ. of Chicago Press, Chicago.
Bullough, W. S., 1954, *Practical Invertebrate Anatomy*, St. Martin's (Macmillan), N.Y.
Clarke, G. L., 1954, *Elements of Ecology*, John Wiley & Sons, N.Y.
Crowder, William, 1931, *Between the Tides*, Dodd, Mead, N.Y.
Davis, C., 1955, *The Marine and Fresh Water Plankton*, Michigan State Univ. Press, East Lansing.
Davis, H. S., 1938, Instructions for Conducting Stream and Lake Surveys, U.S. Dept. Comm., Bur. Fisheries, *Fishery Cir.*, 26, 1–55.
De Latil, P., 1955, *The Underwater Naturalist*, Houghton Mifflin, Boston.
Ekman, S., 1953, *Zoogeography of the Sea*, Sidgwick and Jackson, London.
Hardy, A. C., 1956, *The Open Ocean*, Collins, London.
Hausman, L., 1950, *Beginner's Guide to Fresh Water Life*, Putnam, N.Y.
MacGinitie, G., and N. MacGinitie, 1949, *Natural History of Marine Animals*, McGraw-Hill, N.Y.
Maerz, A., and M. R. Paul, 1950, *Dictionary of Color*, McGraw-Hill, N.Y.
Miner, R. W., 1950, *Fieldbook of Seashore Life*, Putnam, N.Y.
Moore, H. B., 1958, *Marine Ecology*, John Wiley & Sons, N.Y.
Morgan, A., 1930, *Fieldbook of Ponds and Streams*, Putnam, N.Y.
Murie, O. J., 1954, *Field Guide to Animal Tracks*, Houghton Mifflin, Boston.
Murray, Sir John, et al., 1944, *The Depths of the Ocean*, Macmillan, London.
Needham, J., P. Needham, et al., 1937, *Culture Methods for Invertebrate Animals*, Comstock, Ithaca, N.Y.

Needham, J., and P. Needham, 1953, *A Guide of the Study of Fresh Water Biology*, Comstock, Ithaca, N.Y.

Pratt, H. S., 1935, *A Manual of the Common Invertebrate Animals*, (Blakiston) McGraw-Hill, N.Y.

Ricketts, E. F., and J. Calvin, 1952, *Between Pacific Tides*, Stanford Univ. Press, Stanford, Calif.

Ridgway, Robert, 1912, *Color Standards and Color Nomenclature*, Published by the author, Washington, D.C.

Smith, R., *et al.*, 1954, *Intertidal Invertebrates of the Central California Coast*, Univ. of California, Berkeley, Calif.

Sverdrup, H. W., *et al.*, 1942, *The Oceans*, Prentice-Hall, N.Y.

Taylor, William R., 1950, Field Preservation and Shipping of Biological Specimens, *Turtox News*, 28 (2):42–43.

Ward, H. B., and G. W. Whipple, 1945, *Fresh-Water Biology*, John Wiley & Sons, N.Y.

Welch, P. S., 1948, *Limnological Methods*, (Blakiston) McGraw-Hill, N.Y.

Welch, P. S., 1952, *Limnology*, McGraw-Hill, N.Y.

Worden, A. N., 1948, *The Care and Management of Laboratory Animals*, Williams & Wilkins, Baltimore.

CHAPTER 2

The Algae

That group of plants collectively known as "algae" consists of seven distinct phyla found in the Subkingdom Thallophyta. Thallophyta (a group which also includes the bacteria and fungi) have simple plant bodies with little cellular definition; they lack roots, stems, and leaves which are comparable in cellular structure to the higher plants. This simple plant body is referred to as the thallus. Unlike bacteria and fungi, the algae carry on photosynthesis and possess photosensitive pigments of various colors. The seven phyla of algae are thus popularly named after their general coloration and may be referred to as the green algae, yellow-green algae, brown algae, red algae, and so on. With one or two exceptions all of the so-called "seaweeds" belong to the algae, as do a large number of pondweeds and scums found growing on damp surfaces or on tree trunks and moist soil. Because of their evolution, photosynthetic chemistry, means of reproduction and general structure, the collection and preservation of algae are essential for general biological teaching as well as for museum and research studies.

MARINE ALGAE

Algal Habitats

In surveying a long expanse of coastline along any of the continents one may be impressed with the seemingly repetitious appearance of headlands, rocky beaches, sandy beaches, bays, and estuaries. To the trained eye, however, tremendous ecological changes occur between the latitudes, especially in light and temperature. These two factors, along with the presence or absence of nutrients, are most important in controlling plant growth and distribution. Some species of plants or animals may seem to occur almost everywhere along a coastline, but generally there is a gradual progression of species, some dropping out and new ones being added as one moves toward the north or south.

The bulk of the marine algae consists of planktonic one-celled plants that are microscopic in size and are found in the upper water layers, in greatest concentrations close to the continental mass. The common seaweeds, on the other hand, are attached to some substrate and grow in a fixed position. An unstable substrate, such as sand or mud, will support very few species of seaweed except in quiet water. On the other hand, solid substrates such as rocks, reefs, pilings, floating docks, and the like, will provide firm attachment for algae. The student is urged to read through the first chapter on marine collecting and physical factors in order to gain a complete picture of plant requirements. One must bear in mind that light is required for photosynthesis, adequate temperature is essential for growth, nutrients are needed, and individual needs for sunlight or shade exist, and thus, generally, marine algae will be restricted to those substrates that provide at least minimal habitat requirements. Probably the most productive collecting site is the protected rocky coast zone, since this provides optimal conditions for a great number of species.

Morphology of the Algae

Generally speaking, the larger species of algae grow in the cooler temperate water, whereas the stature of most species seems to diminish as one approaches the warmer tropical water. There is great variety in shape and form, ranging from the extremes of erect, free-growing plants to those

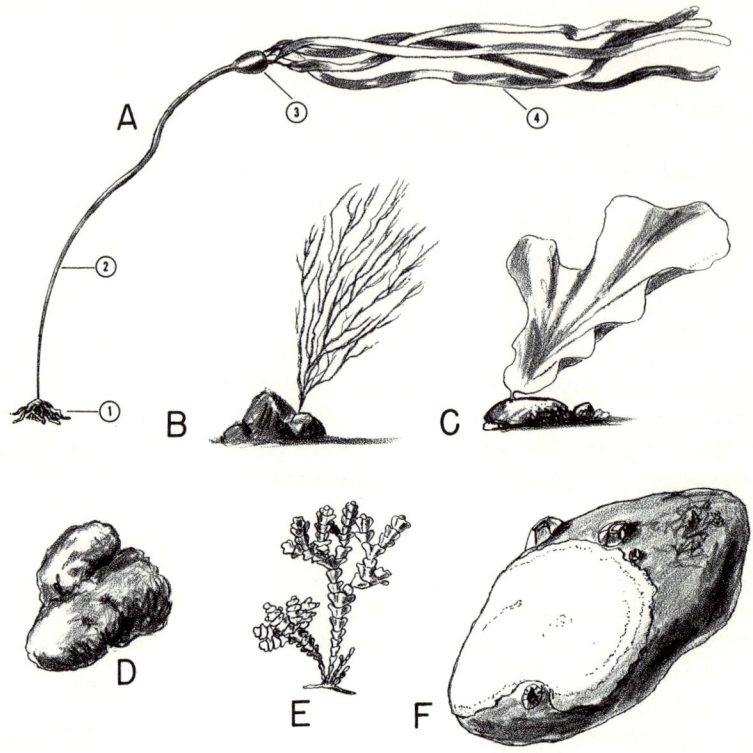

FIG. 2–1. Algal morphology. A. Large kelp. B. Filamentous alga. C. *Ulva*, a sheet form. D. *Culpomenia*, sac-like. E–F. Coralline algae, erect and encrusting. (1) Holdfast, (2) stipe, (3) float, (4) frond.

which encrust on the surfaces of rocks. All of the erect forms have a holdfast for attachment, and may include a stipe (a stem-like structure), a float, and one or more blades (leaf-like portions) as shown in Fig. 2–1A. The holdfast may be a simple, nipple-like growth attached to the substrate or a complex root-like mass. The erect portion may range in morphology from a filamentous, hair-like mass to a large flat blade, or even a sac-like mass, as seen in Fig. 2–1, B, C, and D. Many species of erect algae deposit lime in the thallus, and thus consist of hard, jointed segments (Fig. 2–1E). Quite a few species of both fleshy and coralline (hard, stony) algae grow directly on rocks, in a prostrate, encrusting form and their removal may require the use of a geology pick and chisel (Fig. 2–1E).

Collecting Marine Algae

Equipment. Among the few pieces of equipment required for collecting marine algae are a plastic bucket or a very large, stout plastic bag for transporting specimens, some small plastic bags, rubber bands, a few vials (preferably plastic) for isolating minute specimens, a pocketknife, and a field notebook. A geology or mason's hammer is useful in collecting and turning rocks. Occasionally, a cold chisel is essential for chipping off bits of rock containing encrusting seaweeds. Finally, a protected thermometer is desirable when ecological interests are involved.

Collecting Techniques. Plant specimens of a given species will vary greatly in appearance, depending on the season of the year, the age of a given plant, and the suitability of the habitat in which the plant lives. Whenever possible, collect a series of specimens of each species, attempting to get mature, well-formed plants. One or two specimens will suffice for larger species. Look carefully at the growth of algae surrounding you, so as to include "look-alike species."

In the intertidal and accessible subtidal areas, carefully remove the specimens by slipping a pocketknife underneath the holdfast. Put delicate forms in vials or small plastic bags with a small quantity of water. Carefully select specimens of encrusting algae which grow upon small single rocks or from parts of rocks that can be readily dislodged with a chisel and hammer. Many epiphytes (plants that grow upon other plants) will be found among the algae and should be included. Plants are usually transported to the laboratory for preservation, but occasionally it is essential to temporarily dry or preserve specimens when finished processing is impossible.

When collecting, be sure to collect only enough specimens to satisfy your requirements, to return rocks to their normal position if they are overturned, and, in general, to disturb the habitat as little as possible. There is a great danger not only in overcollecting plant specimens, but in disturbing the habitat for invertebrate animals.

Following severe marine storms great quantities of seaweeds are loosened from their substrata and transported in toward shore. Windrows of weed will occur along the beaches and the tide pools will be filled with loosened specimens. Collecting after storms is very rewarding in that you can get those hard-to-obtain specimens from rocky coasts. Collect on sandy beaches as well as on rocky coasts, for the flora will vary offshore from place to place.

Color and Field Notes. The colors of marine algae will quickly change in the process of dying, drying out, or on exposure to light. Some species of the red and brown algae will turn sufficiently green, if they die slowly in stagnant water, so that even the phylum becomes doubtful. Some preliminary sorting by color into the respective phyla may be essential. The final processing and preservation must take place as quickly as possible to prevent color alteration. When specimens are to be stored in liquid, put them in a dark place to prevent color alteration. Color notes and complete field notes should be taken if the specimens are to be of the greatest scientific value (see Chapter 1).

Preservation

Liquid Preservation. Liquid preservation has a number of distinct uses; a few examples of uses follow. (1) It is desirable to preserve a small number of specimens of each species in liquid, at least until they are identified and especially if the species is quite small. (2) Specimens may be briefly held in 5-percent formalin solution until they can be properly mounted in the laboratory. (3) Large fleshy floats of some of the species of kelp may be suitably preserved in formalin for permanent specimens. (4) Specimens may be preserved for classroom dissection, then soaked for twenty-four hours in fresh water prior to use.

When preserving huge quantities of specimens in small containers, increase the percentage of formalin. Use an FAA for preserving specimens intended for slide-making and histological studies, in the ratio of 93 parts of formaldehyde to 3 parts of glacial acetic acid. One of the drawbacks of liquid preservation is that some species become soft and many will lose their natural coloration, especially if they are exposed to light. Therefore, tightly cork or seal vials containing specimens, along with their field data, and place these vials in a large gallon bottle which is in turn sealed. This double-sealing technique will prevent rapid evaporation and permit simple storage of many vials in a dark place.

Herbarium Mounts. The materials used for herbarium mounts—including papers, methods of attaching specimens to the sheets, methods of constructing and using plant presses, and so on—are discussed in Chapter 5 in considerable detail.

Smaller specimens of marine algae should be mounted directly. Arrange the more rigid specimens on herbarium sheets in such a manner as to show the important taxonomic features. Most small species of algae will "glue" themselves to the herbarium sheets during the process of drying. Cover the exposed surface of the algae with cloth, such as sheeting, muslin, or cheesecloth, to prevent the specimens from sticking to the dryers. These covers are removed when the specimens are dry.

Very flimsy specimens must be floated on to the herbarium sheet in the following manner: Float a specimen in a large pan of sea water and carefully slide a herbarium sheet under it. When the plant is arranged in a suitable fashion slowly lift the sheet from the water, with the plant intact. Drain off the excess water by tilting the sheet, cover the plant with cloth, and press. It may be noted that herbarium sheets are frequently cut into suitable sizes, such as 4 by 6 inches, onto which small specimens are mounted. These are later kept in a card file or, preferably, mounted on a herbarium sheet.

Dawson (1945) recommends a simple method for floating specimens onto cards. Obtain a piece of galvanized sheet metal slightly smaller than the pan used. Bend the corners down to serve as legs and fill the pan so that the water barely covers the surface of the metal. Place the card and the specimen on the metal and depress the metal so that the specimen is floated. When the specimen is in the proper position, the pressure is released gently from the metal, which automatically surfaces the card and fixes the specimen in position. The card is then placed in the press and covered with cloth for drying.

If a plant press is used, be sure that the pressure is light so that undue flattening will not occur. One may also dry specimens by stacking ventilators, blotters, and specimens as described in Chapter 5, then placing a piece of plywood with a moderate amount of additional weight (such as books or rocks) on the top. The pressure on the press or the amount of weighting may be increased as drying progresses, but should never equal that used for the higher plants. Change the dryers daily until the specimens are ready for identification and permanent storage.

Large, fleshy specimens of the red and brown algae should be "field dried" by being spread in the shade until they are dry but still flexible. After the plants are killed in this fashion, they are again soaked in sea water so that they will retain their normal shape, placed on herbarium sheets, covered with cloth, and pressed in the manner described above. Specimens that do not attach themselves to the mounting cards may be treated as the higher plants described in Chapter 5.

Rolled Specimens. Some of the giant species of kelp, which have fronds measuring up to a foot or more in width and measuring from 9 to 90 feet in length, are impossible to mount on herbarium sheets. Small specimens of such species may be selected, and from them representative sections may be mounted in the normal manner. However, intact plants, including the holdfast, may be preserved in the following manner. Field dry the large specimens by spreading them out on newspapers in the shade. Turn them occasionally until they are almost dry and in a leathery state. At that time, roll the specimens, beginning with the holdfast, and secure the roll with a piece of string. Before the roll becomes entirely brittle, wrap each specimen in paper and label it with the appropriate data. Store such specimens permanently in a herbarium locker. Whenever it is necessary to examine or demonstrate a specimen, simply soak it out, preferably in sea water, until it has obtained its lifelike texture and coloration.

Coralline and Other Encrusting Algae

The hard coralline algae and those species encrusting upon rocks present a new problem in preservation. Erect species of coralline algae should be spread out while still fresh, arranged in a suitable position without too much overlapping of the branches, and permitted to partially dry. They are then dipped into herbarium glue on a glass plate (Chapter 5) and attached directly to herbarium sheets or cards where they will complete their drying. The alternative method is to store such specimens, directly or on cards, in herbarium envelopes (described in Chapters 3 and 4).

Select encrusting species of algae attached to rocks which are small enough to be stored in cardboard boxes within the herbarium tray. Chips of rock may be glued to herbarium cards and kept in herbarium envelopes or attached directly to the herbarium sheets.

Slide Techniques. In the identification of algae it is essential to make temporary slides, from time to time, to demonstrate the cellular structure of various parts of the plant. The arrangement of cells and the thickness of the various tissues are frequently characteristic of families, genera, and species. Mount whole fragments of plants in sea water under a cover slip for study. Make cross sections by holding the plant down with one hand and slicing extremely thin sections with a single-edged razor blade. If the specimen is large and fleshy, and therefore difficult to slice, permit it to dry and then slice the section. When the cross section is returned to a slide in a drop of sea water it will resume its natural proportions and be ready for study. Usually no stain is needed, since fresh or dried materials contain their natural pigments.

Specimens for permanent slide mounts should first be killed in 5-percent formalin or in FAA, as stated above, and then washed in fresh water to remove the fixative. Dawson recommends, in his book (1956) and in his course in Marine Botany, the following procedures: stain specimens with acid fuchsin or analin blue while they are on the slide, acidify and wash out the excess stain. Blot away the surplus water and add a 60-percent solution of clear corn syrup. Let this stand for 24 hours in a dust-free place and then replenish the evaporated syrup with an 80-percent solution. For thinner specimens, place a cover slip on at this time and, after labeling, store the slide in a flat position. Thicker specimens may be allowed to stand 24 hours longer then have a slight additional quantity of 80-percent syrup added along with the cover slip. Care must be taken not to dislodge the cover slips or to turn the slide on edge. After a month or two, however, such slides may be considered permanent.

Sass (1951) recommends staining with the self-mordanting hematoxylins or iron hematoxylin, dehydrating with glycerine, and mounting in a glycerine jelly preparation. This preparation is made by dissolving 5 grams of gelatin in 30 cubic centimeters of water and then adding 35 cubic centimeters of glycerine and 5 grams of phenol which has been dissolved in 10 drops of water. This preparation is filtered through coarse filter paper while still warm. To use, place a small quantity of the hardened medium on a slide, warm until melted, and then place the specimen in the medium and add a cover slip. For the more elaborate staining processes used in histological studies, consult any of the excellent texts on plant microtechnique, such as that of Sass (1951) and Johansen (1940).

FRESH-WATER ALGAE

Fresh-water algae are usually small and delicate as compared to the marine species. Therefore, living cultures, when it is possible to maintain them, are often preferred to preserved specimens. Algae are usually cultured until the desired state of growth is obtained; they may then be preserved or dried while attached to the normal substrate or, in the case of some of the larger species, attached to herbarium cards. Usually the first two spring months are the best time for collecting pure colonies of algae; the summer months present mixed cultures consisting of two or more species.

Location and Collection of Fresh-Water Algae

Small aquatic specimens are collected directly into bottles or vials provided with just enough pond water to cover the specimens; containers filled with water will soon stagnate and destroy the samples. Larger specimens, or objects containing colonies of algae, may be placed in plastic bags or rolled into wet (and then dry) newspaper for transportation to the laboratory. Submerged objects in ponds, lakes, and streams provide excellent habitats for algae. Rocks, bits of wood, sticks, and other debris should be removed and examined if they have a green or brown appearance, as this often denotes the presence of algae. These growths can be scraped into vials with a dull pocketknife or fingernail. In the late spring and summer months colonies of floating filamentous algae will appear in great quantities.

Many substrates near ponds and streams, or within humid forests, will harbor flourishing colonies of algae. Greenish soils may be picked up on a piece of tin and transferred into a bottle lid. The bottle is then screwed into the lid and kept in the inverted position while it is transported back to the laboratory. Remove bits of bark containing green algae with a pocketknife. Areas such as water seeps along road banks, bird baths, fish ponds, and the like may also provide sources of less common species of algae.

Culturing Methods

Specimens brought into the laboratory should immediately be transferred to clean culture dishes. The original water should accompany the specimens, or aged aquarium water may be substituted. Water may be boiled to destroy other organisms prior to dispensing. Tap water should not be used as it frequently is contaminated by chlorine or various metals which prove toxic to the algae. Specimens should not be overcrowded and the ratio of surface area to water volume should be favorable for complete aeration. Direct sunlight should be avoided in most cases. Light from a north window proves quite adequate during the growing season, although artificial lighting will enhance algal growth. Nutrients are not essential for short-term culturing, but are required for the maintenance of stock cultures over a long period of time. Dissolved plant-food tablets intended for domestic use may be added in small quantities to the culture medium from time to time. A soil nutrient solution may be prepared (Faridi, 1964) as follows: mix about 200 grams of farm soil for each liter of pond water, boil to sterilize, filter, cool, and dispense for the culture medium. Many highly specific culture media have been developed. Those who are interested may consult the bibliography for literature concerning these media. Biological supply houses (Turtox, Carolina, and others) also provide media concentrates for culturing algae.

Cultures may be started at other times of the year than early spring by collecting the upper crust of mud from dried ponds in the summer and, in the winter, the upper layer of mud from places where algae have been growing during the favorable period (Turtox, 1958). This mud is placed in culture dishes and covered with water. Dormant spores of the algae, if present, will eventually germinate and grow.

Preservation of Fresh-Water Algae

Although fresh-water algae are generally preserved in liquid, larger species may be floated on herbarium cards, pressed, and dried according to the procedure described for marine algae. For liquid preservation pour off the collecting water and replace with 3-percent or 5-percent formalin, or FAA. Add labels containing the field data, cap the vials, and store them in a dark place in airtight jars. The more delicate specimens of algae may be prepared on slides, as described above under marine algae.

DIATOMS AND RELATED FORMS

Diatoms are microscopic one-celled or colonial plants which constitute the most conspicuous class of the phylum Chrysophyta. The members of this phylum have a tendency to deposit silica in the cell walls. The diatom frustule, the cell wall which surrounds the protoplast, is typically divided into two sections or valves. There are two subclasses of diatoms. The members of one class are commonly found in fresh water, brackish or salt water, and on the surfaces of plants, mud or moist soil adjacent to sources of water. The other subclass is predominantly a marine pelagic group which seldom is found on solid substrates. To collect diatoms, tow a 24-mesh plankton net equipped with a terminal collecting vial through the surface water. The sample is then treated with formalin, permitted to settle, and concentrated. Suitable filamentous plants, such as marine algae, fresh-water mosses, and the like should be gathered in quantity and squeezed to remove the free water. The water will carry with it large quantities of diatoms and other microscopic plants. Scrapings from rocks and other structures should also be examined for diatoms. Fresh-water and marine snails that browse and scrape solid surfaces, and certain crabs such as the grapsoid crabs, should be collected, killed, and the stomachs removed, for these are frequently rich with diatoms. Also, the skeletons of diatoms are more or less abundant in the muds of old lake bottoms and are highly concentrated in certain areas of so-called diatomaceous earth.

Diatoms should either be stored in tightly capped vials or prepared on slides. For cytological studies involving various staining techniques see Johansen (1940). Since the skeletal structure of the diatom often proves most important, taxonomically, it is desirable to remove the organic material by cleaning. This may be done by placing some concentrated diatoms on a slide and adding a drop of concentrated nitric acid, boiling by gently heating the slide over a Bunsen burner, and then examining the material on the slide. A mounting medium and cover glass can be added directly to the slide at this point. However, better slides may be made as follows: place concentrated diatoms in about 5 cubic centimeters of nitric acid and let this stand for up to 10 hours. Remove the acid by repeated washings, allowing sufficient time for the diatoms to settle out before decanting the water. Next, replace the water containing the diatoms with alcohol, and transfer a drop of the alcohol-diatom solution to a slide. The streaming action of the alcohol will distribute the diatoms and space them adequately (this reaction may be enhanced by igniting the alcohol). In mounting diatoms, use a medium with a very high refractive index; balsam and similar media are not satisfactory, since the skeleton of the diatom has a refractive index closely approaching that of balsam and will become "lost" when mounted on this medium. Hyrax, Styrax, Storax, and similar media are quite suitable for diatoms. Use the mounting medium sparingly and add a Number 1 cover slip. Diatom specialists use a mounted pig's eyelash to pick up selected diatoms and transfer them to glass slides (Nielson, 1944) when it is desired to isolate distinct species or genera.

MUSEUM AND STORAGE METHODS

Proper storage facilities must be provided in order to conserve not only the collector's time and expense, but the valuable specimens as well. The time has come when repetitious collecting must be avoided in order to conserve and maintain natural plant and animal populations. Specimens preserved in liquid should be tightly sealed, along with internal labels containing field data, and kept in a dark place. Vials should be placed in airtight jars of a convenient size to prevent excessive evaporation. Specimens mounted on herbarium sheets should be carefully labeled and sorted by family, genus, and species. If the collection is sufficiently large, a card file or some other adequate reference should be kept. Standard herbarium cases are readily available at biological supply houses, but can be simply and very inexpensively constructed. These may be made in the home or school workshop for a fraction of the commercial cost, as will be described fully in Chapter 5.

REFERENCES

Dawson, E. Y., 1945, An Annotated List of the Marine Algae and Marine Grasses of San Diego County, California, *Occ. Papers San Diego Nat. Hist. Soc.*, (7):1–87.

Dawson, E. Y., 1956, *How to Know the Seaweeds*, Wm. C. Brown Co., Dubuque, Iowa.

Faridi, M. A., 1964, Some Simple Techniques for the Production of Zoospores in Fresh-Water Algae, *Turtox News*, 42 (2):58–59.

Forest, H. S., 1954, *Handbook of Algae: Special Reference to Tennessee and Southeast*, Univ. of Tennessee, Knoxville.

Fritsch, F. E., 1935, *The Structure and Reproduction of the Algae*, University Press, Cambridge, Vol. I; 1945, Vol. II.

Guberlet, M. L., 1956, *Seaweeds at Ebb Tide*, Univ. of Washington Press, Seattle.

Johansen, Donald A., 1940, *Plant Microtechnique*, McGraw-Hill, N.Y.

Nielson, J. E., 1944, Meet the Diatoms, *Turtox News*, 22 (2):40–43.

Prescott, G. W., 1951, *Algae of the Western Great Lakes Area*, Cranbrook Institute of Science, Cranbrook, Mich.

Prescott, G. W., 1954, *How to Know the Freshwater Algae*, William C. Brown Co., Dubuque, Iowa.

Pringsheim, E. G., 1946, *Pure Cultures of Algae*, Cambridge Univ. Press, N.Y.

Sass, John E., 1951, *Botanical Microtechnique*, Iowa State College Press, Ames, Iowa.

Smith, G. M., 1944, *Marine Algae of the Monterey Peninsula, California*, Stanford Univ. Press, Stanford, Calif.

Smith, G. M., 1950, *The Fresh-Water Algae of the United States*, McGraw-Hill, N.Y.

Smith, G. M., 1955, *Cryptogamic Botany*, Vol. I, McGraw-Hill, N.Y.

Taylor, W. R., 1937, *Marine Algae of North Eastern Coast*, Univ. of Michigan, Ann Arbor.

Taylor, W. R., 1957, *Marine Algae of the Northeastern Coast of North America*, Univ. of Michigan Press, Ann Arbor.

Tiffany, L., and M. Britton, 1952, *The Algae of Illinois*, Univ. of Chicago, Chicago.

Turtox Service Leaflet No. 6, 1958, General Biological Supply House, Chicago.

CHAPTER 3

The Fungi

The fungi constitute a group of primitive plants devoid of chlorophyll which are included in the Subkingdom Thallophyta of the plant kingdom. Without photosynthesis the fungi must manufacture their food by breaking down organic matter such as dead plant and animal tissue, or by parasitic activity. The subgroups of plants included under the fungi by various workers are not consistent. However, for simplicity's sake the bacteria and lichens will be included here, along with the molds, yeasts, mushrooms, puffballs, and so on. The fungi occur the world over and live not only in terrestrial situations, but in marine and fresh-water habitats as well. Bacteria have been collected from some of the highest points of habitation in mountain regions and in some of the deepest trenches found in the Pacific Ocean. The habitat of the fungi is extremely variable, but the species are always located where the source of food is. Only certain fungi are collected and "preserved" in the typical sense of the word. For the bacteria and many of the molds the only truly suitable preparation of various specimens is in a living culture. Therefore, certain institutions working with these forms culture and reculture each species year in and year out, in order to have them constantly on hand.

BACTERIA AND MOLD

Work with these plants generally falls into areas of research, diagnosis of medical and plant disease, and classroom participation. Living organisms rather than prepared museum specimens are used. Thus, the bacteria and molds do not fall into the strict objective of this book in that work with them embraces culturing and rearing rather than preservation and curation. In spite of this fact, brief coverage will be given here, because (1) the bacteria and molds are collected and (2) many biological techniques and types of equipment are involved. Coverage here will embrace only a small fraction of microbiological techniques and is intended to be of value for those wishing to "explore" the bacteria and molds with little equipment or previous training. There is a vast literature concerning techniques in microbiology, with dozens of good texts and manuals that give long and valuable discussions of the methodology of this science for those who wish more complete coverage.

Problems in Culturing

Bacteria and molds obtain most of their food from the organic material in dead plants and animals, or as parasites, and are often highly specific as to the kind of food required. Therefore, the type of nutrient provided must be one of the first considerations in culturing. Various nutrients are prepared in a liquid or solid form in what is called the "medium," as discussed below. Temperature is another critical factor. Parasitic and semiparasitic forms are adapted to live only at the body temperatures of their hosts, whereas the temperatures of those species using dead tissue for food will range tremendously. Although optimal temperatures will enhance and encourage growth, adverse high or low temperatures may cause these plants to become dormant, to encyst or form protective bodies, or even to be killed outright. Certain growths will flourish at below room temperature, at room temperature, or at elevated temperatures within incubators, depending on the species involved. Desiccation quickly interferes with growth processes and may be due either to insufficient moisture in the medium or to the air surrounding the medium, being too dry. For some species free oxygen is toxic, for others it is absolutely essential. The pH (degree of acidity or alkalinity) also must be carefully controlled. The fantastic metabolic and reproductive rates of growing bacteria suggest that excreta will also be plentiful. Metabolic wastes of these plants soon contaminate the culture medium and retard growth or kill the species growing thereon, and thus necessitate reculturing for continued growth. Finally, pure cultures must be developed and maintained. This means that all other species of molds and bacteria must be excluded through a completely aseptic technique of handling.

Some Characteristics of Identification

The molds have a complex morphology and, hence, many characteristics by which they may be identified. The bacteria, on the other hand, are extremely small and rather simple in morphology; because of this one identifies them not only by structure, but also by behavior and colonial appearance. A large vocabulary exists for the description of these plants. *Bergey's Manual of Determinative Bacteriology* is perhaps the most complete and generally used source of bacterial identification. Simple bacterial morphology generally falls into categories of rods, spheres, or spirilla, and the individuals may be found singly or in pairs, clusters or chains. After a reasonable period of growth a bacterial colony will have its own distinct color (occasionally, odor), distinct form, elevation, and margin. Colonial forms may be punctate, filamentous, irregular, rhizoid, spindle-shaped, and so on. The elevations may be flat or raised, convex or umbonate, and so on. Margins may be smooth, undulate, lobate, curled, or still others. When inoculations are streaked across the surface of the solid medium or stabbed down into it, the streaks and stabs of bacteria will develop their own characteristic appearance. They may cause liquefaction of the medium in characteristic ways or the margins of the streaks may provide telltale clues to identification. The specimen's reaction to free oxygen and to the positive or negative acceptance of various diagnostic stains also helps in bacterial identification.

Some Common Media

Because of the wide menu of bacteria and molds a great number of different nutrients must be provided. No single nutrient is in itself all-purpose, though some are more commonly accepted and used than others. Nutrients are usually prepared in two ways: as a broth or as an agar. The broth is prepared and used in cotton-stoppered test tubes; the agar, which is a solid medium, may be dispensed in Petri plates or in tubes (butt and slant), as shown in Fig. 3–1. Agar, a gelatinous material, has the physical property of solidifying at a temperature above the human body temperature. Therefore, it remains solid under incubation at high body temperatures. In addition, the temperature at which the agar solidifies is low enough to permit bacteria

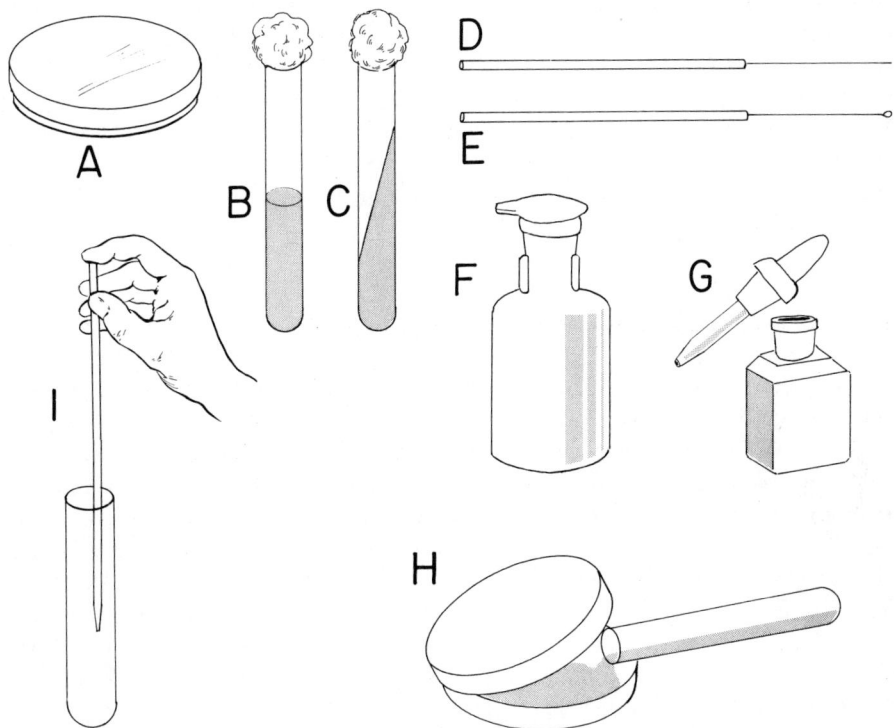

FIG. 3–1. Equipment used in bacteriology. A. Petri dish. B. Butt-agar tube. C. Slant-agar tube. D. Transfer needle. E. Wire loop. F. Dropping bottle. G. Stain bottle. H. Method of opening the Petri dish to transfer agar. I. Transferring with the oral pipette.

to be stirred into the medium just prior to solidification without killing the bacteria. The medium is prepared in a large flask, cotton-stoppered and sterilized, and is then dispensed into smaller sterile holding vessels or into the culture tubes and plates. Those tubes which are intended for agar slants should be placed on the laboratory table with the stoppered end resting on a 1-inch board, and allowed to cool and solidify.

Concentrated agar and broth without nutrient are available for those who wish to make their own medium by adding special ingredients. However, many supply houses provide long lists of prepared media (Carolina Biological Supply, for example, lists several hundred) which are dehydrated and ready for final preparation. Complete instructions for the mixing of these media are on the containers and will not be discussed here. Almost any suitable nutrient may be used as a medium, however. Bouillon cubes, for example, may be dissolved in about three times the required amount of water, poured into test tubes to a depth of about 2 inches, and sterilized at 15 pounds pressure for 20 minutes. This makes a good general-purpose broth for many classroom experiments. Likewise, potato may be cut to resemble the agar slant and pushed to the bottom of a test tube on top of a small piece of cotton. The vial is then stoppered, sterilized, and used as one would use a slant agar tube.

Equipment, Cleaning, and Sterilization

Equipment. Petri plates, used for holding agar, consist of top and bottom halves, or valves, which overlap and exclude contaminating dust and bacteria. Generally, these plates are wrapped and sterilized and maintained in sterile packages or containers until ready for use. To use the Petri plate, heat the medium within a culture tube by placing it in a water bath. When the medium is liquefied, permit the bath to cool to 45° C., remove the tube, dry off the outside, remove the plug, and flame the mouth of the tube. Finally, lift the upper Petri valve (see Fig. 3–1H) and pour the content of the tube into the Petri dish. Lift the upper valve only enough to permit the free entry of the tube, being careful not to touch the inside of the Petri dish. When the agar is solidified, inoculation may take place.

Several types of culture tubes are used. Straight-edge, lipless tubes measuring between 100 and 150 millimeters in length by 20 to 25 millimeters in diameter and supplied with cotton stoppers are very adequate. Special screw-cap tubes are also available and are often desirable in that they maintain the moisture content of the medium somewhat more adequately than the cotton-stoppered tubes. The medium is transferred to the tube either by means of the oral pipette (Fig. 3–1I) or by an automatic pipette which delivers a predetermined volume of medium. In filling, be careful not to contaminate the upper sides or top edge of the tubes. Stopper the tubes (leave screw caps loosened slightly), place them in tube holders, and sterilize.

The wire needle and wire loop are used to transfer bacteria and mold (Fig. 3–1, D and E). Platinum wire is preferable for these instruments, because it withstands repeated heating without excessive burning, corrosion, or flaking. Other wires, such as copper or iron, may be used, although they will be less satisfactory. Very simple needles can be made by forcing a piece of 22-gauge wire into the end of a $3/16$-inch wooden dowel. The wire should measure about 3½ inches in length and the handle should be 5 to 7 inches long. Better needles may be prepared by placing the desired wire in the drilled end of a metal dowel. The end is then crushed in a vise to secure the wire in place. The use of the needle will be discussed below.

If a commercial incubator is not available, a satisfactory one can readily be made from an old refrigerator. Obtain any old refrigerator, preferably an apartment-size model which has not been subjected to rotting in the out-of-doors for a long period of time. The motor and other material should be removed from the under part of the box; the box may be sawed off so as to remove the motor vault. Check the door and, if necessary, repair the seal. Clean out the inside of the box thoroughly, washing the shelves and other items that will be retained. Secure about 15 feet of 18-gauge lamp cord, one electrical plug, one porcelain light socket without a pull-chain, an electrical outlet box, and a thermostat switch. You may have to obtain the last item from a biological supply store or from a large farm merchandising store such as Sears. The thermostat should be of the bi-metal type. The chick incubator thermostat, which sells for about a dollar, is perfectly adequate.

To assemble the unit, first, attach the electrical outlet box to the center bottom of the refrigerator (see Fig. 3–2). Next, mount the thermostat between ⅔ and ¾ of the way up the side of the icebox, depending upon the location of the shelf. Drill a small hole through the side of the icebox to admit the rubber-covered electric wire. Wire the box, one wire going directly to the thermostat and the other directly to one pole of the light receptacle. Secure a short length of single-strand lamp cord (simply remove the other strand) and complete the relay from the thermostat down to the light receptacle as shown in Fig. 3–2B. Attach the light receptacle to the electrical outlet box and equip the receptacle with a 60-watt light bulb. The size of the light bulb will ultimately have to be determined by experimentation. Finally, drill a ½- or ¾-inch hole through the top center of the box (Fig. 3–2A). Secure a stick thermometer (reading in degrees centigrade) and, after wetting the tip of the thermometer, genty pass it through a one-hole rubber stopper. Push the rubber stopper down into the hole in the icebox so that the mercury end of the themometer projects into the box and calibrations at least above 20 are exposed above the box.

This box may be used like any incubator. Careful adjusting of the thermostat to the desired temperature (usually 37° C.) will give very satisfactory results. Do not open the door more than absolutely necessary, as cold air will shock the bacteria or mold being incubated.

Many kinds of biological stains are used in bacteriological work. Two

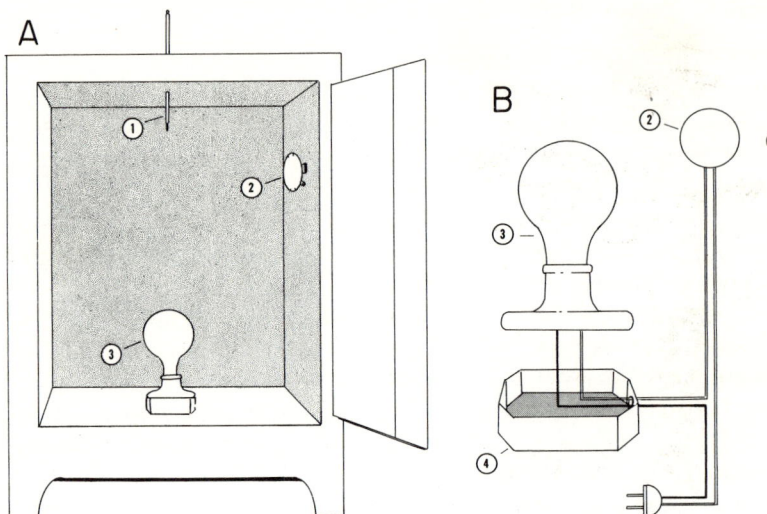

FIG. 3–2. A homemade incubator. A. Modified icebox. B. Wiring diagram. (1) Thermometer, (2) thermostat, (3) light bulb, (4) electrical outlet box.

convenient types of bottles are generally used for dispensing these stains: the dropper bottle with an automatic lid or the dropper bottle containing a pipette (Fig. 3–1, F and G). Other convenient items will be plastic, self-dispensing bottles to hold sterile water, alcohol, and the like.

Cleaning Glassware. Test tubes and plates containing cultures should all be sterilized at 15 pounds pressure for 20 minutes prior to cleaning. After sterilization, remove the culture medium and wash all glassware in a soap solution or in a 1-percent solution of trisodium phosphate. Rinse thoroughly in tap water and dip into a 2-percent solution of hydrochloric acid to remove the film. Rinse again in tap or distilled water and stack to drain and dry.

Sterilization. The autoclave is the standard unit for sterilizing laboratory equipment. Materials are wrapped in brown paper or placed in special containers, and are then subjected to steam heat which must reach 15 pounds pressure for at least 20 minutes. A pressure cooker may be substituted for classroom experiments, if it is used under the same conditions. Dry-heat sterilization may be achieved in an oven heated to 160° F. for 1½ hours.

Collecting Some Bacteria and Molds

Bacteria and molds may be found on almost any surface where some organic material is available or where they have been deposited by contamination or transport through the air. Their spores are air-borne and settle out on almost any surface, from whence they can be collected. A mixture of both mold and bacteria will be obtained in the initial cultures that have been collected.

Sterilized, cotton-tipped swabs are excellent for collecting bacteria. On moist and semimoist surfaces these swabs may be applied directly; they should be moistened with sterilized water when applied to dry surfaces. Rub the swab back and forth over the surface being sampled, then streak the swab across an agar plate or slant as shown in Fig. 3–3, A, B, and C. Be sure to apply the same surface of the swab to the culture medium as you applied to the collecting surface. The principle involved in streaking the swab across the agar surface is to release bacteria or mold spores in ever-decreasing numbers; hence, the swab is directed back and forth, as illustrated.

When inoculating or transferring a culture to a Petri dish (Fig. 3–3D) set the dish on a table, lift the lid only enough to permit the introduction

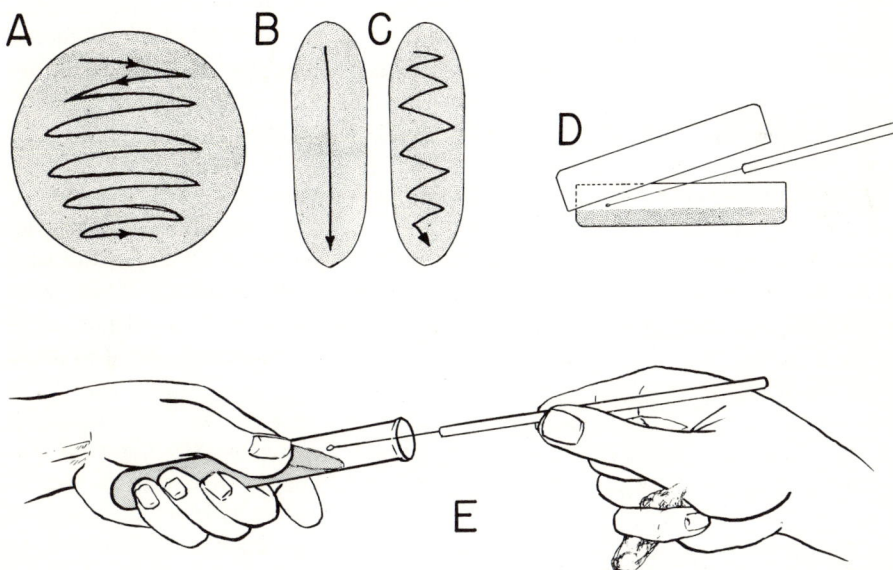

FIG. 3–3. Bacterial techniques. A. Streaking the agar plate. B–C. Two methods of inoculating the slant tube. D. Opening the Petri dish for inoculation. E. Inoculating the slant tube while holding the cotton stopper with the little finger.

of the swab or wire loop, streak the inoculum back and forth, and close the valves after the process is completed. Never lift the upper valve off in such a way as to draw air currents into the dish. Avoid touching the lips of the dish and always burn the swab in a flame or plunge it into a dish of disinfectant. The temperature for incubating the culture (body temperature, room temperature, and so on) will depend somewhat on the source of the bacteria. Turn the Petri dish upside down when incubating; this prevents water droplets from collecting on the lid and falling into the bacterial or mold colonies. To inoculate a slant tube of agar, contaminate the swab from the desired source, hold the slant in a horizontal position with the left hand, and remove the cotton stopper by grasping it with the little finger of the right hand. Run the mouth of the slant tube through the flame once or twice and then streak the inoculum across the surface. Reflame the mouth of the tube and replace the cotton stopper. Now, while still holding the swab, destroy the cotton head either by flaming or in disinfectant.

Air-borne bacteria and mold may be sampled by opening a Petri dish with nutrient agar and exposing this to the air for ½ hour or more. Replace the lid and incubate one plate at room temperature and one at body temperature (37° C.).

The wire loop is used to transfer liquids. The wire loop is flamed to a cherry red, cooled, and then dipped in the desired liquid for the transfer. Following the transfer, it is again flamed to a cherry red before being set down on the work table. Milk, pond water, tap water and the like may provide flourishing cultures of bacteria. Household articles—as tables, sinks, doorknobs, desk tops, and so on—may also be sampled. Soils and decaying organic material of all types will be rich in both molds and bacteria. Any body surface may provide interesting cultures. Fingertips may be applied directly to the agar surface, and the swab is useful for oral contaminations and contaminations around the ears, hands, feet, and so on. In classroom work, do not permit students to culture materials from cuts and infections, since serious dormant infection may be harbored thereafter in the laboratory. Molds may be collected by using moist squares of bread in place of a swab and culturing these in the dark at room temperature. Their growth pattern and morphology are best observed when they are cultured on an agar. Sabouraud's Dextrose Agar is very satisfactory for culturing the more common kinds of mold.

Pure Culture Techniques

Objectives. One seldom collects a pure culture of bacteria in nature. Rather, he may collect dozens of species intermingled in the original sample.

Because of the minuteness of these species it becomes apparent that they must be isolated for identification. The pure-culture technique permits the isolation of each species so that it can be tested in a pure form, yielding "pure" results concerning its identity. Under collecting samples of bacteria it was recommended that the inoculum be streaked across the plate or slant (Fig. 3–3, parts A, B, and C). Although this method releases many bacteria initially, it eventually scatters single specimens across the face of the agar plate as diagrammatically shown in Fig. 3–4A. On the other hand, if the initial sample is taken into a broth culture, the mixed species of bacteria grow at random throughout the culture and, hence, must be isolated.

Streak Plate. If the initial sample was cultured on an agar plate or slant, provide enough new sterile agar plates so as to have several more than the apparent number of species evident. In transferring these species always assume that they are pathogenic (disease-producing). Flame the wire loop in a burner until it is cherry red, cool it, and then pick up part of an isolated colony by carefully opening the Petri plate, as described above. Remove the needle from the initial plate, being careful not to touch any other growths or the edge of the plate itself. Replace the lid of the initial culture, carefully introduce the needle into the clean culture plate, and streak in one of the methods shown in Fig. 3–4, A–C. Close the plate, flame the needle, and then label the plate. Repeat this process with all of the apparently different colonies on the original culture plate. Finally, incubate the plates for 24 to 48 hours until the new colonies develop a strong growth. If it becomes apparent that more than one colony is present in any of the new cultures, repeat the isolation process until pure colonies have been obtained for all species.

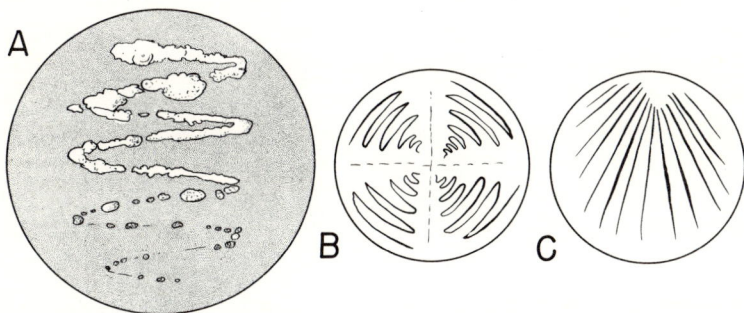

FIG. 3–4. Agar plates. A. Bacterial colonies growing along a streak. B–C. Alternate methods of streaking agar plates.

48 Biological Techniques

Pour Plate. The pour-plate technique (Fig. 3–5) may be used with either broth cultures or agar cultures, but is probably most useful for the former. Obtain one tube of sterile, distilled water and four sterile agar tubes with sufficient agar for one plate each. Melt the agar and permit it to cool to a temperature of 45° C. Flame a wire loop, cool, and transfer one loop of the stock culture to the distilled water. Swish the needle through the water, remove, and flame. Mix the culture into the water by rolling the tube back and forth between your hands at least a dozen times. Mark the agar tubes "1," "2," "3," and "control." Be sure to flame the loop before and after each consecutive transfer, and flame the mouth of each agar tube before and after each transfer so that no contamination will result from the removal or replacement of the cotton stopper. Now, transfer two loops of the water solution to tube 1 and rotate this tube for mixing. By like token, transfer two loops to tube 2; mix and transfer two loops to tube 3. Finally, unstopper and flame each tube, one at a time, and pour it in a sterile Petri dish which has a corresponding number (1, 2, 3) as shown in Fig. 3–5. Pour the control tube into a sterile Petri dish and incubate all of the plates from 24 to 48 hours. Depending on the degree of dilution required, one of the plates

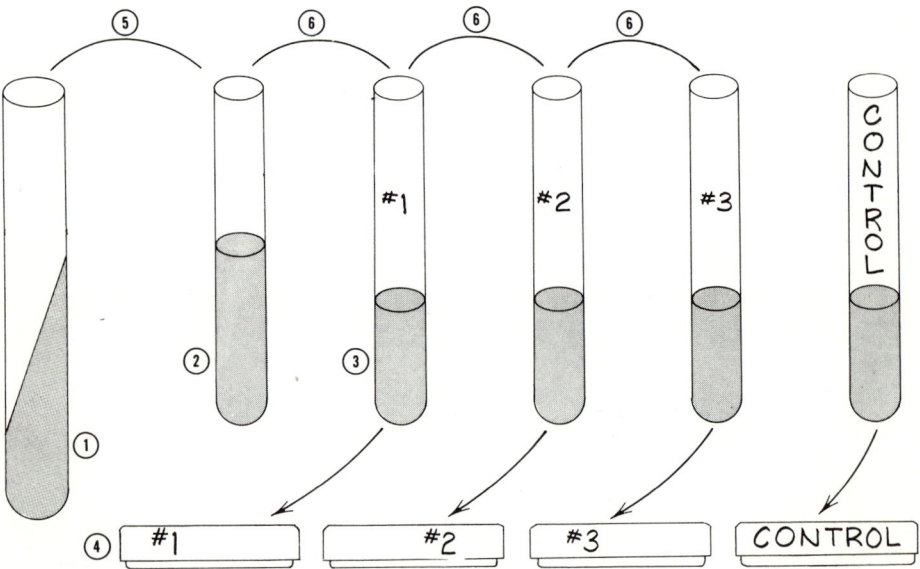

FIG. 3–5. The pour-plate method of diluting bacterial samples.

should have colonies sufficiently isolated from one another to permit these to be recultured in Petri plates. This reculturing technique follows the method described for the streak-plate technique.

Some Staining Techniques

A large number of biological stains may be used in studies of bacteria and molds, but only a few of the most common and most useful are discussed here and are formulated in Appendix C.

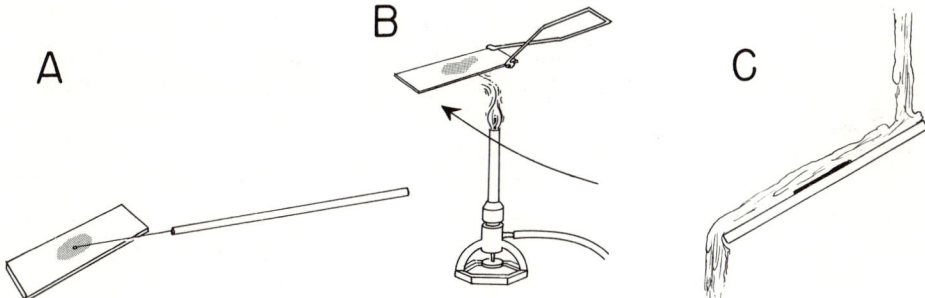

FIG. 3–6. The smear and stain technique (see text).

Simple Smear and Stain Technique. Crystal violet and methylene blue stains are useful for determining the simple morphology of the bacteria. Prepare several clean slides, as directed above, and transfer two loops of sterile water to the center of one slide. Next, flame the loop and transfer a loopful of the bacterial sample; mix this and spread it upon the slide with the loop as shown in Fig. 3–6A. Hold this slide with a wire clip and pass it across a gas flame two or three times at a medium speed. By flaming the underside of the slide (Fig. 3–6B), heat will be conveyed to the culture and will both evaporate the water and fix the bacteria to the slide. Flood the fixed bacteria with one of the stains and permit it to stand for the required amount of time: crystal violet, 5 to 60 seconds; methylene blue, 1 to 2 minutes. Finally, wash the slide, directing a gentle stream of distilled water above the fixed bacteria as shown in Fig. 3–6C. When the stain no longer streams from the bacterial mass, drain the slide, blot it gently with soft tissue paper, and study the slide under the oil-immersion microscope.

Spore Stain. A combination stain technique using carbo-fuchsin and methylene blue prepares bacterial smears so that vegetative cells appear blue and spores are red.

Make a smear and flame as described above. Cover with carbo-fuchsin and heat over a low flame until steaming occurs. Add stain as needed and continue steaming for about 5 minutes. Wash in water and submerge the slide in 5-percent acetic acid for 2 seconds only. Wash again in water and cover the smear with methylene blue for 30 seconds. Finally, wash, blot dry, and examine under an oil-immersion microscope.

Gram's Stain. Because of their different physiological and cytological properties bacteria may be easily separated into two groups, depending on the way they react to a combination stain of crystal violet, safranin, and Gram's iodine. Those species which retain the crystal violet and thus appear dark blue are called "Gram-positive"; those which lose the violet and thus appear red are called "Gram-negative." It is best to work with known species of bacteria until the "reading" of the results becomes routine. Otherwise, there may be some difficulty in discriminating between a light violet and a dark red. Fix a smear to a slide in the usual manner and stain with crystal violet for 1 minute, then wash for a few seconds. Next, apply Gram's iodine for 1 minute and wash. Destain with 95-percent alcohol for about $\frac{1}{2}$ minute until the free stain has been removed. Wash in water and then stain with safranin for 10 seconds. Finally, wash, blot dry, and examine under the oil-immersion microscope.

Culture Destruction or Storage

Culture Destruction. Regardless of the species supposedly involved, destroy all cultures in the autoclaves or pressure cooker at 15 pounds pressure for 20 minutes. This precaution is just a matter of good technique and will prevent dangerous mistakes should occasional pathogenic material be included.

Culture Storage. The growth rate of any culture will quickly drop off, owing to the accumulation of metabolic waste. Therefore, it is necessary to reculture about once a week. The metabolic activity of a culture may be greatly reduced by the low temperatures within a refrigerator; under such conditions the need for reculturing may be reduced by several weeks. Many species may be stored for longer periods of time in a refrigerator when clear mineral oil is added to the slant tube containing them. The oil greatly restricts the flow of oxygen to the bacteria and thus inhibits metabolic activity.

Microscopic Study of Molds

A number of papers in the literature, including that of Hedgcock and Spaulding (1906) and Merilh (1939), deal with the growth of molds for microscopic study and herbarium use. A simple culture-chamber slide method, developed by the Turtox Service Department for the temporary or permanent preparation of molds for microscopic study, is as follows:

A simple but effective moist-chamber slide method for growing thin "sections" of mold colony that lends itself well to photography and observation at any growth stage consists of the following steps: liquefy a tube of agar and spread a thin streak of the melted medium about one inch long parallel to the long axis and near the center of a standard microscope slide which has been previously flamed. As soon as the agar on the slide has congealed, sow spores of the mold to be studied with a flamed bacteriological loop and apply a flamed coverslip. The degree of pressure applied on the slip and the thickness of agar originally streaked on the slide are the factors which control the amount of spread and the depth of the preparation. A well-made culture slide provides plenty of air space for development of the organism and does not necessitate observation close to the edge of the coverslip. The coverglass is anchored to the slide by placing a drop of heated paraffin on each of the four corners in such a manner as to cause an overflow onto the slide. The completed culture slides are then inserted into the slots of a Coplin jar, and the jar is subsequently placed inside of a wide-mouth screw-cap jar containing about an inch of *distilled* water. This sealed moist-chamber can be opened repeatedly for removal of culture slides and for their inspection under the microscope. A mature slide culture can be "fixed" and made into a permanent mount.[*]

LICHENS

Lichens are symbiotic plants composed of fungi, which make up the supporting body and provide the form, and of algae, usually belonging to the green phylum or blue-green phylum. Lichens grow abundantly and are characteristic plants found in the forest. They may be seen as "Spanish moss" hanging in a bearded form from trees or growing attached to bark, branches, mosses, or directly on the ground. In contrast, lichens are also characteristic plants of barren regions such as deserts, where they are found

---

[*] "The Culture and Microscopy of Molds," *Turtox Service Leaflet No. 32*, pp. 1–4, 1959. Quoted by permission of the General Biological Supply House, Inc., Chicago, Illinois.

growing on rocks, or the tundra where they are one of the dominant plant forms. Lichens are often referred to as "soil builders" because of their petrolytic action which disintegrates rocks and converts them into crude soils. Most lichen genera are world-wide in their distribution (Mozingo, 1961).

Lichens grow in three dominant forms of morphology, as shown in Fig. 3–7. The *crustose* form grows directly on the surfaces of soil, rocks, bark, and other suitable substrates. Typically, it does not have a sheet-like appearance and often is so granular that it resembles the surface of the substrate upon which it grows. So perfectly are these plants hidden, in some cases, that their presence is unsuspected by amateur collectors. A second form, call *foliose*, is more or less leaf-like in appearance. Any of the above substrata may support the foliose form. This morphotype may be broadly or narrowly branched, but is pigmented by the algae only on one surface and is completely pale on the under surface. The third form is the *fruticose* which stands erect and often is highly branched. The branches may be cylindrical or rather flattened in cross section, but exhibit the alga growing around the branch, rather than on one surface as in the preceding form. All in all, the lichens may be extremely beautiful in their pigmentation and are sought after for decorative as well as scientific uses.

Collecting Lichens

Collect and transport specimens in plastic bags. A knife may prove useful for dislodging pieces of bark containing lichens, for digging up soil-dwelling

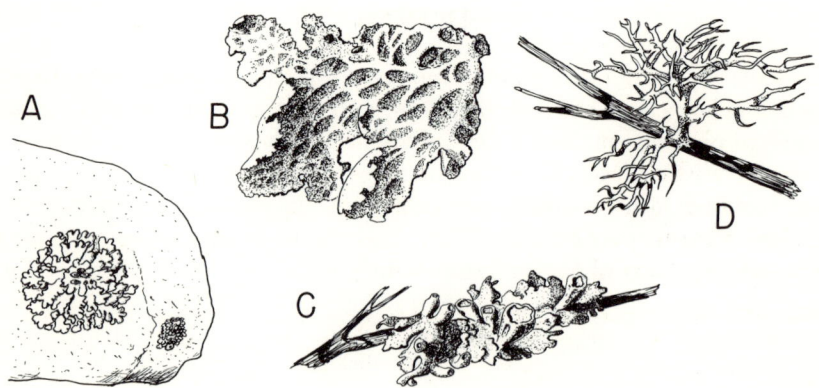

FIG. 3–7. Lichen morphology. A. Crustose. B–C. Foliose. D. Fruticose.

forms, for cutting branches with lichens, and so on. A geology pick and chisel will be essential for chipping off rocks containing crustose lichens. The field notebook should always be carried along for recording of habitat data (see Chapter 5).

Those species of lichens which grow on trees, bushes, mosses, and the like, are removed from the substrate when possible or collected attached to small portions of the substrate. Be careful to get at least some of the rhizoids, the root-like structures which grow on the under side of the lichens and attach them to their substrate. When digging lichens directly from the soil wash them to free the rhizoids from the soil. A thorough washing in the field will greatly simplify the cleaning preparation in the laboratory. Substrata other than soil are often desirable, in small quantities, in order to show the specimen as it was growing in nature.

Those species which grow directly on the surfaces of rocks are more difficult to collect. By carefully looking over the terrain you may find rock specimens small enough to be collected directly. Others may be conveniently sampled with a coal chisel and hammer. Place the chisel in such a way as to fracture off a thin section of rock. Wear glasses or goggles to protect your eyes from rock and metal chips. Look for rock fissures where thin flat layers of rock containing lichens may be pried loose.

Preservation of Lichens

Bulk Methods. Large quantities of lichens are often preserved in bulk for classroom study or when all of the lichen forms from one ecological habitat are being collected for comparison with those of other such habitats. For these objectives, use bulk preservation techniques. Wash dirt, insects, and other debris from the lichens and sort out those specimens to be preserved. Wrap the entire collection, along with a label containing the field data, in a large piece of cheesecloth and tie with a piece of string. Dry the specimens in a warm room and then store them in closed cardboard boxes which include paradichlorobenzene or naphthalene to control insect pests. One of the nice things about lichens is that the form and color will be well preserved by this method. When ready to study the specimens, soak the entire collection, cheesecloth and all, and the lichens will soon appear amazingly life-like and fresh. They may be redried and remoistened as frequently as necessary.

Museum Techniques. Lichens are stored in special envelopes which, in turn, are kept in convenient boxes or pinned to herbarium sheets. Dry foliose and fruticose lichens and place them in labeled envelopes. If these are bulky, place them between newspapers and lightly press by setting a book or board on top of a small stack. Never subject lichens to heavy pressure as the resulting distortion of the specimen will render it of less value. Small specimens may be glued to pieces of herbarium sheet cut to fit inside the storage envelope. Crustose forms growing on bark, branches, rock, and so on, may also be glued to small squares of herbarium sheet and kept in envelopes, unless they are too bulky. If the bulk makes some specimens inconvenient, put them in small boxes, along with their data, and keep these in the herbarium cabinet.

The method of folding envelopes and making standard storage boxes and other equipment used for lichens is described in Chapter 4.

To study dried lichens, simply soak them in water for about ½ hour. The specimens will absorb sufficient water to regain their bulk and may be sectioned or otherwise studied. Be sure to fumigate and maintain small quantities of paradichlorobenzene or naphthalene flakes with lichen specimens, as they are subject to attack by insect pests.

MUSHROOMS AND OTHER HIGHER FUNGI

This section deals with those plants commonly referred to as mushrooms, coral mushrooms, shelf fungi, bird's nest fungi, puff balls, and the like (Fig. 3–8). For the most part, these forms are merely fruiting bodies, or reproductive bodies, which appear seasonally as offshoots projecting from the main body of the plant (mycelium) which grows within soil, dead wood, and so on.

Taxonomic Features

Unlike the woody species of the higher fungi, which undergo little distortion as a result of drying, the fleshy members of this group may suffer great morphological changes and color alteration due to drying. It is essential, therefore, to make notes concerning the morphology and color of the fungi as well as field notes concerning the locality of the collection.

Morphology. The morphology of the common mushroom form (Fig. 3–8A) consists of the cap (pileus), stem (stipe), and the gills which bear spore-producing bodies. In addition, an annulus, or ring, may girdle the stem and a volva, a sac-like membrane, may receive the base of the stem. After the mushroom has been carefully extracted from its substrate, field notes should at least record the presence or absence of the typical body

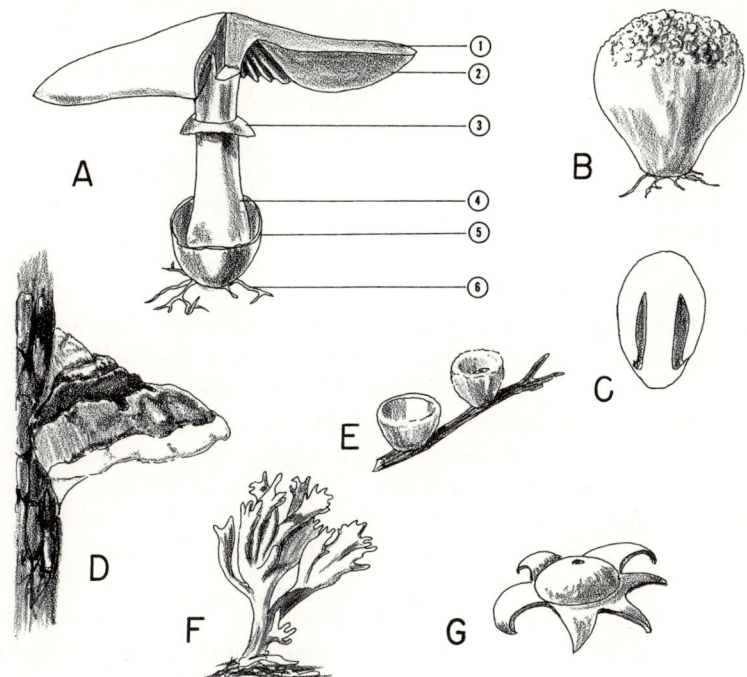

FIG. 3–8. Morphology of higher fungi. A. Mushroom morphology. (1) Cap, (2) gill, (3) annulus, (4) stem, (5) volva, (6) "roots" or rhizoids. B. Puffball. C. Cross-section through a young mushroom. D. Bracket fungus on tree bark. E. Bird's-nest fungi. F. Coral fungus. G. *Geastor*.

structures and describe the attitude of the plant itself. Sketches may be essential to denote the position of the cap, with reference to the stem of the more delicate species. The cap may be turned down, directed outward or even directed upward like the walls of a funnel. The manner in which the gills attach to the cap and/or the stem, the position of the annulus with reference to the length of the stem, and the overall shape of the basal portion of the stem where it projects into its substratum must also be noted. The surface texture of the cap and stem may vary from waxy to coarse or porous. Furthermore, it may be smooth or bare, have hair-like, flake-like, or scab-like projections, or conversely, be sculptured and eroded in various patterns. For an excellent series of diagrams relative to these and other characteristics, see Snell and Dick, *A Glossary of Mycology* (1957).

Color. Almost all identification of mushrooms will involve color; hence, accurate records on that subject must be kept. The color of both young and mature specimens should be recorded (see Chapter 1) by means of published color standards, by special color standards presented in some mushroom books, or by recording colors with water-color paints or colored pencils. Note carefully the color of the stem, cap, gills, and spores. The gills change color drastically with age, and may produce spores that appear in a color other than that of the gill. Therefore, spore prints must be made, as described below.

Other Characteristics. Such characteristics as taste and odor, the presence of a milky sap, and the ability to glow in the dark occasionally prove of taxonomic importance. Thomas (1948) describes taste as "acrid (biting, peppery); astringent (puckery, bitter); disagreeable but not acrid; mild; sweetish." Professional collectors generally record these features in their field notes at the time of collecting (using caution not to swallow any portion of mushrooms being tasted). When describing the locality of a collection be sure to indicate the surroundings of the mushroom, as to whether they were dark and dank or exposed and well lighted. The nature of the substratum supporting the specimen should also be recorded, as well as whether the specimen grew in large clusters, in rings, or in solitude. The season and prevailing weather conditions may also be of some importance.

Collecting the Higher Fungi

Depending on the locality, mushrooms and other higher fungi may appear continuously all year long, although spring, summer and autumn will produce the greatest number of species. Careful records should be kept concerning the weather conditions preceding and during the appearance of different species in order to record the seasonal succession of the fungi.

A box or picnic basket should be used as a collecting receptacle (Burt, 1898; Thomas, 1948; Smith, 1949), as the fleshy species will be crushed and destroyed if they are collected in a plastic bag or similar receptacle. In addition to a basket or cardboard box with rope handles, take along some smaller boxes in which to place delicate forms and some soft paper and string for wrapping individual collections. Paper toweling or premoistened newspaper may be quite suitable for this purpose. Burt (1898) recommends recording field data on small slips of paper and including these in each bundle in order to keep one collection separated from another. Use a large knife or trowel for digging specimens from the soil. A wood chisel or knife may be used for removing specimens from bark and other wooden substrates. Be careful to collect all of the structures surrounding the base of the mushroom where it enters the soil; do not cut the stem at the ground line. Collect large series and attempt to get all stages of development from the button stage on up to fully mature specimens. Spore prints (discussed below) may be initiated at the time of collecting or in the laboratory. If there is any delay in field work or in processing specimens in the laboratory, keep the specimens cool, but do not freeze them. Freezing distorts the tissue and renders the specimens unusable. Finally, record color notes and other field notes before leaving the field. Do not rely on your memory for field data.

Preservation

Remarks. Woody specimens present no real problems in preservation; they are simply dried, fumigated, and stored in boxes or other suitable containers along with their field data and identification slips. Fleshy fungi, on the other hand, cannot be satisfactorily preserved. In liquid preservation (with FAA or 10-percent formalin) the form is retained, while overall shrinking occurs. However, the color quickly disappears and, ultimately, the flesh becomes macerated and will break down. On the other hand, heat drying will retain many of the colors, although considerable shrinking and distortion will follow. Heat drying, nevertheless, is quite permanent and proves to be the only suitable method for museum collections. Frozen specimens are almost always reduced to a disintegrated mess upon thawing. Freeze-drying techniques (see Chapter 13) are suitable for museum displays, but render specimens too fragile for taxonomic work.

Spore Prints. Spore prints should always be made of specimens intended for taxonomic work. Select mature specimens with expanded caps and cut off the stem as close to the cap as possible (Fig. 3–9A). Place the cap, gill-side down on a piece of white paper (index card or herbarium paper) and cover this with an inverted glass or dish (Fig. 3–9B). The dish will prevent drafts from blowing spores as they fall during the 6- to 12-hour waiting period. The walls of the gills are lined with minute cells (basidia) which in turn produce the colored spores (Fig. 3–9C). If the gills appear whitish or very light in color put a piece of black paper under one quarter or one half of the gill area (Fig. 3–9D); an accurate record may be obtained in this fashion. When the cap is removed, apply several thin coats of artist's spray fixative (used for pencil and charcoal drawings) as directed on the can. This will protect the spores. The print then is included in the packet or herbarium sheet with the mushroom specimens. Figure 3–9E shows an actual spore print made in this manner.

58   Biological Techniques

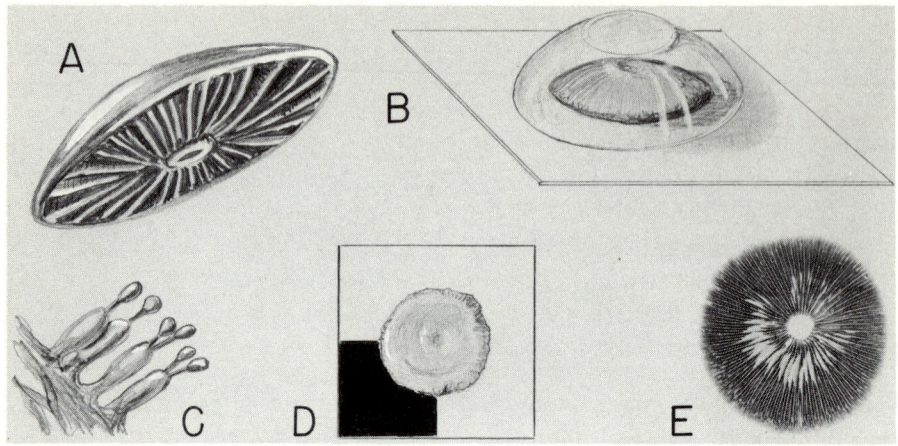

FIG. 3–9. Spore print technique. A–D. See text. E. An actual spore print of a mushroom.

Burt (1898) recommends making spore prints in the field; his technique is slightly modified here. Place the freshly cut cap on a long slip of paper large enough to go under and over the top of the cap. Slide the "sandwich" into a small plastic bag (freezer or sandwich bag) and secure with paper clips. Place the mushroom cap gill-side down in the collecting basket. By this method spore prints will be ready shortly after you have returned to the laboratory.

Liquid Preservation.   Liquid preservation is suitable when specimens are to be held briefly for classroom use, or for liquid displays when these specimens may be replaced periodically. FAA is a suitable preservative for such use, although no preservative is adequate for long-term use. Research is necessary in this area to determine if some other solution might be used which will maintain specimens in a firm condition and, if possible, retain their color.

Drying Techniques.   Mushrooms are best dried for permanent storage. Drying in sunlight or at room temperature is not satisfactory (Burt, 1898) because decay and the attack of insect larvae will be extensive before the specimens are cured. Specimens should be placed on screen trays or shelves made with ⅛-inch screen or window screen. The screen trays are placed above some source of heat, such as a hotplate or other electrical heating element at a distance where the heat will not burn or destroy the specimens.

Dr. Stuntz (in personal communication at the University of Washington) says that heat over 40° C. damages the tissues of mushrooms for taxonomic work. He recommends that temperatures between 35° and 37° C. be used for drying, and that a dilute, 3-percent KOH solution be used to moisten dried specimens when tissue sections are required.

Methods for field drying may be adapted from those outlined in Chapter 5, under "Field Drying Techniques." A very convenient drying chamber is described for the starfishes in the chapter on echinoderms.

Prepare specimens for drying by carefully cleaning away dirt and debris and placing them directly on screens. Place a data sheet nearby, or attach it to the specimens with a string. Large specimens will have to be split lengthwise (Fig. 3–10, A and B) and spread on the screens to dry. Mushrooms are usually fragile and resist bending when first collected, but become pliant during the process of drying (Burt, 1898), so that they can be placed in a desirable way. The caps may be bent down on the stem or spread in a lateral fashion, as shown in Fig. 3–10, C and D. If placed in a moisture chamber or near an open window, mushroom specimens will quickly reabsorb enough water so that final positioning may be completed. Mushrooms must never be pressed, but they may be held flat with a minimum amount of pressure during the final drying process.

Small specimens which are properly placed may be kept in envelopes (described in Chapter 4) and stored in shoe boxes or attached to herbarium sheets. Large specimens, along with their field data, should be placed in small open boxes and kept in trays within the herbarium cabinet. Large woody and fleshy mushrooms which have been dried should be subjected to fumigation to destroy any insects. Place specimens in an air-tight cabinet for 24 hours, along with a dish containing a small quantity of carbon bisulphide. Fumigation can also be accomplished by using excessive amounts of paradichlorobenzene for three or four days. Add small quantities of paradichlorobenzene to all fungi collections in the herbarium cases. The woody fungi are most attractive to insect pests and should be observed during biannual fumigation to ensure that pests have not invaded the collection.

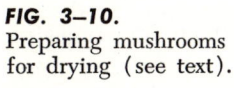
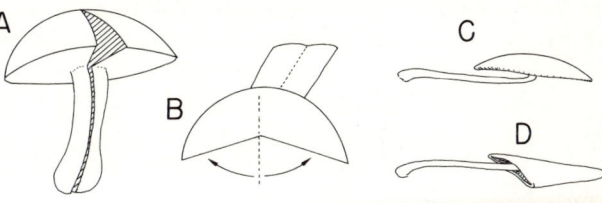

FIG. 3–10. Preparing mushrooms for drying (see text).

REFERENCES

Alexopoulos, C. J., 1952, *Introductory Mycology,* John Wiley & Sons, N.Y.

Barnett, H. L., 1955, *Illustrated Genera of Imperfect Fungi,* Burgess Publishing Co., Minneapolis.

Bergey's Manual of Determinative Bacteriology, 1957, Williams & Wilkins, Baltimore.

Bessey, E. A., 1950, *Morphology and Taxonomy of Fungi,* Blakiston, Philadelphia.

Birkeland, Jorgen, 1949, *Microbiology and Man,* Appleton-Century-Crofts, N.Y.

Burt, E. A., 1898, On Collecting and Preparing Fleshy Fungi for the Herbarium, *Botanical Gazette,* 25:172–187.

Christensen, C. M., 1943, *Common Edible Mushrooms,* Univ. of Minnesota, Minneapolis.

Christensen, C. M., 1955, *Common Fleshy Fungi,* Burgess Publishing Co., Minneapolis.

The Culture and Microscopy of Molds, 1959, *Turtox Service Leaflet No. 32,* pp. 1–4, General Biological Supply House, Chicago.

Eifert, V., 1952, *Exploring for Mushrooms,* Illinois State Museum, Springfield.

Fitzpatrick, H., and W. Ray, 1955, Some Common Edible and Poisonous Mushrooms, *Cornell Extension Bull.,* 386, Ithaca, N.Y.

Funder, S., 1953, *Practical Mycology Manual for Identification of Fungi,* Hafner Publishing Co., N.Y.

Hedgcock, G. G., and P. Spaulding, 1906, A New Method of Mounting Fungi Grown in Cultures for the Herbarium, *J. of Mycology,* 12:147.

Henrici, A. T., 1947, *Molds, Yeasts and Actinomycetes,* John Wiley & Sons, N.Y.

Krieger, L. C. C., 1947, *The Mushroom Handbook,* Macmillan, N.Y.

Merilh, E. L., 1939, A Method of Cultivating and Mounting Molds, *Turtox News,* 17(2):50.

Mozingo, Hugh N., 1961, "Lichens" in *The Encyclopedia of the Biological Sciences* (Peter Gray, ed.), Reinhold, N.Y., pp. 559–560.

Nearing, G. G., 1947, *The Lichen Book,* published by the author, Ridgwood, N.Y.

Ramsbottom, J., 1953, *Mushrooms and Toadstools,* William Collins Sons, N.Y.

Sarles, W. B., *et al.,* 1956, *Microbiology: General and Applied,* Harper & Brothers, N.Y.

Smith, A. H., 1949, *Mushrooms in Their Native Habitats,* Sawyer's Inc., N.Y.

Smith, A. H., 1951, *Puffballs and Their Allies in Michigan,* Univ. of Michigan Press, Ann Arbor.

Smith, G. M., 1955, *Cryptogamic Botany, Vol. I: Algae and Fungi,* McGraw-Hill, N.Y.

Snell, H., and E. A. Dick, 1957, *A Glossary of Mycology,* Harvard Univ. Press, Cambridge, Mass.

Stanier, R. Y., *et al.*, 1957, *Microbial World,* Prentice-Hall, Englewood Cliffs, N.J.
Stuntz, D. E., 1963 (personal communication, Univ. of Washington, Seattle).
Thimann, K. V., 1955, *The Life of Bacteria,* Macmillan, N.Y.
Thomas, S. J., and T. H. Grainger, 1952, *Bacteria,* McGraw-Hill., N.Y.
Thomas, W., 1948, *Fieldbook of Common Mushrooms,* Putnam, N.Y.

CHAPTER 4

The Mosses and Liverworts

This chapter will include a brief discussion on the so-called "club mosses" (Tracheophyta), in addition to the true mosses and the liverworts which are members of the phylum Bryophyta.

THE MOSSES

Introductory Remarks

The true mosses, found the world over, are moisture-loving plants which possess root-like, stem-like, and leaf-like structures in the gametophyte generation, but which lack the vascular tissue of the higher plants. In the mosses the sexes are separate and gametes develop in large reproductive organs, the antheridia and archegonia. These plants experience alternation of generation, the gametophyte generation being conspicuous, in contrast to the vascular plants. The sporophyte is parasitic and is borne on top of the female gametophyte (Fig. 4–1). On the other hand, the club mosses, which are members of the true vascular plants, differ in having vascular tissue, a conspicuous sporophyte generation and an inconspicuous gametophyte generation. In reproduction the club mosses (which go by such names as ground pine, ground fir, ground cedar, and so on, because of their appearance) reproduce in a manner almost identical to that of the ferns. This group bears reproductive cones, usually on upright stalks. The shape of the gametophyte, its posture, branching, leaf structure, leaf margin, cellular structure, and the nature of the sporophyte and its capsule, all are important characteristics of this group. It is essential, therefore, not only to collect mosses at random, but to attempt to get mature gametophytes, with their gametangia well developed, in one season of the year, and to collect specimens in which the capsule of the sporophyte at a point of maturity during the opposite season. The same is true of the club mosses, for specimens gathered without the fruiting cones will be of less value taxonomically. Students beginning to study mosses are encouraged to read *How to Know the Mosses and Liverworts*, by H. S. Conard (1956).

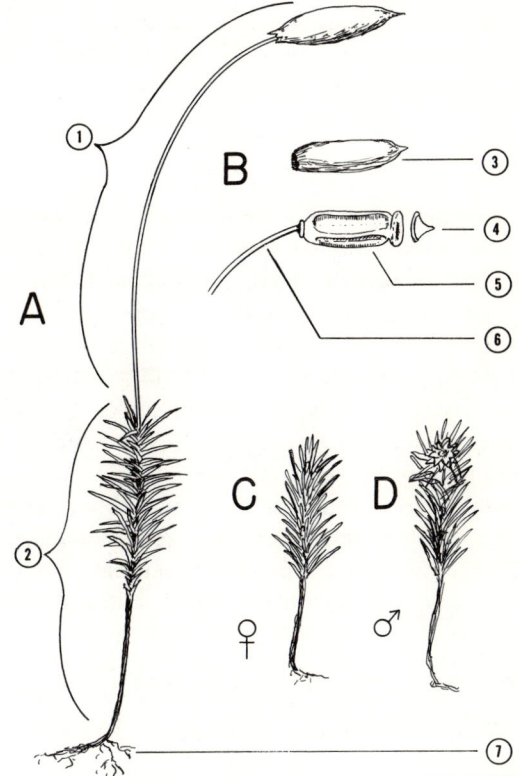

FIG. 4–1. Simple moss morphology. A. Two generations growing together. B. Capsule morphology. C. Female gametophyte. D. Male gametophyte. (1) Sporophyte generation, (2) gametophyte generation, (3) the cap, (4) operculum, (5) sporangium, (6) stalk, (7) rhizoids.

Collecting Methods

Provide yourself with a vasculum, a large plastic bag, or knapsack in which to carry specimens. Use small plastic freezer bags or sandwich bags and rubber bands to isolate individual specimens, along with their field data. Take along a pocketknife and a trowel for dislodging specimens attached to stones, bark, and the like, or for digging specimens from the soil. Be careful to dig up specimens so as to obtain the root-like rhizoids. Select only specimens that are mature and which possess reproductive structures. Several field trips throughout the year may be essential if the collector is to obtain local specimens in peak condition. Collect a large number of specimens whenever possible. Carefully wash mud and debris away from the rhizoids and basal portions of the gametophyte. Specimens that grow upon rocks, bark, and so on, have a habit of collecting soil which should be washed away. When collecting free-floating mosses from ponds, select specimens that are actively growing rather than dead or dormant specimens. Isolate each collection and insert a field note and number referring to the collecting site and data in the field notebook.

Preservation of Specimens

Arrangement and "Pressing." Specimens to be stored in envelopes or on herbarium sheets are carefully washed, blotted partially dry, and arranged on several thicknesses of newspaper folded in quarters. Position the specimens to display all of the essential body parts to their best advantage. Highly branched forms must be spread out, to avoid multiple overlapping at any one point. Conard (1945) recommends stacking papers with specimens in piles of about 12 inches each, and then adding a light weight of about 2 pounds to "press" the specimens. Do not add enough weight to literally "press" the specimen, but only enough to keep it in a two-dimensional plane. When leaflets and other structures are flattened against the main body of the plant the taxonomic characteristics are difficult to analyze. If specimens become misshapen they may be dipped in water for a few moments, until they are pliable, and then "repressed" in the desired posture.

Envelope Storage Technique

Most collections of mosses are kept in envelopes or packets, along with their field data. These envelopes are in turn stored in boxes 4 by 6 by 17 inches, or they are attached to herbarium sheets.

Select a good grade of 20-pound bond paper with at least 50-percent rag content, in the standard 8½- by 11-inch size. If boxes (shoe boxes or the like) are used for storage, determine an envelope size that will conveniently fit into them. Standard envelopes are usually 5½ inches wide and from 3¾ to 4½ inches high. The older style of envelope, still in use, can be made by folding the bottom upward to the desired height of the envelope, folding the top flap down, and creasing the ends backwards and underneath (Chamberlain, 1903). (See Fig. 4–2, A through C.) This envelope has the slight disadvantage of requiring considerable bending, which may damage the specimen on reopening. Advocates of this style, however, feel that this envelope will not open by itself and release the specimen.

A second style of envelope frequently in use is made from the same quality of 8½- by 11-inch sheets of paper, as follows: fold the lower margin up to the desired height of the envelope (Fig. 4–2A); next, fold the side margins in to produce the desired width of the envelope; finally, fold the top flap down as shown in Fig. 4–2, D and E. This envelope may readily be opened for examination of the specimens and will not release the specimens spontaneously.

For folding large quantities of envelopes a preruled board with sharp edges will prove handy. Rule the board to indicate the position of the paper for each of the successive folds, numbering each ruled line 1, 2, 3, and so on, to indicate the sequence of folding. Place the board on the edge of a table so that the paper may hang over when folded along the edge of the board. A substitute for this apparatus is the sharp cutting edge of a paper cutter, which also has half-inch rulings suitable for folding envelopes.

Write the scientific and common name of the specimen, along with its field data, on the upper flap of the envelope. (Consult Chapter 5 for field label data and techniques, as well as the rubber-stamp technique, which should prove useful in this instance.) Half-size envelopes are frequently made to hold very small specimens. These, in turn, are stored within the larger envelope. Envelopes containing specimens are arranged in taxonomic order by family, genus, and species and stored in convenient boxes. Many excellent collections are kept in shoe boxes, while larger herbariums have special boxes made 4 by 6 by 17 inches which fit conveniently into herbarium cases. The alternative to this method is to affix the envelopes to herbarium sheets with pins (Fig. 4–2F), staples, or glue. The former method is preferred, in that the envelopes can readily be moved from one sheet to another.

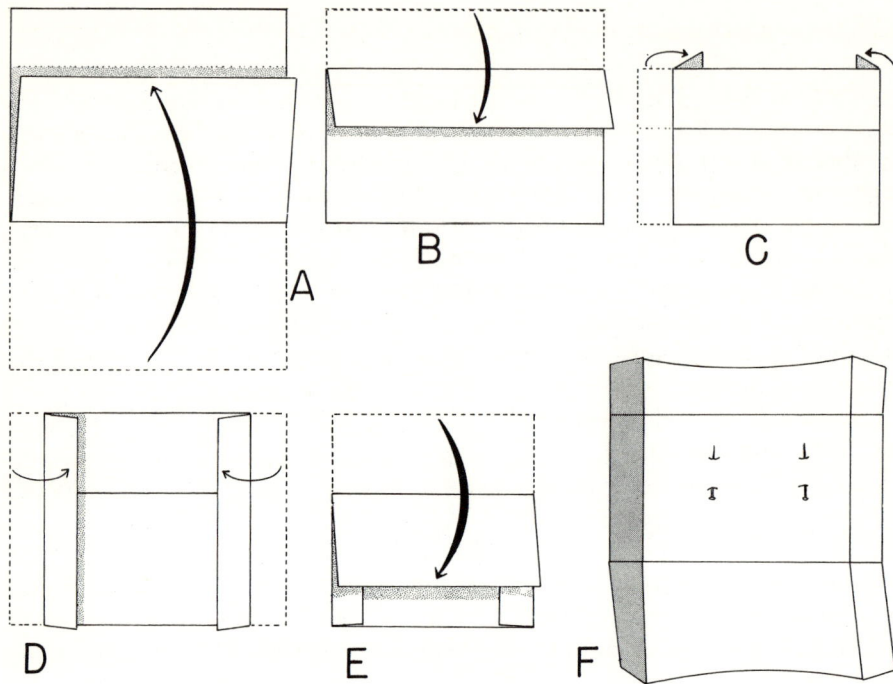

FIG. 4–2. Two methods of folding envelopes for storing all types of botanical specimens, either in filing boxes or by pinning them directly to herbarium sheets (see text).

Herbarium Sheet Mounts

A second common practice is to mount moss specimens directly on herbarium sheets. For this technique follow the instructions in Chapter 5. The glue-plate method will probably prove most suitable. Direct mounts have the advantages of making specimens more easily accessible and more readily viewed. However, it is impossible to adequately study such specimens under the microscope or to boil or steam them in order to manipulate the various parts of the foliage. Because of their fragile nature mosses must be protected in genus folders (Chapter 5).

Bulk-Dry Methods

If it is desirable to preserve large quantities of mosses for classroom observation, or to preserve the larger club mosses in a natural posture, the bulk-dry method will be convenient. Mosses are cleaned in the usual manner and carefully arranged on a piece of cheesecloth, along with a slip containing the field data. The cloth is then wrapped in such a manner as to contain the moss but not to distort its shape. Tie the corners of the cheesecloth together or secure the bundle with string. Allow the specimens to dry and store them in a pasteboard box with some paradichlorobenzene. For larger club mosses record the collecting data and scientific name on a waterproof label (1 inch by 2 inches) and tie this to the stem of the fresh moss. Moss specimens so prepared should be carefully placed in a cardboard box and permitted to dry. When they are dry, some paradichlorobenzene is added and the box is tied shut and stored in a dark place. These dry specimens are quite adequate for casual observation or they may be subjected to soaking, steaming, or even dipping in boiling water, to permit their manipulation in the classroom.

Liquid Preservation

Moss capsules and other structures preserved at their prime in liquid are useful for taxonomic or classroom dissections. For normal preservation use FAA. Chapter 5 describes a solution that will to some extent preserve the natural green color of the plant, if this is desired.

LIVERWORTS AND HORNWORTS

At first glance liverworts are inconspicuous plants produced as a flat thallus or as a central axis with simple or multilobed leaves. Hornworts resemble the flattened thallus of the liverworts, but tend to produce erect, spindle-shaped sporophytes from which the common name is derived. As in the cycle of the mosses, the gametophyte generation of liverworts and hornworts is most conspicuous, whereas the sporophyte generation tends to be less conspicuous in appearance. Alternation between the asexual and sexual generations is the common mode of reproduction. However, simple asexual reproduction by budding within the gemmae cups or by fragmentation is also common. Most of the liverworts possess root-like rhizoids, which are produced from single cells; these provide a means of attachment to the substrate. Both generations are photosynthetic, and thus have a lusty green color.

Habitat and Collecting

Liverworts prefer moist, shaded areas usually associated with some source of water. Stream and lake banks, moist road banks in forested areas, and similar habitats, will support the terrestrial species. Look carefully among grasses and other vegetation for liverworts growing in any of these situations. Where the ground is moist and damp, lift up bushes and overhanging branches to inspect for liverworts. Ponds and shallow lakes are a favorite habitat for the aquatic species. Proskauer (1961) suggests that hornworts require a temperate microclimate with acidic substrates mostly of soil, but occasionally on bark.

Use plastic freezer bags of about one-quart capacity for collecting. A knife, trowel, or dip net (see Chapter 13 for construction) are used for removing specimens from their substrate or dipping them from water. Place the specimens in plastic bags, inflate the bag, then seal it with a rubber band. When placed in a carrying box, the inflated bag will prevent crushing when other specimens are added. Take careful field notes concerning the exact microhabitat, dominant species of plants associated with the habitat, and light and moisture conditions. Include a copy of this information or a reference number to these data in the field notebook in each plastic bag.

Collect well-formed, mature specimens showing ripened reproductive bodies whenever possible. Liverworts are easily kept alive in the aquarium or laboratory terrarium and may be cultured until reproductive stages appear and mature. This is the only suitable way of getting excellent specimens, unless numerous field trips can be made. If kept cool and provided with some moisture, specimens will last many days in their plastic bag containers and can be transported from very long distances. Provide an aquarium with pond water for aquatic species; for terrestrial species provide a sloping surface in a terrarium containing a small quantity of water. A mixture of peat moss and loam makes an excellent substrate in which to plant liverworts. Place a sheet of glass over the aquarium or terrarium to maintain the humidity level, and keep the specimens in somewhat subdued light such as that from a north window.

Preservation Techniques

Liverworts should be pressed between specimen sheets and dryers with only moderate pressure, following the techniques described in Chapter 5.

Should specimens tend to stick to the specimen sheets, place them directly on herbarium paper and cover them with cloth, as described for the marine algae. Either the gummed-cloth or herbarium-paste method (both discussed in Chapter 5) should be used for attaching specimens to standard herbarium sheets. The herbarium label, containing the scientific name and field data, is attached to the lower right-hand corner of the sheet. Herbarium sheets are stored in genus folders. An alternate (and preferable) method is to store the pressed material in moss envelopes, as described above. In pressing, be sure to apply only enough pressure to hold the plant in a two-dimensional plane, but not enough to crush it or totally flatten it.

Liquid Preservation

For many purposes liquid preservation is quite suitable. It is advisable to preserve some specimens in liquid, even though the bulk of these are to be dried for herbarium sheets or envelopes. Several solutions are quite suitable for this work. (1) A 4-percent formalin solution, with or without a small quantity of copper sulfate, should be used for general preservation. (2) FAA (see Appendix C) is very suitable for general preservation and should be used for specimens intended for microscope slides. Again, a small quantity of copper sulfate may be added if desired. (3) A more complex formula for maintaining the green coloration of plants is presented in Chapter 5.

REFERENCES

Bodenberg, E. T., 1954, *Mosses: A New Approach to the Identification of Common Species*, Burgess Publishing Co., Minneapolis.

Chamberlain, Edward B., 1903, Mounting Moss Specimens, *The Bryologist*, 6:102–103.

Conard, H. S., 1944, *How to Know the Mosses*, William C. Brown, Dubuque, Iowa.

Conard, H. S., 1945, The Bryophyte Herbarium. A Moss Collection: Preparation and Care, *The Bryologist*, 48(4):198–202.

Conard, H. S., 1956, *How to Know the Mosses and Liverworts*, William C. Brown, Dubuque, Iowa.

Grout, A. J., 1947, *Mosses with a Hand Lens*, published by the author, Staten Island, N.Y.

Jennings, O. E., 1951, *Manual of the Mosses of Western Pennsylvania and Adjacent Regions*, Univ. of Notre Dame Press, Notre Dame, Ind.

Proskauer, J., 1961, "Hepaticae" in *The Encyclopedia of the Biological Sciences*, (Peter Gray, ed.), Reinhold Publishing Corp., N.Y., pp. 472–474.

Richards, P. W., 1950, *Mosses*, Penguin Books Baltimore.

Watson, E. V., 1955, *British Mosses and Liverworts*, Cambridge Univ. Press, N.Y.

CHAPTER 5

The Higher Plants (Tracheophyta)

The phylum Tracheophyta contains a number of subphyla with diverse species which are all related by the common characteristic of having tracheids (a type of vascular tissue) or tracheid derivatives. To most of us the tracheophytes are the most familiar and important plants. Some of the plant groups involved, their characteristics, and the techniques required for their study will be discussed.

SUBPHYLA OF THE HIGHER PLANTS

Club Mosses

The club mosses (Lycopsida) resemble the true mosses and some of the gymnosperms, or conifers, in both appearance and foliage. The plant consists of a rhizome (Fig. 5–1F) which bears roots and gives rise to erect stems. In season the erect stems will bear cones, or strobili. These cones bear both kinds of sporangia and thus, popularly but incorrectly, are both "male" and "female." The entire plant, including rhizomes and roots, should be collected and dried in the manner of true mosses (see Chapter 4).

Horsetails

The horsetails, subphylum Sphenopsida, are primitive plants that are widely distributed. They grow profusely along some highways, on wet banks near lakes, and in moist forested areas. Horsetails somewhat resemble the ferns in their gross structure and life history. The plant consists of a rhizome which gives rise to roots and to stems bearing needle-like leaves. Some of the stems give rise to spore-bearing cones. When released, the spores develop into the bisexual gametophyte generation. Collect all plant parts and, if necessary, make special trips to get immature and mature cones. The gametophytes may be reared in the same manner as fern prothallia.

Ferns

Ferns represent one large class of plants belonging to the subphylum Pteropsida. This subphylum also includes the conifers and the true flowering plants. In collecting ferns, gather more than just the frond, or leaf (see Fig. 5-1, A, B, and C); get the entire stipe, the rhizome, and some of the roots. Collect some fronds bearing reproductive bodies (sporangia) and mount these so that the sporangia may be seen on the herbarium sheet. The sexual, or gametophyte, generation of ferns may be reared on agar, with plant food added, or on clean bricks which are kept moist with a solution of plant food and water. These are best preserved in liquid or on permanent microscope slides.

Conifers

The conifers represent the class Gymnospermae of the subphylum Pteropsida and include such familiar plants as pines, cedars, and firs. In collecting throughout the year obtain both the "male" and "female" cones, as well as needles, terminal branches, growth buds, and the like, from the same plant. Take notes on the nature of the bark, the height of the tree, the altitude at which it occurs, the nature in which the cone comes off the branch, and other things of taxonomic use. Large cones are dried; smaller ones may be preserved in liquid or pressed, along with the main plant specimen.

Flowering Plants

The flowering plants, the Class Angiospermae, of the subphylum Pteropsida, should be carefully collected so that all plant structures, includ-

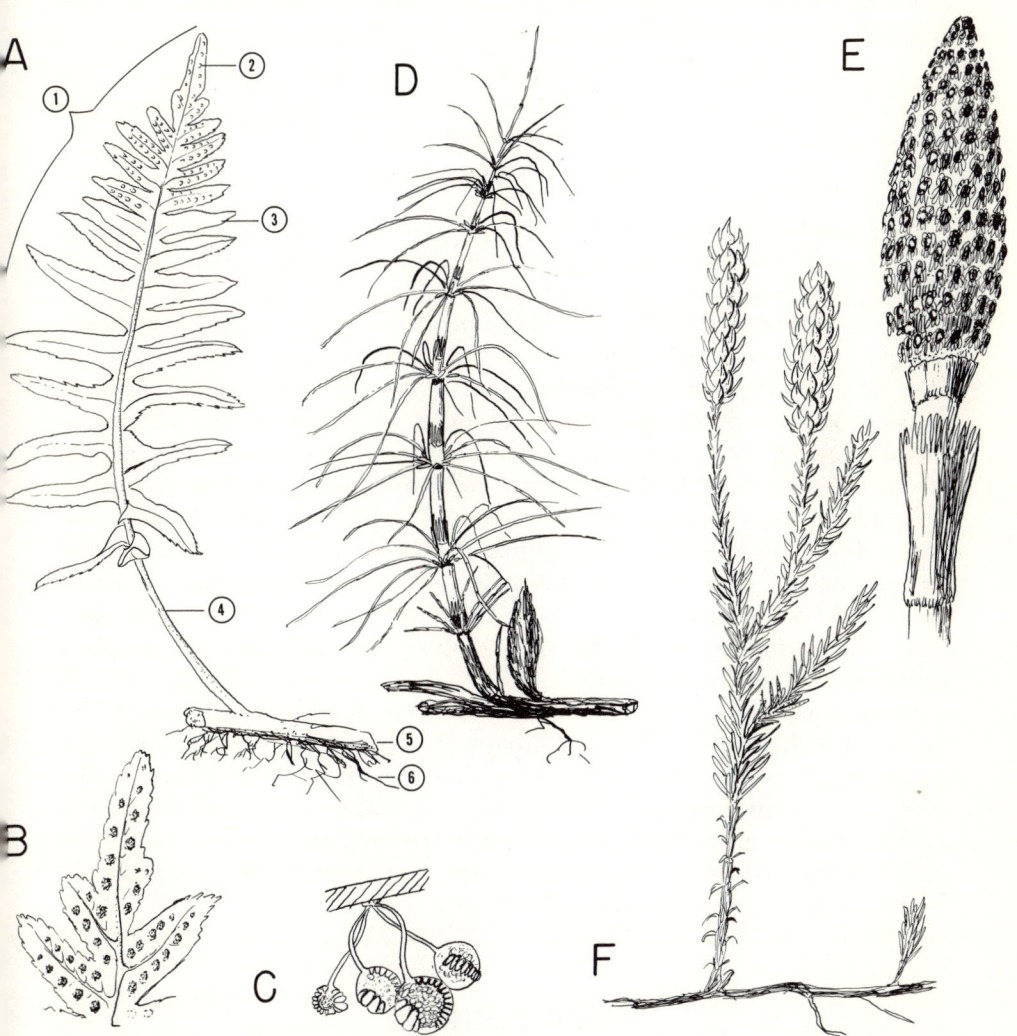

FIG. 5–1. Some higher plants. A–C. Fern morphology. D–E. Horsetails. F. Club moss. (1) Frond, (2) sporangia, (3) pinna or leaflets, (4) stipe, (5) rhizome, (6) rhizoids or roots.

ing fruits and seeds, are obtained. The general description of collecting methods which follows gives many suggestions for the treatment of this material.

PURPOSE OF THE HERBARIUM

Plant collections maintained by botanists vary in both size and purpose. Students who are interested in plant taxonomy in general, or the taxonomy of some family or genus of plants in particular, may make very extensive collections. When carefully identified and labeled, the specimen becomes an important reference for illustrating the particular species characteristics and for checking future identifications. With reference collections new students may learn the procedures of plant identification and determine the accuracy of their own work by referring to the herbarium collection. In taxonomic collections field data, plant measurements, and plant characteristics are an important part of the preserved specimens. The collector must be careful to get all of the essential plant portions which serve as taxonomic features and to cross reference his collection in such a way that all forms of collections may easily be surveyed. Complete specimens with ample field data lend themselves to studies of ecology, plant life histories, and plant geography, as well as taxonomy.

Many teachers maintain small plant collections of selected material to illustrate such things as different groups of plants, types of leaves, flowers, fruits, seeds, cones, or other features that may be of importance in the teaching of plant sciences. Such collections should be made with the utmost care to provide all of the taxonomic parts required and to demonstrate the proper technique for herbarium specimens. Poorly prepared specimens are of little value in teaching on any level.

FIELD METHODS AND MATERIALS

Collecting Tools

A number of small tools will be required for field collecting. A pocketknife, hand pruning shears, field notebook, 6-foot steel tape, some form of trowel, a vasculum, and a plant press make up the essential list. The trowel is used for extracting the roots of small plants; the steel tape may be used to take data on height of the population. The vasculum consists of a metal cylinder with a sliding door, usually worn on a strap over the collector's shoulder, into which plant specimens are placed. However, large plastic bags are more versatile and suitable than the vasculum, for they prevent more fully dehydration of specimens between the time of collecting and the time of pressing. A collector should carry several small plastic or paper bags for fruits, bark, and other specimens.

Methods of Collecting

Collecting does not start in the field. In order to collect usable plant specimens you must know what essential plant parts are needed from each group of plants. Thus a study of plant keys and literature (see the characteristics given above, or Fogg, 1940, and others) will ensure more accurate field collecting.

When collecting small herbs, pick out plants of medium size which show both mature flowers and (if possible) mature fruits (immature flowers and fruits are of little value). Get the complete plant, including the roots, stems, leaves, and floral and fruiting portions. Collect extra flowers and fruits for identification. Carefully dig up the plant and wash the dirt from the roots so that the root mass may be included on the finished pressed specimen. Specimens up to 3 feet in length may be handled in their entirety, provided they are not too thick and woody, by bending the stem once or twice in order to make the plant fit the herbarium sheet. Collect many specimens of smaller plants and several of the medium and larger ones to ensure that all of the essential characteristics will be displayed on the herbarium sheet. Collect within an area repeatedly, if necessary, to obtain mature flowers and fruiting bodies. Take ample field notes and measurements to complete the record.

When collecting larger plants, choose portions with flowers, fruits, and a representative of leaf types. Leaves on the terminal branches of trees often differ from those down low on the trunk, so that several samples must be taken. Samples of bark and fruits may be put into plastic bags, along with a field number, and later may be incorporated with the finished specimen. If you are looking for cones, seeds, or fruits of various plants, make sure those specimens picked up from the ground actually came from the plant in question. In some plants the flowers, leaves, and fruits appear at different times; thus, tag a particular tree or plant and collect from the same specimen throughout the year until all of the essential parts are obtained. Higher branches may be collected with the aid of a 40- or 50-foot rope. Fill a small cloth bag with dirt and tie this securely with a piece of string. Tie the end of the string to the rope in such manner that if the bag becomes entangled in the tree the string will break and release the rope. Next, carefully toss the weighted end of the rope into the outer branches of the tree and, when the rope is entangled, pull the specimen down from the tree.

When collecting club mosses, horsetails, or ferns make sure that root specimens, complete stems, and fruiting bodies, as well as sterile portions, are collected.

Put specimens into the press as quickly as possible, using the technique described below, making sure that the field data or a field number is placed on each specimen sheet. Coarse brush that will not quickly wilt may be kept in the order of collection by attachment to a stout cord with slip knots.

In the tropics the almost constant rain and extreme humidity make the drying of specimens and the maintenance of specimens almost impossible without artificial heat and other facilities. Schultes (1947) found that collecting by normal methods in the Amazon drainage area of Colombia was impossible and resorted to temporary preservation of plant specimens. His method consisted of pressing plants for 24 hours between sheets of folded newspaper, after which they were dipped for a few moments in a strong formaldehyde solution consisting of two parts of commercial 40-percent formaldehyde and three parts of water. The dripping specimens were then replaced in their newspaper folders and put back into the press. The specimens, press and all, were then wrapped in an airtight rubberized bag and shipped to a laboratory where the plants were pressed and dried in the conventional manner. Schultes found that when shipping time was less than 15 days a solution of one part full-strength formaldehyde to two parts water was strong enough to prevent rotting. Moore (1950) reports excellent results from preserving bundles of plants in a 1-percent solution of hydroxy-quinoline sulphate, presumably in plastic bags, and then later pressing the specimens. He also found that palm fruits and flowers could be preserved in the same concentration and transported to the museum in that way.

The Plant Press

The plant press is an indispensable tool by means of which fresh plant specimens are pressed and dried in preparation for permanent mounting. The many styles of plant presses fall into two chief categories, the field press and the laboratory press. These are available through biological supply houses, but a student can quickly make his own press at very little cost.

Laboratory Press. The laboratory press is of heavy construction and is intended to be used in the laboratory or carried in an automobile. It derives its pressure for pressing plants from two threaded bolts, one at either end of the press. Figure 5–2 shows the detail of construction of the laboratory press. In constructing this press, obtain the following items: two pieces of $3/4$-inch waterproof plywood measuring 21 inches by 12 inches; four pieces of 1- by 2- by 12-inch clear fir or other suitable wood (finished board actually measures $3/4$ inch by $1 1/2$ inches by 12 inches); two $1/2$- by 12-inch carriage bolts (not machine bolts), each equipped with one large washer and one large-size wing nut.

To make the top board of the press, attach two pieces of the 1- by 2-inch wood to both ends of the top board by means of a good grade of waterproof wood glue and nails or wood screws. When the glue has dried, drill $5/8$-inch holes through both ends of the top board and the cleat, at a point $3/4$ inch from the end, as illustrated in Fig. 5–2A.

In a machine shop have the threads on the carriage bolts extended down the shank for a distance of 8 inches (Fig. 5–2B). Next, drill $1/2$-inch holes through both ends of the bottom plywood which will be perfectly in line with the holes drilled in the top board. Put each carriage bolt in its hole and with a hammer drive the square shoulder of the bolt into the plywood until the bolt head rests flush against the wood surface. Cut away the center of the 1- by 2-inch cleat, so that it will fit firmly over the bolt without leaving any excess space, and then glue and nail this in place, as illustrated in Fig. 5–2C. The top board should now easily slide down the two bolts and fit uniformly over the bottom board. If, in the future, the carriage bolt should begin to turn, simply drill a hole through the bolt with a small metal drill and drive a nail into the bolt, as illustrated in Fig. 5–2C. The press is now ready for use.

Field Press. The lightweight construction of the field press permits the collector to take it with him into the field and to press plants immediately after collecting. Thus, the press eliminates the wilting and disarrangement of floral parts that might occur. Two easily constructed presses will be described.

Figure 5–3, A and B, show a lightweight press constructed of wood lath. To make this press, obtain some thin, strong, finished lath (this will measure about $5/16$ by $1 3/8$ inches) and cut fourteen pieces 12 inches long, and 8 pieces 18 inches long. Glue and fasten these pieces at right angles, as illustrated. Use either copper rivets and washers or short box nails, which are bent over, for securing the lath pieces. Use two nails or rivets at each inter-

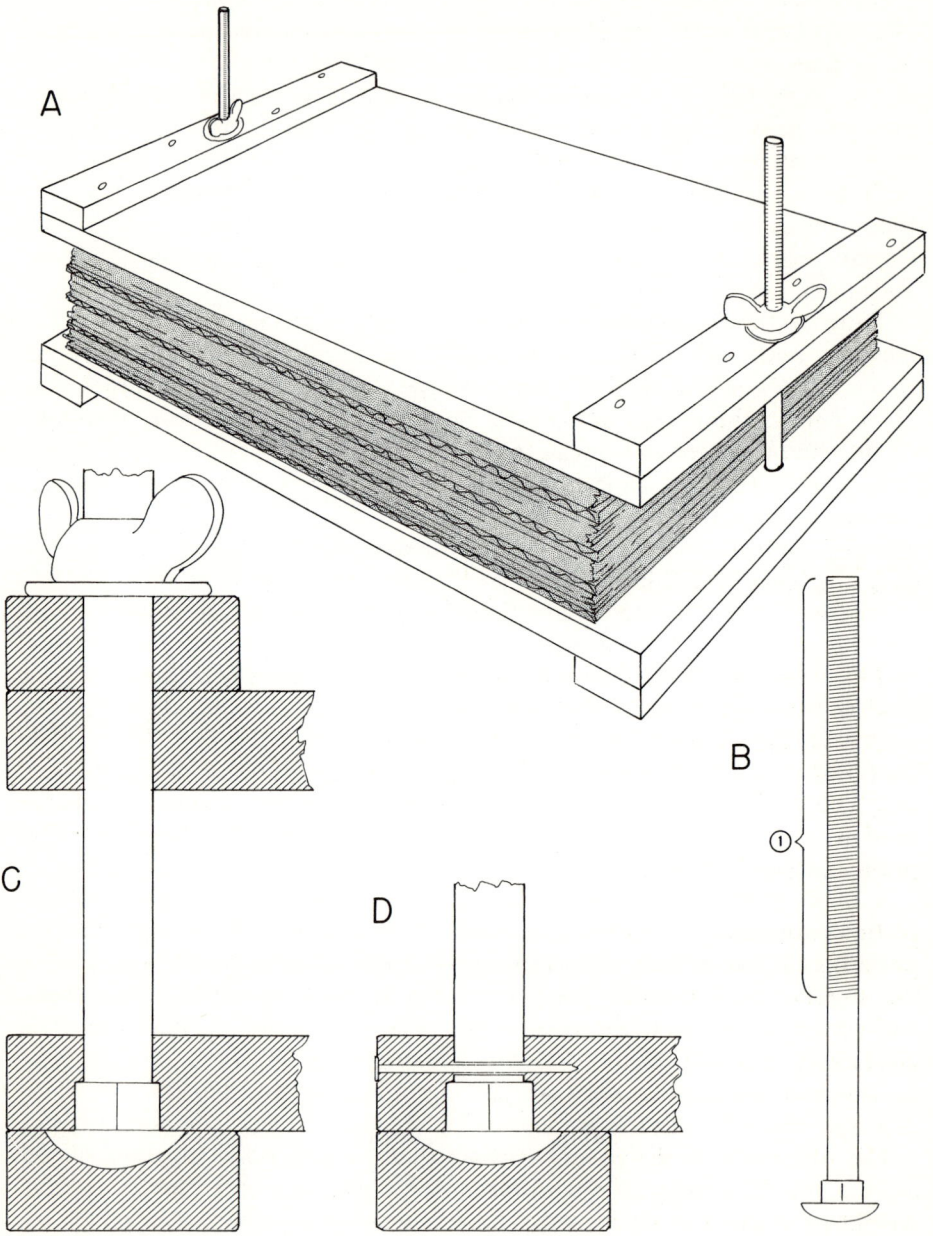

FIG. 5–2. Methods for building a laboratory plant press. See text; (1) cut threads 8 inches down the length of the bolt.

The Higher Plants (Tracheophyta)

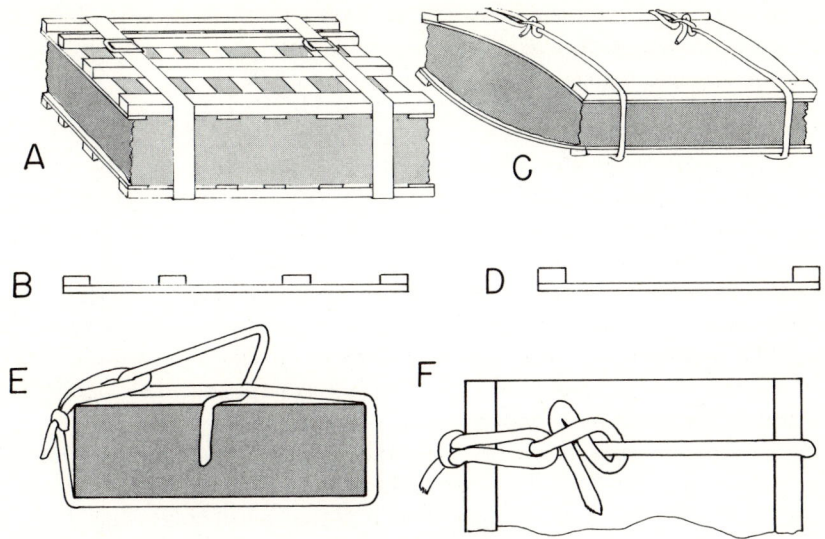

FIG. 5–3. Field presses. A–B. Lath construction. C–D. Plywood construction. E–F. Method of tying a field press with rope.

section when attaching the two outside 18-inch lath strips. Use one nail or rivet at each intersection when attaching the two long inner strips.

Pressure for pressing plant specimens may be obtained by means of two pieces of rope tied as illustrated in Fig. 5–3, E and F. As an alternative use two web cargo straps, ¾ inch wide and provided with a buckle, which are used for car-top carriers and are available at any automotive store. Use this press as described below.

A press that is much simpler in design and construction and yet completely satisfactory is that illustrated in Fig. 5–3, C and D. To build this press obtain two pieces of ¼-inch exterior plywood 18 by 12 inches, with the surface grain running lengthwise. Obtain four cleats, 18 inches long, and attach one to each long side of the plywood boards, as illustrated. The cleats may measure up to ½ by 1½ inches and should be attached to the plywood by wood glue and nails or short wood screws. Provide two ropes or cargo straps, as described above, and the press is ready to use.

The difference in the construction of the two presses follows two trends of thought. The all-lath construction originally provided the only light-weight press and was thought to leave a lot of open space from which water vapor could escape from the drying plants. If corrugated ventilators (described below) are used, there is no need for latticework construction to carry off the water vapor. Thus, the plywood press is equally light, more easily constructed, and will bend somewhat to provide pressure on the outside of the plant mass as well as along the center.

Using the Plant Press. Obtain some form of botanical dryers which absorb moisture from the plant specimens, and some ventilators which permit the circulation of air between each group of plants being pressed. A fine-grained absorbent felt dryer or blotter, 12 by 18 inches, may be purchased, but folded newspaper is just as satisfactory. Likewise, commercial corrugated cardboard ventilators or aluminum ventilators, 12 by 18 inches, may be purchased but can easily be made from large pasteboard boxes. The corrugated ventilators permit air to circulate between the dryers and draw moisture off more readily. Finally, provide specimen papers made from single sheets of newspaper folded over.

Put a stack of ventilators, dryers, and specimen papers in your press when going into the field. Press specimens shortly after collecting, so that wilting will not distort their appearance. After a specimen has been collected, put a ventilator and a dryer, in that order, on the bottom board of the press. Open a specimen sheet and put this on top of the dryer. Arrange the plant in the way you wish the finished specimen to appear, remembering to show all of the floral parts, leaves, stems, and roots, so that every taxonomic feature can be observed. Arrange the roots on the lower part of the folder and, if necessary, bend the stem so that the bulk of the plant is distributed uniformly. Do not let several parts overlap, causing unnecessary thickness in any one area. Show both sides of leaves and both sides of large disc-shaped flowers such as daisies. When the corolla is tubular in nature, split the tube of one or two flowers to show the inner parts. Add extra flowers and fruits that may later be stored in a paper envelope attached to the herbarium sheet. These will be useful for dissection and microscopic study. On the inside of the specimen sheet write the field data or, at least, a field number that corresponds to data kept in the notebook. The specimen sheet will remain with the plant until it is finally mounted on a herbarium sheet. Hold the plant in place and fold over the other half of the specimen sheet, add another dryer and a ventilator, and, finally, put the press together again and retighten the straps.

When the press is full of specimens, place it on the ground and, while kneeling on it, tighten the press straps. Change the dryers every 12 to 24 hours to reduce drying time and prevent molding. Wet dryers must be placed in the sun or dried artificially so that they will be ready for reuse. When the dryers are changed in the laboratory, specimens may be transferred from the field press to the laboratory press, if this is desirable. When specimens are dry, stack them in their specimen sheets, between two ventilators, and tie them in bundles of about 50 plants. These are now ready for identification and mounting.

When working with thick succulent plants such as the cactus, halve the flat segments lengthwise and scoop out the inside so that only the surface is pressed. Cylindrical sections may be cut to appropriate lengths, split down on one side and hollowed out, rolled out flat, and pressed. The tops of small barrel cacti should be cut off, with a minimum of flesh attached, scooped out, and pressed as a disc. The side of the barrel is treated as a cylinder, split, cleaned, and flattened. Large flowers may be cut in half lengthwise and the fleshy base reduced before pressing. Otherwise, the receptacle may be cut off just below the petals and the flower pressed with all the petals radiating like those of a daisy. Occasionally, the folding of the spines will become difficult, but this may be achieved by putting a dryer and a ventilator board over the cactus and simultaneously pressing downward and to one side. This motion will fold the spines over and permit the bulk of the cactus to be reduced.

Some flowers, such as the iris and many aquatic forms, will stick to the specimen papers or dryers. In such cases the flowers may be placed directly on a herbarium sheet and covered with waxed paper or very fine cotton cloth. After drying in the usual manner, the waxed paper or cotton cloth is very carefully peeled away. The stem and leaves must be strapped down after pressing so that the flower is not broken.

Many of the conifers lose their needles, even though they may be covered and given the best of care. Sharp (1935) recommends a method which is slightly modified here. Cut fresh specimens to fit the herbarium sheet and then put them on a plate of glass covered with partially diluted herbarium glue (see mounting methods below). Carefully lift the specimens and put them on the herbarium sheet, being careful not to have an excess of glue. Cover these with several dryers and press until dry. Another method, recommended by Wherry (1949), takes advantage of one of the clear plastic sprays, sold under the name of Kraylon. He recommends that the conifer be dried in the usual manner and then sprayed where the needles attach to the stem. The specimen is then mounted on a herbarium sheet and may be covered with cellophane, although this is unnecessary.

Field Drying Methods. It is often imperative to hasten the drying of pressed material, both to free the presses for more specimens and to prevent the specimens from becoming moldy or rotten. In sunny weather field presses may be set in the sunshine over pavement or hot sand, or suspended by screw eyes in the sunshine. In very damp climates, however, it may be essential to use artificial heat for drying. Specimens that are carefully dried with artificial heat, not exceeding 115° F., will show no appreciable difference in appearance from those dried normally. If electricity is available in the field, place your specimens in a dryer constructed of a pasteboard box, a light bulb and receptacle, and an extension cord, as described in Chapter 13 under "Field Drying." Place the plant presses on edge over the source of heat so that hot air will continuously circulate through the ventilators and remove moist air. Gates (1950) describes in detail the construction of a more elaborate electrical dryer for herbarium specimens designed primarily for museum use. However, an inexpensive controlled-heat dryer may be made by providing a box, over which the presses may be placed, and providing this with a coil-heating element or a bank of electric lights and a chick incubator thermostat (which costs around one dollar).

When electricity is not available, the heat from pressure gasoline lanterns or stoves, or from kerosene lanterns, serves quite well. Place the plant presses on edge on some sort of rack or suspend them by means of rope and screw eyes over a Coleman lantern and drape a sheet of muslin skirt around the presses to funnel the hot air up through the ventilators. In using this technique be extremely careful to provide space for air to enter underneath the sheet and keep the plant presses far enough away from the lantern to prevent burning. If kerosene lanterns are used, be sure the fuel is clean; otherwise, soot will be deposited throughout the ventilators.

Field Notes and Data

Students should get into the habit of taking field notebooks with them and recording accurate data while collecting. Although the amount of field data recorded may depend on the intended use of the specimens, only those specimens which have complete data will be of universal value in botany. The recognition of new taxonomic characteristics often makes the splitting of species into respective subspecies impossible unless accurate notes have been kept. Keeping of locality data, arranged according to serial numbers, is described in Chapter 1. In addition to locality data, however, the student should record information concerning the species itself and its surroundings: the type of plant association (meadow, woodland, and so on), associated plants (sagebrush, etc.), altitude, the slope of the ground, the nature of the soil and its moisture content, and other pertinent factors. Insofar as the species itself is concerned, record the average size of specimens, the color and pattern of the flower, the nature of the bark for trees, and any other detail that the preserved specimen will not give after color has faded. The recording of actual color notes for new species, or for species whose identi-

The Higher Plants (Tracheophyta)

fication depends on color, is discussed in Chapter 1. Color photographs are not very suitable, unless some color standard is included in the picture to serve later as a norm in the laboratory.

IDENTIFICATION, MOUNTING, AND PRESERVATION

Identification

Whenever possible, collect extra material for immediate identification. Transfer species names to the proper specimen sheets during the first change of dryers on the plant press. If immediate identification is not possible, identify specimens prior to mounting them on herbarium sheets. One or two flowers may be removed from the specimen for dissection if extra flowers and fruits have not been pressed. Place the flowers in a bath of boiling water from one to three minutes or hold them over the spout of a steam kettle until they are thoroughly relaxed. Return the dissected parts to paper envelopes and affix these to the herbarium sheet along with the main specimen. If extremely large numbers of plants must be handled and mounted, it is best to postpone identification until after mounting.

Mounting Pressed Specimens

Mounting is a process whereby pressed and identified specimens are attached to herbarium sheets for both protection and permanent storage. From one to many specimens representing only one species may be mounted on a single sheet. Arrange the specimens so that the bulk is more or less uniformly distributed around the sheet (Fig. 5–4E). Herbarium paper comes in a standard size, 11½ by 16½ inches; it should be of a good quality, preferably a high-rag-content paper that will not yellow with age. The standard weight approximates that of manilla folder stock or Bristol ledger, but heavier 5-ply mounting cards are also obtainable.

There are four acceptable methods of mounting. The first method consists of fastening the specimen down with narrow strips of herbarium tape. This is a white cloth tape, heavily gummed, and measuring from 1 to 1¼ inches in width. Apply tape straps amply so that the plant will be flexible and yet firmly attached to the sheet. Plastic and celluloid tapes are not suitable, as they are not permanent. Tape may be made by using thin strips of cloth to which herbarium paste has been applied.

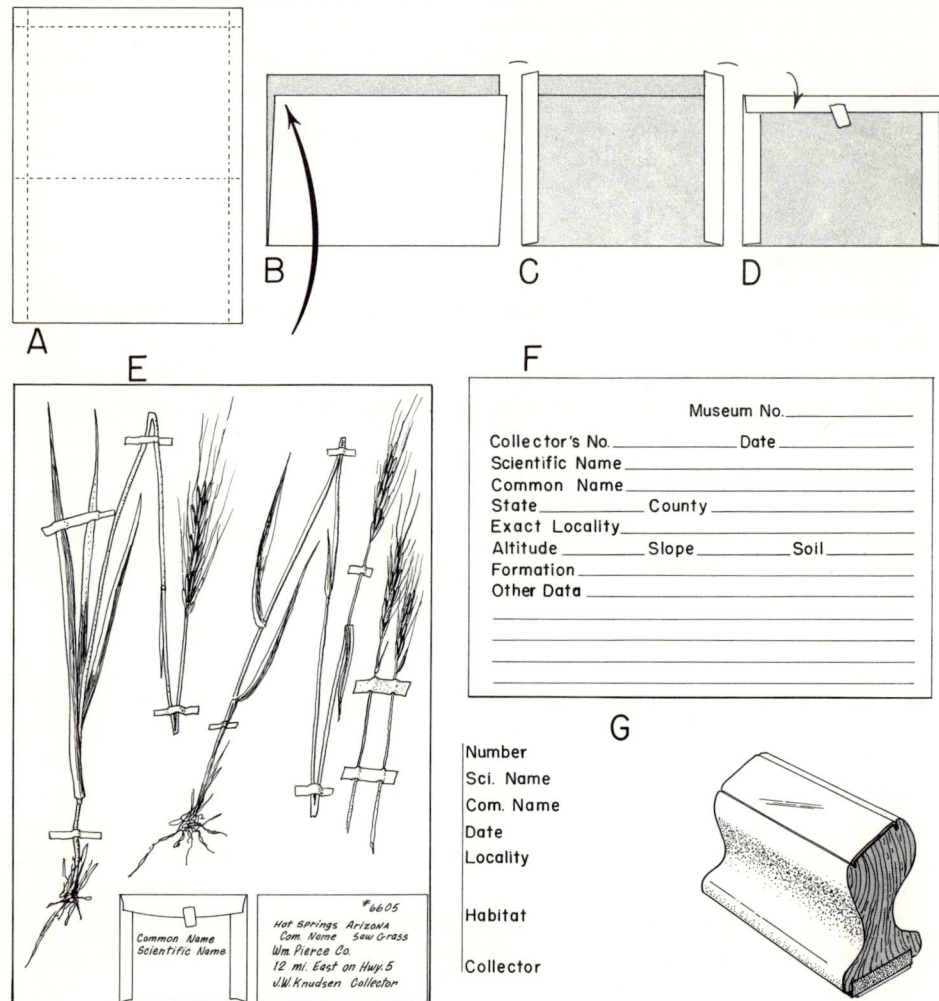

FIG. 5–4. Herbarium sheets and mounting techniques. A–D. Folding a storage envelope for the herbarium sheet. E. Mounting specimens on a herbarium sheet with cloth tape (see text). Note storage envelope and label at the bottom of the herbarium sheet. F. A herbarium label modified from the U.S. Forestry Service label. G. A simple, rubber-stamp label suitable for museum work.

The second method of mounting consists of turning the specimen face down and applying glue with a brush to all of the exposed surfaces. The specimen is then positioned on the herbarium sheet and firmly pressed into place with a soft cloth. Hold very flexible specimens by the stem or roots, dangling the floral portion downward, and apply a herbarium sheet against the specimen. When the specimen is partially attached, place the sheet on a table and gently affix the specimen with a soft cloth. Be careful not to get excess glue on the herbarium sheet in places where it will stick to neighboring sheets. Use a good grade of herbarium glue or paste or various white glues or acorn glues, such as Elmer's Glue. This last-named may be diluted with one part of water to two parts of glue or with equal parts of water and glue. Most workers do not recommend rubber cement, although the writer has specimens in excellent condition that have been mounted for over twelve years with rubber cement.

The third method consists of spreading diluted Elmer's Glue (equal parts of glue and water) or other glue on a glass sheet and placing the specimen on the glass until the undersurface is thoroughly coated. Then the specimen is applied to the card or the card to the specimen, as described above. This method is extremely quick and very satisfactory. Cloth straps to hold the plant in position will not be necessary unless the glue used is untrustworthy.

The fourth method consists of using a new plastic mounting medium that takes the place of gluing or strapping, and is extremely useful when large numbers of specimens must be mounted. The mounting material, developed by Archer (1950), is made with the following formula: 800 cubic centimeters of toluene, 200 cubic centimeters of methanol, 250 grams of ethyl cellulose (standard 7 cps.), and 75 grams of Dow Resin 276 V-2. This is mixed to the consistency of a heavy syrup and is applied from a squirt oil can or polyethylene squirt bottle. The material is now available in small quantities from Carolina Biological Supply. In this method specimens are paced on herbarium sheets and, if necessary, weighted down. The medium is squirted over stems or leaves in the form of straps, as in the cloth strap method, in sufficient number to firmly affix a plant. After drying, the plant may be permanently stored.

Seed or Fruit Collections

A collection of seeds and fruits will be useful both in teaching and in taxonomy. Fleshy fruits must be preserved in liquid, as described below. Place small dry fruits or seeds in cellophane envelopes or paper envelopes and firmly affix them to the herbarium sheet containing the specimen from which the fruits were collected. Larger fruits or large quantities of small fruits and seeds should be dried and stored in paper bags or small boxes; include the field data. Write the field number and species name on the bag or box. Staple the bag shut and store along with the mounted specimens in the herbarium case, if space permits.

If a seed collection is desirable for demonstrating different types of seeds to beginning students, two satisfactory display methods are available. First, write the scientific name of the specimen on a 3- by 5-inch card and put this in a small, screw-cap vial, along with a small sample of seeds. Arrange the vials in a glass-top box, Riker mount, or holding case made by drilling holes of an appropriate size in a board. Internal labels should be included, even if external labels are used, so that identification of the specimens will not be confused. A second method of mounting is to use prepared seed-mounting cards which are available from biological supply houses. These are made of heavy cardboard stock and have up to 25 circular depressions ranging from $\frac{1}{8}$ inch to $\frac{3}{8}$ inch deep. Seeds are placed in the depression, the name is printed beneath the depression, and a glass top is attached. Seed mounts can be made by drilling a series of uniform holes in wood $\frac{1}{4}$ or $\frac{3}{8}$ inch thick which has been covered with white, glazed paper. A piece of cardboard is glued to the bottom of the wood and a glass is taped to the top after seeds and labels have been added. Coin collectors' cards treated in the same manner may be suitable for very small seeds.

Pollen Collections

Quantities of identified pollen prove beneficial for general teaching, plant taxonomy, and for diagnosing the gut contents of herbivorous animals or the fossil records left in old peat bogs. To collect large quantities of pollen, tie small paper bags around flower heads until the pollen is released. Dry the pollen in a desiccator equipped with silica gel, and store it in vials or preserve it in liquid.

To make slides of pollen, collect flowers and bring them into the laboratory. When they are mature, shake a small quantity of pollen onto the slide and add a drop or two of alcohol. When the pollen is partially dry, add additional alcohol until the grease is floated away from the pollen. When the alcohol is evaporated, wipe the grease from the slide with a swab dipped in alcohol. Add a drop of a weak aqueous solution of basic fuchsin to the pollen and stir with a needle. After a moment withdraw the excess stain with blotting paper. Warm the slide, add a drop of hot glycerine jelly, and stir with a needle. Finally, add a cover slip before the glycerine jelly has solidified. For an alternative method add a few drops of a 60-percent white corn syrup solution to the pollen, stir, and let stand overnight.

When part of the water has evaporated add a few drops of an 80-percent corn syrup solution and cover with a cover slip. The corn syrup will dry around the outside in three or four days. However, do not store the slides on edge for many weeks or the pollen will migrate to one edge. For a monographic treatment of pollen and pollen studies see Wodehouse (1935).

Cone Collections

The cones of club mosses, horsetails, and the conifers should be collected as part of the taxonomic collection. If necessary, make special trips to obtain cones of all degrees of maturity and of both sexes. For taxonomic work, smaller cones are usually left attached to the foliage and are mounted on the herbarium sheet, extra cones being stored in envelopes on the sheet. Store medium-sized cones in envelopes, paper bags, or small boxes, with the field number and specimen number clearly recorded on the container and, if possible, on the cone itself. For large cones, either attach heavy-grade tags by means of strings or paint one of the cone leaves (megasporophyll) white and record the field number and specimen number. These may then be stored in bags or in a large box and kept near the herbarium sheets.

Younger developing cones should be preserved in liquids for purposes of dissection or tissue sectioning and slide preparation. FAA or other preservatives described below are suitable for this.

Wood and Bark Collections

Wood and bark collections are often essential in taxonomy, field botany courses, general biology courses, and so on. Figure 5–5, A and B, illustrate some of the important macroscopic aspects of wood and bark.

Young Wood. To demonstrate the young bark and wood of various trees select, preferably, material that is already cut down or that is scheduled to be cut down. Specimens of an average diameter of approximately 2½ inches should be sought. If the wood is not cured, permit it to dry several months so that cracks and flaws will not occur. With a saw make the cuts illustrated in Fig. 5–5, C and D, and affix a label as shown in Fig. 5–5E. Make the top and side cuts before reducing the wood to its finished length. This will afford a greater degree of safety, if you are using a table saw, or room for putting the wood in a vise, if you are using a handsaw. Cut the wood to length and sand the four surfaces with a belt sander or sandpaper block. Use increasingly finer grades of sandpaper until a smooth finish is achieved. Apply a furniture wax to the exposed surfaces.

88 Biological Techniques

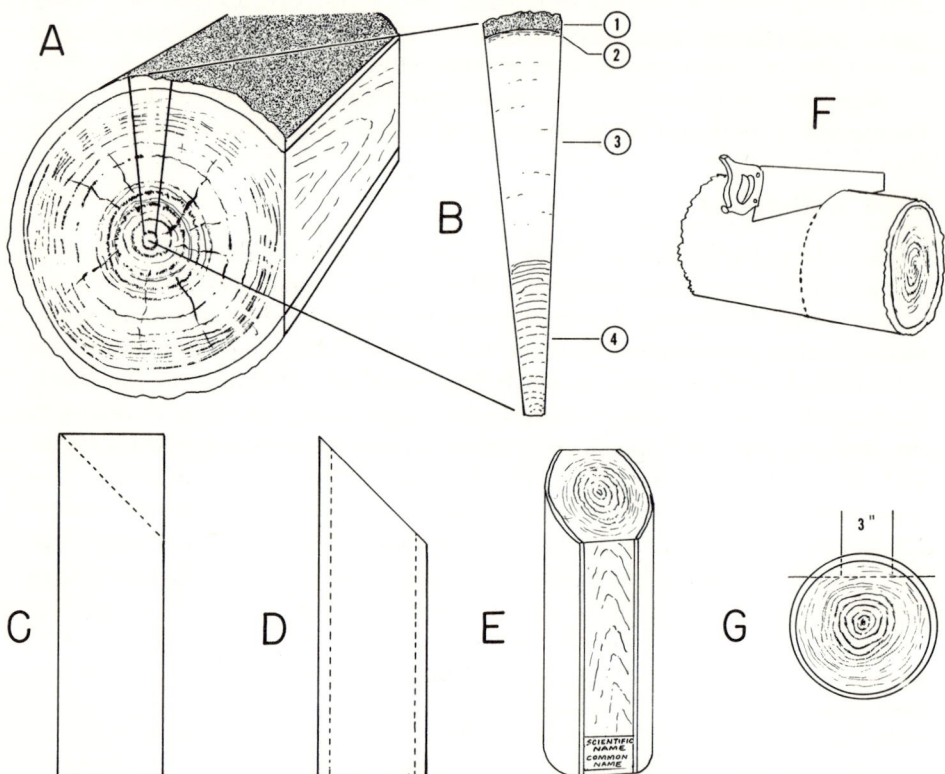

FIG. 5–5. Collecting wood samples. See text for details. A–B. Wood morphology. C–E. Making wood and bark samples from trees of a small diameter. Cut on dotted lines. F–G. Making bark samples from larger logs. Cut on dotted lines. (1) Bark, (2) cambium, (3) sapwood, (4) heartwood.

Mature Bark. Preferably, select specimens of mature bark from recently fallen, sound, and relatively dry logs. Cut off a piece of log 6 to 8 inches long and section out the bark sample, as shown in Fig. 5–5, F and G. The cut surface needs no sanding, but should carry a label bearing the name, specimen number, and collecting number. If such samples are glued permanently into the bottoms of display boxes (see Appendix A), place the label on the lower end of the bark specimen. Specimens used for taxonomic purposes may simply be removed from the tree and kept in the herbarium in paper bags or boxes with the field data.

Leaf Collections

The leaves of higher plants differ tremendously in over-all shape and complexity, nature of the margin, nature of the tip and the base, and venation. Consistent variations are, of course, extremely useful in plant identification and from them a beginning botany student must learn to recognize various leaf characteristics. Of great value, therefore, is the demonstration of these leaf characteristics.

Leaf Characteristics. To make teaching displays of leaf characteristics, either mount separate leaf types on small botany cards or make a composite mount of leaf characteristics on full-size herbarium sheets, as illustrated in Fig. 5–6. For the latter, select heavy sheets of white poster board or art board (11½ by 16½ inches), heavier than standard herbarium paper. On a piece of scratch paper the size of the herbarium card rule off the areas that will be occupied by each leaf type. Consult any taxonomic work for the many examples of leaf margins, tips, and bases not illustrated in Fig. 5–6. With a finished layout you may determine sizes of leaves needed for your display. Collect and press characteristic leaves of the proper size, in sufficient number to make as many displays as required. When the leaves have been pressed and dried, mount each with glue (see above); type, cut, and mount labels describing each leaf. Finally, protect the display mount with Saran wrap held in place with tape, glue, or paper clips. Keep the display cards in a species folder or herbarium portfolio.

Venation Techniques. When it is necessary to demonstrate or study the intricate patterns of leaf venation, permanent mounts of leaves with the mesophyll removed are useful. Mount venation specimens directly on index cards or herbarium sheets with herbarium glue (see above).

During the spring excellent specimens may be obtained underneath piles of leaf litter from those leaves that were next to the soil mass and acted upon by bacteria. Carefully select undamaged specimens which show the venation pattern and keep these moist. In the laboratory brush away dirt and debris or float the specimen in water and gently agitate it. Press the specimen between a mounting card and a piece of waxed paper, or mount it directly on a card. Many workers suggest that leaves kept in water at room temperature for long periods of time will have the mesophyll reduced by bacteria, thus exposing the venation.

A rapid method for reducing the soft leaf tissues is as follows (the source

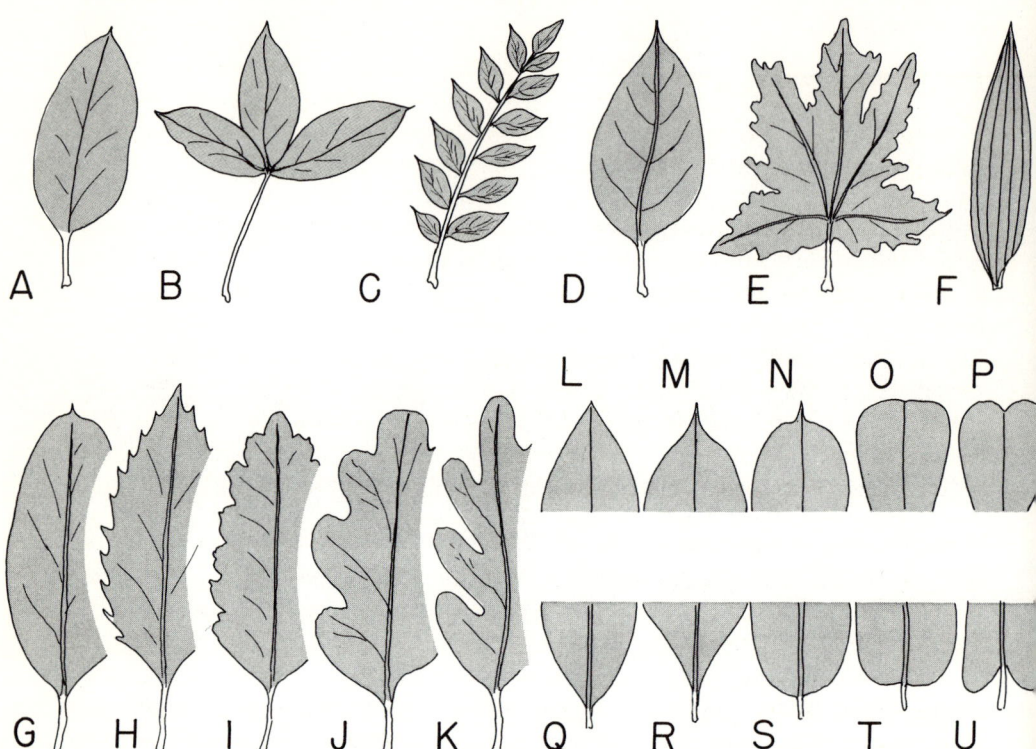

FIG. 5–6. Some types of leaf morphology. A. Simple leaf. B. Compound leaf, palmately branched. C. Compound leaf, pinnately branched. D. Simple venation. E. Palmate venation. F. Parallel venation. G. Entire margin. H. Serrate margin. I. Dentate margin. J. Lobed margin. K. Cleft margin. L. Acute tip. M. Aristate tip. N. Cuspidate tip. O. Truncate tip. P. Emarginate tip. Q. Acute base. R. Cuneate base. S. Rounded base. T. Truncate base. U. Emarginate base.

of this method is unknown to the author). Mix 57 grams of sodium carbonate ($NaCO_3$) to 500 cubic centimeters of water in a beaker. Next, add 23½ grams of calcium oxide (CaO) and, after mixing, boil for 20 minutes and then filter the solution. Select leaves to be treated and simmer these (do not boil) for one hour or more, until the epidermis and mesophyll are readily removed. Remove the leaf from the solution, dip it in clear water, and use a brush and dissecting needle to remove the soft portions. Finally, press and mount the specimen. If necessary, float the leaf onto a mounting card, cover it with waxed paper, and press it.

Liquid Preservation

Preserved specimens are useful in general laboratory studies when fresh material is unavailable and dried specimens are unsuitable.

Preservation for General Use. When flowers, fruits, stems, leaves, and other plant parts are required for classroom study, preservation in a 4-percent formalin solution is satisfactory. For large, fleshy specimens, or when many specimens are placed in one jar, use a 5- to 6-percent solution. To facilitate speedy penetration of plants, especially those with waxy coats, replace from 20 to 50 percent of the water with alcohol. A general disadvantage with the formalin or formal-alcohol preparation is that color will be bleached or removed entirely. Specimens should be soaked in fresh water for several hours prior to classroom use.

Preparation for Chlorophyll Retention. On many occasions the retention of green color is extremely useful in the study of small organisms such as fern prothallia. A satisfactory solution for retaining color (*Turtox Service Leaflet 3*) is made by adding 20 grams of phenol c.p., 20 grams lactic acid Specific Gravity 1.21, 40 grams of glycerine Specific Gravity 1.25, 0.2 grams cupric chloride, 0.2 grams cupric acetate, and 20 cubic centimeters of distilled water. Specimens are stored in this solution until required.

Color Preservation. No method for the color preservation of flowers or fruits is everywhere satisfactory. Color retention by any method is not wholly satisfactory for taxonomic purposes and, therefore, one should use color notes made at the time of collection (see Chapter 1). Many dried plant specimens will retain natural colors for very long periods of time, but exposure to light, humidity, temperature, chemical action, and other factors may, gradually or quickly, alter color. Scully (1937) developed solutions that give satisfactory results for many flower and fruit colors. To use this method, place specimens in vials containing a 5-percent copper sulfate solution for 24 hours to set the color. Wash the specimens several times and put them into a second solution of 16 cubic centimeters of sulphuric acid, 21 grams of sodium sulphite, and 1000 cubic centimeters of water. The material is maintained in the solution in stoppered vials. According to Scully, delicate pinks and blues or some of the deep reds and purples will fade, whereas many of the colors, especially the yellow range, will show very little alteration.

Preservation for Histology. When specimens are intended for tissue studies and slide preparation, preserve them in a formal-acetic-alcohol solution made of 6 parts formaldehyde, 50 parts 95-percent ethyl alcohol, 2 parts glacial acetic acid, and 40 parts distilled water. Specimens may be stored in this solution indefinitely for general use or for histological techniques.

Natural Display Methods

Plants in natural history displays are often dried in their natural three-dimensional positions. The author frequently utilizes plant material in displays of insects which requires the specimens to appear natural, unwilted, and with normal color. Not all plants—especially, not all flowers—lend themselves to such displays. Therefore, be highly selective, taking small compacted leaves and flowers in preference to large delicately constructed leaves and flowers.

Initial drying may be achieved in two ways. (1) Attach the plant to a base and stand it upright in a container such as a small pasteboard box. Very carefully sift in extremely fine-grained, clean, white sand, filling all of the available spaces around the plant until it is covered. The sand should be added in such a way that the petals, leaves, and branches are not bent out of their normal position. Place the filled container in an oven and heat to 115° F. until dry. Next, remove the sand, gently blow away any dust attached to the plant, and mount the plant in the display. If you anticipate the fading of plant parts, add water-color pigments with an airbrush (see Chapter 24). (2) Substitute very fine-grained silica gel for sand, with or without heat. This compound has such an affinity for water that it will quickly dry the plant, leaving it in its natural posture and form. With larger fleshy plants some heating may be necessary to evaporate the water from the silica gel. The silica gel may be dried in a hot oven before being reused.

LABELS AND STORAGE

Labeling and Labels

Specimens are of little value without labels that reflect the data recorded in the field notebook. Labels for preserved specimens should be written with waterproof India ink, dried for about 10 minutes, and then placed inside the container, along with the specimen. Labels for herbarium sheets, usually attached to the lower right-hand corner, measure about 3 by 5 inches. The label should include: (1) the locality, including state, county, and distance from the nearest town; (2) date; (3) field number or collector's number; (4) scientific name; (5) common name; (6) habitat, which may include the slope, the soil, the altitude, associated vegetation, etc.; (7) collector's name. Such labels are usually printed to the collector's specifications and headed with the name of the institution or the collection represented, as shown in Fig. 5-4F. However, for small collections labels may be made by means of a rubber stamp applied directly to the herbarium sheet (Fig. 5-4G). The following headings should be arranged in a column: locality, date, field number, scientific name, common name, habitat, collector. Be sure to leave sufficient room between headings to permit the transcription of adequate field data. The label is stamped with black ink about 5 inches in from the lower right-hand margin.

Storage Techniques

Problems Involved. Light, moisture, insect infestation, and rough handling of specimens represent the common elements which damage herbarium specimens. Specimens that are properly cared for and handled will not wear out or disintegrate. After identification, file the members of a particular genus in one or more herbarium folders and store these in a herbarium cabinet. The folders are usually made of manilla and are slightly larger than the herbarium sheet. When handling individual sheets, be careful not to bend or twist the specimens; pick up each sheet individually and set it aside when working through a stack. With proper handling one need not cover specimens with thin plastic. For general classroom use, however, covering the specimen is often advantageous. Always keep some fumigant such as paradichlorobenzene or naphthalene flakes inside the herbarium cabinet. This is not designed to kill insects that have already infested the specimens, but to prevent infestations from starting. The cautious use of carbon bisulfide may be required for existing infestations. Pour a small quantity of carbon bisulfide in a jar and place this on a top shelf in the herbarium cabinet. Close the cabinet for 24 hours and then remove the carbon bisulfide and air out the herbarium room.

Arrange your specimens either alphabetically or taxonomically within the cabinet. The keeping of a card file is not essential for a small collection but is highly desirable in large herbaria. Such a file cross-references the collection and makes the specimens more readily available for study.

Storage Cabinet. The storage cabinet must be light-, dust-, and insect-proof, and provide convenient pigeonhole units for the specimens. Serviceable cabinets may be made for a fraction of the market price. The one described herein is a single unit with 18 storage compartments, made of ¾-inch plywood and sealed against insect infestation.

The parts of this cabinet are made up of the main box and the base (see Fig. 5–7). Cut four pieces of ¾-inch plywood (A-A or B-B grade) 36 by 18 inches; these make up the vertical portions. Next, cut a top and bottom piece of ¾-inch ply 45 by 18 inches. Following this, cut ⅛-inch grooves, 3⁄16-inch deep, in the vertical sections to receive the Masonite shelves, as shown in Fig. 5–7, A and B. These are cut on 6-inch centers. Only the inner sides of the end boards are grooved, whereas both sides of the inner vertical partitions are grooved. When this is finished, glue and nail this portion of the main box and then add a ¼-inch plywood back (37½ by 45 inches) with nails and glue. Make sure the plywood back is cut perfectly square and use it to square up the main box.

Next, cut 15 shelves out of ⅛-inch Masonite, 17¾ by 14 5⁄16 inches. Finally, make up a base of 1- by 3-inch wood, as shown in Fig. 5–7, C and D. This should be recessed 3 inches from the front of the main box, to provide a toe board. This can be attached to the main box or merely set underneath it, as the weight of the cabinet will hold it in place.

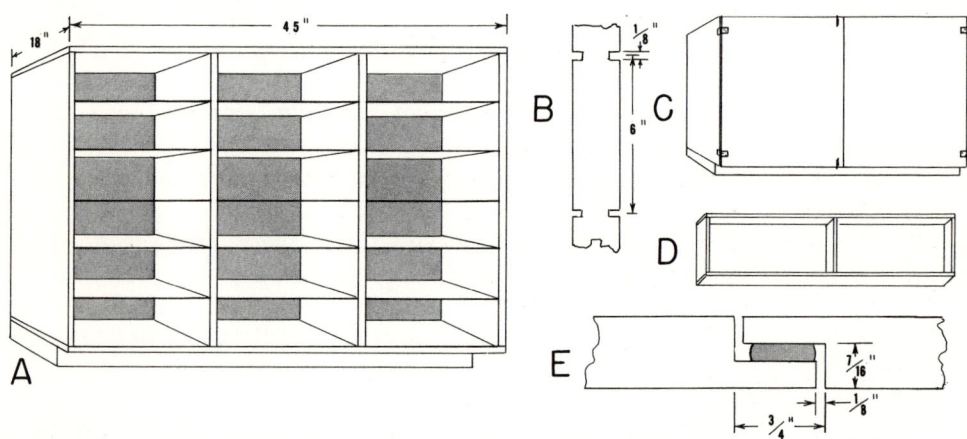

FIG. 5–7. Methods of making a herbarium storage cabinet. See text for full instructions. A. Cabinet without doors. B. Central partition showing grooves for shelves. C. Cabinet showing door hardware. D. Construction of the base. E. Cutaway view of the door showing a sponge rubber or sponge plastic seal between the overlapping doors.

The two doors must measure 22¾ by 37¾ inches; they are made of ¾-inch plywood. This dimension will permit a ¾-inch overlap, as illustrated in Fig. 5–7E. When the doors are completed, remove the shelves and varnish all surfaces with one or two coats of clear spar varnish. Sand the front edges of the main box and varnish this to produce a smooth surface. When the varnish is thoroughly dry attach a strip of self-sealing ⅛- by ½-inch foam rubber or foam plastic on the front outside edges of the main box. This foam will receive the doors and create an airtight cushion. Do not put foam on the inner vertical partitions. Finally, put a strip of foam on one of the doors where they overlap (Fig. 5–7E). Two pairs of hinges and two catches, like those illustrated in Chapter 13 under "Insect Storage Cabinets," should be installed to complete this unit.

REFERENCES

Archer, W. Andrew, 1950, New Plastic Aid in Mounting Herbarium Specimens, *Rhodora,* 52(623):298–299.

Armstrong, Margaret, 1915, *Field Book of Western Wild Flowers,* Putnam, N.Y.

Baerg, Harry, 1955, *How to Know the Western Trees,* William C. Brown, Dubuque, Iowa.

Blackburn, B. C., 1952, *Trees and Shrubs in Eastern North America,* Oxford Univ. Press, N.Y.

Bowers, N. A., 1942, *Cone-bearing Trees of the Pacific Coast,* Pacific Books, Palo Alto, Calif.

Canada Department of Resources and Development, 1949, Forestry Branch, Native Trees of Canada, *Canada Dept. Res. and Devel. Bull.,* 61.

Cobb, Boughton, 1956, *A Field Guide to the Ferns,* Houghton Mifflin, Boston.

Core, Earl L., 1955, *Plant Taxonomy,* Prentice-Hall, Englewood Cliffs, N.J.

Cuthbert, Mabel, 1948, *How to Know the Fall Flowers,* William C. Brown, Dubuque, Iowa.

Cuthbert, Mabel, 1949, *How to Know the Spring Flowers,* William C. Brown, Dubuque, Iowa.

Durand, H., 1949, *Fieldbook of Common Ferns,* Putnam, N.Y.

Fernald, M. L., 1950, *Gray's Manual of Botany,* American Book, N.Y.

Fogg, John M., 1940, Suggestions for Collectors, *Rhodora,* 42(497):145–157.

Frye, T. C., 1934, *Ferns of the Northwest,* Binford & Mort, Portland, Ore.

Gates, Burton N., 1950, An Electrical Drier for Herbarium Specimens, *Rhodora,* 52(618):130–135.

Graves, A. H., 1956, *Illustrated Guide to Trees and Shrubs,* Harper & Brothers, N.Y.

Harlow, William M., 1942, *Trees of the Eastern United States and Canada,* McGraw-Hill, N.Y.

Harrar, Elwood S., and J. George Harrar, 1946, *Guide to Southern Trees*, Mc-Graw-Hill, N.Y.

Jaeger, E. C., 1940, *Desert Wild Flowers*, Stanford Univ. Press, Stanford, Calif.

Jaques, H. E., 1946, *How to Know the Trees*, William C. Brown, Dubuque, Iowa.

Matthews, F. S., 1955, *Field Book of American Wild Flowers*, Putnam, N.Y.

Moore, H. E., Jr., 1950, A Substitute for Formaldehyde and Alcohol in Plant Collecting, *Rhodora*, 54(653):123–124.

Parsons, Frances Theodora, 1961, *How to Know the Ferns*, Dover Publications, N.Y.

Porter, C. L., 1959, *Taxonomy of Flowering Plants*, Freeman, San Francisco.

Preston, R. J., Jr., 1948, *North American Trees*, Iowa State College Press, Ames, Iowa.

Preston, R. J., Jr., 1940, *Rocky Mountain Trees*, Iowa State College Press, Ames, Iowa.

Schultes, Richard E., 1947, The Use of Formaldehyde in Plant Collecting, *Rhodora*, 49(578):54–60.

Scully, Francis J., 1937, Preservation of Plant Material in Natural Colors, *Rhodora*, 39(469):16–19.

Sharp, A. J., 1935, Preparing Gymnosperms for the Herbarium, *Rhodora*, 37(453):266–268.

Swingle, Deane B., 1946, *A Textbook of Systematic Botany*, McGraw-Hill, N.Y.

Taylor, T. M. C., 1963, *The Ferns and Fern-allies of British Columbia*, A. Sutton, Victoria, B.C.

Templeton, Bonnie C., 1932, Methods of Preserving Cacti, *Desert*, 3:127.

Turtox Service Leaflet No. 3, 1959, General Biological Supply House, Chicago.

United States Department of Agriculture, 1949, *Trees: Yearbook of Agriculture*, U.S. Dept. of Agriculture.

Wherry, Edgar T., 1949, A Plastic Spray for Coating Herbarium Specimens, *Bartonia*, 25:86.

Wodehouse, R. P., 1935, *Pollen Grains, Their Structure, Identification and Significance in Science and Medicine*, McGraw-Hill, N.Y.

CHAPTER 6

The Protozoans

The phylum Protozoa constitutes a complex of several very old animal groups. These animals are amazingly complex, especially for their size, and are as diversified and well-distributed ecologically as the arthropods and vertebrates. They have invaded most habitats, from marine and freshwater to terrestrial and parasitic. Whether they should be called one-celled or acellular may remain debatable. Nevertheless, these animals occupy a fascinating taxonomic position and are of interest because of their morphology, physiology, and general behavior. Most techniques for collecting, other than the plankton net, yield relatively few specimens and necessitate some method for concentrating or culturing the catch. Generally, prepared specimens are much less suitable for study than living specimens. It is often more practical to re-collect and culture specimens for classroom use than to go through the elaborate processes of staining and sliding such extremely small organisms. Parasitic animals and groups such as the Foraminifera and Radiolaria are exceptions, as are type specimens used in taxonomy.

COLLECTING PROTOZOANS

General Remarks

In collection of protozoans, especially the fresh-water species, many samples should be taken from each microhabitat in a pond or lake. In collecting bottom plant debris, for example, any minor change in the degree of light or shade, the plants associated with this debris or contributing to it, the nature of the substrate below the debris, the depth of the material, and other factors may be of sufficient importance to attract entirely different populations. Seasonal variations of light and temperature also influence the appearance and disappearance of species. Furthermore, each species is usually cyclic and will fluctuate between low and high numbers of population over a period of a few weeks. Thus, multiple factors may influence the presence or absence of protozoans in all of the collecting places.

Fresh-Water Protozoans

When collecting fresh-water protozoans, provide yourself with a large number of clean bottles with lids. Since culturing is the general rule for these specimens, attempt to collect from as many microhabitats as available. The open surface waters of lakes, ponds, and marshes, permanent or temporary mud puddles, and the like, will be rich with species of flagellate protozoans, some of the actinopod protozoans (such as *Actinosphaerium* and *Lithocolla*), and many species of ciliates. Use a fine plankton net of Number 20 mesh (which contains 173 meshes per inch). Provide the net with a small terminal vial to receive the specimens. Tow the net behind a skiff or cast it from shore and retrieve it by means of a long hand line. In pools too small for active netting, dip up water in a convenient container and concentrate the specimens by pouring the water through the side of a plankton net. To retrieve the specimens from the net either wash them down into a terminal vial or reverse the net and wash off the filtering surface into a jar of water. In shallow water containing large quantities of vegetation, dip out quantities of greenish looking water, for it may be rich in flagellate protozoans such as *Euglena*. Look for surface scums and collect these by submerging a bottle so that one edge is approximately a millimeter below the surface of the water. The surface layer with the scum will float into the bottle while the bottle is moved along the pond's surface. Such collections may be concentrated through the side of a fine plankton net and then killed or cultured directly.

With a large cooking ladle or tin-can lid, carefully scrape off the upper ⅛ inch of surface mud or sand from pond and lake bottoms and examine this in the laboratory. Depending on the size of granules making up the substrate numerous amoeboid species, ciliates, and other protozoans will be present. Pond vegetation (living, dead, or decomposing) will also attract amoeboid, flagellate, and ciliate forms directly to feed on the vegetation or indirectly to feed on the bacteria living there. Moist soils adjacent to streams and ponds (and elsewhere) also harbor protozoans.

Marine Protozoans

Marine specimens are less frequently cultured than fresh-water specimens, owing to difficulty in maintaining perfect environmental conditions. You must be prepared, therefore, to work immediately with living marine specimens or to preserve them for later study. Preservation, however, will cause distortion of naked protozoans. Formalin will attack the shells of foraminiferans, and alcohol may cause the precipitation of those salts found in sea water. Any of these problems may render all or part of a protozoan sample valueless. In other words, it is difficult to preserve bulk samples of marine protozoans so that all of the specimens will be usable. You should either plan to collect some particular type of protozoans (such as Foraminifera or Radiolaria), or else preserve parts of each sample by different methods.

Marine protozoans may be found in many of the same situations as those described for fresh-water collecting, above. The uppermost layer of sand or mud from all depths of water is usually populated with living protozoans, including Foraminifera and Radiolaria, and contains the shells of such protozoans which have been deposited on the bottom. Collect samples of sediment (see Chapter 1 for collecting devices) and preserve these as follows: Place the sediment sample in a bottle; permit ample time for the specimens to sink to the bottom if water is present. Next, pour off the superfluous water and place this with 95-percent isopropyl alcohol. A solution of 3-percent neutral formalin or buffered formalin (Appendix C) may be used, but this is dangerous, inasmuch as the formalin may become acidic and attack the shells of Foraminifera. By adding a small quantity of rose bengal stain (1 gram of dry stain per liter of distilled water), you will stain the protoplasm of all protozoans that were alive at the time of preservation. Therefore, it will be possible to separate and make a census of the living and nonliving specimens.

The plankton net is also useful for collecting pelagic protozoans, including the Radiolaria and the smaller Foraminifera. This material should likewise be preserved in alcohol and may be treated with rose bengal. Large Foraminifera may be obtained by rubbing marine grass and coarse specimens of marine algae over a sieve submerged in a bucket of water. Select a sieve fine enough to trap most of the vegetation, but coarse enough to permit the protozoans to crawl through to the bottom of the bucket. Wash the sieve in the water and concentrate the water by pouring it through the side of a plankton net or through a filter. Kill the specimens in alcohol and preserve in 70-percent alcohol. Some invertebrate filter-feeders and gleaners—such as crabs, clams, and the like—frequently contain large quantities of protozoans and diatoms in their digestive tracts. The gut content of such invertebrates may be preserved directly in alcohol and studied at a later date. When red tides occur, collect the protozoans involved either by means of a plankton net or directly in quart containers. Since those species which cause red tide seldom contain calcium carbonate in their skeletons, they may be poisoned by adding a few drops of formalin or alcohol to containers of water. Allow the specimens to settle to the bottom (about ½ hour) and pour off the superfluous water. Preserve in 3-percent neutral formalin or wash and store in 70-percent alcohol.

Parasitic Protozoans

1. The body surfaces and cavities of fresh-water, marine, and terrestrial invertebrates and vertebrates may all harbor parasitic protozoans. Small nodules on the skin or gills of fishes and amphibians often harbor protozoans. Tease such nodules apart on a clean glass slide with a few drops of isotonic saline. Such nodules may be prepared as temporary slides or smears, or sectioned, using histological techniques.

2. The oral chambers and gill chambers of vertebrate animals frequently harbor amoeboid protozoans. For example, remove some material from around the gums of your own teeth and mix this with saline on a clean slide. Examine this under the microscope with weak illumination for *Endamoeba gingivalis*.

3. The digestive tracts of almost all vertebrates and many invertebrates provide good hunting for protozoans. The lower small intestine and the large intestine, along with the cloaca, harbor many species.

4. Fishes, amphibians, and reptiles may all have characteristic species of ciliates or amoeboid species. Section the gut into small pieces and examine each in turn, both in the general content and along the intestinal wall. Temporary slides are most suitable but permanent smears may be produced.

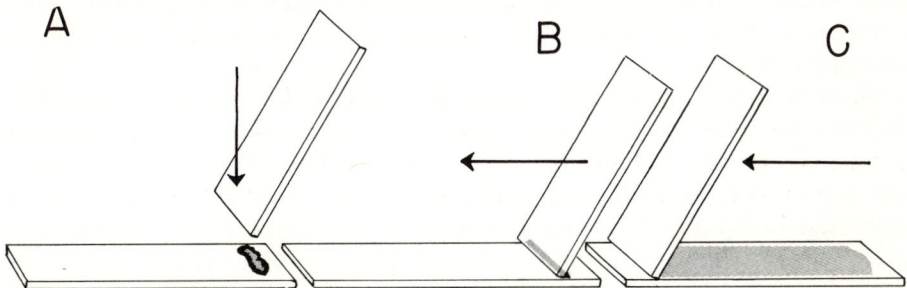

FIG. 6–1. Making a blood smear. See text for details.

5. Nodules on the liver, spleen, and kidneys should all be examined, as well as the content and wall of the urinary bladder.

6. The blood of amphibians, reptiles, and mammals may be rich with haemogregarines or others in the red blood cells, or flagellate protozoans within the lymph. Blood samples should be taken from freshly killed specimens, made into a thin smear, stained, and studied under the microscope. Make a thin smear by putting a fresh drop of blood on the end of a scrupulously clean slide. Place a second slide down into the drop of blood and after a second push this to the far end of the initial slide (Fig. 6–1, A, B, and C); stain with Wright's stain (see Appendix C).

SOME CULTURE METHODS

Care of Native Cultures

Field collections may be maintained in the original jars or directly subcultured. Do not overcrowd jars with too much vegetation. Stacking dishes or finger bowls are preferable to jars, in that they provide a high ratio of surface area to volume. Stack finger bowls and cover the top bowl with glass. Be sure that all glassware is meticulously clean and rinsed before collecting or subculturing. Raw or filtered pond water or stream water should be used in preference to tap water which is often contaminated with various metals. Place about ½ inch of pond water in a finger bowl and add some vegetation. Examine cultures under the dissecting microscope for the presence of protozoans. If desired species are abundant, subculture immediately by pipetting them into culture dishes. Usually, however, several species are present in low numbers and require up to two weeks of reproduction before they are numerous. Protozoans show distinct cyclic population effects in culture and will fluctuate periodically from high to low populations.

Physical conditions of the culture dishes must be carefully controlled. Bright light, with occasional sunshine, is required for euglenoid forms; it is tolerable to many ciliates but detrimental to most amoeboid forms which are bottom dwellers. The last-named species are best cultured away from windows, preferably in rooms facing north. Moderate room temperatures are ideal for fresh-water forms, whereas parasitic and marine species require special temperatures. If temperatures tend to approach 90° F., put all culture dishes and jars in a large water bath to reduce temperature fluctuation, since high temperatures are lethal to protozoans. Other critical factors, such as pH and mineral content of the water, may prove important. Native cultures require little or no food until after several weeks have passed. Aquatic protozoans feed on decaying vegetation and bacteria, or are carnivorous and use other species of protozoans for food. Feed as directed below. Finally, many species of protozoans and other invertebrate animals such as rotifers, water fleas, flat worms, and the like, may prove detrimental to cultures by controlling the populations of protozoans. With a fine pipette or a filter, remove such organisms from culture dishes.

Reculturing Techniques

Provide clean finger bowls with about ½ inch of water. Although filtered pond water may be satisfactory, distilled water or treated pond water may prove superior. Treat pond water by boiling it for a few minutes to kill unwanted protozoans and bacteria, cool, and then shake vigorously in a half-filled bottle to charge the water with oxygen. Protozoans may be introduced in one of two ways. Transfer bits of vegetation or small quantities of culture water containing the specimens to the new culture. Pest-free cultures are made, however, by using a fine pipette and transferring specimens with the aid of the dissecting microscope. To make such a pipette (Fig. 6–2) rotate a ¼-inch glass tube in a flame until it becomes extremely soft. Next, pull on the two ends of the tube to sharply reduce the tube diameter, and make a bend in the tube before it regains a solid state. Cut the tube at the desired point and attach an 18-inch length of rubber tubing or a rubber pipette bulb. Suction is provided orally with the rubber tubing or manually with the bulb. Make a shoulder on the end of the glass tube to receive the pipette bulb by heating the tube until the glass melts and pressing the end

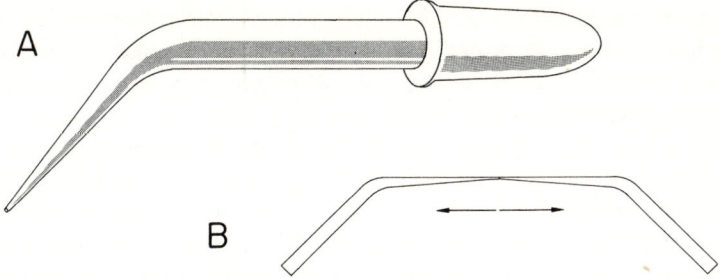

FIG. 6–2. A micro-pipette. A. Pipette with bulb. B. Drawing out the glass tubing.

straight down on a preheated metal surface such as the top of a Bunsen burner stand.

Fresh-Water Amoeboid Protozoans. Free-living and shell-bearing amoeboid protozoans such as *Amoeba, Difflugia, Pyxidicula,* and *Arcella* are collected among live and rotting vegetation in the bottoms of small lakes, ponds, marshes, etc. When these species are detected in native cultures, transfer them to subculture dishes prepared and kept in subdued lighting at room temperature. A hay infusion has long been used as a food source (Hyman, 1925) but hay seeds, wheat kernels, or rice may also be used. Boil the food source for 5 minutes and add no more than four or five 1-inch pieces of hay per culture dish along with, possibly, a few pieces of grain. Bacteria will be transferred when you are pipetting amoebas into these cultures and will thrive on the vegetable matter, thereby producing a source of food. The decaying vegetable matter will also serve as food, but should be discarded if mold begins to thrive in the cultures. Specimens will become noticeably abundant after two to three weeks and need reculturing after about seven weeks.

Ciliate Protozoans. Many species of ciliate protozoans in mixed or pure cultures are used in the classroom—most notably *Paramecium*. The food of these ciliates varies from bacteria which grow upon decaying plant or animal matter to living protozoans. *Paramecium* cultures are prepared as described in the previous section, but are kept in moderately bright light (avoid direct sunlight) where bacteria flourish. Transfer specimens by means of a pipette. *Paramecium* populations will reach their peak in from 10 days to 2 weeks (*Turtox Service Leaflet No. 4,* 1959) at which time reculturing is advisable. Predacious ciliates may be kept in *Paramecium* cultures if new supplies of *Paramecium* are added periodically.

Euglenoid Protozoans. *Euglena* and other similar protozoans can easily be cultured by the Turtox method (*Turtox Service Leaflet No. 4,* 1959). Use clean, wide-mouth, gallon bottles or battery jars filled with boiled pond water or tap water and add a handful of wild grass or hay to each. Age this culture in a sunny place for a week and then inoculate with a *Euglena* culture collected as described above or acquired from a biological supply house. Some additional food (hay or wheat) must be added every few weeks to maintain the culture for a period of several months. Room temperature, with bright light and occasional sunlight, proves best for these protozoans.

Parasitic Protozoans of Insects. Some common parasites, such as *Gregarina*, can be "collected" by collecting field crickets and keeping these in a dark container with such food as dry dog biscuits, dry vegetation to hide in, and a moist sponge for water. When the digestive tract is dissected upon a slide in saline, *Gregarina* can readily be detected in a large percentage of the specimens. *Trichonympha* and other termite flagellates are best maintained by keeping termites in a dark, covered container. Moist wood, preferably wood that has already been partially worked by the termites, should be used as a food source. Hold a large worker termite by the thorax and gently apply pressure toward the terminal end of the abdomen. This will usually cause the termite to defecate and thus release hundreds of protozoans. Collect the drop of feces on a slide with some saline solution, mix, add a coverslip, and study under the microscope. Frequently these flagellates will go into temporary shock and may remain inactive for a brief period of time. Trager (1934) has worked out two media which may be used for pure cultures of these species.

Literature on Culturing. Lengthy discussions of culture media for freshwater, marine, and parasitic protozoans are presented by Kudo (1954), Galtsoff *et al.* (1959), Turtox Service Department (*Leaflet No. 4,* 1959), and others.

STUDY METHODS FOR LIVE PROTOZOANS

Temporary Slides

Wet Mount. Wet mounts are used for temporary examination of cultures or for classroom study. Clean a slide and coverslip and transfer a few drops

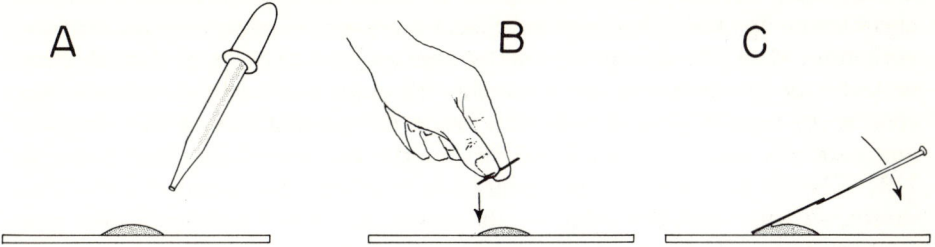

FIG. 6–3. Methods of making a wet-mount slide. See text.

of culture material to the slide. Select material either from the bottom (including some bottom debris), from the lighted side of the culture jar, or from the surface of the culture water. Grasp the coverslip between the thumb and first finger and lower it to the slide so that it rests at the edge of the drop of culture media. Finally, support the upper edge of the coverslip with a dissecting needle and lower the coverslip into place (Fig. 6–3). In the process, air will naturally be expelled and few air bubbles will be included in the slide. Examine under the compound microscope. When working with larger protozoans, support the coverslip with cotton fibers or short lengths of hair to prevent crushing.

Fecal Smears. Fecal material may be obtained from the cages of host animals (frogs, rats, and others), or the intestinal content may be removed during dissection and examination for parasites. Human fecal samples should be collected in clean containers. Here, 1-pint cylindrical containers used for custom-packed ice cream are desirable, in that they are waterproof and disposable. The stool should be firm if the trophozoite stages of amoeboid protozoans are desired. Diarrhetic stools contain predominantly cyst forms. Examine the fecal material within an hour after collecting, as the trophozoite forms quickly die out, losing their taxonomic characteristics.

Place a few drops of saline appropriate to the host animal (Appendix C) on a clean slide. Add to this a small quantity of fecal material and mix this uniformly to make a thin smear. Lower a coverslip as directed above, and seal the coverslip with a soft wax mixture. This mixture is made by adding equal parts of paraffin and Vaseline, melting it, and applying the liquid with a cotton swab. The soft-wax seal is optional, but is recommended when pathogenic species are being examined. Study under the oil-immersion microscope.

Hanging-Drop Technique. This technique is used to study a limited number of protozoans over long periods of time. Secure a concave slide, a clean coverslip, and a fine pipette. Select a few protozoans from the culture, with the aid of the dissecting microscope, and place these in a small drop of water in the center of the coverslip. Examine the coverslip microscopically to be sure the protozoans you desire are present. Invert the coverslip and carefully lower it into the cavity of the concave slide (Fig. 6–4A, p. 112). The drop should not be so large as to contact the bottom of the concavity. Finally, seal the edge of the coverslip with Vaseline or soft wax (Appendix C). This preparation will last for 8 hours or more of continuous observation and greatly restricts the movement of the protozoans. The protozoans may be stained with an appropriate vital stain before they are mounted.

Use of the Microscope

For most protozoans the standard compound microscope with 50, 100, and 430 magnifications is quite adequate. For the smaller species, especially the parasitic forms, the oil-immersion microscope is used, giving 100, 430, and about 950 magnifications on the three objectives. Controlling the light is very important in locating protozoans. Frequently, light which is too bright will completely obscure protozoans, whereas subdued light enhances their outline. Learn to alter the substage mirror from time to time in order to sharpen the image of specimens under the standard microscope. The use of a substage condenser greatly facilitates study of protozoans and is recommended. Learn to scan a slide under the lowest usable power. This requires that you concentrate and bear in mind the many shapes and forms of protozoans as you scan. When you develop an "eye" for seeing protozoans, low-magnification scanning will save considerable time. Move the slide back and forth under the objective so that you systematically cover the entire area of the coverslip. When you locate a specimen, go to the higher ranges of magnification for study. Guard against overheating or dehydrating specimens. Either seal the coverslip in soft wax or periodically add a drop of water at the edge of the coverslip. Stain may be introduced to slides while they are on the microscope, as discussed below.

Slowing Protozoan Movement

The faster-swimming flagellate and ciliate protozoans move too rapidly for observation of contractile-vacuole movement, digestion, locomotor movement, and the like. Several standard methods for slowing protozoan movement are now used, including placing a small amount of cotton or lens paper that has been teased apart on the slide, adding a drop of culture in the center of this material, and supplying a coverslip. The objective is to create small "cells" surrounded by fibers in which the protozoans are contained. Although this method is somewhat satisfactory, the use of methyl cellulose or related alcohol compounds is much more effective. Put some methyl cellulose on a toothpick or dissecting needle and make a ring about ¼ inch in diameter on the slide. Place a drop of culture containing protozoans into this ring and mix the methyl cellulose with the culture water. Finally, add a coverslip so as to exclude air bubbles, and study under the microscope. The very viscous methyl cellulose greatly retards the movement of the protozoans but does not readily interfere with their metabolic activities. One should be most cautious in the use of this compound, however. If the microscope objective becomes coated, clean the lens immediately, for a dried coat of methyl cellulose is very difficult to remove and may damage the finish of the lens.

Vital Stains

Vital stains may be used in dilute concentrations to stain living organisms without interfering with their metabolic activity. Some of these stains are selective for certain structures or materials within the bodies of organisms. Some experimentation may be required to determine the quantity of vital stain necessary. There are three general methods of applying the stain: (1) Add a small quantity of vital stain to a drop of culture medium on a microscope slide, stir with a dissecting needle, add a coverslip, and examine. (2) Add stain after the wet mount has been under examination. Place a drop of stain at one end of the coverslip and "pull" this under the coverslip by applying a piece of paper toweling against the moisture at the other end of the coverslip. With absorption of water from under the coverslip, the stain is forced in and around the specimens. Specimens will show various degrees of staining, depending on the amount of contact they have made with the stain. (3) Add small quantities of stain to the culture until the organisms absorb the desired amount. Then make wet slides in the normal manner. This last technique may be useful for mixing different strains of *Paramecium,* in order to observe conjugation, for example; the different-colored mating strains make the process of clumping and pairing off more obvious.

In addition to staining the cellular content of the organism, vital stains may be used to detect enzyme activity in food vacuoles or to contrast contractile-vacuole activity. For example, feeding *Paramecium* yeast that has been stained with Congo red has been recorded in the literature and is becoming a popular study in laboratory classes. Stiles (1961) suggests mixing 35 milligrams of Congo red with ¼ of a cake of yeast and boiling this in 10 cubic centimeters of water for 10 minutes. When the mixture cools, small quantities are fed to the *Paramecium*. The relative pH can be identified by the color of this vital stain, for pH that ranges from neutral to alkaline registers a brilliant red, whereas pH below 3.0 turns this stain bright blue. Between these levels, purple hues result. Hence, the pH shift in digestion can be observed from the point of food intake to the time of waste discharge. Contractile vacuoles may also be observed by adding fine, particulate carbon to the wet mount. This enhances the appearance of the contractile vacuole and makes it more easily observed. Some of the more common vital stains (which are all mixed with water) are methylene blue, methylene green, aniline yellow, Janus green, neutral red, safranin, and crystal violet. For general laboratory use an assortment of ten vital stains under the trade name "Parstains" is available, along with complete instructions for mixing.

Stain-Kill Methods

Protozoans and other small organisms are readily stopped and killed by the introduction of strong stains or stains containing alcohol or other poisonous compounds. Alcoholic stains may cause some distortion if they are too strong. A large number of stains may be used for study of specific structures. For example, iodine stains may show up flagella, cilia, and nuclei; aceto-carmine, methyl green, and crystal violet also contrast the nucleus; Bismarck brown and methylene blue may cause trichocyst discharge from ciliates, and so on. These stains are added to wet mounts that have already been prepared. Pull the stain under the coverslip, as described above, and observe through the low-power microscope during the staining process. Stain, if too dark, may be removed from the culture water by pulling enough fresh water through to dilute the existing stain. Difficulty may arise in pulling or streaming the specimens from under the coverslip. Specimens may be trapped, however, with cotton fibers, or the speed of streaming may be controlled by the rate at which water is absorbed from under the coverslip. For the many additional stains available, consult any good handbook on biological staining techniques (see Appendix B).

PERMANENT SLIDES

For the most part, cytological and histological techniques are beyond the scope of this text. There are many complex methods of staining and counterstaining to show subcellular structures. For animals other than the protozoans such methods are beyond the scope of collecting and preserving. However, in work with the so-called one-celled animals familiarity with a few basic techniques of cytological work is essential. Only one method of fixing and one of staining will be presented here. For more complex methods, consult such authors as McClung (1950), Lee (1950), Gurr (1953), and others (Appendix B).

Smears

Protozoans are most frequently fixed directly on slides, unless they are available in large quantities. In the latter case they may be concentrated by a slow centrifuge, fixed, stained, dehydrated before being infiltrated with the mounting medium, and transferred to the permanent slide. Smear techniques may be carried out on either the slide or the coverslip. Many media containing protozoans will cause them to adhere to the glass surface. One must avoid drying in all processes of mounting protozoans, in order to eliminate distortion.

Free-Living Protozoans. On the slide (or coverslip) cover an area slightly smaller than the coverslip with Mayer's albumen and allow it to dry until sticky. Next, place a small drop of culture containing flagellates, ciliates, or other protozoans on the albumen and distribute the culture evenly. From this point the author has found the method of Cable (1957), using Schaudinn's fixative and iron-alum hematoxylin, very satisfactory and somewhat less complex than other procedures. This combination of fixative and stain is very satisfactory for all protozoan work. Cable recommends, before smears are made, numbering five slides consecutively with a carborundum or diamond pencil and having ready a 100-cubic-centimeter beaker about two thirds full of Schaudinn's fixative heated to between 60° and 75° C.

Fix immediately by standing the slide on end in a beaker of fixative. Hold the smears apart with pieces of applicator placed across the top of the beaker. Fix for at least 5 minutes. Pass through 30% and 50% alcohol, 2 minutes each. Treat with 70% iodine-alcohol 3 to 5 minutes. Pass through fresh 50% and 30% alcohol 2 minutes each. Wash with two changes of distilled water. Mordant in 2% aqueous iron alum (ferric ammonium sulphate) for 30 minutes or longer. Wash rapidly twice in distilled water. Place in 0.5% aqueous hematoxylin for 30 minutes or longer. Wash twice in distilled water, 1 minute each. Destain as follows: Place all five slides at one time in 1% aqueous iron alum. At the end of one minute, transfer slide number 1 to distilled water, at the end of two minutes, slide number 2, and so on so that the staining time of the various ones will have ranged from 1 to 5 minutes. Change the water at once and let stand 2 minutes. Intensify in saturated aqueous lithium carbonate one minute. Note that the smears become decidedly bluish in color. Dehydrate by passing through 30%, 50%, 70% and 80% alcohol 2 minutes each, 95% 3 minutes, and absolute, 5 minutes. Clear in xylol (3 to 5 minutes). Mount in thin xylo-damar or Permount, using No. 1 coverglasses. Let dry and examine with oil immersion lens and compare staining with the duration of the destaining process.

Surface Scums. McClung (1950) recommends the following technique for surface scums:

If the culture bears a surface scum in which the Protozoa may be more or less entangled, drop a clean coverglass on the scum, then lift it up with forceps, keeping it horizontal, and a section of the scum will adhere to the undersurface. Tilt the cover and drain off excess water on a piece of filter paper, then drop, scum side down, on the surface of the fixative which has been placed in a convenient container.

From this point Cable's method, as cited above, should be followed.

Fecal and Intestinal Smears. Material from these sources are self-adherent and need no albumen. The more or less liquid intestinal content may be placed directly on the slide, whereas harder fecal material should be mixed with a drop or two of saline on the slide. With an applicator stick work the material into a thin uniform smear and transfer it to a fixative, before it has time to dry, as directed above.

Foraminifera and Radiolaria

These shelled protozoans should be collected, killed in 95-percent isopropyl alcohol, and treated with rose bengal, as directed above (p. 99). When the specimens have settled to the bottom of the container they may be stored in 70-percent alcohol.

The skeletons of foraminiferans and radiolarians are found in various earth deposits and make up over half of the surface area of the ocean bottom. Sverdrup *et al.* (1949) show that the foraminiferan *Globigerina* makes up a bottom ooze covering over 47 percent of the floors of the Atlantic, Pacific, and Indian oceans, an area of 126.4 million square kilometers. Radiolarian oozes cover about 6.9 million square kilometers. These deposits of tiny skeletons are almost pure in composition and may measure hundreds of feet thick.

Isolating Specimens. Samples of these oozes and other preserved bottom sediments should be processed to remove the mineral content and to isolate the protozoan tests. Very fine mud sediments and coarse sand grains must be eliminated by various methods. One such method is screening through 40-, 80-, and 200-mesh screens to remove the fine muds, followed by the tedious procedure of picking out specimens under the microscope. Secrist (1934) removes clay particles and other extremely minute material by directing a jet of water against the side of a beaker at such an angle as to obtain a whirling motion in the container. This causes the fine material to be momentarily suspended so that it can be removed, leaving the coarser residue at the bottom. This cyclonic movement of water within the beaker also tends to deposit the heavier sand grains in the center; the lighter material, which remains waterborne longer, is distributed toward the outside. Thus, specimens are partially separated from the mineral content. With heavy liquids the differential weight of the mineral content and the animal tests may also be taken advantage of. Schenck and Adams (1943) list carbon tetrachloride, bromoform, ethylene dibromide, and potassium mercuric iodide dissolved in water as being among the most useful liquids. Carson (1933) states that the use of heavy liquids works well only for samples in which the protozoan tests are not filled or pyritized (which would make them as heavy as the mineral content). When dried samples are introduced into carbon tetrachloride the tests float, while the remainder of the sediments sink to the bottom, thus permitting easy separation. Carson's method with bromoform is as follows: dilute bromoform with absolute alcohol to a point where the quartz crystal taken from the sand sample will just sink. This solution is placed in a funnel provided with a valve at the bottom, and the dry sample is added to this solution and stirred. After a few moments the heavy concentrations sink to the bottom. If considerable sand remains on the surface, continue stirring and add small quantities of absolute alcohol until the desired degree of separation occurs. The heavy concentrate is drawn off onto filter paper and the alcohol-bromoform solution is recovered. The concentrate should be washed in two or three changes of alcohol to further remove the bromoform. Next, the floating portion is drawn off onto filter paper and likewise washed with alcohol. These practices are primarily applicable to foraminiferan samples. Many of these heavy liquids are poisonous and should be used with extreme caution.

Dry-mounting Foraminiferans. Special slides are obtainable or may be made for the purpose of dry-mounting foraminiferans. The slide consists of two pieces of laminated cardboard, the upper piece (Fig. 6–4B) is perforated with a round hole about ½ inch in diameter; the lower piece beneath this perforation is painted dull black. If the black surface is not impregnated with a water-soluble glue, a clear, permanent (but soluble) blue must be provided. Foraminiferans are isolated from temporary wet-mount slides without coverslips by pushing the desired specimens to one side with a probe and then picking them up on the moist tip of a fine

112 Biological Techniques

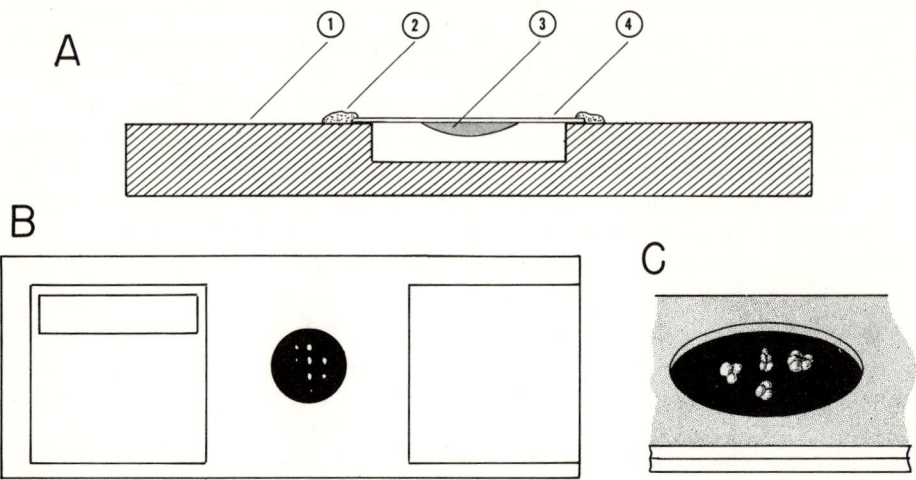

FIG. 6–4. Other slide techniques. A. Hanging-drop technique. B–C. Foraminifera-dry-mount slide. (1) Concave slide shown in cross-section, (2) soft wax seal, (3) hanging drop with specimens, (4) coverslip.

water-color paintbrush, size 000. The moisture from the brush temporarily activates the glue and fastens the specimens down. When possible, mount several specimens of the same species, each in a different attitude to show all characteristics. The glue quickly adheres to the foraminiferan and will hold permanently or until it is remoistened. The specimen data are recorded on the ends of the slide, and the slides are, in turn, stored in standard slide boxes. Generally, it is not essential to place a clear glass slide over the cardboard preparation. However, this may be useful in very moist climates. For more elaborate mounting systems, consult Schenck and Adams (1943) who detail the operational procedures of commercial micropaleontologic laboratories and present an excellent bibliography.

Other Techniques. Tests of radiolarians and foraminiferans may be mounted on standard microscope slides with coverslips. Heat-dry samples of foraminiferans, place in xylene for one week, and mount on slides with Permount, using either concave slides, spun cells, plastic rings, or other supports for the coverslip as directed in Appendix B. Dried specimens may be treated for a week in 70-percent alcohol and mounted directly in Turtox CMC-10. Radiolarians may be mounted directly from water or alcohol in Gray and Wess's mounting medium. Alternately, they may be dried and stored in Euparal Essence for several days and then mounted in Euparal.

REFERENCES

Cable, R. M., 1957, *An Illustrated Laboratory Manual of Parasitology*, Burgess Publishing Co., Minneapolis.
Carson, Carlton M., 1933, Paleontological Notes: A Method of Concentrating Foraminifera, *J. of Paleontology* 7:439.
Cushman, J., 1948, *Foraminifera: Their Classification and Economic Use*, Harvard Univ. Press, Cambridge, Mass.
Galtsoff, P. S., *et al.*, 1959, *Culture Methods for Invertebrate Animals*, Dover Publications, N.Y.
Gurr, Edward, 1953, *A Practical Manual of Medical and Biological Staining Techniques*, Interscience Publishers, N.Y.
Hall, R. P., 1953, *Protozoology*, Prentice-Hall, Englewood Cliffs, N.J.
Hyman, L. H., 1925, Methods of Securing and Cultivating Protozoa, I. General Statements and Methods, *Trans. Amer. Micr. Soc.* 44:216.
Hyman, L. H., 1931, Methods of Securing and Cultivating Protozoa, II. Paramecium and Other Ciliates, *Trans. Amer. Micr. Soc.* 50:50.
Jahn, T. L., 1949, *How to Know the Protozoa*, William C. Brown, Dubuque, Iowa.
Jones, Ruth McClung, 1950, *McClung's Handbook of Microscopial Technique*, Paul B. Hoeber, Inc., Medical Book Dept. of Harper & Row, N.Y. (Now published by Hafner, N.Y.).
Kirby, H., 1950, *Materials and Methods in the Study of Protozoa*, Univ. of Calif. Press, Berkeley, Calif.
Kudo, Richard R., 1954, *Handbook of Protozoology*, Charles C Thomas, Springfield, Ill.
Lee, Bolles, 1950, *The Microtomist's Vade-Mecum*, J. B. Gatenby, ed., Blakiston, Philadelphia.
McClung's Handbook of Microscopial Technique (see entry under Jones above).
Schenck, Hubert G., and Bradford C. Adams, 1943, Operations of Commercial Micropaleontologic Laboratories, *J. of Paleontology*, 17(6):554–583.
Secrist, Mark H., 1934, Technique for the Recovery of Paleozoic Arenaceous Foraminifera (Paleontological Notes), *J. of Paleontology*, 8:245–246.
Stiles, Karl A., 1961, *Laboratory Explorations in General Zoology*, Macmillan, N.Y.
Sverdrup, H. U., *et al.*, 1949, *The Oceans*, Prentice-Hall, Englewood Cliffs, N.J.
Trager, William, 1934, The Cultivation of a Cellulose-Digesting Flagellate, *Trichomonas termopsidis*, and of Certain Other Termite Protozoa; in a thesis submitted for Ph.D., at Harvard.
Turtox Service Leaflet No. 4, 1959, General Biological Supply House, Chicago.

(See Chapter 1 for other References.)

CHAPTER 7

The Sponges

The sponges, phylum Porifera, are primitive multicellular animals found predominantly in marine waters. They have a cellular grade of construction, since well-developed tissues or organs are lacking. Owing to their primitive development it is difficult to distinguish between individuals and colonies. The name Porifera acknowledges the multitudes of pores, or openings, which allow water to enter the body cavity of the sponge. Collar cells equipped with flagella line the body cavities and are responsible for creating a water current through the body of the sponge and for collecting food.

The taxonomy of sponges may be based on: (1) the gross morphology of the colony or individual (asconoid, leuconoid, syconoid, encrusting, erect, branched, globosed, and so on); (2) the color of the living animal; (3) the substrate—rock, mud, limestone, or mollusk shells; (4) the nature of the skeleton. The skeleton may be made of hard, glass-like calcareous or siliceous spicules or fiber-like spongin. Spicules vary tremendously in relative location, size, sculpturing, and gross shape (see Hyman, 1940). The phylum Porifera is separated into three classes on the basis of the nature and type of skeletal material. The identification of sponges, therefore, depends in part on the analysis of this material.

COLLECTING SITES AND METHODS

Fresh-water sponges are found in clear lakes, ponds, and streams in relatively shallow water. They may be collected by hand or with a dredge net (Chapter 13). Remove rocks, heavy pieces of vegetation, or other debris and examine all surfaces for encrusting sponges. Obtain specimens with gemmules which will appear as light-colored bodies scattered throughout the sponge colony. If necessary, collect several times during the year in order to obtain these important reproductive and taxonomic structures. Preserve and dry portions of sponge attached to their substrates, as well as additional material carefully removed with a pocketknife or spatula.

The majority of the marine sponges are found attached to rocks in quiet waters and unprotected and protected rocky habitats, or attached to pilings, floating docks, or other suitable substrata. Most species are to be found in shallow water, although few are found above the zero tide level except on floating substrates. In extremely deep water endemic sponges occur on the softest of muds, supporting themselves by long growths of spicules down into the mud. Collect marine sponges by hand during low tide, placing specimens in wet, nonmucous-secreting algae until the time of preservation. It is essential to keep the specimens cool and moist and to preserve them as soon as possible. Skin-diving will provide many additional specimens. Be sure to turn boulders and look into crevices for sponges growing in highly protected microhabitats. A knife will be essential to remove the sponges from their substrates in most instances.

The biological dredge (Chapter 1) is indispensable when collecting in the intermediate and deep-water habitats. Sponges growing where either water current or water action influences the substrate will be found on rocky or rubble-covered bottoms rather than on sand or mud. In deeper water, however, sponges may be obtained from all substrates.

Killing, Preservation, and Storage

Liquid Preservation. Sponges may be dried or preserved in liquid. For liquid preservation, wash specimens in water to remove debris and preserve in 75- to 95-percent alcohol. The weaker solution is suitable for small specimens, but large specimens or specimens crowded in a single bottle should be preserved in 95-percent alcohol. Change the alcohol after 24 hours on large or crowded specimens. Do not use formalin for preservation, as it will destroy calcareous spicules and break down the sponge "tissue." Neutralize formalin if it must be used and transfer specimens as quickly as possible into strong alcohol. Specimens are stored in the museum in strong alcohol. Small specimens should be kept in alcohol-filled, cotton-stoppered vials, along with their field data. These vials, in turn, are placed in a larger museum jar filled with alcohol. Large specimens are kept in single jars, with the field data included.

Dry Preservation. Dried sponges are very suitable for taxonomic purposes and cost less to prepare, but are not too useful for beginning biology students. If sponges are dried, additional small individuals or portions should be preserved in alcohol. Sponges may be dried, with or without their substrate, by being washed and placed in a warm place out of the sun. Before drying large specimens, attach labels by means of a string threaded through the body of the sponge. When they are thoroughly dried, sponges should be kept in small cardboard boxes and supported with tissue paper. Field data and taxonomic data should be included with each specimen. Paradichlorobenzene or naphthalene flakes may be added to dry containers, though this is usually not necessary.

Fixation for Histology

Sponges intended for "tissue" studies must be fixed in the field with Champy's Fixative or other standard fixatives containing osmic acid. Wash specimens to free them of debris and place them directly in the fixative for about 6 hours. Subject the fixed specimens to running water for an additional 6 hours and transfer them to strong alcohol. Do not leave specimens in the fixative, as any water-bearing solution will break down the tissues. On returning to the laboratory, remove the specimen's spicules, stain, and section as described in handbooks of microscopic technique (Gatenby and Beams, 1950, and others).

Skeletal Material

Sponge skeletal material consists of fibrous spongin or of siliceous or calcareous spicules. Megascleres are larger spicules which make up the loose inner mass of support; microscleres are peripheral and make up small dermal spicules. Spicules must be studied independently and occasionally *in situ* to show the spicule arrangement.

Spongin. Sponges containing spongin should be washed in fresh water and dried. Beat the sponge to loosen the cells mechanically, rewash, stain with safranin or eosin, and dry. Remove small portions and mount in Permount or balsam slides.

Siliceous Spicules. Select a sample piece of sponge and remove the soft tissue by boiling it in several drops of concentrated nitric acid on a slide over a flame. When it is dry, add a mounting medium (Permount, balsam, or either) and a coverslip, and store in a horizontal position until dry. A better preparation with less debris is obtained by treating sponge overnight in a test tube containing 6 or 7 cubic centimeters of nitric acid or Clorox. In the morning, carefully pour off the acid, leaving the spicules at the bottom, and wash with several changes of water. Permit the spicules to settle to the bottom after each washing. Exchange the water for alcohol and transfer a drop of the alcohol–spicule mixture to a slide. When completely dry, add mounting medium and a coverslip.

Calcareous Spicules. The composition of spicules in unknown sponges may be determined by treating a small portion of the sponge in acid. Calcareous spicules will react with the acid: their "boiling" testifies to their composition. Treat calcareous spicules as described under siliceous spicules, but use a 10-percent potassium hydroxide (KOH) solution in place of the nitric acid. After the treatment, wash, transfer to alcohol, and prepare the slides, as described above.

Spicule Arrangement. To determine the spicule arrangement *in situ*, remove a small but undisturbed sample of sponge and dry. When it is completely dry, wet with xylene, transfer to a xylene–balsam mixture, and then transfer to a slide. Mount in balsam or Permount, using broken pieces of a slide or coverslip to prop up the cover glass if necessary.

Transport and Culture

Transport sponges loosely packed in wet mucous-free seaweed or in small volumes of aerated water, as described in Chapter 1. The specimens must be kept cool, supplied with oxygen if in liquid, and free of polluting or suffocating materials such as mucus or excreta from other animals. Overcrowding must be avoided.

The requirements of sponges to be cultured in the aquarium are relatively simple. Place the sponges on a suitable substrate well above the bottom debris to prevent the clogging of pores and to facilitate feeding. Circulate running seawater, or pond water in the case of fresh-water sponges. If running water is not available, change the water in the aquarium frequently, in order to introduce new food. Aerate the aquarium to prevent stagnation and maintain the water temperature at or slightly below that of the native environment. See Needham *et al.* (1959), Wilson (1907 and 1911), and others for special culture techniques.

REFERENCES

Gatenby, J. B., and H. W. Beams, 1950, *The Microtomist's Vade-Mecum (Bolles Lee)*, Blakiston, Philadelphia.

Hyman, Libbie H., 1940, *The Invertebrates: Protozoa through Ctenophora,* McGraw-Hill, N.Y.

Jewell, M., 1959, "Porifera," in *Ward and Whipple's Freshwater Biology,* ed. by W. T. Edmondson, John Wiley & Sons, N.Y., pp. 298–312.

Needham, James G., Paul Galtsoff, Frank Lutz, and Paul Welch, 1959, *Culture Methods for Invertebrate Animals,* Dover Publications, N.Y.

Wilson, H. V., 1907, A New Method by Which Sponges May be Artificially Reared, *Science,* 25:912.

Wilson, H. V., 1911, Development of Sponges from Dissociated Tissue Cells, *Bull. U.S. Bur. Fish.*, 30:1.

(See Chapter 1 for other References.)

CHAPTER 8

The Coelenterates and Ctenophorans

The phylum Coelenterata contains a diverse group of animals, from the point of view of general appearance, skeletal structure, life history, and means of reproduction. The coelenterates are found mostly in marine water, as both sessile and pelagic forms. A few species, however, are fresh-water dwellers and occur the world over. Figure 8–1, A through L, illustrates specimens referred to in this chapter. The phylum Ctenophora, frequently confused with the jellyfishes, differs considerably from that group, not only in anatomy (the nervous system in particular), but in that ctenophores must be relaxed and preserved by methods somewhat different from those used for the coelenterates. Being highly contractile, many coelenterates will require narcotization prior to preservation.

PHYLUM COELENTERATA

Fresh-Water Hydra

Collecting. The general habitat of the hydra is the small fresh-water lake or pond and small moving streams and rivers, especially where they

120   Biological Techniques

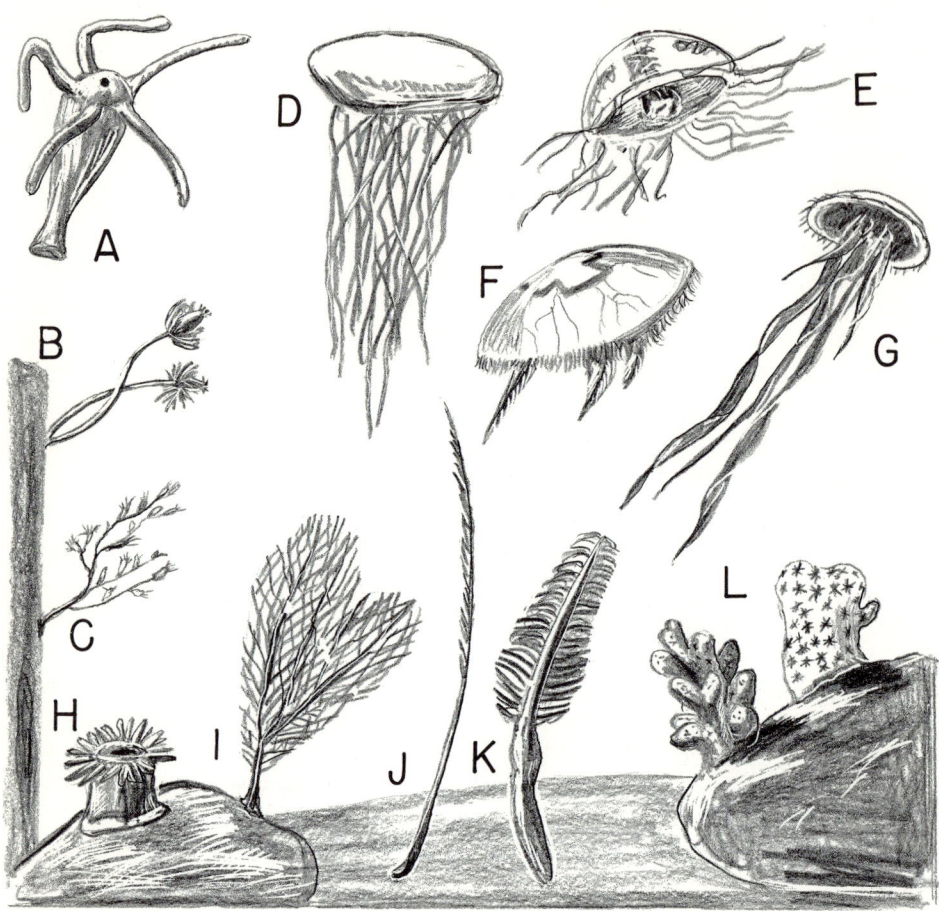

FIG. 8–1.   Types of coelenterates (see text). A. A hydra. B. *Tubularia.* C. *Obelia* colony. D. *Physalia.* E. *Gonionemus.* F. *Aurelia.* G. Large jellyfish showing oral tentacles. H. Anemone. I. Sea fan. J. Sea whip. K. Sea pen. L. Hard coral.

flow from lakes. In particular, shallow waters about one foot in depth produce the best light and temperature combination for hydra. Specimens may be seasonal where the winters are severe, but are generally available through the late spring, summer, and autumn. In western Washington hydra may be collected all year round, although they are contracted and inactive during the winter months. Locating hydra for the first time may require careful searching. These animals live habitually on the surfaces of rocks (especially in slow-moving streams or where moderate wave action is experienced in lakes), or on aquatic vegetation such as *Elodea,* sedges, grasses, and the like. In one little stream where the author collects, these animals are so numerous for a distance of about 200 feet that they can be seen as a continuous grayish mass on the surfaces of almost all rocks, and are easily discernible from 20 or 30 feet away. Where hydra are scarce, look carefully on the surfaces of rocks for grayish or brownish forms which are either expanded or contracted. Should this prove difficult, place rocks in jars or plastic buckets and allow them to stand 5 or 10 minutes. If hydra are present they will expand and be readily detectable. When you develop an "eye" for hydra you can spot them clinging to aquatic vegetation. Pick quantities of vegetation and place it in jars of pond water. After a reasonable length of time, hold these jars up to the light and examine the vegetation for hydra. When you locate a few animals, search diligently on all likely substrates and you will probably find large concentrations of specimens.

Hydra may be transported to the laboratory attached to their normal substrate in a small quantity of water. Take care not to let the specimens become too warm. Generally, when hydra float upon the surface of the water, the amount of oxygen is insufficient.

Evicting Hydra. Hydra may be dislodged from their substrate by squirting them with a jet of water from a pipette. To transfer them to a new culture, pick them up quickly with a pipette which has a moderately large opening. Be sure to expel them with equal rapidity; otherwise, they will cling to the inside of the pipette. When hundreds or thousands of hydra cling to rocks or vegetation, bring this material into the laboratory, place it in a fairly dark area, and allow it to stand several hours or overnight. The hydra will dislodge themselves and float to the surface as the available oxygen decreases in the container. They can then easily be removed with a small aquarium net or pipette and transferred to a new culture dish.

Fixation and Preservation. Transfer large numbers of hydra to Petri dishes with about 3 millimeters of water. After allowing them to become fully expanded, pour hot Bouin's fixative (50° C.) over them. After a few moments pour off the diluted fixative and add fresh Bouin's fixative for 30 minutes. Following this, wash in several changes of 30-percent alcohol, move the specimens through 50-percent alcohol, and store in 70-percent alcohol. In this procedure, be prepared to fix the specimens as soon as they become fully expanded; prolonged waiting may cause them to partially contract and remain that way. When specimens are few, treat them individually so that each may be caught at the peak of its expansion.

Preserved hydra should be placed in vials of alcohol, along with their collecting data, and kept in a large airtight jar of 70-percent alcohol. These specimens may be used on temporary slides, stained with methylene blue, neutral red, or other standard laboratory stains, and provided with supports for the coverslip (Appendix B). For permanent slides, stain with borax carmine or alum hematoxylin, dehydrate through absolute alcohol, transfer to xylene, and mount in balsam or Permount (see Appendix B).

Study of Nematocysts. The stinging cells of hydra may be discharged if living specimens are placed on a slide and covered with a 1-percent solution of nitric acid or a 2-percent solution of glacial acetic acid to which a small quantity of methylene blue has been added. Pennak (1953) recommends placing specimens on a slide with Hertwig-Schneider's fluid (consisting of 1 part .02-percent osmic acid and 4 parts 5-percent glacial acetic acid).

Colonial Hydroids

Colonial hydroids appear as small weed-like colonies. These represent the asexually reproductive generation in the life history of alternation of generations. Special colonial forms bud off the medusoid generation (jellyfish-like animals) which sexually give rise to the swimming plannula larva. It is the job of the larval form to seek out the proper habitat and attach itself. It then gives rise to the hydroid colony (Fig. 8–1, B and C). The location of the habitat is, to a large extent, on a trial-and-error basis; those larvae attaching in an unfavorable situation may begin to grow, but will quickly die out.

Collecting and Seasonal Studies. Hydroids, being filter feeders that require large quantities of microscopic organisms, live in the upper layers of ocean water and are generally attached to rocks, pilings, and other suitable substrates. They may be found offshore by pulling in giant brown kelp and examining the fronds and holdfasts. It will be noted that hydroids will grow in great profusion on certain levels of the pilings and corresponding levels of rock substrates. They may be collected in moderate numbers in protected rocky coast zones, but are most abundant in bays and estuaries. In the latter situation, floating docks should be examined, preferably from underneath, for the log floats harbor tremendous quantities of invertebrates, including hydroids. When collecting, use a pocketknife to free the hydroid from its substrate. Place specimens directly into a clean container of fresh sea water and keep them cool while transporting them back to the laboratory.

Hydroids and other sessile invertebrates may also be collected by suspending wood blocks in the water for them to attach to. In following a long bay or estuary you will note that the species composition varies tremendously as you proceed from the open ocean to the point farthest away from the ocean. The species composition not only varies with the distance away from the ocean, but also with the depth of the water. Physical oceanographic conditions such as pH, dissolved oxygen, temperature, light, turbidity, and unnatural pollution may contribute to regulating the distribution of various hydroid species. Prepare a rope (¼-inch hemp) with a weight and a series of wood blocks held in place by knots tied above and below, as shown in Fig. 8–2. These blocks may be made from 2- by 2-inch fir wood, about 6 inches long and drilled with a ⅜-inch hole through the center. Arrange the blocks either at 1- or 2-foot intervals. Tie the string of blocks to a floating dock so that the first block settles at 1 foot in depth below the surface. Retrieve these blocks every 28 days, and either sort out the specimens, directly preserving each in its proper fashion, or preserve the entire block of invertebrates in 7-percent formalin.

Narcotizing, Killing, and Preserving. Place small colonies in jars of clean sea water, in a cool semidark area. When the polyps are fully expanded add small quantities of epsom salts (see Appendix D). Continue to add small quantities every 20 to 30 minutes for several hours, being careful not to disturb the polyps. After 3 or 4 hours probe one or two polyps with a dissecting needle and when they prove insensible to touch, carefully add enough formalin to produce a 5-percent solution. If narcotizing compounds are not available, kill specimens with formalin and maintain in 5-percent neutral formalin or 70-percent alcohol after going through intermediate changes of 30 percent and 50 percent.

Bianco (1800) worked out a method which is still used today for killing expanded specimens.

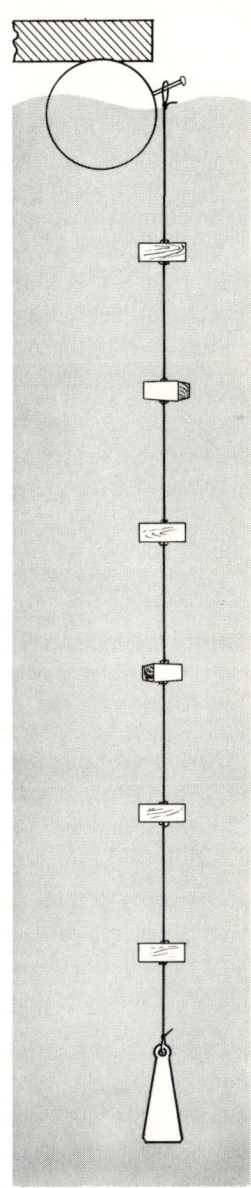

FIG. 8–2. Wooden blocks suspended from a floating dock in order to collect settling marine organisms.

Place specimens in a beaker of sea water and, when they are fully expanded, siphon off all water except enough to cover the specimens without disturbing them. With great caution, poor in a solution of corrosive sublimate (heated to 50° to 60° C.), then cool the solution, wash in several changes of fresh water, and preserve as above. (Consult Appendix C for the use and neutralization of corrosive sublimate.) Good results may also be obtained by following Bianco's method, using hot Bouin's fixative (50° to 60° C.) as the killing agent. If slides are desired, stain with borax carmine, dehydrate, and mount on slides in balsam or Permount with coverslip supports as directed in Appendix B.

Medusae

The small medusas of the Hydrozoa and the jellyfish forms of Scyphozoa will be treated simultaneously. A number of methods are available for specific groups of medusae, but as Bianco (1899) states, different species within a single genus may require different techniques. Obviously, some experimentation is required in testing the different methods listed. For general use, most of the medusae may be preserved directly in sea water with 5-percent neutral formalin.

Collecting. Medusae are pelagic forms found in most strata of water from the surface down to the aphotic, or lightless, zone. Bays, harbors, and estuaries tend to swarm with medusae during the summer months. Because they are extremely soft, they can easily be destroyed in general collecting nets such as tow nets, plankton nets, bathy-pelagic nets, and the like. Whenever possible, it is best to dip specimens out directly with a net or a bucket. Regardless of the means of collection, however, transfer specimens immediately to containers of sea water. The best-preserved specimens are those which are narcotized and preserved while in a living condition rather than those which are dead when preserved.

Various Narcotizing and Preserving Techniques. The small medusae from hydroid colonies, such as *Obelia*, or small scyphozoan jellyfishes, may best be narcotized with ethyl urethane (Galigher, 1934) or with Chloretone or epsom salts. Place specimens in a shallow dish of sea water, add a few crystals of urethane or Chloretone, and let stand for several hours. With epsom salts, add crystals periodically, as described under colonial hydroids. When the medusae are insensitive to probing, pipette out most of the water and kill them with 5-percent formalin. Store specimens in this solution or run them up through the alcohols to 70 percent.

Bianco (1899) recommends killing these same small forms with a copper corrosive solution, washing in fresh water until the precipitate has vanished, and then storing them in alcohol. In another method, designed to kill quickly, add small quantities of acetic acid to dishes containing the specimens, and immediate transfer specimens to a chrom-osmic mixture for 15 minutes, washing and preserving as above.

For members of the Siphonophora (*Velella, Physalia,* and others), or larger scyphozoans, fairly good specimens may be obtained by direct preservation in 5-percent formalin. Store specimens in 5-percent formalin or 70-percent alcohol. Always consider the fluid content of the animal when measuring the preservative. For better specimens, narcotize with menthol crystals (Appendix D) or epsom salts, as directed above.

Large jellyfishes become distorted when allowed to sink to the bottom of a fixing bottle. The author uses a transparent plastic disc tied to a piece of nylon fish-leader to support such specimens. The free end of the fish-leader is threaded up through the subumbrellar surface of the jellyfish by means of a long needle, and out through the upper surface of the animal. The animal is thus suspended in sea water so that the tentacles can hang down freely and naturally. It is then narcotized and preserved in the above manner. This same technique is excellent for holding specimens in large display jars and cylinders. The fish-leader is held firmly in place by the jar lid.

Anemones

Collecting. Anemones are either sessile, attached organisms or sand-dwelling burrowers. They are capable of some locomotion, creeping along on the pedal disc or even swimming. Sand-dwelling species are simply dug from the substrate; attached forms may be collected along with part of their substrate or by slipping the blade of a very dull knife under the pedal disc. The specimens should be kept in clean sea water and should not be overcrowded with other organisms. Deep-water species are obtained with the biological dredge (Chapter 1).

Narcotizing and Preserving. Sea anemones must be expanded in sea water, and then narcotized, before they may be preserved. The author has had outstanding success with many species of sea anemones, using clove oil and Chloretone. Regardless of the method, however, it is best to treat each anemone in a separate jar of sea water, and to begin narcotizing in the evening so that specimens may stand all night under the influence of the particular drug. Several large drops of clove oil were added to each quart jar (¾ full), once each hour for 3 hours, beginning at 8 o'clock in the evening. The amount of clove oil was increased somewhat during the last addition. By 8 o'clock the next morning specimens should be well expanded (provided they were expanded before the treatment began) and completely insensitive to formalin. Probe the tentacles lightly to determine if the animal is sensitive to touch. If it is not, add enough formalin to make a 5-percent solution. The Chloretone method is essentially the same as the clove oil method. Add a moderate pinch of Chloretone crystals each hour for 3 or 4 hours, in slightly increasing amounts, during the evening. Permit the specimens to stand overnight, test them for sensitivity by probing the tentacles, and preserve in formalin when they become insensitive.

Epsom salts have long been used to anesthetize sea anemones. The writer finds that this drug is less effective if it is used as directed for Chloretone above; thus, epsom salts should be added in small amounts every *half hour*, for a period up to 24 hours, until the animals are insensitive to touch. One problem here is that animals tested prematurely may contract the entire body and fail to reexpand under the effects of epsom salts. Therefore, gently tease one or two tentacles, rather than probing the entire animal. When the specimen is totally insensible, preserve it in 5-percent formalin. Gohar (1937) found that some anemones began to macerate before they were completely narcotized. Thus, he suggests beginning the epsom salts treatment with specimens placed in 50 to 100 times their own volume of sea water. During this process add three drops of 1-percent formalin for every 100 cubic centimeters of water, every 15 minutes, and double the volume added each hour until the specimens are insensible; then preserve in 5-percent formalin.

Another technique that will work for sea anemones is to let expanded specimens stand in containers of water at room temperature, from 24 hours to several days. Such specimens will gradually become insensitive as the oxygen is used up and as the water becomes polluted with excreta. This process may be hastened by beginning with chilled sea water which was previously boiled to remove the free oxygen. A second method of hastening the process is to add some organic material to further pollute the water and deplete the oxygen.

Some sea anemones intended for histological work should be expanded in beakers of sea water. When the specimens are fully expanded, siphon off most of the water and kill them by pouring in hot Bouin's fixative (50° to 60° C.), hot 20-percent formalin, or hot FAA.

Horny Corals, Sea Pens, Sea Whips, and Others

The gorgonians, or horny corals, may be unbranched, highly branched and plant-like, or highly branched with cross-branching forming the typical sea fan. The body contains a central axis surrounded by an outer coat, the coenenchyme, which contains the polyps. The sea pens, sea pansies, sea feathers, and the like, possess a central horny axis covered with a more fleshy material which supports polyps singly or in clusters.

Collecting. The horny corals grow in shallow to moderately deep water, attached to reefs or isolated stony substrates. Deep-water forms are collected by dredging; shallow species are obtained by skin diving. A geology pick is quite suitable for cracking off chips of rock which support the horny corals. They may also be dislodged by forcing a heavy knife between the "holdfast" and the substrate. Sea pens, sea pansies, and the like, may be obtained at low tide or by dredging. They are found on sandy substrates just below the low tide level, frequently in channels, bays, and estuaries.

Narcotizing and Preserving. The horny corals are very useful when dried. Dip specimens in neutral formalin for 15 minutes and dry them in a warm but shaded place. Prior to drying, arrange the branches so that they will take up the least amount of space in a museum tray, if this is important. Always secure a waterproof tag containing the field data to the stem of each specimen before drying. After the specimens are thoroughly dried keep them in light-proof boxes with a few crystals of paradichlorobenzene. Samples of horny corals and all sea pens and related species should be narcotized and preserved. Permit these specimens to expand normally in containers of sea water and narcotize as described above under "Anemones," or kill them quickly with hot Bouin's, hot corrosive sublimate, or hot formalin.

Color notes should be taken (consult Chapter 1) for the sea pens and related forms before they are preserved. Five-percent neutral formalin or 70-percent alcohol makes a suitable preservative (the latter will alter color, however). Store specimens in sealed jars, preferably in a dark place, or mount for display as directed in Appendix A.

Stony Corals

Collecting. Stony corals are dominant forms in tropical marine waters, occupying the intertidal zone and extending down to considerable depths. At the higher latitudes the number of species decreases markedly, until only one genus, *Balanophyllia,* a solitary coral, is found intertidally and a few other species are found in deeper water extending on to the north. When a rare coral is collected, part of the rock substrate to which it is attached should be chipped away, along with the specimen, by means of a cold chisel and hammer. Where corals abound they may be collected while you are skin diving, with a geology pick or by means of a dredge.

Narcotizing and Preserving. When expanded specimens are required, narcotize and fix as described under "Horny Corals." Use 70-percent alcohol for a preservative or 5-percent neutral formalin. Straight formalin will greatly damage the skeleton of any coral.

Dried coral specimens are the most useful in spite of the fact that they are fragile. There are three general methods for preparing dried coral; the second method is perhaps the best. To prepare corals by the first method, simply wash freshly collected coral to remove mucus and foreign debris. Next, place them in a well-ventilated area to dry, but avoid constant sunlight. When the specimens are thoroughly dry, store in cardboard containers along with their field and taxonomic data. This technique is especially useful for those species with very small polyps (feeding animals). Some fleshy specimens prepared by this method will have a faint odor that may be offensive and may attract insect pests; fumigate these specimens periodically with paradichlorobenzene.

The second and most useful method for preparing dried coral involves the removal of all tissue prior to drying. Place specimens in containers of water for three to four days. Following this, wash the specimens under a hose to remove rotted tissue and debris. Next, place the specimens in a 10-percent solution of household bleach for one day, or until the specimens are whitened, rinse them thoroughly in several changes of fresh water, and then permit them to dry.

The third method requires the least amount of work but is more expensive and may cause some specimens to become yellowish. Place freshly collected corals in a 10- to 20-percent solution of household bleach until all of the tissue has been removed (generally from 3 to 4 days). Rinse specimens thoroughly in several changes of fresh water, and permit them to dry.

CTENOPHORA

Members of the phylum Ctenophora like jellyfishes are collected on surface ocean waters. Larger forms such as *Beroe* should be carefully dipped from the water with a net or a bucket. Specimens should be isolated and preserved as soon as possible. If they are dead when captured, preserve them immediately. Since most of the larger species will not preserve in formalin, other methods are needed. Sea walnuts (*Pleurobrachia*) are exceptions to the formalin rule and should be narcotized with menthol crystals and placed in 5-percent formalin when insensible.

Bianco (1899) recommends killing larger ctenophores by dropping them in either the chrom-osmic mixture or in a corrosive sublimate-copper solution from 15 to 60 minutes. After washing, move these specimens up through the alcohols and harden them in 70-percent alcohol. He suggests that a small, inverted glass tube may be inserted in the gastric cavity to keep it distended and to keep the animal afloat while it is hardening.

REFERENCES

Bianco, S. L., 1899, The Methods Employed at the Naples Zoological Station for the Preservation of Marine Animals, *Bull. U.S. Nat. Mus.*, 39(M):1–37.

Galigher, A. E., 1934, "The Essentials of Practical Microtechnique," published privately.

Gohar, H. A. F., 1937, The Preservation of Contractile Marine Animals in an Expanded Condition, *J. Mar. Biol. Assoc.*, 22(1):295–299.

Pennak, Robert W., 1953, *Fresh-Water Invertebrates of the United States*, Ronald Press, N.Y.

Smith, F. G. W., 1948, *Atlantic Reef Corals*, Univ. of Miami Press, Coral Gables, Fla.

(See Chapter 1 for other References.)

CHAPTER 9

The Lower "Worm" Phyla

This chapter includes the lower phyla of bilaterally symmetrical animals which lack the true body cavity (coelom). Those involved here are the acoelomate animals, the Platyhelminthes, or flatworms, and the Nemertea, or ribbonworms, which have no body cavity but rather have the internal organs surrounded by a parenchyma. The second major group, the pseudocoelomate animals, have the internal organs in a false body cavity which lacks the peritoneum found in the higher animals. This group is represented by the phyla Entoprocta, Aschelminthes, and Acanthocephala. The Entoprocta are minute fresh-water and marine worms (formerly included in the Bryozoa; they are very rare and not treated herein); the Acanthocephala are parasitic spiny-headed worms. Four of the five classes found in phylum Aschelminthes may be considered worm-like, but the rotifers are not generally considered to be worm-like animals. This fact accounts for the quotation marks in the chapter title.

PHYLUM PLATYHELMINTHES

Class Turbellaria

Diversity. Turbellaria are free-living or commensal flatworms which are nonsegmented, possess cilia on the epidermis, have no true suckers, and have the mouth (when present) on the ventral surface. This class possesses five orders usually distinguishable by the nature of the intestine. The Acoela lack the intestine completely, the Rhabdocoela have a straight intestine, the Alloecoela have a more or less straight intestine with short branches, the Tricladida (the order which includes the familiar planarians) have a basically three-branched intestine, and the Polycladida, as the name implies, possess many-branched intestines. The lower orders of these worms possess very small individuals ranging from .3 to 4 or 5 millimeters in length, whereas some very large individuals are found in the Tricladida and Polycladida, groups ranging up to 500 millimeters (Storer and Usinger, 1957).

Collecting Fresh-Water Species. Collect fresh-water turbellarians in lakes, ponds, marshes, streams, and eddies or back runs of larger rivers where the water is a foot or less in depth. These animals are photonegative and thus are inactive during the day and are found in places of hiding. The undersides of rocks (except those firmly buried in mud), submerged logs or boards, lily pads, and the like may harbor specimens. Algal turfs and scums or rocks are frequented by smaller specimens. Standing pond vegetation should also be examined. One method of obtaining small individuals is to collect large quantities of pond or stream vegetation into buckets and store them in a dimly lighted area. As the oxygen tension changes turbellarians will be seen on the sides of the bucket or on the surface of the water. Replacing a large quantity of the water with freshly boiled and cooled water may hasten the process.

Shallow streams may be baited with pieces of raw beef or liver by placing the meat between rocks or other debris where turbellarians may occur. The bait should be checked every half hour or, if it is left overnight, the hiding places in the vicinity of the bait should be examined for specimens. A simple trap (Fig. 9–1) can be made by cutting out both ends of a small tin can, inserting a small piece of ¾-inch board, and flattening the can by stepping on it. Next, drill a hole through the middle of the can and a second hole for a tie string, as shown. Cut a cube of meat about ¾-inch square, slip this into the center of the can, and secure it by pushing a nail or a piece of wire

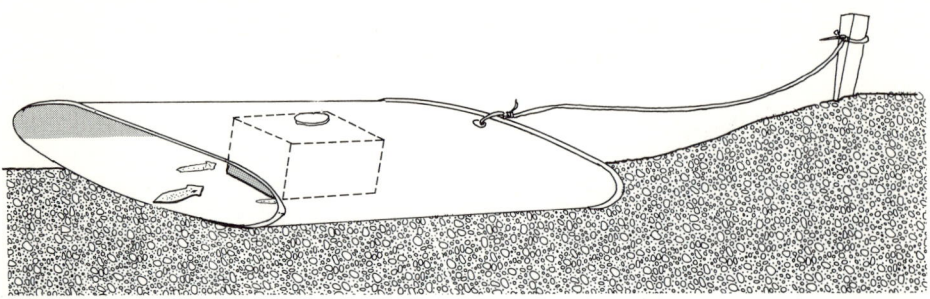

FIG. 9–1. A planaria trap (see text).

through the holes. Place such traps in likely collecting sites and tend them periodically or after each night's trapping period. This trap is especially useful for attracting crayfish, when they are present in the habitat. The author has found that planarians can be collected in the winter by breaking the ice and turning rocks where specimens will be found in a state of contraction.

Collecting Marine Specimens. Many of the methods mentioned above are adequate for marine collecting. The prime requisite of the larger triclads and polyclads is a suitable hiding place. Thus, crevices where one rock covers another, places where rocks cover coarse sand or gravel substrates, clusters of barnacles and mussels clinging to rocks or pilings, holdfasts of algae, discarded clam and snail shells, gill chambers of some arthropod animals, and the like, may all harbor these animals. When turning rocks pick off the obvious specimens and observe the rock periodically to detect the smaller hidden species. Be sure to return rocks to their normal position and to disturb the habitat as little as possible in the interest of conservation. Loosen specimens by sliding your fingernail or the blade of a dull knife under them.

Culturing Turbellarians. Both marine and fresh-water turbellarians turn up as uninvited guests in well-stocked aquaria which have some algae present. Clean wide-mouth jars or finger bowls filled half full with water are suitable culturing vessels. If room temperature fluctuates, place vessels in a large water bath. Provide small stones or other debris for hiding places and use pond water or well-aged tap water for fresh-water specimens. Partially decaying pond weed or natural algal growths may provide suitable food for smaller specimens. Small pieces of meat, crushed insects, cut earthworm, or the like, may be introduced about once a week, but should be removed after the main feeding has been completed. Some species require small crustaceans such as copepods and *Daphnia* for food. The water must be changed every three to four days; the lighting should be kept somewhat subdued. Marine specimens are cultured more or less in the same manner. For specific techniques on culturing consult Galtsoff *et al.* (1937).

Study of Living Specimens. Small turbellarians should be placed on concave or flat slides in a large drop of water, supplied with a coverslip, and studied under the microscope. Long-term studies are best carried out by supporting the coverslip with short pieces of hair and sealing the coverslip with soft wax (Appendix C). The hanging-drop technique (Chapter 6, p. 105) is quite useful. Pennak (1953) suggests placing small turbellarians between two coverslips and sealing the edges with vaseline or soft wax. The advantage here is that the specimen may be turned over and viewed from either side.

Narcotizing, Fixing, and Preserving. Turbellarians are either prepared on slide mounts or preserved in 70-percent alcohol or 5-percent formalin. In either case, specimens must be well expanded before they are fixed and preserved. This presents little difficulty with the small individuals, but may require narcotizing or other methods for larger specimens. FAA (Appendix C) is one of the best all-around fixatives, and may also be used for temporary storage. Gilson's fixative (Appendix C) is also widely used. However, specimens should be kept in Gilson's fixative for 24 hours or longer to ensure thorough fixation. A 5-percent formalin solution (mixed with sea water for marine specimens) is quite suitable for general preservation where specimens are not intended for slide making.

Fix specimens as follows: transfer small specimens in a pipette with a minimum amount of water to the bottom of a dry Petri dish or a glass plate. The water should be just sufficient to permit the animal to expand. When it is thoroughly expanded, squirt on hot (50° to 60° C.) fixative with a pipette. After several minutes pick up the specimen with a thin spatula and transfer to a vial of fixative for 3 or 4 hours (FAA) to 24 hours (Gilson's fixative). Store in 5-percent formalin or move specimens up through the alcohols to 70 percent.

Moderate-sized turbellarians should be transferred to a glass plate with a small amount of water. When they are fully expanded, drop a coverslip on top of the specimens. The more muscular specimens may be covered with part of a microscope slide. The volume of water should be small enough so as not to float the slide or coverslip, but rather to create suction which holds the worm in an expanded position. Add FAA to one end of the coverslip and pull this under and around the specimen by absorbing water, with a piece of paper toweling, at the opposite end. Again, be careful not to introduce enough liquid to float the coverslip and allow the animal to contract. Specimens should be treated up to ½ hour with additional quantities of FAA, added when necessary. Finally, lift the coverslip and slide the specimen into a container of fixative. Treat as above. An older method for dealing with very large turbellarians was to place them between two microscope slides which were subsequently bound with rubber bands. This preparation was then placed in a container with fixative for 2 to 5 hours, depending on specimen size, and then transferred to a proper fixative and treated as above.

Narcotizing probably gives the best results with large specimens. Place specimens in a small dish with water from their normal habitat. Add crystals of Chloretone, chloral hydrate, or menthol in small amounts for up to 3 or 4 hours. When the animals are insensitive, place the specimen on a glass plate, cover with a glass slide, and fix as directed above.

Lavoie (1958) found the following method very suitable for large polyclad turbellarians: Transfer specimens to a glass Petri dish with no excess water, and place in the freezing compartment of a refrigerator for about 20 minutes. Remove from the freezer and immediately fix with an FAA solution, prepared as follows: 90 parts of alcohol (95 percent), 5 parts of formalin. Heat this solution and pour directly over the specimen. Lavoie reports little shrinking or loss of natural color, and no histological damage with this method.

Slide Preparation. Lightly pigmented specimens may be cleared and mounted in Permount or balsam, or lightly stained and mounted. Borax carmine (described below) is a good general stain. Pennak (1953) recommends orange G and Delafield's hematoxylin for whole mounts of rhabdocoels (and probably other forms). Darkly pigmented specimens should be bleached in undiluted hydrogen peroxide, after which they are run up through the alcohols to 70 percent, stained, and mounted.

Galigher (1934) suggests feeding planarians on liver pulp ground with carmine prior to fixation with 10-percent formalin. Specimens are then dehydrated, cleared, and mounted unstained.

Borax-Carmine Stain. Transfer specimens from 70-percent alcohol into borax carmine and let stand until the specimens are thoroughly permeated, sometimes one or more days. Following this, destain in 70-percent acetic alcohol (Appendix C) until a bright, translucent pink appearance is obtained. Use the dissecting microscope every few hours, for as long as two days, to determine the progress of destaining. Finally, wash in neutral 70-percent alcohol, dehydrate in successive washes of alcohol through 95-percent and absolute alcohol. Next, clear in neutral beechwood creosote, clove oil, or xylol and mount in Permount or balsam with supported coverslips. An alternative method is to mount directly from the 70-percent neutral alcohol in Turtox CMC-10 mounting medium. This medium permits objects to be transferred directly from water or alcohol; it will both mount and clear specimens, but should be sealed with asphaltum upon drying.

Turtox CMC-S is a stain-mounting medium designed for small invertebrates and is excellent for fixed flatworms. Living specimens may be killed, stained, and mounted when placed in this medium on a slide with a coverslip. Tissues with similar refractive indexes sometimes become uniformly colored, since this medium has no destaining process. This process can be overcome somewhat in smaller specimens by mixing 1 part of CMC-S with 1 part of CMC-10 before using. Ring with asphaltum when dry.

Class *Trematoda*

Collecting Adults. Trematodes are parasitic animals which are non-segmented, lack cilia, possess a cuticle in place of the outer epidermis of the turbellarians, and possess suckers for attachment. They are most common as parasites among vertebrate animals, especially those associated with water. Their life cycles are quite complex, a multiplicity of larval forms frequently involving intermediate hosts. Trematodes may be found on their hosts in nasal chambers, gills and gill chambers, in stomachs (rarely), commonly in the small and large intestines, cloaca, lungs, and urinary bladder, or hiding under scales of fishes. Gills of fishes and amphibians should be placed in a saline solution (Appendix C) with some Chloretone crystals, set aside, and examined later under the dissecting microscope. Trematodes hiding there become relaxed and may be "combed" out of the gills with a dissecting needle. They will be found by carefully examining the gills and the bottom of the dish.

The complete digestive tract of the host should be examined. Remove the entire tract and place it in a dish of saline, then cut it into short sections, split each section, and examine it for trematodes. Large trematodes will be quite obvious, whereas small ones will hide among the villae of the intestine and must be located with the dissecting microscope. When adult specimens are located, remove them by dislodging the suckers; the fingernail or a dull scalpel may be forced against the worm in such a way as to dislodge the sucker without injuring the worm. Transfer worms to dishes of saline to clean them prior to fixation.

Cut lungs of possible host animals open in saline and examine under the dissecting microscope. Likewise, puncture the urinary bladder, draining it of its contents, cut open, and examine in a dish of saline. Also examine the urine drained from the bladder. Finally, look for nodules under the scales of fishes, and elsewhere, which may harbor adult or larval nematodes.

Fixing and Narcotizing. Trematodes are almost always prepared for microscopic slides in that the internal anatomy is of paramount taxonomic importance, whereas external anatomy is of little value except for general display of large specimens. Therefore, all adults are fixed while under the pressure of a coverslip or slide, with or without narcotizing, depending on their size. Follow all of those techniques presented for the Turbellaria when working with the Trematoda.

Slide Preparation. Borax carmine will yield excellent results when used with the Trematoda. A mixture of Turtox CMC-S and CMC-10 will also give fair results. Follow the instructions under "Turbellaria." Hematein, described for tapeworms below, is also very useful.

Collection of Eggs. Adult trematodes are almost always ready to expel eggs from the uterus. Isolate specimens by species in individual dishes with saline and allow to stand. Heating to room temperature or other adverse conditions usually cause the adult to expel tremendous quantities of eggs, thus making the adult easier to study under the microscope. Collect eggs with a fine pipette (Chapter 6, p. 102) and store in 10-percent formalin in small screw-cap vials. Be sure to include field data and the identification or field number of the adult and the host.

Collecting Larvae. Monogenetic trematodes have more or less direct life histories, whereas the digenetic trematodes may have as many as six distinct larval forms. Some of the larvae, in turn, are parasitic and may infect one or more hosts during the life cycle. The miracidium hatches from an egg released in water, and locates an intermediate host, usually a snail. After penetrating the snail, it metamorphoses into a sporocyst stage which may, in turn, produce either the daughter sporocysts or the redia stage. Either sporocyst or redial stage gives rise, ultimately, to the cercaria stage. This latter form is free-swimming and leaves the snail in order to penetrate the final host, to penetrate a second intermediate host where it encysts, or to encyst on a piece of vegetation. The encysted form, known as the metacercaria, is eaten by the final host along with the vegetation or the second intermediate host.

Place adult worms removed from the final host in a dish containing an appropriate saline (Appendix C). When eggs are expelled, transfer them to the water used by the final host (pond water or sea water). Observe the water every few hours for miracidia by placing the culture dish over a dark surface and shining a light through from the side. Swimming miracidia may be pipetted into formalin or FAA and stained and mounted as adult specimens; the Turtox medium is useful here. To obtain the next sequence of larval forms, collect snails from appropriate habitats. Most fresh-water snails from slow-moving rivers, streams, ponds, and lakes are potential hosts. The majority of the marine snails living where wave action is severe will have little or no infection, whereas those found in back bays and estuaries, either in or at the edge of the water, or living on muddy bottoms (moon snails), are more likely to be infected. Snails from the back bay at Newport, California, for example, are infected with over twenty different species of trematode larvae. Select larger mature snails, as these are more likely to be infected.

Snails can be treated in one of two ways. Place them in finger bowls with the habitat water, and observe daily for cercaria being expelled. This is the best way to get mature, characteristically formed cercaria. The other method for obtaining cercaria, and the only means of getting sporocysts or redia, is to dissect the snail. Gently crack the shell without crushing the soft parts of the snail. Pick away the shell or unwind the animal from the shell case. Those coiled portions occupying the tip of the shell (the gonad and liver) are the best hunting grounds. Place the evicted snail in a dish of saline and gently tease apart the gonad and liver. Many animals will be so infected that the gonad will consist of a membranous sac crawling with sporocysts, redia, and cercaria. Look through the dissected debris and the water on the bottom of the dissecting microscope. Remove larval forms to a clean dish of saline with a small pipette.

One of the important taxonomic features of cercaria is the location and makeup of the excretory system. Waste products are collected by flame cells which are distributed in a characteristic fashion. These cells are mapped and reduced to a formula as a means of identification. Although it is quite difficult to see these cells in mounted specimens, they can easily be observed in living forms. For the study of living forms consult the "Turbellaria," above. Vital staining techniques may also be used for trematodes. Place a drop of vital stain on a slide, add a few cercaria in water, and cover with a coverslip. The animals will absorb the stain and show some differential structuring.

Preservation of Larvae. Larvae are stained and prepared in the same manner as adults. Place larvae in small quantities of water on a glass plate and fix with hot FAA, without coverslips, as described above. Transfer specimens to a vial for staining. Permit these specimens to settle to the bottom after each successive wash, or gently spin them to the bottom with a hand centrifuge. Mount in balsam or Permount. Excellent results may also be obtained by mixing Turtox CMC-S with an equal amount of Turtox CMC-10 mounting medium. Place living larvae on a slide, blot away extra water, and add only enough mounting medium to fill out the coverslip. Place a coverslip on top, avoiding air bubbles, and set the slide aside to dry in a dust-free place. When the specimen is dry, seal with asphaltum, gold size, or other sealing agent. This medium will kill, stain, clear, and mount the specimen.

Class Cestoidea

Collecting and Preserving Cestodarians. Members of Subclass Cestodaria are very primitive tapeworms which differ from true tapeworms because they lack segmentation, have no true scolex but rather use a rosette-like structure for attachment, and have a ten-hooked larval form. Furthermore, they are found in the digestive tracts of very primitive cartilaginous fishes, the rat fishes or *Chimaera*. In Puget Sound these common fishes all possess one or two primitive tapes. There seems to be a crowding effect, for when numerous these specimens are quite small. When collecting rat fishes beware of the poisonous spines on the pectoral fin. Remove the digestive tract and carefully cut it lengthwise in a dish of saline. The primitive white tapes can readily be spotted and easily removed to a dish of saline. Squirt the specimens several times with a pipette to remove foreign food material. Place in the cool part of the refrigerator to relax, or narcotize with Chloretone (see "Turbellaria," above). When they are fully expanded, carefully siphon off most of the water and fix the specimens by pouring hot (60° C.) FAA over them. Either store in 5-percent solution of formalin and sea water or run them up through successive changes of alcohol to 70 percent. Smaller specimens may be relaxed and fixed under glass slides as described for the Turbellaria, above. These, however, lose the characteristic twisting and curling of the body, but are useful for study of the internal anatomy.

Collecting True Tapeworms. The subclass Eucestoda is comprised of segmented worms possessing a scolex for attachment and larvae bearing only six hooks. There are a number of different larval forms, depending on the order of tapeworms involved. Tapeworms occupy the digestive tract, most frequently the small and large intestines, of all of the vertebrate animals. Moreover, both herbivores and carnivores become infected with tapeworms. Most tapeworms shed mature segments and never become extremely large, measuring up to 30 inches or so; others, which do not shed their segments, may grow from 10 to 20 feet long. Eviscerate the host animal and carefully open the digestive tract, looking continuously for chains of tapeworm segments. If no large tapes are found, cut the tract into shorter sections and submerge in saline. Smaller tapes may appear as very thin whitish threads a few inches in length. When you locate a tapeworm, do not extract it, but carefully locate the scolex (at point of attachment). Observe this under the dissecting scope, and with a dull scapel blade or probe attempt to dislodge the hooks or suckers. Tapeworm bodies are seldom strong enough to support even their own weight and will break very readily. The scolex is essential for identification and must be secured. Float very fragile tapes away from the intestine and pick up in a glass tube provided with a suction bulb. Larger tapes may be removed with the forceps or fingers after they are dislodged, and should be transferred to saline. Wash debris away from the specimens.

Larval tapeworms may be lodged in almost any organ of the body. They are common in the liver, spleen, lining of the body cavity, and in the fat layers under the skin, especially of marine mammals. Larval tapes may appear sac-like, pigmented or white, fibrous, or cyst-like, and are, in general, difficult to recognize. Consult Wardle and McLeod (1952) or textbooks on parasitology for the appearance of larval tapes.

Preservation of Tapeworms. Since internal structures are of paramount taxonomic importance, adult and larval tapeworms are generally prepared on slides. Both, however, may be fixed and stored in 5-percent formalin or even displayed against black glass as described in Appendix A. Transfer small tapeworms to a glass plate with a large open-mouth pipette. Arrange these specimens so they will conveniently fit under a coverslip, blot away excess water, add a coverslip to hold them in place, and fix with FAA. Place the FAA at one end of the coverslip and pull through by absorbing the water at the other end with a piece of paper toweling. Stain with borax carmine, as described under "Turbellaria." Large specimens are cut into sections small enough to go under coverslips. The "head" (scolex), "neck," and first few segments are kept intact, some immature segments are preserved, some mature segments from farther down the chain are preserved, and some of the terminal segments which should be ripe are also secured. After cutting them into convenient lengths, place specimens on a glass plate, cover with coverslips or slides, and fix with FAA as described above. More "muscular" and robust tapeworms have to be relaxed in their entirety in order to render them flat enough and thin enough for taxonomic uses. Place the entire animal in a dish of saline and put this in the refrigerator (not the freezer) until it is thoroughly chilled. Next, lift the specimen to a glass plate which also has been chilled. The plate should be long enough to accommodate the entire tapeworm or sections of at least 16 inches. Arrange the specimen in a straight line and blot the water away from the anterior end. Next, gently hold the anterior end and gently pull the opposite end out on the glass plate, in order to stretch the specimen. Remove excess water from the distal end so that it will not contract again, cover the desired sections of the tapeworm with glass slides, and fix by adding FAA for 30 minutes. Specimens may be stained in borax carmine (see p. 134) or with hematein.

Staining with Hematein. Dehydrate the specimen by treating with 50-percent and 30-percent alcohol for 30 minutes or longer with each. Wash in distilled water, add stain (Appendix C), and let stand for one hour. Following this, wash in two changes of distilled water for 10 minutes each. Destain in 30-percent alcohol until the color changes from blue to medium pink. Next, treat in ammonia water for about 20 minutes to intensify the color. Wash rapidly in three changes of distilled water, move through 30-, 50-, 70-, 85-, and 95-percent alcohol for 25 minutes each (or longer with large specimens), wash in two additional changes of 95-percent alcohol and move to absolute alcohol, clear, and mount (see Appendix B).

PHYLUM NEMERTEA

The ribbonworms are long, slender animals that belong in the division Acoelomata along with the Platyhelminthes because they lack a body cavity. They possess a complete digestive tract with an anus; they are often striking in appearance owing to their colors and the long eversible proboscis which is extended periodically from the anterior end. Nemerteans are chiefly marine, but also occur in fresh water and damp areas.

Collecting Nemerteans

Marine nemerteans are found mostly under rocks between tides or below the tide level, more commonly on sandy or rocky beaches which are protected in bays, but also along protected rocky coasts (see Chapter 1). They may occasionally be encountered among mussel beds, kelp holdfasts, and the like. Specimens are usually very colorful and measure up to 3 feet long and $\frac{3}{8}$ inch wide. One usually does not set out to collect nemerteans, but, rather, picks up occasional specimens as they are found. When the worm is located, grasp it to prevent its escape but do not pull it. Carefully check to see if the worm is partially buried or completely free of any substrate.

If it is buried, remove sand and debris, being careful not to stretch the worm unduly. When it is free of the substrate, place the specimen in a small plastic bag with some water rather than mixing it directly with other specimens in a bucket.

The writer has not collected fresh-water nemerteans. Pennak (1953, p. 142) suggests that they are very "spotty" in distribution. "Weedy ponds, masses of filamentous algae, the undersides of lily pads, and the general substrate of littoral areas are preferred habitats, but often nemerteans may be taken from only one restricted area in a pond and are apparently absent elsewhere." Sample such areas and examine materials in water under the dissecting microscope.

Narcotizing and Preserving

It is desirable to have nemerteans extended to their full length and to have the proboscis protruding from the body, if possible. Place the specimen in a container of sea water in a cool, dimly lighted place, and add small quantities of Chloretone for up to 3 or 4 hours until it becomes insensible. Should you stimulate it prematurely and cause it to contract, revive the worm in clean water and begin the process again when it has expanded. Menthol or epsom salts produce fair results also. If the specimen is in a pan, gently coil it back and forth so that it will fit into a jar or vial when preserved. Siphon off most of the water and kill the worm by adding strong, 10-percent formalin or hot (60° C.) FAA. Store marine specimens in 5-percent formalin, or wash in water and store in 60-percent isopropyl alcohol or 70-percent ethyl alcohol. Pennak (1953) recommends killing and storing fresh-water nemerteans in 80-percent alcohol. Smaller specimens should be placed in vials with their preservative and a waterproof label containing the field data (Chapter 1) and scientific name of the specimen. Vials, in turn, are sealed in a jar which also contains some of the preservative. Larger specimens are placed directly in jars with their field data.

PHYLUM ASCHELMINTHES

This phylum includes several groups of animals which appear quite unalike and which were classed as separate phyla at one time. Nevertheless, the six classes (Rotifera, Gastrotricha, Kinorhyncha (not treated herein), Nematoda, Priapulida (see Chapter 10), and Nematomorpha) have been placed together under the name Aschelminthes, because all possess a false body cavity (pseudocoelom) and have a complete digestive tract with a posterior anus.

Class Rotifera

Collecting Rotifers. Rotifers are aquatic animals of two general types: the pelagic (free-swimming) and the benthic (bottom-dwelling). Lakes, ponds, marshes, brackish water, and the like should be sampled. Clear water and stagnant water (which is usually fertile and highly congested with microorganisms) should be sampled. Free-swimming specimens are obtained with the standard plankton net (125 meshes) or the coarse plankton net (75 meshes) which is towed through the water (Chapter 1). Specimens should periodically be transferred from the collecting vial attached to the tip of the plankton net to a clean bottle of water. Bottom-dwelling specimens are best collected along with plant debris. Sample different kinds of plants from various depths, including rotting vegetation, and keep separated in clean jars two-thirds full of water.

Place the jars with plant debris near a north window in the laboratory. Push the plant debris toward the side of the jar away from the light. Rotifers will migrate toward the light and can be pipetted into a vial.

De Beauchamp (1912) recommends collecting both wet and dried mosses and scrapings from mud surfaces of dried pools. These are placed in culture dishes and supplied with pond water which was previously boiled, cooled, and shaken well in a jar to add oxygen. Eventually, rotifers will emerge from dormant stages or from eggs contained in the moss or mud surfaces. These are then transferred to a vial for narcotizing. Rotifers vary in abundance seasonally throughout most of the United States, having two population peaks which probably correspond with nutrient cycles: April–May and September–October.

Culturing Rotifers. Some genera of rotifers, such as *Philodina,* are predaceous on small protozoans and thus are very easily cultured by maintaining them in *Paramecium* cultures as directed in Chapter 6. Other rotifers feed on microscopic plants, bacteria, or protozoans and may do well in cultures inoculated with *Euglena,* desmids, or diatoms, or fed on bacteria which may be grown on powdered egg, powdered malted milk, or other nutrients. Water should be changed frequently in cultures with foodstuffs, by filtering through the side of a plankton net or by centrifuging the specimens out of the culture medium. See Pennak (1953) for details concerning some culture techniques.

Narcotizing and Fixing. In preserving any group of organisms one must bear in mind the taxonomic characteristics which will be needed for identification after fixation and preservation. Some features of the rotifer (Fig. 9–2) are found in the soft parts of the anatomy; others are derived from the lorica, a shell-like covering which is quite hard and may be highly ornamented. Specimens containing the lorica may be killed in 5-percent formalin without narcotizing, since the shell characteristics are better seen in contracted specimens (Harring and Myers, 1922). In most cases, however, thoroughly expanded specimens are preferable, to exhibit such things as the cilia pattern on the corona, the nature of the foot and toes, whether the ovary is single or paired, and the structure of the mastax with its jaw-like trophi. Furthermore, soft tissues may have to be removed from the trophi to expose the structure and dentition.

Harring and Myers (1922) recommend narcotizing specimens placed in a watch glass which can then be observed under the dissecting microscope. Add a few crystals of Chloretone (in place of coccine hydrochlorate, which is no longer available) to the specimen and observe it until the cilia almost stop moving. Quickly add a few drops of 1-percent osmic acid directly on the animals and mix well. Wash through two or three changes of water, from 10 to 15 minutes each, to remove the osmic acid. If the specimens are darkened, bleach in a weak solution of potassium hydroxide, and rewash. Store in 2- to 5-percent formalin, or run up through 30- and 50-percent alcohol and store in 70-percent alcohol.

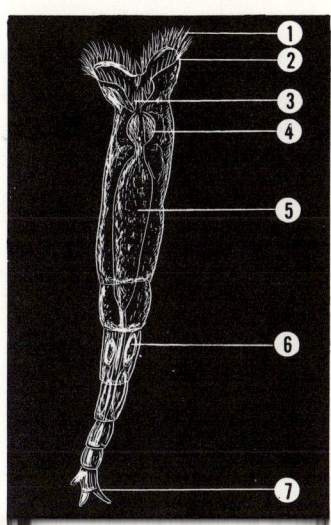

FIG. 9–2. Rotifer anatomy: (1) cilia, (2) trochal disc, (3) mouth, (4) mastax with jaws, (5) stomach, (6) ovaries, (7) toes on foot.

Other compounds have given good results in relaxing specimens. A freshly made, 1-percent solution of hydroxylamine hydrochloride or a 2-percent solution of strychnine sulphate (*Turtox Service Leaflet No. 2*, 1959), 2-percent benzamine lactate, 2-percent butyn or 2-percent hydroxylamine hydrochloride (Pennak, 1953), or 1-percent neosynephrin hydrochloride (Myers, 1942) have been used.

Temporary Slide Mounts. Harring and Myers (1922) suggest the following: Secure depression slides with the concave measuring about 10 millimeters in diameter by $\frac{1}{2}$ millimeter in depth, and place a drop of liquid containing preserved specimens on the inside edge of the depression. Lower a large, square coverslip down over the drop, and then push the coverslip so it covers the concavity. Place the slide on a microscope and tilt the microscope back sharply for examination. The specimens will gradually settle to the lower edge of the concavity until they touch both coverslip and slide. They can then be turned and positioned by moving the coverslip. Seal the slide with soft wax or vaseline to prevent evaporation.

Permanent Slides. Fairly satisfactory permanent slides may be made with glycerin jelly in what is a standard procedure for many small invertebrates. Transfer specimens to a 10-percent glycerin and 70-percent alcohol solution in a concave slide and store in a dust-free place for about five days. The water will gradually evaporate and the glycerin permeate the specimens completely, thus leaving them in pure glycerin. Next, add a piece of glycerin jelly about the size of a wooden match head to a slide, and warm the slide over a flame to melt the jelly. Transfer a rotifer by means of a bristle into the melted glycerin jelly, lower a circular coverslip, and position the specimen, while observing with the compound microscope, by moving the coverslip from side to side. Never use a pipette when transferring specimens from glycerin (Harring and Myers, 1922), but rather use a mounted pig's bristle (a flattened toothbrush bristle glued to a matchstick will do; Pennak, 1953). A specimen may be balanced on the bristle and lifted from the glycerin for transfer. When the slide is thoroughly hardened, coat the edges of the coverslip with Murrayite cement or other waterproof and alcohol-proof cement. Label the slide with field and taxonomic data and store in a flat position.

Another very simple technique that will give fair results involves the use of Turtox CMC mounting medium. This is a nonresinous, hard-drying medium into which living or preserved specimens may be placed directly. The CMC-S medium stains specimens as well as mounting; the CMC-10 does not contain stain. For rotifers, carefully relax, fix, and wash specimens before mounting. Place only enough medium on a clean slide to fill the space under the coverslip and transfer specimens directly into this. Add chips of broken No. 2 coverslips around the specimen for support, and lower the permanent coverslip into place. Orient specimens as described above and dry the slide in the flat position. Seal the edges of the coverslip when the medium is dry with Murrayite or asphaltum. If specimens are stained too darkly on the initial attempt, mix various proportions of stained and unstained media for better results, or add small quantities of dried, water-soluble stain (Eosin, orange G, fast green, etc.) to CMC-10 to achieve your own results. The problem of specimen shrinkage is always a difficult one with soft-bodied rotifers; shelled species present no problem. It may be possible to diffuse the mounting medium into the specimens, as described for glycerin above.

Mounting Trophi. The work of Pennak (1953, p. 186) is followed directly here.

Since genus and species identifications are often dependent upon the anatomical details of the trophi, it is sometimes essential that permanent or semi-permanent mounts of these jaws be made. The following method, requiring some skill and practice and much patience, is modified from Myers (1937). Place a drop of 1:10 Chlorox, or similar caustic alkali, just inside the concavity of a shallow concavity slide, and place a similar drop just outside the concavity. Then place a 22 mm. square coverslip on the outside drop and push it over the concavity until it is almost in contact with the inside drop. Next, place the rotifer in the inner drop by means of a bristle. If [the] coverslip is pushed slowly over the inner drop, the rotifer will be drawn under. And by working the coverslip over the concavity by short pushes, and adding small quantities of solution after each advance, the rotifer will be forced into the acute angle formed by the edge of the concavity and the undersurface of the coverslip. The edges of the coverslip should then be carefully dried and painted with vaseline or Murrayite. The position of the rotifer should be noted and in about a half hour it will have dissolved except the trophi. Such trophi slides will last for several to many months, and if a vaseline seal is used the trophi may be orientated by tapping or moving the coverslip slightly.[*]

Class Gastrotricha

Collecting and Culturing. Gastrotrichs (Fig. 9–3) are marine and fresh-water invertebrate animals associated with vegetation, decaying plant matter and other organic debris. They are less well-known from marine waters, fewer marine species having been described to date. Nevertheless, marine forms as well as the fresh-water species may be collected.

Techniques used for protozoans and rotifers are also used for gastrotrichs.

---

[*] Robert W. Pennak, *Fresh-Water Invertebrates of the United States.* Copyright 1953, The Ronald Press Company.

146 Biological Techniques

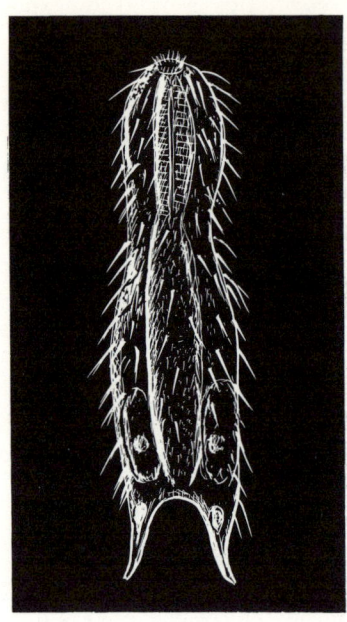

FIG. 9–3. A gastrotrich.

Collect plant material strewn about the bottom, including some that is decaying, and transfer to clean finger bowls in the laboratory. Carefully scrape the upper surface of bottom muds, especially those possessing light growth of green algae. Gastrotrichs are also numerous on quiet, sandy bottoms just above and below the water mark, especially in the fresh-water habitat, where they are found moving freely among the sand grains. Place mud and sand samples in clean finger bowls provided with pond water that has been previously boiled and cooled to exclude other organisms. Hunt through old protozoan and rotifer cultures, as well, for gastrotrichs.

Living specimens are preferable for study, and may be maintained by methods suggested for rotifers.

Narcotizing, Fixing, and Preserving. The majority of gastrotrichs may be pipetted to a slide with a minimal amount of water and fixed by squirting with hot (60° C.) FAA fixative. Some, however, require expansion and should be narcotized. Therefore, unless specimens are numerous, narcotize all gastrotrichs to ensure good specimens. The methods for narcotizing rotifers are suitable for gastrotrichs, and the 2-percent osmic acid solution or FAA works well as a fixative. Store specimens in 5-percent formalin or run them up through 30- and 50-percent alcohol, and store in 70-percent alcohol. Slides are made with glycerin jelly or with Turtox CMC mounting medium, as described for the rotifers.

Class Nematoda

The roundworms, class Nematoda, like the insects, protozoans, and vertebrate animals, have successfully invaded most available habitats. They are common in marine and fresh water, in soil or snow, as parasites or free-living animals. Barnes (1963, p. 152) sums it up:

> . . . they occur from the polar regions to the tropics in all types of environments, including deserts, hot springs, high mountain elevations, and great ocean depths. They are often present in enormous numbers. One square meter of bottom mud off the Dutch coast has been reported to contain as many as 4,420,000 nematodes. An acre of good farm soil has been estimated to contain from several hundred million to billions of terrestrial nematodes. A single decomposing apple on the ground of an orchard has yielded 90,000 round worms belonging to a number of different species.

Similarly, nematodes parasitize all manner of plant and animal material. Plant roots, bulbs, and stems are often so heavily infected as to render agriculture impractical. This writer has found nematode larvae and adults as common guests in the bodies of vertebrates and of large numbers of invertebrates, especially crustaceans. Chandler (1957, p. 365) sums it up by saying:

> . . . probably every species of vertebrate animal on the earth affords harborage for nematode parasites, and Stoll (1947) estimated 2000 million human nematode infections in a world harboring 22,000 million human inhabitants, a tribute, as he said, to the variety and biological efficiency of nematode life cycles.

Nematode Characteristics. Nematodes are usually small, cylindrical, straight worms which generally taper toward the anterior and posterior ends. The body is covered with a tough cuticle which is very hard to penetrate with preservatives. This cuticle is generally plain, but may be sculptured or ornamented. The symmetry of nematodes is often thought of as being biradial or triradial because of the distribution of mouth parts, sensory organs, and other structures. Possibly, the more or less uniform external appearance of small nematodes has caused most students to shy away from them taxonomically. Internal characteristics are highly important, including the reproductive system, digestive system, and excretory system. However, many nematodes may easily be studied internally, even while alive, thus permitting their identification. Biometrics (ratios between the body length, width, and position of various organs), of ultimate importance here, has been more or less standardized by Cobb (1913) who has worked up a set of symbols and formulas for measurements.

Nematode Study. For general work nematodes are best stored in vials and studied in temporary microscope slide mounts, either as preserved or as living material. Stored specimens may be cleared with glycerin and stained to exhibit internal structures. This procedure may eliminate the need for making permanent microscope slides. Vialed material, however, is subject to drying unless it is carefully maintained, and is not fully suitable for beginning students. Larger worms are generally dissected for internal studies or sectioned serially and mounted on slides by standard microtoming techniques (see other sources). Small nematodes may collapse and become distorted when mounted by more or less conventional methods. However, reasonable success may be achieved with the methods mentioned herein. When working with live parasitic nematode larvae in the tissues of intermediate hosts (such as *Trichinella spiralis*) use extreme caution and work with rubber gloves, as self-infection is quite possible.

Collecting Parasitic Nematodes. Nematodes may inhabit any vertebrate host. Marine and fresh-water fishes, frogs, reptiles, birds, and mammals are all suitable hosts. However, unless a particular form of research is being conducted, limit classroom studies to laboratory animals or such vertebrates (fishes or frogs) as are prolific, abundant, and in little danger of being overcollected. Look for parasites as soon as the host has been killed (see the appropriate chapter for methods of treating host animals). Open the body cavity of the host and, in turn, open all parts of the digestive tract. The stomach, intestine, and cloaca may all harbor adult worms. Occasionally the body cavity is invaded by adults and frequently harbors larval nematodes encysted as welts upon the internal organs (liver, spleen, digestive tract, body wall, mesenteries). Remove large worms first and then study individual sections of the host's digestive tract or other organs, submerged in an appropriate saline, for smaller ones (Appendix C). Worms should be transferred with a pipette or forceps to saline until ready for fixation. Also, remove the intestinal content to a dish, mix with saline, and examine under the dissecting microscope for parasites. Carnivorous and herbivorous invertebrates, as well as some filter feeders, should be examined in the same manner. Always keep nematodes isolated and labeled as to the part of the body in which they were found. Do not mix specimens from several parts of the digestive tract, as the exact locality of the collection is extremely important. Do not overlook the lungs and urinary bladder as likely sources of nematodes. Treat specimens as directed below.

Collecting Soil and Mud Nematodes. Any damp soil around various types of vegetation; the mud bottom of ponds, lakes, and marine shores, sand from beaches, both above and within the intertidal zone or in fresh-water areas; and soil beneath dung or other excreta harbor numerous species of free-living or larval, parasitic nematodes. The first method for separating specimens requires the least amount of equipment but the most work. Break up lumps of soil and dilute small quantities in a flat dish. Using substage lighting, hunt out nematodes and remove them with a pipette or fine dissecting needle (mount a size 000 insect pin in a match stick).

A second method of collecting nematodes is to wash the sample through a sieve made of window screen to remove coarse particular matter. Next, wash the material through brass testing sieves of 24 meshes, 50 meshes, 100 meshes, and if available, 150 meshes. (A fine plankton net may be substituted for the last dimension.) Examine microscopically the washings from each successive sieve and remove specimens with a fine pipette.

The Baermann Funnel is used as a standard tool for removing soil nematodes, is the most satisfactory, and requires the least amount of effort. To construct this funnel (Fig. 9–4) set up a standard ringstand with a supporting ring, place a large funnel in the ring, and equip the tip of the funnel with a piece of rubber tubing and a spring clamp or screw clamp. Next, fashion a piece of window screen to form a shelf that will sit part way down in the funnel, as shown in Fig. 9–4B. Fill the funnel with warm (50° C.)

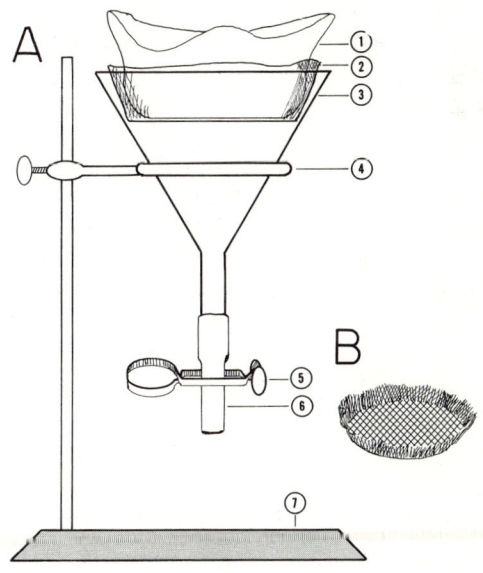

FIG. 9–4. The Baermann funnel for removing small nematodes. A. The funnel setup. B. The screen. (1) Cloth containing a soil sample, (2) screen shelf in place, (3) funnel, (4) ring stand, (5) clamp, (6) rubber tubing, (7) ring-stand base.

water so that it covers the screen by ½ inch to 1 inch. Place the soil in a piece of muslin or cotton cloth and set this on the screen so that it is covered by the water. Note, the screen may be eliminated by tying the soil within a cloth and suspending this in the funnel by means of a wire or string. After about ½ hour remove a small quantity of water into a dish by loosening the clamp. If the nematodes are numerous, withdraw only the bottom portion of water, but if they are few, centrifuge all of the water from the funnel to concentrate the nematodes. Treat specimens as directed below.

Collecting Plant Nematodes. Plant nematodes may be removed with the Baermann Funnel. These nematodes will be numerous in bulb or root nodules, which may appear partially decayed, from decaying vegetation, and the like. Bits of cut apple or other material may even be placed directly on the soil as a trap for such specimens. Macerate the plant material by teasing it apart with dissecting needles and then treat in the same manner (described above) as a soil sample.

Fixing Small Nematodes. Very small to small nematodes (those which will conveniently fit under a coverslip) may be fixed in one of two ways. In the first method, pipette the specimens onto a sheet of glass with a minimal amount of water and fix by squirting them with hot FAA (60° C., Appendix C) which has ten parts glacial acetic acid instead of the standard two parts. This will increase the rate of permeation and, hence, fix the internal tissues more quickly.

The second general method is to pipette specimens into a concave slide depression and warm this above a burner flame until the worms stop wiggling. Do not boil or overheat. Remove excess water and fix by flooding with FAA.

Fixing Medium and Large Nematodes. Specimens too large for slide preparation should be preserved in 5- to 10-percent formalin (depending on the size) or 75-percent alcohol. Smaller specimens may be dropped directly into FAA and transferred to the desired preservative, or they may be held between two glass slides and flooded with FAA. An alternate method for medium specimens, and a necessary one for large specimens such as *Ascaris*, is as follows: Straighten the specimens and place them on a piece of cheesecloth. In the meantime, bring a pan of water to a temperature near boiling. Roll the cheesecloth and grasp both ends with forceps; lower the worms into the near-boiling water, keeping them fully extended; and, after a moment, remove them. Next, place the specimens in a pan of FAA or 5-percent formalin to fix for 24 hours.

Liquid Preservation. Place specimens in appropriate-sized vials or jars with collecting data, including the name of the host, the collecting locality, and the place in the body from which the worm was taken. Vials should be cork-stoppered or fitted with screw caps and sealed in an airtight museum jar with an additional volume of preservative. Use 5-percent formalin (stronger, if worms are crowded) or run up through the alcohols and store in 70-percent alcohol.

Glycerin Jelly Mounts. Mount as described for rotifers (p. 144).

Other Slide Mounts. The author's parasitology students have prepared dozens of small nematode slides with Turtox CMC-S mounting medium, most of which were very satisfactory and showed no collapsing. This is a nonresinous medium with stain. Larger, robust specimens which require much more mounting medium are least satisfactory in that they absorb too much stain. Very small nematodes (up to 12 millimeters in length), however, are not overstained, but are rendered with the internal anatomy easily discernible. With this medium specimens may be fixed and mounted, or even placed within the mounting medium while alive. The medium will kill them, stain them, and clear them, as well as mounting them. The author has had success in adding dry stain (methylene blue, fast green, orange G, eosin, etc.) in varying quantities to the colorless Turtox CMC-10 mounting medium. When the stain is dissolved in this clear medium, the specimen is added and treated as above.

Stained balsam mounts may be quite difficult to prepare, because specimens frequently collapse, thus rendering the internal anatomy difficult to analyze. Although staining may be omitted, a suitable stain such as alum cochineal may be used. Specimens are dehydrated, cleared, and mounted like any other specimens; however, extreme caution must be used. When moving specimens up through the alcohols, for example, use gradations of no more than 10-percent alcohol. After washing in absolute alcohol for several changes, very slowly add clove oil, cedarwood oil, or other clearing agents, a drop at a time, so as to increase the amount slowly. By like token, add small quantities of the mounting medium to the clearing agent over a long period of time, permitting it to infiltrate the specimen slowly. When it is in pure mounting medium, transfer the specimen to a slide, add extra medium and a coverslip. Use coverslip supports (Appendix B) if needed.

Class Nematomorpha

Collecting. The horsehair worms, or gordians, spend their larval stages in insects, such as beetles or grasshoppers, but live briefly as adults in slow streams, ponds, and lakes, swimming slowly in an undulating manner.

These worms may measure 26 inches or more in length and slightly more or less than $1/16$ inch in diameter; with their dark-brown coloration they thus appear as slowly undulating horsehairs. The author has encountered these specimens in southern California, always by chance and in the midsummer months. One should be observant, for these animals could be encountered in any shallow body of water, either near or away from vegetation. One specimen collected was hidden among willow roots in a pond, but was finally detected because of its persistent undulating.

Preservation. Specimens may be put directly in 5-percent formaldehyde along with their field data. Pennak (1953) recommends 70- to 90-percent alcohol as a fixative and preservative for general use, and hot saturated mercuric chloride containing 5- to 10-percent acetic acid for histological work. In the latter case, specimens would be placed in a pan and flooded with the hot preparation.

PHYLUM ACANTHOCEPHALA

Adult spiny-headed worms are common parasites in fishes and some birds, but may be found in all vertebrate animal groups. Larval forms are parasitic in aquatic crustacea or in the larvae and adults of beetles, roaches, and other insects. Adult worms and encysted preadult forms possess a proboscis which is eversible and is armed with various numbers and combinations of hooks for attachment to the host. The digestive tract is lacking; the sexes are separate, with well-developed and unique reproductive systems. The body is covered with a very tough cuticle which is almost impermeable to preserving solutions. Internal structures are diagnostic for identification, as is the proboscis. It is essential, therefore, to preserve specimens with the proboscis everted.

Collecting Procedures. Acanthocephalans are routinely obtained during collecting of other parasites from the digestive tracts of vertebrates. In Washington, the common source is the Pacific tomcod (*Microgadus proximus*), which is heavily infected; also, occasional rock bass or ducks obtained by local hunters harbor acanthocephalans. Remove the visceral mass and carefully examine the stomach, small intestine, and large intestine by sectioning the digestive tract, placing sections in finger bowls with an appropriate saline (Appendix C), splitting the side of the tract carefully, and examining. Adults will have the proboscis firmly embedded in the gut wall. Usually the worms can be extracted by grasping them between the fingers and pulling gently. Other worms may dislodge themselves when left alone in the saline dish. Remove worms to dishes of clean saline where they will alternately contract and extend the proboscis. Leave them in the solution long enough to free the body of debris. A small water-color paintbrush can be used for cleaning. Be sure to keep specimens from various parts of the host's digestive tract isolated, as its location is significant in the ecology of the worm.

Fixing and Preserving. Always attempt to kill spiny-headed worms with the proboscis extended. After they are killed, specimens of all sizes may be preserved in liquid such as FAA or 5-percent formalin. However, smaller specimens are usually prepared on slides. Thus, in addition to ensuring that the proboscis is extended, fix these worms in a position such that they will fit under a standard coverslip.

Several methods may be employed to kill the worms with the proboscis everted. For small to medium-sized worms suitable for slide-making, transfer each specimen with a moderate quantity of saline to a sheet of glass. Position the worm so that it will fit under a coverslip. When the proboscis is everted, quickly drop a microscope slide on top of the worm. Usually, the suction created between the glass plate and the slide by the saline will maintain enough pressure to keep the proboscis everted. Otherwise, slight additional pressure could be applied to the slide during the killing. Prepare FAA as described in Appendix C, except that 10 parts of glacial acetic acid, rather than 2 parts, should be added. Put this solution on one side of the slide and pull it around the specimen by absorbing water on the opposite side with a piece of paper toweling. Maintain pressure on the worm during this procedure.

A second method is to place worms in the refrigerator until they are thoroughly chilled and expanded. (Owing to the cold marine temperatures in Washington, the author has found it necessary to further chill the worms by placing them in the freezer to render them insensible, being careful not to freeze the solution.) Next, siphon off the water and pour hot (50° to 60° C.) FAA solution over the worms.

A technique that works fairly well for very large acanthocephalans is to chill them as directed above. Then inject, with a fine hypodermic needle placed in the posterior part of the body cavity, sufficient FAA solution to extend the worm and force the proboscis to be everted.

Slide Mounts. For quick mounting the author has had excellent success with the Turtox CMC-S mounting medium as discussed under "Nematoda" (p. 151). Worms are placed directly on the slide, mounting medium is added, supports for coverslips are added, if necessary (see Appendix B), and a coverslip is lowered into place. Often, this medium will dry and cause air pockets to form along the sides of the coverslip. These must be refilled with a mounting medium until this tendency is overcome. Ring the slides with Murrayite or asphaltum to insure permanency, and label with data on the host and taxonomic information.

Mounts with Permount or balsam and stained with Ehrlich's acid hematoxylin (following Cable, 1957) work very well. Cable recommends pricking the body wall with a fine pin to speed the movement of stains and alcohols into and out of the body. For this procedure wash specimens in 50-percent and then 30-percent alcohol for 30 minutes or more, depending on their size, and stain in Ehrlich's acid hematoxylin for several hours. Next, wash in 30-percent alcohol and move up through 50- and 70-percent alcohol, allowing at least 30 minutes for each washing. Finally, destain in 70-percent acid alcohol until the specimens are pink, and then wash in 70-percent alkaline alcohol until they appear blue. Finally, move specimens through 85-percent, 95-percent, and two changes of absolute alcohol for 30 minutes each, and clear and mount with Permount or balsam as directed in Appendix B.

REFERENCES

Barnes, Robert D., 1963, *Invertebrate Zoology*, W. B. Saunders, Philadelphia.

Cable, Raymond M., 1957, *An Illustrated Laboratory Manual of Parasitology*, Burgess, Minneapolis.

Chandler, Asa, 1955, *Introduction to Parasitology*, John Wiley & Sons, N.Y. (Also published by Chapman & Hall, London, in 1957.)

Chitwood, B. G., and M. B. Chitwood, 1950, *An Introduction to Nematology*, Monumental Printing, Baltimore.

Cobb, N. A., 1913, New Nematode Genera Found Inhabiting Fresh Water and Non-brackish Soils, *Jour. Wash. Acad. Sci.*, 3(10):432–444.

Dawes, B., 1956, *The Trematoda*, Cambridge Univ. Press, London.

De Beauchamp, P. M., 1912, Instructions for Collecting and Fixing Rotifers in Bulk, *Proc. U.S. Natl. Mus.*, 42:181–185.

Galigher, A. E., 1934, "The Essentials of Practical Microtechnique," published privately.

Galtsoff, P. S., et al., 1937, *Culture Methods for Invertebrate Animals*, Dover, N.Y.

Harring, H. K., and F. J. Myers, 1922, The Rotifers of Wisconsin, *Trans. Wisc. Acad. Sci., Arts and Letters*, 20:553–662.

Lavoie, M. E., 1958, The Preparation of Polyclad Whole Mounts, *Turtox News*, 36(1):45–46.

Myers, F. J., 1937, A Method of Mounting Rotifer Jaws for Study, *Trans. Amer. Micros. Soc.*, 56:256–257.

Myers, F. J., 1942, The Rotarian Fauna of the Pocono Plateau and Environs, *Proc. Acad. Nat. Sci. Phila.*, 94:251–285.

Pennak, Robert W., 1953, *Fresh-Water Invertebrates of the United States*, Ronald Press, N.Y.

Rothschild, Miriam, and Teresa Clay, 1952, *Fleas, Flukes and Cuckoos*, William Collins, London.

Stoll, N. R., 1947, This Wormy World, *J. Parasitology*, 3(1):1–18.

Storer, Tracy I., and Robert L. Usinger, 1957, *General Zoology*, McGraw-Hill, N.Y.

Turtox Service Leaflet No. 2, 1959, General Biological Supply House, Chicago.

Wardle, R. A., and J. A. McLeod, 1952, *The Zoology of Tapeworms*, Univ. of Minnesota Press, Minneapolis.

CHAPTER 10

The Higher Worm Phyla

The animals covered by this chapter have been grouped together on the artificial basis that the techniques for collection, narcotization, and preservation are similar and thus should be treated collectively. (While many of the higher invertebrate phyla are put out of taxonomic order by this arrangement, the author feels that the simplicity of presenting biological techniques should be paramount, even at the expense of the phylogenetic arrangement.)

PHYLUM ANNELIDA

Class Polychaeta: Sandworms, Tube Worms, and Others

Polychaete worms make up a very important part of the marine fauna as free-living and tube-dwelling species. The latter species lend the beautiful colors of their radioles (plume-like tentacles) to the underwater landscape. Frequently, the entire surface of a submerged rock will appear orange, red, and yellow from the expanded radioles of tube worms, only to revert to its drab green when the plume worms become startled and withdraw into their tubes. Polychaetes are frequently omitted in collections or are badly preserved when collected. It is best not to collect contractile forms unless you have adequate time to carefully narcotize and preserve the specimens. Members of this class are almost exclusively marine; about seventeen known species are from fresh-water sources (Pennak, 1953).

Collecting. Polychaete worms are found in and upon mud and sand, under and upon rocks, on pilings and floating objects, attached to marine algae, and as commensals associated with other animals. Thus, these worms occupy all types of beach as well as deeper water habitats. When collecting intertidally you will notice a marked difference in distribution of different species between the high- and low-tide levels. Therefore, all levels must be sampled if a complete faunal list is being made.

A sand screen is essential for collecting in mud and sand substrates. An ideal screen for this purpose may be constructed as shown in Fig. 10–1. Select fine-grained, knot-free, soft wood, such as pine or cedar, in a standard 1- by 3-inch or 1- by 4-inch size. Cut two pieces 16 inches long and two pieces 22½ inches long. Nail them together, as shown in Fig. 10–1B, and add corner braces made by cutting a piece 2 by 2 inches diagonally. A

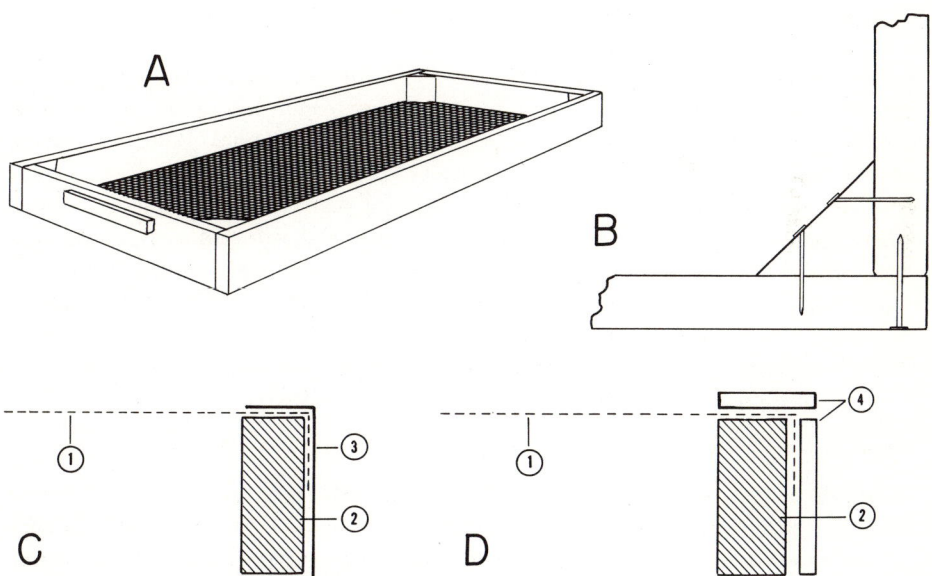

FIG. 10–1. Construction of a beach screen. A. Finished screen. B. Diagram for bracing and nailing the corners. C. The use of sheet metal to protect the screen. D. The use of wooden lath to protect the screen. (1) Screen, (2) side board, (3) sheet-metal guard, (4) lath.

screen with 12 meshes to the inch is a good average size. Heavier brass screen (ordered from hardware stores) is preferable to window screening. Cut the screen so that it will extend up the sides of the box at least 1 inch (Fig. 10–1, C and D). Tack the screen in place and then add metal or wood cleats. Pieces of heavy gauge galvanized iron bent and nailed in place, as shown in Fig. 10–1C, are preferable, but wooden cleats, applied as shown in Fig. 10–1D will do quite well. Nail on two short pieces of wood for handles at either end of the screen (Fig. 10–1A). If all wood surfaces are painted before and after construction, the screen will last for many years. A screen must always be washed thoroughly in fresh water and dried following collecting trips.

To screen for polychaete worms, place a shovelful of mud on the screen, wade into the water, and slosh the screen up and down. This will permit most of the sand and fine mud to go through the screen, but will maintain amazingly small worms long enough for them to be collected with a forceps. Standard sieves may be used in the same manner, using different sizes of screen for different sizes of worms. Select and screen mud from all levels and of all compositions of sand and gravel.

Examine the undersides of rocks, found in all substrates, for free-living polychaete worms, and hunt through mussel beds, barnacle beds, and the like. Search through kelp holdfasts, as these harbor many species. Examine kelp fronds, coralline algae, and the like, for small tube-building polychaetes. The sides of rocks may contain numerous chitonous or calcified tubes of polychaetes. These must be removed with a pocketknife or chisel and hammer, or else the entire rock must be collected. The fouling organisms on pilings and floats are also likely hiding places for polychaete worms. Commensals may also be taken from around the tube feet of starfishes. The biological dredge, orange-peel bucket, or other bottom samplers (Chapter 1) are useful for collecting deep-water species. Fouling organisms such as kelp holdfasts may be quickly examined by submerging them in a bucket of very weak formalin solution. Worms and other organisms will become irritated, leave their hiding places, and sink to the bottom of the bucket.

Narcotizing. Although researchers frequently preserve in bulk, specimens intended for display, classroom study, or even taxonomic work, should be thoroughly relaxed before preservation. The author has experimented with large numbers of the scale worm *Halosydna,* various species of *Nereis,* the shell-binder worm *Thelepus,* the iridescent worm *Hemipodia,* the plume worm *Eudistylia,* and the proboscis worm *Glycera.* Various narcotizing agents—clove oil, menthol, alcohol, epsom salts, Chloretone, benzocain, chloroform, and others—were tried. Results are not always consistent, even with the better techniques, but some are much more reliable than others. Bear in mind that the volume of water containing the specimen will have an effect on the time and efficiency of the narcotizing agent. Specimens treated overnight are perhaps simplest to deal with, in that you do not disturb them by constantly probing and pinching. Occasionally, specimens are completely insensitive to severe pinching and probing but, nevertheless, contract when placed in formalin. Perhaps the best results are obtained by the alcohol method, fair results being obtained with the menthol, Chloretone, or clove oil method. When combined with the alcohol technique, menthol gives good results.

For the alcohol method, place specimens in a container of sea water large enough to permit expansion. Add 75-percent alcohol, drop by drop, until the worms become insensible. Stop the procedure when enough alcohol has been added to equal about 10 percent of the water or when the sea water becomes slightly milky. Test the worm by gently probing and, finally, by pinching the posterior end with forceps. If the worm continues to be active after the alcohol is introduced, test the worm each hour until the specimen is insensitive. When the worm no longer responds, stretch it out on a rough-sawed board and add a small quantity of FAA (Appendix C, use Bouin's Fixative for histological specimens) to the specimen. After a few moments stretch the specimen out in a pan of FAA for 24 hours. Preserve as directed below.

Tube worms and all free-living forms may also be treated with menthol. Place specimens in finger bowls two thirds full of sea water, and sprinkle menthol crystals on the surface and cover. This will produce beautifully expanded worms, though 12 hours to 24 hours may be required for complete narcotizing of the specimens. When relaxed, tube worms can be withdrawn from their tubes and fixed as directed above. After they have been in menthol a few hours, alcohol can be introduced, as directed above, with excellent results.

A few drops of clove oil placed in a finger bowl or Petri dish containing specimens tends to render specimens insensible more quickly than any other compound. However, clove oil has two drawbacks: (1) it will irritate specimens that contact it directly and cause local contraction, and (2) insensible specimens often react to FAA. Probably a clove oil–alcohol combination would work quite well.

Epsom salts give mixed results, some specimens being completely narcotized, whereas others periodically revive. One large plume worm in the author's collection revived off and on over a period of 72 hours, and finally died in a well-expanded condition.

The author finds that proboscis-bearing worms such as *Glycera* or *Nereis* often die with the proboscis fully expanded and the jaws exposed, when permitted to stand in a bucket of seaweed and other specimens in the laboratory. The oxygen is used up as the water putrifies. The addition of some organic material (the gut mass of a sea cucumber or any similar material) will hasten this process. Preserve specimens as soon as they are narcotized or dead.

Preservation and Storage. If time does not permit anesthetizing, specimens should be placed in a bottle of sea water to which enough formaldehyde is added to make a 10-percent solution. After a week or more in this solution, they may be transferred to either 5-percent formalin, or, preferably, 70-percent alcohol. Some specialists strongly prefer 50- to 70-percent isopropyl alcohol for polychaete worms.

Small specimens are isolated in vials, along with their field data. The vials are filled with preservative, are cotton-stoppered, and are maintained in a large airtight jar also filled with preservative. Larger specimens are kept in a cool, dark place when not in use.

Slide Mounts. If slide mounts of small polychaete worms are desired, treat as follows: Anesthetize the specimen and/or place the specimen on a sheet of glass and add a small quantity of FAA or Bouin's Fixative. Cover with the fixative and let stand overnight. Next, wash in two changes of 50-percent alcohol and one of 70-percent alcohol, for 30 minutes each. Stain in borax carmine for 1 or more hours, and then destain in 70-percent acid alcohol (Appendix C). Move the specimen up through the alcohols to 95 percent and then treat in one of the following ways: (1) Place in Euparal Essence to clear, and mount in natural Euparal. (2) Dehydrate and clear in Terpineol, and mount in Permount or other similar media. (3) Wash in two changes of absolute alcohol, clear in xylene, and mount in Permount or similar media. (4) As an alternative, remove specimens from 70- or 95-percent alcohol and mount in Turtox CMC-10 (see Appendix B). Due to the thickness of polychaete worms, a depression slide, a gold size cell, plastic ring, or other coverslip supports may prove useful (see Appendix B).

Class Oligochaeta: Earthworms and Related Species

Collecting Aquatic Specimens. Ponds, lakes, and slow-moving streams and rivers (both above and below the water line) are favorite habitats for small oligochaetes. These worms live in mud and sand, around decaying vegetation, under the loosened bark of submerged logs and branches, under submerged boards or pieces of paper, and in similar habitats. Any of these objects should be carefully retrieved from the water by hand or with a dredge net (Chapter 13) and examined. The upper $\frac{1}{8}$ to $\frac{1}{4}$ inch of mud or sand should be carefully scraped into a container and examined, a small portion at a time, under a dissecting microscope. Specimens may be maintained in aquariums in the laboratory for moderate periods of time.

Collecting Earthworms. Worms are usually obtained by digging the surface layers of moist soil. The species composition is likely to change from place to place as the nature of the soil (grain size, chemical composition, pH, organic content) changes. Collecting in domestic areas will not necessarily produce a true picture of the species composition, in that many worms are transported into an area with plants. Digging near buildings, where grass is lush and green, under manure piles, and in other such areas will produce many specimens. Specimens may be taken at night following a very heavy, prolonged rainstorm or an all-day lawn sprinkling. Under such conditions worms come partially out of their burrows, but they are very sensitive to bright light and heavy vibrations. Nevertheless, they may be approached, grasped firmly, and extracted in great numbers. During the wet season any kind of ground litter, from old boards to rotten paper, may harbor earthworms.

As a fishing enthusiast the author has obtained earthworms between two poles of an alternating current circuit (Fig. 10–2). Such an electrical shocker is easily built. Obtain two metal rods about 15 inches long, 10 feet of rubber-covered lamp cord and a plug, an extension cord, two pieces of wood 6 inches long with a groove cut in one side, and some plastic electrician's tape. Cut grooves in the handle boards, as shown in Fig. 10–2B. Split the two strands of wire apart for a distance of 3 feet, bare about 3 inches of the wire on each strand, and tape one to the end of each metal rod (Fig. 10–2B). Finally, insert the rods in the handles and completely wrap with electrician's tape.

To use this shocker, select a moist area such as a garden, and stick the two rods down in the ground about 20 inches apart. Only after the rods are placed in the ground should you plug the cord into the alternating electric current. After a few seconds (to a few minutes) worms will begin to emerge from the soil. Be sure you unplug the cord before moving the shocker to a new area, as this rig is very dangerous when improperly handled.

Narcotizing. Various degrees of narcotization may be used for small aquatic oligochaetes. Specimens may be placed on a wet glass plate, per-

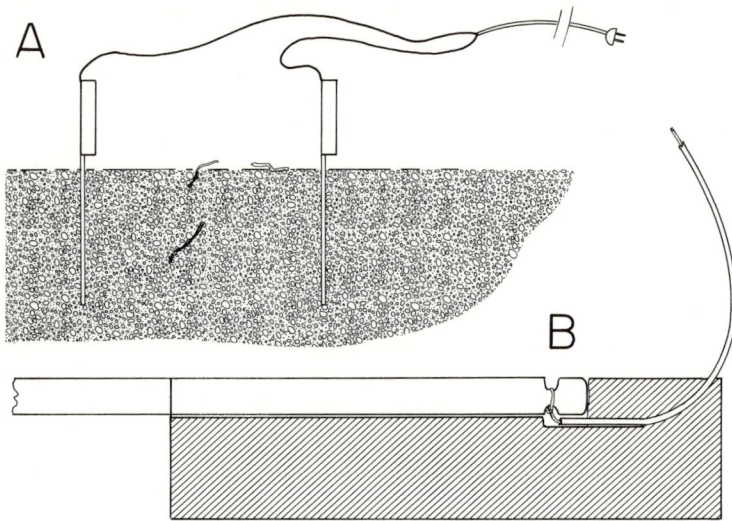

FIG. 10–2. Electric stimulater for collecting earthworms (see text for details). A. Apparatus at work. B. Cross-section of handle.

mitted to crawl and thus stretch out, then killed with hot (60° C.) FAA or Bouin's fixative. By another method, they may be placed in appropriate containers of fresh water, supplied with a few drops of chloroform (which will sink to the bottom if introduced beneath the surface of the water), and covered with a glass plate until anesthetized. Alternately, epsom salts may be introduced in small, but increasing, quantities until the specimens are insensitive. Chloretone added initially to the container will eventually anesthetize the specimen. When the specimens no longer respond to mechanical stimuli, such as probing, siphon off the narcotizing solution and replace with FAA or 5-percent formalin.

Earthworms are best prepared by placing them in water and adding, drop by drop, sufficient alcohol to make a 10-percent solution. When fully anesthetized they may be fixed in 5-percent formalin. To prepare internal organs in the best manner for dissection, the Carolina Biological Supply Company injects earthworms with chromic acid, which hardens and thoroughly fixes the internal organs. The following instructions were kindly supplied by the company:

Inject specimens with a 1-percent chromic acid solution by introducing the hypodermic needle in a posterior direction about one inch behind the clitellum (the conspicuous glandular swelling). One injection will usually suffice to render specimens turgid. Next, straighten the specimens, roll them into a bundle, and submerge them in a 1-percent chromic acid solution for a period of four hours. Following this, wash the specimens in running tap water from twelve to sixteen hours, or until the chromic acid has been displaced, then run them up through various alcohol solutions of increasing strength and store them at 85-percent alcohol. Specimens may lose some of their color and become partially flattened by this process. However, many of the internal structures which are difficult to locate such as the nephridia, will stand out with clarity.

Preservation and Storage. Oligochaete worms may be kept in 5-percent formalin or moved up through the alcohols to 75 or 85 percent. Store as directed for polychaetes.

Slide Mounts. When fresh-water oligochaetes are mounted, follow the procedure listed for polychaete worms, above.

Class Hirudinea: Leeches

Collecting. Fresh-water leeches may be collected in large numbers in lakes, ponds, and slow-moving water, attached to the undersides of rocks, submerged structures, or aquatic vegetation. They are occasionally taken on host species which they parasitize, such as fishes, turtles, and frogs. Marine leeches are less numerous and are usually collected incidentally. The author has found them most abundant in moderately deep water, attached to rocks or other debris. Also, some species of marine fishes seem to habitually possess leeches, but, again, these are taken incidentally. Terrestrial leeches, it is often said, will find you if they are present.

Narcotizing and Fixing. Small leeches may be placed on a glass plate and, when expanded, covered with a microscope slide. Immediately, FAA should, be added to the edge of the slide so that it will run around the specimen. Do not flood the plate and thus float the slide away from the specimen. After 20 minutes move specimens into FAA, 5-percent formalin, or 70-percent alcohol. Many narcotizing agents for leeches are listed in the literature. Pennak (1953) suggests carbonated water, Chloretone, chloroform vapors, chloral hydrate, and other compounds. The author has always used Chloretone, as described for oligochaetes, with great success. Following narcotization, specimens may be placed directly in preservatives or fixed under slides with FAA or Bouin's fixative.

Preservation, Storage, and Slide Mounts. For these topics follow the instructions given under the class Polychaeta, above.

MINOR PHYLA

Phylum Phoronida

Phoronid worms are nonsegmented animals which bear tentacles upon the U-shaped lophophore which also surrounds the mouth and anus. Barnes (1963) relates that "the phoronids are a small group consisting of two genera and approximately 15 species of worm-like animals. All members are marine and live within a chitinous tube that is either buried in sand or attached to rocks, shells, and other objects in shallow water. . . . A few species bore into mollusk shells." Phoronids range in size from a few millimeters to 12 to 15 inches. In the author's experience these specimens have been collected incidentally during dredging in shallow waters off the coast of California or while skin diving in bays and estuaries in the Gulf of California. The shape of the lophophore is the best means for recognizing sand-dwelling phoronids under water. Try narcotizing these specimens as directed for polychaete worms, above.

Phyla Sipunculoidea, Echiuroidea, and Priapuloidea

These three phyla were once placed together in the phylum Gephyrea because of their superficial characteristics. They may all be preserved in the same manner.

Sipunculoidea. Sipunculid worms are moderately large animals found in sand or mud from the intertidal zone down to extreme depths or, in protected rocky coast zones, wedged between rocks which make up the substrata. The specimens are obtained by dredging (Chapter 1) or are collected by hand by turning rocks or looking for their tentacles protruding from hiding places. When collecting in rocky places remember that tremendous damage may be done to the general habitat if rocks are not returned to their normal position.

Echiuroidea. Members of this phylum inhabit muddy and sandy bottoms of tropical and temperate marine waters. They are sausage-shaped and may vary in color from greenish-gray to red. They possess a proboscis, but lack tentacles around the mouth. These worms are collected by dredging or by using the orange-peel bucket on appropriate marine bottoms (see Chapter 1).

Priapuloidea. Priapulid worms (also recognized as the class Priapulida of the Aschelminthes) are superficially segmented and measure up to 3 inches in length. They dwell in muds, but are extremely restricted in geographic distribution; according to Storer and Usinger (1957) they have been found from the Massachusetts and Belgian coasts northward, from Patagonia south to Antarctica, and from Tomales Bay, California, south into Mexican waters.

Narcotizing and Preserving. These phyla are contractile and will withdraw the proboscis or tentacles, as the case may be, when preserved. They may be preserved in 5- to 10-percent formalin solution, but should be relaxed and expanded first. The *British Museum* collector's guide (1954) recommends menthol crystals or alcohol for narcotizing. Place the specimens in finger bowls containing sea water and sprinkle liberally with menthol crystals after they are fully expanded. After an hour or two begin to introduce, drop by drop, enough alcohol to make a 10-percent solution. Otherwise, begin directly with the alcohol treatment without the menthol. Very carefully touch the contractile parts after a few hours, to determine if the animals are still sensitive. Should they contract immediately, wash them in fresh, cool sea water, place them in a finger bowl of clean sea water until they are expanded, and begin anesthetizing all over again. Alternately, try Chloretone or epsom salts, as directed for polychaete worms above. Specimens may be stored in 80-percent alcohol or 5-percent formalin. Change the preservative solutions after 24 hours.

Phylum Chaetognatha

The arrow worms are common members of the marine pelagic fauna in tropical and temperate waters. They may be collected by towing a coarse plankton net through the water (Chapter 1). Frequently empty the contents of the plankton net into a finger bowl, and carefully isolate the fragile arrow worms into clean, fresh water. Preserve specimens directly by adding enough formaldehyde to make a 5-percent solution, and keep the specimens stored indefinitely in this solution.

REFERENCES

Barnes, Robert D., 1963, *Invertebrate Zoology*, W. B. Saunders, Philadelphia.
British Museum of Natural History, 1954, *Instructions for Collectors, No. 9A: Invertebrate Animals Other Than Insects,* London.

Buchsbaum, Ralph, 1948, *Animals Without Backbones*, Univ. of Chicago Press, Chicago.

Carolina Biological Supply Co., *Catalogue No. 34, 1963–1964*, Burlington, N.C.

Crowder, William, 1931, *Between the Tides*, Dodd, Mead, N.Y.

MacGinitie, G., and N. MacGinitie, 1949, *Natural History of Marine Animals*, McGraw-Hill, N.Y.

Needham, J., P. Needham, *et al.*, 1937, *Culture Methods for Invertebrate Animals*, Comstock, Ithaca, N.Y.

Pennak, Robert W., 1953, *Fresh-Water Invertebrates of the United States*, Ronald Press, N.Y.

Pratt, H. S., 1935, *A Manual of the Common Invertebrate Animals*, Blakiston (McGraw-Hill), N.Y.

Storer, Tracy I., and Robert L. Usinger, 1957, *General Zoology*, McGraw-Hill, N.Y.

(See Chapter 1 for other References.)

CHAPTER 11

The Mollusks and Brachiopods

Mollusks and brachiopods belong to distinct phyla which were lumped at one time, owing to the similarity of the brachiopod shell to that of some clams. They are combined here, not for taxonomic reasons, but for the sake of convenience. The Latin word *mollis* means "soft" and refers to the soft, unsegmented bodies of these animals. Nevertheless, we tend to characterize mollusks by their hard shells and think of nudibranchs, octopi, and squids as being irregular or unusual representatives of the phylum. The mollusks are very successful, having invaded most possible habitats, including relatively high mountains and some deserts. The five classes are highly divergent in appearance and habits and, to some degree, in habitat. Such extremes of divergence make it difficult for beginning students to link a swift and sensitive animal as the squid with the seemingly headless clam. These forms are, however, not only related taxonomically but, also, to some degree in methods of preservation.

A word of caution. Collecting mollusk shells has long been a popular hobby for private individuals and shell clubs. This in itself is excellent, and we must credit much of the classification of the mollusks and the compilation of faunal lists to such individuals or groups. However, the many books dealing with mollusks, catalogues and lists of shells for sale or exchange, and "nature guides" which tell some of the details of collecting and curating mollusks say little about the danger of overcollecting. Myra Keen (1958) is to be commended for her fresh statement urging sensible collecting and collecting procedures. Those of us who have been fortunate to collect in distant lands and on remote islands can easily detect the difference between the richness of fauna and flora there and on our own beaches. Thus, the reader is urged to collect only small series of specimens and to leave the habitat as unmolested as possible.

Most states have laws, or are developing laws, governing collecting in general and the collection of "shellfish" or mollusks in particular. All three states along the West Coast, for example, have distinct rules for scientific collecting and partial laws covering mollusks; the State of Oregon, which owns most of its ocean beaches, restricts collecting on the vast majority of its beaches to those individuals who have permits. The State of California now has a collector's permit and a student collector's permit which cover all forms of terrestrial and marine plant and animal life and must be renewed annually. To those of us who are thinking toward the future, these laws are a step in the right direction, for they will ensure the continuance of our marine fauna and flora. Write to your own state's fish and game department to ascertain state or local restrictions pertaining to the collection of mollusks.

PHYLUM MOLLUSCA

Class Gastropoda: Snails, Slugs, and Related Forms

Characteristics. Snail classification is based on the soft parts, as well as on shell characteristics. The higher groupings (orders, families, genera) almost always refer to characteristics other than the shell for their definition. Myra Keen (1958) states:

> . . . the shell alone is an unsatisfactory basis for classification, as the same form (for example, a limpet-like shape) may appear in what are obviously unrelated stocks. The radula seems to be the best single guide to relationships, but the gills (ctenidia), heart, and nervous system are also useful. One does not, of course, have to make an anatomical study of every specimen in order to discover a name for it. The shells are very convenient when it comes to identification—after the classification or pigeon-holing has been worked out by the zoologist.

For general classroom studies the shells will suffice to demonstrate different species, but for advanced courses and for people doing serious taxonomic work at least a portion of the collection should be preserved in liquid. If the shells are to be cleaned and dried, preserve the soft parts along with adequate data so they may be matched up with the shells, if for no other reason than to yield the radula.

Field Data. In addition to making field notes as to the locality of collection, note carefully the environmental conditions (see Chapter 1) under which the specimen was obtained. Be on the lookout for such things as mating, egg-laying, or the arrival of newly metamorphosed snails. Record the various foods which snails are using (if necessary preserve algae or other food for later identification) and note the method of feeding. Snail growth and snail migration may be studied by marking and releasing animals in the field. You may occasionally observe predators or commensals associated with snail or other animal populations, and these observations should be recorded. Such field notes, along with a preserved specimen for proper identification, will be most welcome in the files of specialists should you have no personal use for them.

Habitats and Collecting.
1. Marine collecting. Gastropods are found in all marine situations. A tremendous wealth of specimens occurs intertidally. Low tides, therefore, are best for hand collecting. Night collecting, when safe (see Chapter 1), is often more productive than daytime collecting, especially in bays and estuaries, for most mollusks are nocturnal in their behavior.

When collecting on rocky beaches along the open ocean check some of the following places: (1) The spray zone above the high-tide level where limpets (Fig. 11-1), turban snails, and others feed on the microscopic growths of algae. Look especially in small crevices where splashed water will accumulate and return to the sea. Moving down in the intertidal zone check (2) the tops, sides, and undersides of rocks for limpets and a host of other snails, turning rocks back to their normal position when finished. (3) In tide pools look among the fronds of algae and collect not only the large colorful snails but the minute species as well. Nudibranchs will be abundant in both the protected and partially protected rocky coast zones clinging to algae or other objects. Gloves and a geology pick are handy for handling rocks, and a dull knife is often essential to slip beneath the shell and foot of an unsuspecting snail to pry it loose. Generally, collecting techniques mentioned in Chapter 1 are suitable.

On sandy beaches in bays and estuaries a small amount of rock rubble will usually be present. One should screen sand and mud, collect around seaweed and rock deposits, and look for tracks of snails in the sand while wading out in the water. The moon snail and olive snail (Fig. 11-1, G and

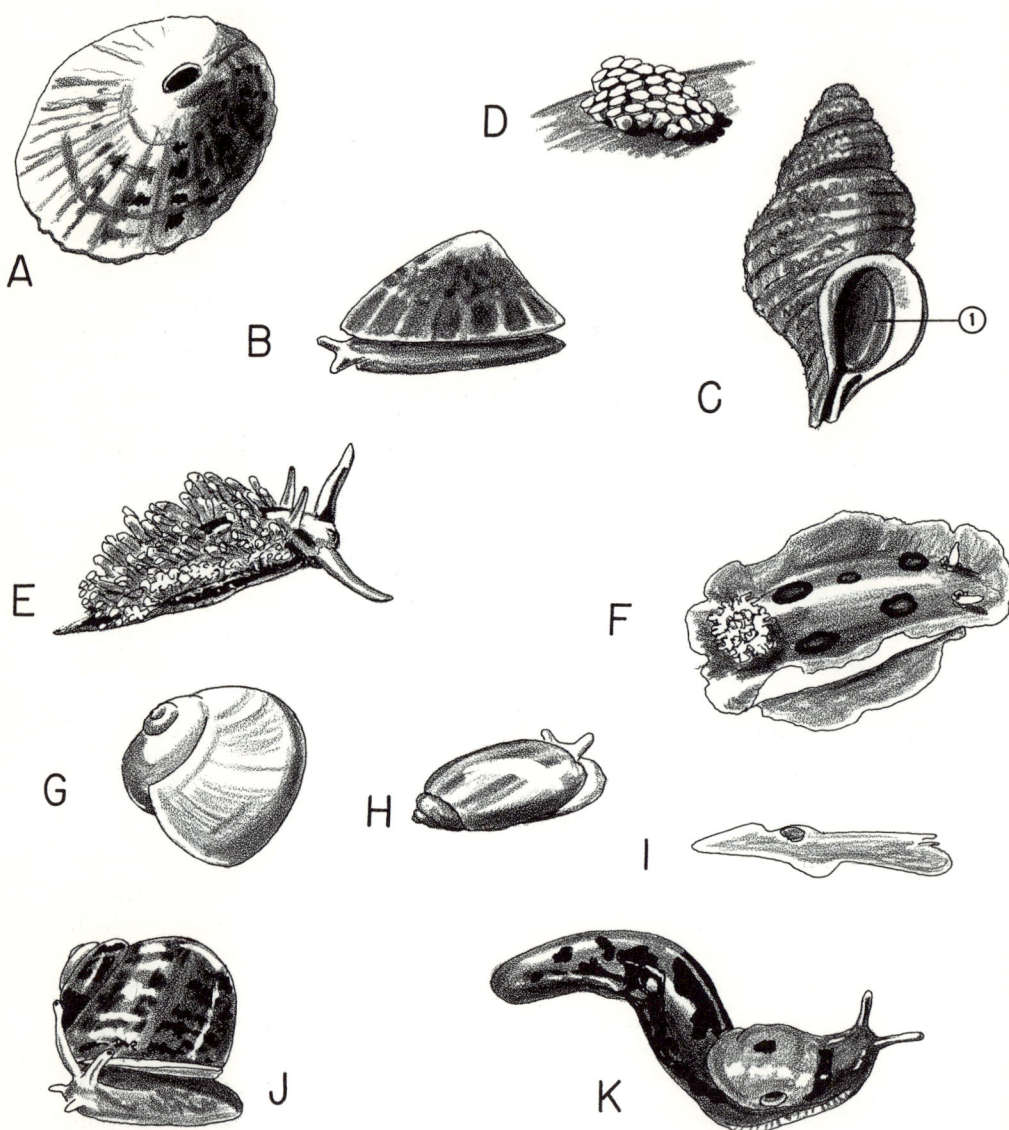

FIG. 11-1. Various types of snails. A–B. Limpets or hatshell snails. C. Snail with operculum (1). D. "Sea-corn," or snail eggs. E–F. Nudibranches. G. Moon snail. H. Olive snail. I. Heteropod, a transparent pelagic snail. J. *Helix,* the common land snail. K. *Limax,* the land slug.

H) often remain buried but make horseshoe-shaped impressions or little furrows in the surface of the sand as they move underneath. A large number of oyster drills and whelks are also found in this habitat. Traps and baits, described below, may also be set for snails.

In the offshore waters skin diving and dredging are excellent means for bringing up deep-water mollusks. For dredging techniques and dredge construction, consult Chapter 1.

2. Fresh-water collecting. Snails are found in all types of fresh-water situations. In lakes, ponds, and slow-moving streams they may be found on muddy bottoms, especially near the edges of the water, attached to rocks, sunken branches and logs or other debris, and dwelling on fresh-water plants. Other than direct observation, the best means for locating fresh-water snails is to collect various kinds of debris and examine them on the shore. The dredge net and dip net (Chapter 13) are very useful for collecting plants and other debris. Carefully pull logs or branches from the water and examine both the upper and under surfaces, as well as under loose bark. Be on the lookout for the switchback trail left by the radula of a feeding snail or for gelatinous masses of eggs which signify that snails must be close by. Again, nocturnal collecting with a gasoline lantern and dip net is desirable, in that most snails are nocturnal. Specimens should be placed in jars of water and transported to the laboratory for proper preservation or culturing. Fresh-water snails may easily be kept in the laboratory aquarium when they are supplied with the proper food. Clams are found in many of the snail habitats described. Look for empty shells as a sign of their presence.

3. Terrestrial collecting. The land snail *Helix* (Fig. 11-1) is cosmopolitan in distribution; it may be found in very high mountains and deserts as well as in gardens and fields. Garden slugs and other species are also cosmopolitan. In the Pacific Northwest gigantic specimens of *Limax* make up part of the typical rain-forest fauna. *Limax* is almost always present during the daytime, except on days that are unusually dry or brilliant. On cloudy days these specimens can be found especially where leaf litter, fallen logs, and ferns and other shrubs are present. *Helix, Limax,* and related snails are also active at night and can be picked from vegetation, walls of buildings, sidewalks, along pathways in the woods, near streams or ponds, or in other damp places. Collect these species directly into a plastic bag with a small amount of vegetation to prevent crushing.

4. Traps and baits. Carnivorous snails are readily attracted to the baits of crab and shrimp traps and fish traps. Any kind of funnel trap, crab trap, fish trap, salamander trap, or the like, may be adapted for snail collecting. Meat baits, such as pieces of fish, should be securely tied in the trap. Leave traps out for 24 hours or more, as snails are slow to move into them. Many workers place suitable baits in a gunny sack filled with rocks, deposit this at the low-water mark on one night, and revisit it the next night when numerous snails will be found hiding on or around the bait.

Preparation of Dry Shells. The removal of animal tissues from the shells is undertaken (1) to permit easy storage and quick reference without the fuss of liquid preservatives, and (2) to eliminate the large number of jars and volume of alcohol that would be necessary to preserve all of the specimens in a taxonomic collection. Usually, smaller snails are preserved in liquid or placed in 5-percent formalin for 24 hours, then washed externally, and dried. Large snails are cleaned by boiling or by other methods.

1. Boiling technique. For general use, prepare some boiling water and lower the specimen into it from 2 to 5 minutes. The heat quickly kills the animal and loosens the flesh where it is attached to the shell. A wire basket made of ¼- or ½-inch mesh hardware cloth (Fig. 11–2A), or even a soup strainer, is indispensable for introducing and removing the specimens from the boiling water. Those snails with glassy, hard shells on the exposed surfaces may develop minute cracks; among these are the cowries, some of the *Murex* species with glassy shell openings, some of the cones and olive shells. Place these animals in warm water and very gradually bring them to a boil; cool slowly, never permitting them to contact cold water or cold objects until

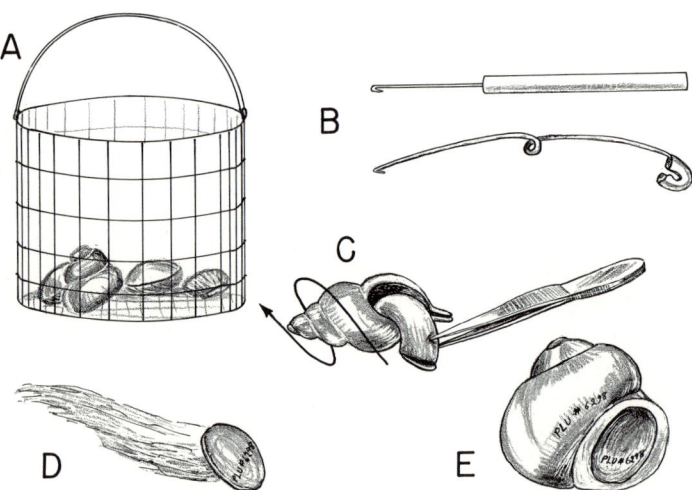

FIG. 11–2. Techniques of cleaning snail shells. A. Wire basket. B. Hooks. C. Twisting to remove soft parts. D. Operculum attached to cotton. E. Operculum glued in the cleaned snail shell with museum numbers intact.

they themselves are cool. Some of the very thin-shelled land snails should also be treated by this method.

When the specimens have been boiled and cooled, remove the soft parts by grasping them with a forceps or a hooked wire, crochet needle, or safety pin (Fig. 11–2B). With a firm grasp on the tissue, and on the shell, carefully pull the specimen out of the shell. Here, a little experience pays off in that you will develop a "touch." The tissue may adhere firmly to the shell, then suddenly become loosened, at which time you may jerk part of the animal free of the shell only to break off the soft terminal portion. Work close to some solid object, such as a table top, which your hand will strike the instant the tissue releases. Once the tissue has loosened, twist the shell in its normal direction so that you literally "unscrew" the soft part from the shell (Fig. 11–2C). If part of the tissue remains in the shell, treat as directed below.

2. Other methods. Other methods may be more time-consuming and, occasionally, less pleasant. Specimens may be boiled, or killed in fresh water, placed in a jar two thirds full of water, sealed, and permitted to rot over a period of a week or more. Upon reopening the jar, the specimens must be washed and the contents thoroughly shaken from the shell or washed out with a stream of running water. Specimens should then be washed in soap and water to remove the odor, then dried on a blotter.

For those people living near sandy beaches, specimens may be killed by placing them for 12 hours in fresh water or by boiling, and placed in a wire basket or a wide-mouth bottle with a screen top. They are then lowered to the bottom from a pier or float, or suspended beneath a floating dock. Small crustaceans (amphipods or isopods) will usually remove the soft parts in a week or less. A similar method is as follows: Scoop out a depression in the sand above the high-water mark, place the killed animals in this depression, cover with a small board and then heap mounds of drift algae on top of the shells. Crustaceans will leave the drift algae and reduce the animal tissue in a week or so.

3. Retrieving broken flesh. Should the "tail end" of the snail break off in the shell, this may be removed by filling the shell with water and shaking it vigorously, attempting to float the tissue out as the shell is drained. Directing a forceful stream of water into the shell sometimes will also dislodge this tissue. Many workers recommend placing formalin in the aperture and curing the tissue for 24 hours, rinsing the shell and drying it with the tissue intact. The procedure recommended for overcoming any resulting smell is to stuff cotton well back into the aperture of the shell. This technique, however, is not fully satisfactory, except for small specimens. The author prefers the method of placing the shell so that it will not tip over, and then introducing some 10-percent potassium hydroxide (KOH) down into the spiral. This will eat away the tissue. KOH will affect the brown outer layer (the periostracum) of some shells, but usually will have no effect on the other parts of the shell. After 24 hours, wash out this solution by directing water into the aperture, rinse thoroughly, and dry. This treatment may be repeated should you suspect that more tissue remains in the shell.

4. The operculum. Many snails possess the operculum, which serves as a door and is attached to the side of the foot. When the animal retreats into its shell the operculum closes the aperture for protection. The operculum is either chitinous or hard and calcareous like the rest of the shell. The operculum must be saved, and should be removed from the soft parts of the animal at the time of cleaning. Keep the operculum with the shell while drying and cleaning. Record the field number or museum number on both shell and operculum; then glue the operculum to a piece of cotton (Fig. 11–2D), which is inserted in the mouth of the shell (Fig. 11–2E).

5. Removal of the periostracum. Some collectors prefer to remove the outer, brown periostracum that occasionally occurs on snails and other mollusks. However, this layer is frequently ornamented, forming tufts of hair-like protrusions which in themselves are of taxonomic importance. While removal of the periostracum may be desirable to show the underparts of the shell, the practice should be restricted to only a few specimens in a series intended for taxonomic purposes. The periostracum is removed by placing specimens in a 10-percent solution of potassium hydroxide (KOH).

6. Coating shells. It seems to be a fad among amateur collectors to apply varnish or other substances to their shells. This creates an artificial appearance and should be avoided. The natural luster may be returned to the shell by rubbing it lightly with mineral oil (not olive or vegetable oil which may become rancid on the specimens) to enhance the appearance of the shell without altering the surface. By the same token, the treatment of shells in acid to produce a luster should be avoided, because acid alters the surface and lessens its taxonomic value.

Preparation of the Radula. The radula is a rasp consisting of many rows of transversely orientated teeth. The number of teeth per row, their shape and occurrence can be reduced to a formula; thus, the radula is extremely useful for identification. The radular teeth are fastened to a belt-like sheet of tissue (Fig. 11–3) which draws back and forth over a cartilaginous "tongue" like a shoeshine rag pulled over the tip of a shoe. The radula is used to grind up plant and animal tissue, or even to drill through the shells of other mollusks.

The radula is generally mounted on a slide, in either a stained or an unstained condition. The radula may be freed from surrounding tissues by

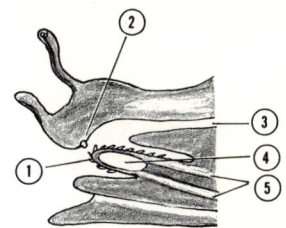

FIG. 11–3. Cross-section through the head of a snail. (1) Radula, (2) jaw, (3) esophagus, (4) radular support, (5) radular muscles.

treating it in potassium hydroxide at room temperature until it is fully cleaned (up to about 24 hours), or by gently boiling it in a KOH solution. For small snails, remove the entire anterior part of the body for treatment. Remove unnecessary tissue in larger snails, in order to reduce the cleaning time. When it is clean, wash the radula in several changes of water and place between two slides supported with index cards or other material to prevent crushing. Bind these slides together and wash in 35-, 50-, and 70-percent alcohol for 20 minutes each. The radula may then be completely dehydrated by treating on up through absolute alcohol, then cleared and mounted. The alternative is to stain in borax carmine, dehydrate, clear, and mount by the following steps: Transfer the radula and its restricting slides to a vessel of stain (Appendix C) for 1 hour. Destain in acid alcohol until the specimen is a light pink. Dehydrate up through 95-percent and absolute alcohol, and clear in xylene. Remove the restricting slides, place the radula on a clean slide with the teeth upward, add Permount or a similar mounting medium, and, finally, place a cover glass over this. Dry in a flat position. Note that coverslip supports (Appendix B) will probably be necessary.

A more simplified technique for radular mounting is as follows: Wash the specimen thoroughly after treatment in KOH, transfer to 35-percent acid alcohol, after a few moments wash in several changes of 35-percent alcohol, and then mount directly in Turtox CMC-S or CMC-10 mounting medium (see Appendix B). Either of these media will clear, stain, and mount the specimen, all in one operation. The slide should be cleaned, ringed, and labeled.

Narcotizing Specimens. Fully relaxed specimens are often required for study and dissection. All snails will contract badly when preserved under normal conditions. Some of the aquatic snails, however, are incapable of projecting very far beyond the shell aperture and relaxation is of little value. On the other hand, many snails, including land snails and slugs, nudibranchs, and sea slugs, can only be of value when fully relaxed prior to preservation. One must recognize that no single procedure is applicable to all snails, nor will a technique always work consistently, even with the same species treated under more or less the same general conditions. The procedures included herein, however, are usually quite satisfactory.

1. General techniques. For many of the heavy-bodied, fresh-water and marine snails the following four methods yield some results.

Possibly the best general method is treatment with alcohol. Place the specimens in a container with their habitat water and add, drop by drop, enough alcohol to make a 10-percent solution. This should be introduced over a period of several hours—or days, if necessary. By very gently probing the expanded snail one can determine whether or not it will react in the preservative. When fully insensitive, the snail may be fixed in 10-percent formalin or FAA. Plunging specimens into concentrated glacial acetic acid gave only poor results for the author. Possibly the alcohol method can be improved upon in the following way: When the animal appears insensitive, slowly introduce small quantities of formalin so that it is gradually poisoned. As soon as it is dead, enough formalin should be added to make a 10-percent solution. Gradually introducing formalin without first using alcohol also gives fair results for some species.

An alternate method that is occasionally quite good, but somewhat difficult to gauge, is based on epsom salts. Place the specimens in containers of water from the habitat and slowly add crystals of epsom salts every 10 or 15 minutes, increasing the amounts after each hour. The crystals should be dropped in carefully, so as not to disturb the expanded specimens. Some specimens are easily relaxed in this way and may be transferred to 35-percent alcohol or 5- or 10-percent formalin for killing with no contraction. On the other hand, it has been the author's experience that many species require several days of treatment—and, even then, they have occasionally contracted.

Many workers suggest placing specimens in water and gradually heating them so as to kill them in an expanded condition. The author has found this method quite unsatisfactory in most cases.

The freezing technique of Gohar (1937), which he recommends for tritonid snails and nudibranchs, works quite well for many species of marine snails. (Fresh-water snails were not experimented with.) Place the specimens in a jar containing water, and when they are fully expanded, place them in the freezer. Most of the specimens will remain expanded and, thus, are killed by the reduced temperature. After freezing, chip away or melt away most of the ice and place the specimen, encased in ice, in strong formalin which will rapidly penetrate the body and fix the tissues. The author tried clove oil, Chloretone, menthol (recommended by many), chloroform, and other compounds, but with inconsistent results, both for fresh-water and marine (tritonid) snails. The large sea slugs should be treated either by methods mentioned above or as directed for nudibranchs below.

2. Narcotizing of Nudibranchia. The nudibranchs and related animals are possibly among the most beautiful of all the snails, and yet the most disappointing when preserved. These animals are brilliantly colored, adorned with external gills and other projections (cerata), and lack a shell. They display brilliant reds and oranges, blues, greens, yellows, opaque whites, or opalescent colors. Some of the Doris-type nudibranchs are fairly easily preserved, whereas the more highly ornamented forms such as *Hermisenda* (see Fig. 11–1, E and F) or *Aeolis* are so delicate that they can scarcely be collected from the sea water without losing most of the cerata. Many workers recommend narcotizing with menthol and then (with or without narcotizing) plunging the specimens directly in concentrated glacial acetic acid. The author's experience here is quite disappointing, resulting only in badly contracted and dismembered specimens. For the more delicate forms the most promising techniques are the use of alcohol and then formalin, the gradual poisoning with formalin alone, or the freezing technique of Gohar, as described above under "general techniques." The author has also obtained fair results by adding menthol to the water, covering the vessel for 12 hours, and then slowly adding formalin, a few drops at a time, over a period of many hours. Another method which gave excellent results in some cases, but was not fully consistent, was the use of Benzocain, a new local anesthetic available at any drugstore. A moderate pinch of this should be added to the culture jar. After 6 to 12 hours the specimen is usually completely relaxed and may be slowly poisoned by adding formalin, drop by drop, or plunged directly in 5-percent or 10-percent formalin. Five-percent formalin is an adequate preservative for nudibranchs; although alcohol will harden the tissues and make them somewhat more firm it will probably alter the color.

3. Narcotizing of land snails and slugs. Many methods have been recommended for preserving of land snails such as *Helix* and land slugs such as *Limax*. One method found throughout the literature is that of placing the specimen on the end of a stick until it is fully expanded, then plunging it directly into boiling water. The author finds that even the best specimens prepared in this way are quite poor. The antennae are almost always contracted, the body shortens badly, and a tremendous quantity of mucus is discharged, indicating a significant reaction on the part of the animal. The author has experimented with menthol, clove oil, alcohol, chloroform, epsom salts, and direct injections with massive quantities of

10-percent formalin or FAA. Each of these methods gave only partially successful results. (After preserving more than a hundred large *Limax* simply to recheck the techniques from the literature, the author became so coated with mucus that a renaming of the land slug as "S*Limax*" seemed in order.)

The best technique is as follows: Place a moderate number of individuals in an appropriate jar, submerge the jar in a bucket of water, and place the cap on the jar (under water) so as to exclude all air. After 24 hours or more, the specimens can be preserved in formalin. Boiling the water beforehand is not essential, although slightly less time may be required for complete relaxation if the water has been boiled. Including some tobacco (such as a 2-inch length of cigarette used by Wells, 1932, and others) worked very well, but did not improve upon the plain-water method. The author placed six large *Limax*, measuring 4 to 7 inches, in a tall, straight-sided jar, and placed the jar on its side after the animals were sealed inside. Twelve *Helix* were similarly placed in a tall, straight-sided, half-pint jar. The *Limax* required 24 to 36 hours before they could be put directly in 10-percent formalin. It is best to test one or two specimens before the entire batch is pickled. *Helix* in previously boiled or unboiled water is almost always ready in about 24 hours. With this procedure, all of the animals expand fully, extend the antennae and the foot, and show no signs of local contraction.

Preserving Specimens. Specimens are frequently killed in the field in 5- to 10-percent formalin. Specimens should be washed in fresh water and transferred through 35- and 50-percent alcohol, and stored in 70-percent alcohol. Formalin solutions, unless neutralized, will act upon the shells and ultimately destroy them. Alcohol, on the other hand, will not harm the shell in any way, though it may act upon colors deposited in the soft tissues. No preservative will maintain animal colors adequately. Neutral formalin is perhaps the best preservative for maintenance of color. Shell colors are usually not altered by various preservatives. Specimens intended for histological work should be fixed directly in Bouin's fixative or FAA (Appendix C).

Inducing Egg Laying. Frequently, the eggs of fresh-water snails are desired for classroom work or embryological studies. The method used by Mattox (1952) is simple and yields rather consistent results, if the snails are not overworked. Remove snails from the laboratory aquarium and place them in the refrigerator (not in the freezer) overnight. In the morning, they should be removed to room temperature. Upon warming, specimens will usually begin egg deposition as with the onset of spring.

Class Pelecypoda: Clams, Oysters, and Related Forms

Characteristics. Clams, oysters, pectins, and the like, are very well-known and need little introduction. These specimens occur predominantly in the marine environment, but also are found in estuaries, large fresh-water rivers, streams, lakes, and ponds. These animals are usually bilaterally symmetrical and possess a pair of shells (valves) which protect the soft parts (Fig. 11-4D). Soft parts—the gills, foot, siphons, and so on—are of major importance in placing pelecypods initially in their proper orders and families; shell characteristics become important at the family, genus, and species level. Therefore, in serious taxonomic collections or for classroom studies, intact animals should be preserved. Figure 11-4, A, B, and C, shows some of the important characteristics of the shell. The ligament forms a hinge which opens the shell; the cardinal and lateral teeth, when present, align the shells. The position and number of muscle scars, and the configuration of the pallial line, which denotes the point where the soft tissues of the mantle attach to the shell, are important. Externally, the umbo de-

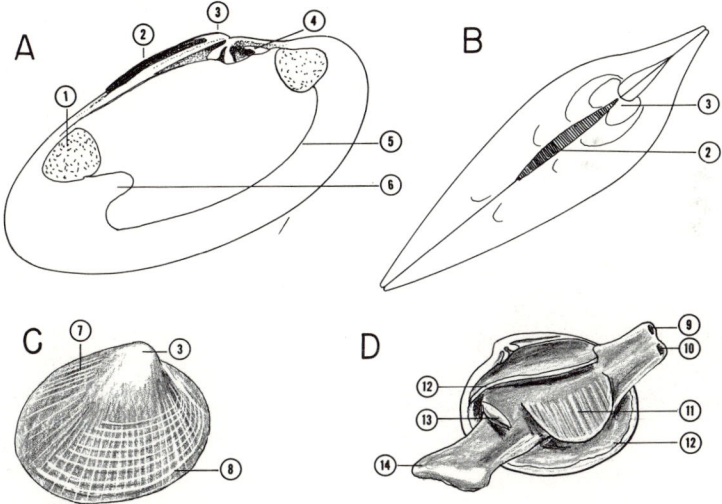

FIG. 11-4. Some aspects of clam anatomy. A. Internal shell characteristics. B. Dorsal shell characteristics. C. Shell sculpturing. D. Internal soft anatomy. (1) Muscle scar, (2) hinge ligament, (3) umbo, (4) teeth, (5) pallial line, (6) pallial sinus, (7) radiate sculpturing, (8) concentric sculpturing (note, where this crosses radiate sculpturing it may be referred to as cancellate sculpturing, (9) excurrent siphon, (10) incurrent siphon, (11) gill, (12) mantle, (13) oral palps, (14) foot.

notes the oldest part of the shell. The shell may be covered with a brownish periostracum, which may be smooth or "hairy" in appearance. Pelecypods display many types of sculpturing, in addition to color patterns that are of importance. Chief of these are the radial, concentric, and cancellate types. The last-named is a combination of both radial and concentric lines which cancel out each other. Some species of clams vary greatly in degree of sculpturing, color pattern, hairiness, and other features; some of these may be correlated with the age of the specimen, whereas others may simply be true variations. When variations are apparent, a selective but adequate series of specimens should be collected. Some of the burrowing clams found in rocks and wood possess pallets (calcium carbonate ornaments) attached to the siphon end of the body, which are taxonomically useful.

Habitats and Collecting
1. Marine habitats. Pelecypods are found from brackish water to bays and estuaries, and on out to both the unprotected sandy beaches and unprotected and protected rocky beaches of the open ocean. Many of the species common to the protected and unprotected rocky beaches may also be found in large inland bodies of marine water, such as the Puget Sound Narrows, where currents are swift. Like the snails, clams and other pelecypods are distributed on the basis of the substrate and local conditions of the sea water. With every change in the nature of waves, currents, oxygen content in the water, temperature, and the like, there will be a change in pelecypod distribution. On exposed headlands and on pilings and rocks (Fig. 11–5A) mussels are found firmly attached by byssal fibers. Jingle shells are found lower down on rock surfaces, with one valve firmly cemented to the substrate. Frequently this inner valve may be loosened with a knife, although occasionally a hammer and chisel will be essential for loosening the specimen. Pectins, other scallops, oysters, and the like, also attach to protected rocks. Look for holes in the rocks, especially on the sandstone formation, for many species of burrowing clams (Fig. 11–5D) may be collected with a geologist's pick or hammer and chisel.

Collecting is best at low tide. Specimens are usually dug by hand and located individually, either by the hole left in the sand by the siphon, or by an occasional squirt of water. However, sand, gravel, and mud may be screened with a coarse-mesh screen (built as directed in Chapter 10) to obtain small individuals. Some collectors take the razor-like clam *Tagelus* with a bent wire. This clam has a double siphon and leaves two small holes in the mud where it occurs. A piece of straight clothes-hanger wire, 16 inches long, with $\frac{3}{8}$ inch of the end bent at right angles to the shank, may be inserted with the end tab aligned with the two siphons. This wire is

The Mollusks and Brachiopods   181

FIG. 11–5. Various pelecypod habitats. A. Mussels. B. Jingle-shell clam. C. Pecten attached to rock. D. Burrowing clam in rock. E. *Tagelus* in mud. F. Horse clam. G. Butter clam. H. *Teredo* burrowing in wood.

thrust quickly down into the mud for its full length, turned 90°, and retracted, usually with the clam intact. The wire passes between the shells on the way down, but after turning will snag one of the two shells, and thus retract the animal. Many clams will be found under rocks or between rocks which are distributed over mud and gravel bottoms. Always look for submerged wood, old pilings, and the like, for species of burrowing "ship" worms which are really burrowing clams of the genus *Teredo* or *Bankia*. These animals burrow thoroughly through the inner parts of pilings, old logs, and even ship sides, but show little sign externally. However, small holes are present where the siphons enter the water. They are so numerous that an old log can be split with one blow of a geology pick to expose hundreds of specimens for preservation. In deeper water the biological dredge is very effective for bringing up all forms of mollusks (see Chapter 1).

2. Fresh-water collecting. Fresh-water specimens are collected by hand or by means of a rake or dredge net (Chapter 13). They can easily be located in the shallow waters of lakes, ponds and slow-moving rivers, and streams. Juveniles and small species are found associated with plant material. Such material is collected and carefully sorted through in white pans to locate the specimens.

Preparation of Shells. Almost all sizes of pelecypods may be prepared as dried shells. Place specimens in fresh water and bring to a boil, continue to boil for 2 or 3 minutes (longer for large specimens), and cool slowly. Upon heating, the large adductor muscles usually become loosened, the shell opens automatically, and the soft parts are thus easily removed. The little muscle tissue that adheres to the shell may be pulled away with a forceps. At this point shells are generally closed and tied with string or fastened with tape while drying. One specimen should be left open, however, so that internal characteristics may be observed. Closed shells may be opened at any time with a short period of soaking in fresh water. The author prefers to leave fouling organisms attached to the shells; however, if the collector so desires, they may be removed with a pocketknife or by brushing with a stiff-bristled (not a wire) brush. Pennak (1953) recommends the application of varnish or Vaseline to the periostracum of fresh-water pelecypods. A little mineral oil may be used for other shells, but other coatings are not recommended (see "Gastropoda," p. 174).

Narcotizing and/or Preservation. Marine pelecypods may be killed by soaking in fresh water. They may then be preserved in 75-percent alcohol after being moved through 35- and 50-percent alcohol. Another method for preserving all pelecypods is to place them in habitat water until the shells open. Next, insert a wooden plug $3/16$ inch thick, and preserve the animals by running them up through 35-, 50-, and 75-percent alcohol. Other methods that work very well are the heat-kill method, narcotizing with alcoholized water or menthol, or the freeze-kill method. For a detailed discussion of these methods and of preservatives consult "Gastropoda," above.

Collection and Treatment of Glochidia. Fresh-water clams have a unique reproductive cycle in which fertilized eggs are maintained in the gill chambers of the female clams where they develop into glochidia, or larvae. These larvae are later expelled and become parasitic in fish gills, but they ultimately leave their host and metamorphose to normal adult clams. Thus, during the breeding season the gills of females are almost always swollen with glochidia. Pennak (1953) in writing about North American fresh-water pelecypods says "the Unioninea are short-term breeders and are gravid between April and August; the Anodontinae and Lampsilinae are long-term breeders, the eggs being fertilized in midsummer and carried until the following spring or summer." These specimens are best preserved in alcohol and used for wet-mount slides when desired. However, stained or unstained mounts may be made in Turtox CMC-S or CMC-10 (Appendix B). An alternate method is to transfer them directly from 70-percent alcohol to borax carmine stain, destain, dehydrate, and mount as described for the Gastropoda.

Class Amphineura: Chitons

Characteristics. Amphineurans, the chitons, are so named for the double nerve trunk which extends down both sides of the body. These animals are remarkably snail-like, with tentacles and anterior mouth and a large foot for locomotion. The dorsal part of the body is covered with a soft mantle which protects gills on both sides of the body and holds the shell in place. The shell is divided into eight individual plates (Fig. 11-6, B and C) which overlap and are embedded laterally in the mantle. The mantle and the plates may be simple or highly ornamented, drab or very brilliantly pigmented. Usually the plates are exposed, but some species possess a mantle which completely covers the plates dorsally. Chitons are usually preserved intact in liquid.

Collecting Chitons. Chitons are extremely numerous in the intertidal

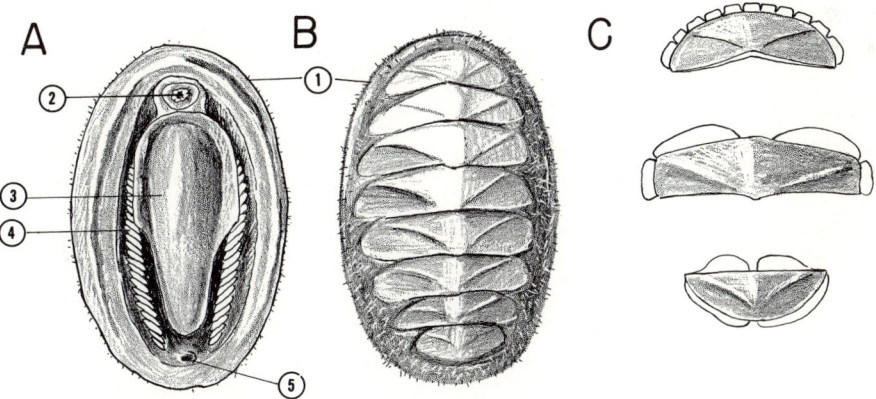

FIG. 11–6. Chitons, class Amphineura. A. Ventral view. B. Dorsal view showing the eight plates surrounded by the mantle. C. Representative plates, anterior, middle, posterior. (1) Mantle, (2) mouth, (3) foot, (4) gills, (5) anus.

zone, but are restricted to solid substrates such as rocks or pilings. They occur from the splash zone into the subintertidal zone where they may be dredged. Chitons attach very firmly to their substrate, but may be loosened with a dull pocketknife or pulled free from the rocks by hand if they are taken by surprise. Specimens should be placed in a bucket of sea water with the foot flat against the bottom, in an attempt to prevent them from curling up.

Narcotizing and/or Preservation. The main requisite for preserving chitons is to keep the foot and body flattened. This may be achieved by narcotizing or by binding and preserving. Most collectors simply place two specimens together, foot to foot, either in the field or in the laboratory, and bind these with strips of cloth or tape. String or rubber bands are not recommended as they distort the specimens. Specimens so bound may be dropped directly into 10-percent formalin or 35-percent alcohol for killing. All specimens are washed and treated in 35-, 50-percent alcohol, and finally, are stored in 70-percent alcohol. When the bindings are removed specimens will remain flat and in perfect condition. Menthol, epsom salts, and the freeze-kill method have all yielded fairly good results with amphineurans, when used as directed for gastropods (p. 176). The author finds that specimens placed in clean sea water with several drops of clove oil added become relaxed in about $\frac{1}{2}$ hour. If specimens move out of the water, simply return them until they are fully relaxed. Specimens narcotized by all methods had best be bound before preservation.

There are two methods of working with chiton shells. The shells may be dissected from the mantle, numbered with India ink (from anterior to posterior), and stored in small boxes. The second method is to preserve bound specimens in 10-percent formalin for 24 hours, wash in fresh water, and dissect away the main portion of the animal, leaving only the outer rim of the mantle to keep the shells intact. Such specimens should be kept in small boxes within insect-proof cabinets, inasmuch as dermestid beetles may attack the mantle.

Preparation of the Radula. The radula is less important in identifying chitons than in snail identification: thus, it is infrequently prepared on permanent slides. The technique of preparation begins with the removal of the head and anterior portion of the body containing the radula, and is completed in the manner described for snails.

Class Scaphopoda: Tooth or Tusk Shells

The tooth shells are generally restricted to mud flats occurring at the lowest intertidal levels of certain bays, but, more typically, at subintertidal depths down to 15,000 feet (Storer and Usinger, 1957). These unusual animals occupy tall, cone-shaped shells which are open at both ends. The mouth and foot occur inside the large aperture and extend down into the mud or sand. Scaphopods are generally obtained by dredging across muddy and sandy bottoms. They may be killed in fresh water and preserved by moving through changes of 35-, 50-, and 75-percent alcohol in which they are stored. Shells may be cleaned by placing specimens in cool water, bringing the water slowly to a boil, cooling slowly, and then withdrawing the flesh with a forceps.

Class Cephalopoda: Octopi and Squids

Characteristics. Since a large number of cephalopods lack an external and visible shell, identification is based primarily on external and internal soft structures. The number of tentacles, the number of suction cups per tentacle, the relative length of the tentacle, the nature of the mantle, and so on, are features of importance. Shell characteristics, when present, are also considered. Therefore, thoroughly relaxed and carefully preserved specimens are most desirable for taxonomic purposes. Badly contracted forms cannot be measured with accuracy. These animals are quite sensitive and die easily during transportation, necessitating careful handling. Usually, however, when specimens die en route they are expanded and quite useful for taxonomic studies.

Collecting
1. Octopi. The octopus usually has a home den, or hiding place, from which it ventures for feeding activities. These dens are below the zero tide level but may be exposed on extremely low tides. They are found where rocks are piled on top of one another so as to provide numerous spaces underneath. Any kind of container, such as a submerged tin can or bottle, should be drained and examined for small octopi. In Alaska the writer found large octopi (up to 30 pounds) at low tide by looking for trails of broken crab shells. Crustaceans make up a big part of the octopus' diet and, hence, are frequently scattered about the entrance of the hiding place. If the octopus can be reached by hand it is a simple matter to extract it, regardless of its size. Reach in and feel around until the ventral edge of the mantle is located. At this point a stout ligamentous muscle ties the mantle to the main part of the body. Insert a finger into the mantle on either side of this muscle and pull with a uniform pressure. The animal will soon release its grip and may be drawn from its hiding place. Larger octopi will not remain

in a bucket, and should be placed in a plastic bag containing sea water which is in turn carried within a bucket. Inflate the bag partially and tie it closed so as to form a small balloon of air.

2. Squids and other cephalopods. This collective group of animals is considered nectonic, that is, animals capable of directional swimming in spite of average currents. The squid, for example, is one of the few swimming invertebrates which follows a migratory pattern. Because of this, squids are somewhat seasonal in their occurrence at certain latitudes. In Puget Sound, for example, they are most frequently seen in the winter months from November through February, whereas in the Channel Islands off California they are very common during the summer months. Perhaps the best way to collect squid is with the use of a night light (see Chapter 1) set up from shipside in a remote anchorage. In Puget Sound these animals come in to the lighted fishing docks in tremendous numbers for a few weeks each year. Squids and like animals tend to remain on the surface, just beneath the night light, and can easily be dipped from the water. Some of these specimens swim toward the light with such velocity that they sail through the air at a height of 4 or 5 feet and land on shipboard. Another method that is always successful in the Puget Sound area is use of the beach seine, which is employed for catching fishes (see Chapter 18 for construction). A large 50- to 100-foot seine set 200 to 300 feet offshore on a sand and gravel bottom, and then skillfully retrieved to the beach, will usually yield many squids. Squids may also be obtained with large oceanic tow nets measuring 5 or 6 feet in diameter, but this equipment is rare insofar as the average collector is concerned.

Narcotizing and/or Preserving. Squids are best preserved alive. Place specimens in a large container (a bucket or plastic garbage can) big enough to permit swimming. Add enough formaldehyde to make a solution only a fraction of 1 percent. The amount is not important, so long as it is small. The specimens will swim in this, slowly die, and sink to the bottom in a natural position. Next, transfer specimens to the museum jars in which they will be stored, and position them as desired. Add 5-percent formalin for 24 hours, then replace with a fresh 5-percent solution. Place field data and taxonomic information in the storage container and seal with a lid. Small octopi and squid may first be hardened in plastic or wax-bottomed pans. Arrange the tentacles, body, and fins in the desired position, being mindful to bend the tentacles, if necessary, so that the specimen will fit the storage container. By drawing contracted tentacles through your fingers, you can stretch them out and they will remain enlongated. Flood with 5-percent formalin for 24 hours, transfer specimens and their field data to appropriate bottles, add a fresh 5-percent solution, and seal.

Most of the nonshell-bearing cephalopods may be stored in 5-percent formalin, except for large specimens which may go into 8- or 10-percent formalin. Otherwise, the specimens may be transferred through 50-percent alcohol and stored in 70-percent alcohol. Formalin preserves the color better than alcohol, but is less agreeable to work with. Shelled specimens should be stored in alcohol.

Octopi and squids are easily narcotized. Bianco (1899) and others recommend the use of chloral hydrate, used in the same way as Chloretone. The author has had excellent results from the alcohol, epsom salts, and Chloretone methods. The first two techniques were conducted as described for snails (p. 176). All specimens were isolated in appropriate jars with sea water, placed in a semi-dark place, and narcotized as directed. Chloretone crystal was added initially, and again after 1 hour. Check specimens by gently probing the tentacles. Let them remain in the narcotizing solution for ½ hour after they show no reaction. For moderate-sized specimens, 1½ to 3 hours, or more, are required with alcohol; 3 to 6 hours are required with epsom salts, and 2 to 4 hours are required with Chloretone. Before flooding the specimens with a preservative, pipette a small quantity of preservative onto one tentacle. If there is a reaction place the animal back in the narcotizing solution. Should some faint contraction occur during preservation the tentacles may be stretched through the fingers and restored to normal length.

All cephalopods bearing external shells should not be subjected to formalin solutions, which are harmful to the shell. Rather, kill in 35-percent alcohol, transfer to 50-percent, and store in 70-percent alcohol. Bianco (1899) recommends that transparent pelagic forms be put into Kleinenberg's solution for 1 hour. After fixing, transfer to 35-percent alcohol and move on up into 70-percent alcohol.

PHYLUM BRACHIOPODA: LAMP SHELLS

Characteristics and Collecting. The brachiopods are often confused with clams in that they possess two shells which cover the soft parts of the body. They actually resemble bryozoans much more than they do the pelecypods. For one thing, the shells are dorsoventrally located, and there is a median-posterior peduncle, a stalk which attaches the specimen to its substrate. Within the shells a lophophore with tentacles is supported on a shelly loop. These latter structures, and other anatomical features, are not characteristic of the phylum Mollusca.

188 Biological Techniques

Both the shell and soft parts are used as taxonomic features for identification. In the shell such things as the over-all shape, the cross section at various points, whether the two shells are articulated or not, the position of muscle scars, the peduncular opening, and the shape and location of the shelly loop are all of importance. Important features among the soft parts are the lophophore and its tentacles, the presence or absence of the anus, the location of various internal organs, and so on. It is therefore necessary to preserve specimens in such a way as to have expanded internal portions which may be readily examined.

Most of the brachiopods attach to solid substrates and are members of the lower intertidal fauna in the north latitudes, but are found in deeper (and thus cooler) waters as one goes toward the equator. Specimens have been dredged from depths of over 18,000 feet. Some genera, such as *Glotidia*, have a well-developed and expanded stalk which supports the animal on sandy substrates below the intertidal level.

Preservation. The minimal treatment a brachiopod should receive is as follows: Place the specimen in sea water until the shells are agape, insert a wooden plug between the valves, preserve in 35-percent alcohol, transfer through 50-percent and store in 70-percent alcohol. Formalin solutions damage the shell: if they are used, specimens must be washed as quickly as possible, soaked out in fresh water for several hours, and transferred up through the alcohols for final preservation.

A better treatment is as follows: Permit specimens to open and expand in sea water and add small quantities of alcohol, drop by drop, over a period of several hours until a solution of 5 to 10 percent results. Permit the animals to remain in this solution until they are completely insensitive, which may take 6 to 12 hours. At this time, test a single individual by placing it in alcohol. If it does not contract, the remainder of the specimens should then be preserved as described above.

STORAGE METHODS AND EQUIPMENT

Methods for Specimens in Liquid. Small specimens should be put into vials along with a label (printed in India ink and thoroughly dried) which gives the field data and scientific names of the specimens. Fill the vials with preservative, and stopper with cotton. Place these vials in a larger jar, along with some preservative and an internal label, and seal with an airtight lid. Keep specimens in a dark, cool place and check once or twice a year to make sure the preservative level remains constant. Larger specimens will go directly into airtight jars with their respective field and specimen data.

Shell Cabinets. For the proper taxonomic arrangement and storage of shells, some simple cabinet system should be adopted. The cabinet described herein is adapted from a series of shelf measurements given by Webb (1935). The cabinet stands 60 inches high, 30 inches wide, and is 23¾ inches deep, not including the door. This cabinet has 20 drawers: 6 drawers 1⅜ inches deep, 6 drawers 2 inches deep, 6 drawers 2½ inches deep, and 2 drawers 4 inches deep. The cabinet is made entirely of ¾-inch plywood (sound both sides), except for the back which is made of ¼-inch plywood (sound one side). Finish nails and glue are used at all joints. Nails must be set and sealed.

To make this cabinet, cut a 4- by 8-foot sheet of plywood off at 60 inches. Split this 60-inch piece down the center, giving two pieces 23¾ inches wide. These pieces form the sides of the cabinets. Next, set a ¾-inch wide dado blade to cut ¼ inch deep, and dado the top, middle, and bottom areas of the side boards as shown in Fig. 11–7B. After this, cut the cabinet top, bottom, and middle shelf, each 29 inches wide and 23¾ inches deep, with the exception of the middle shelf which is 23½ inches deep.

With a wide dado set to cut ¼ inch deep, dado out the back of the side boards and top and bottom boards, as shown in Fig. 11–7C, in order to make a recess for the ¼-inch plywood back. Finally, cut a toe board out of ¾-inch plywood, which is 3 inches high and 30 inches wide. With a handsaw carefully cut out the front, bottom corner of the side boards (3 by 3¾ inches) as shown in Fig. 11–7B.

Assemble the main box of the cabinet by putting white wood glue in the three dado cuts of one side board, insert the top, bottom, and middle shelves, and nail from the outside in, making sure that the shelves are flush with the front of the box (check again to see that the middle shelf is only 23½ inches deep). Turn this assembly over, put glue on the butt ends of the top, bottom, and middle boards, and nail the other side in place. Next, place the cabinet face down, put glue all the way around the dado cuts which are to receive the back sheet of plywood, and install a piece of plywood cut slightly smaller than 29½ by 56½ inches. The plywood back should be cut extremely square and is used to square the entire cabinet. Finally, nail the toe board in position as shown in Fig. 11–7D.

Now rip 36 pieces of hardwood, ½ by ½ inch by 23⅜ inches long. These are for drawer runners, and should be drilled at appropriate intervals so that they may be nailed in position without splitting. These are installed with wood glue and 1-inch nails, as shown in Fig. 11–7E. The safest and most accurate way to install the drawer runners is to cut four different-sized spacers approximately 24 inches long and measuring exactly 4⅛ inches wide, 2⅝ inches wide, 2⅛ inches wide, and 1½ inches wide respectively. Install the runners on one side of the cabinet at a time. Turn the cabinet on its side and place the 4⅛-inch spacer firmly against the bottom. Apply glue to a runner, put the nails in the small holes, and position the runner firmly against the drawer spacer. Drive the nails in position, holding the runner in place, and set the nails. Use this same spacer for the second drawer up and affix the runner in the identical manner. Following this, use the 2⅝-inch spacer for all six drawers which are located below the middle shelf. When these are in place, put the 2⅛-inch spacer against the middle

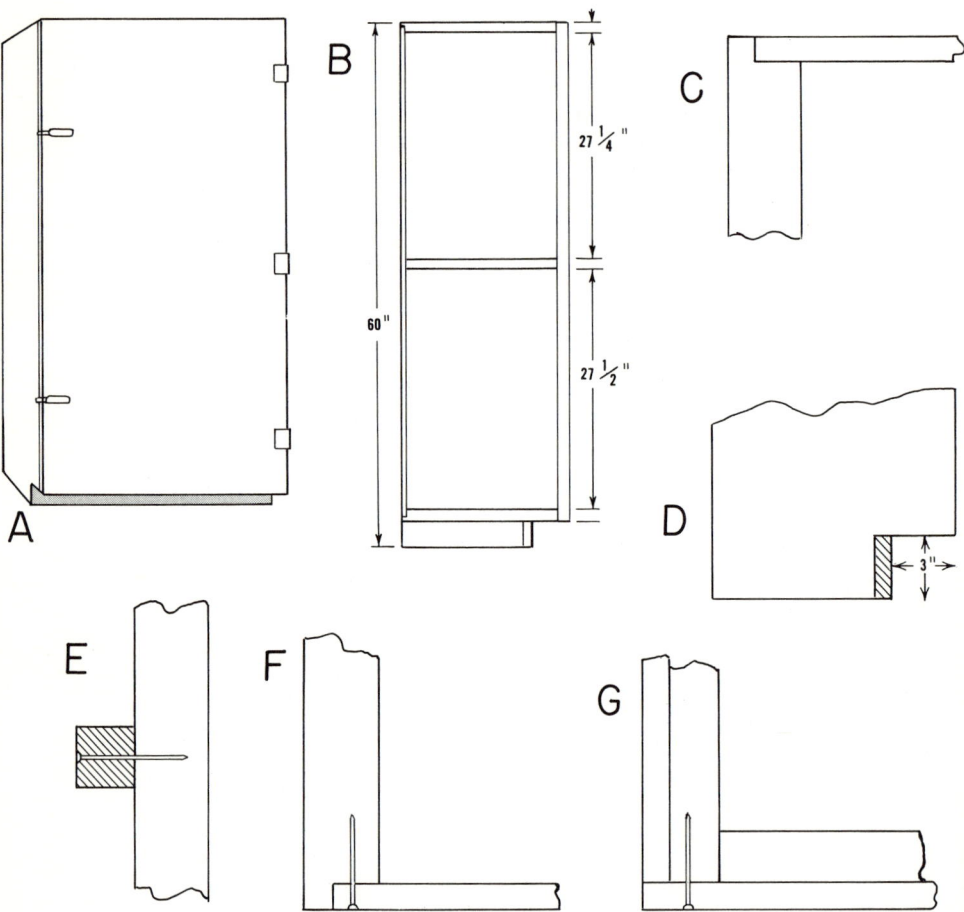

FIG. 11–7. Construction of a shell storage cabinet. A. Cabinet with door hardware. B. Long section through the cabinet with measurements. C–G. See text for details.

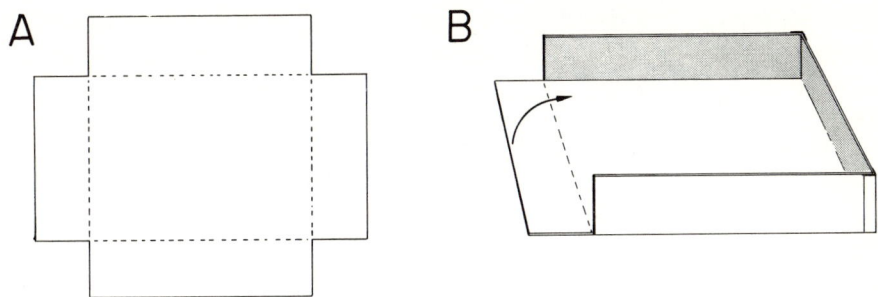

FIG. 11–8. Making individual shell-storage boxes.

shelf and nail the next six (not five) drawer runners in place, using this spacer. Finally, install the last five drawer runners with the 1½-inch spacer. Repeat this process, beginning at the bottom of the other side. When the cabinet has been dried, sanded, and varnished or shellacked, foam rubber or foam plastic (⅛ inch by ¾ inch wide) should be placed all the way around the door opening. Cut a ¾-inch plywood door, 57 by 30 inches. Use three hinges and two door catches of the same type, and install as described under "Insect Storage Cabinets" (p. 000).

All drawers should measure about ⅛ inch smaller than the space provided, and may be constructed of a clear, fine-grained softwood measuring exactly ¾ inch wide, or of ¾-inch plywood. The bottom should be constructed of ⅛-inch masonite. The drawer fronts, backs, and sides will all be ripped to the dimensions given above (4, 2½, 2, 1⅜ inches). The fronts and backs should be cut off at exactly 28⅜ inches in length; the side boards are cut at 22⅞ inches in length. With the dado blade cut a depression ⅛ inch deep and ½ inch wide on the bottom edge of all drawer sides, fronts, and backs. Next, with a ¾-inch-wide dado cut the ends of each front and back piece (not the sides) ½ inch deep, in order to receive the side boards, as shown in Fig. 11–7G. Cut twenty drawer bottoms just a shade under 22⅞ by 27⅞ inches. Assemble the sides, fronts, and backs of each drawer with wood glue and finish nails, turn the assembly bottom-side-up, and with glue and finish nails install the bottom masonite which will square the drawer. Finish the drawers by sanding and applying a coat of shellac or varnish.

Unit Trays. Shells are kept in small cardboard unit trays along with a printed label containing the shell's number (which is also recorded on the shell), its name, and the field data. Small, conveniently sized cardboard boxes may be purchased at any large paper company, or they may be made from poster paper (lightweight glazed cardboard) to any desired dimension and depth, as illustrated in Fig. 11-8, A and B. Cut the cardboard into convenient squares of the right dimensions and place each in turn over a block of wood which has the corners cut out exactly as the finished cardboard should be (Fig. 11-8A). With a sharp pocketknife cut the corners out over the wooden template. Next, place the cardboard over a wooden block and bend the sides down. Finally, fold the sides and tape with masking tape (Fig. 11-8B). When specimens are kept in vials, these vials are in turn placed in such unit trays.

REFERENCES

Abbott, R. T., 1954, *American Sea Shells,* Van Nostrand, Princeton, N.J.

Bianco, S. L., 1899, The Methods Employed at the Naples Zoological Station for the Preservation of Marine Animals, *Bull. U.S. Nat. Mus.,* 39(M):1–37.

Gilbert, M., 1961, *Starting a Shell Collection,* Hammond, Newark.

Gohar, H. A. F., 1937, The Preservation of Contractile Marine Animals in an Expanded Condition, *J. Mar. Biol. Assoc.,* 22(1):295–299.

Hyman, Libbie H., 1959, "Phylum Brachiopoda," in *The Invertebrates,* Vol. V, McGraw-Hill, N.Y.

Keen, A. M., 1958, *Sea Shells of Tropical West America,* Stanford Univ. Press, Stanford, Calif.

Mattox, N. T., 1952, (Personal direction given in the Univ. So. Calif. Laboratory)

Morris, P. A., 1951, *A Field Guide to the Shells of Atlantic and Gulf Coasts,* Houghton Mifflin, Boston.

Morris, P. A., 1952, *A Field Guide to the Shells of the Pacific Coast and Hawaii,* Houghton Mifflin, Boston.

Muir-Wood, H. M., 1955, "A History of the Classification of the Phylum Brachiopoda," British Museum, London.

Pennak, Robert W., 1953, *Fresh-Water Invertebrates of the United States,* Ronald Press, N.Y.

Pilsbry, Henry A., 1939–1948, *Land Mollusca of North America,* 4 vols., Academy of Natural Sciences, Monograph 3, Philadelphia.

Storer, Tracy I., and Robert L. Usinger, 1957, *General Zoology,* McGraw-Hill, N.Y.

Thompson, J. Allan, 1927, "Brachiopod Morphology and Genera (Recent and Tertiary)," Wellington, New Zealand Board of Science and Art, Manual No. 7.

Verrill, A. H., 1950, *Shell Collector's Handbook,* Putnam, N.Y.

Webb, W. F., 1935, *A Handbook for Shell Collectors,* W. F. Webb, Rochester, N.Y. and St. Petersburg, Fla.

Wells, M. M., 1932, *The Collection and Preservation of Animal Forms,* General Biological Supply House, Chicago, Ill.

(See Chapter 1 for other References.)

CHAPTER 12

The Bryozoans

The name Bryozoa is variously used on the phylum or class level, and may include the phylum Entoprocta (according to Osburn, 1950), may be discarded in favor of Ectoprocta (Barnes, 1963), or (by British workers) be referred to as the Polyzoa. These animals are predominantly marine, although a limited number of species make up an important part of the fresh-water invertebrates.

MARINE BRYOZOANS

Characteristics. The living bryozoan animal, or zooid, lives within a secreted housing, the zooecium (Fig. 12–1A). The anterior end, which projects from the housing, the introvert, terminates in a characteristic swelling, the lophophore, which bears soft, ciliated tentacles. This tentacular ring surrounds the mouth and is everted for food-gathering. The zooecium of marine bryozoans is either chitonous (Order Ctenostoma) or calcified (Orders Cyclostomata and Chielostomata). The opening of the zooecium may be naked or provided with a membrane or a trap-door-like operculum. The zooecium itself may be highly ornamented with spines or provided

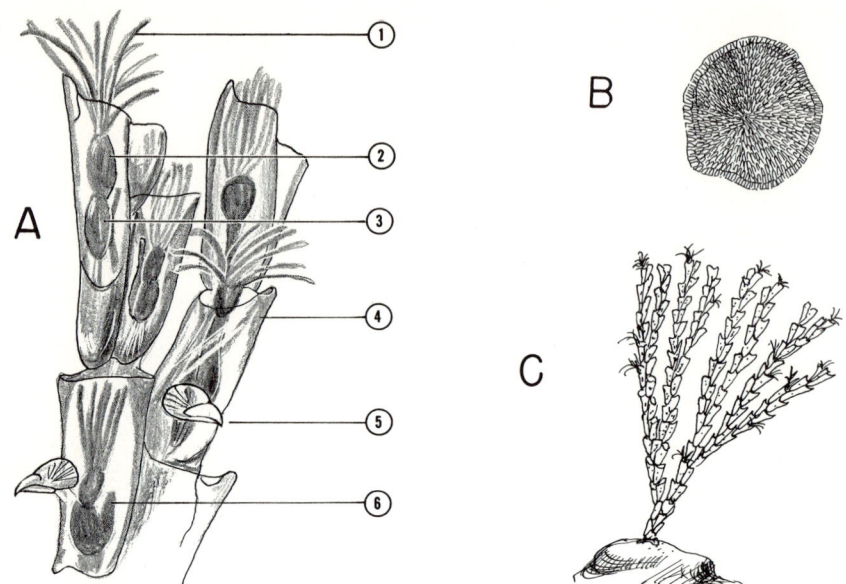

FIG. 12–1. Bryozoan morphology and anatomy. A. Anatomy of *Bugula*. B. An encrusting colony. C. An erect colony such as *Bugula*. (1) Tentacles, (2) pharynx, (3) stomach, (4) zooecium, (5) avicularium, (6) intestine.

with bird-head-like avicularia, which are defense mechanisms. Bryozoans bud off new individuals to form complex colonies which may encrust flat surfaces or erect moss-like structures (Fig. 12–1, B and C); or they may reproduce sexually, giving rise to motile larvae.

For the ctenostome bryozoans, the soft parts as well as the chitonous zooecium are used for identification; the remainder of the marine bryozoans bearing calcium carbonate in the zooecium are identified predominantly by shell characteristics. This latter group may be dried, as directed below; the former should be preserved well enough to demonstrate the internal anatomy. Anesthetizing specimens is recommended for the ctenostomes and is essential whenever good teaching slides are to be made. Finally, the nature of the zooecium determines the preservative used, as discussed below.

Habitat and Collecting. Marine bryozoans are extremely numerous in bays and estuaries (more than 190 species in Puget Sound), in protected and unprotected rocky coast zones, and are abundant in the intertidal zone extending down to relatively deep water. They are generally absent from muddy or sandy bottoms unless some solid structure is present to support them. They attach to the sides and undersides of rocks, pilings, floating docks, boats, marine algae with their holdfasts and fronds, and the shells of marine invertebrates and vertebrates such as barnacles, mussels, turtles, and the like. Collecting is generally done by hand in the intertidal zone and on floating structures; the dredge (Chapter 1) or dredge net (Chapter 13) may be used in deeper waters to obtain objects bearing bryozoans.

Collect intact rocks or shells, cut off bits of piling, and pull up kelp from the bottom, examining carefully the fronds and holdfasts and removing those parts with bryozoans. A pocketknife is handy for dislodging encrusting bryozoans from ship bottoms, floats, and the like; erect forms may be plucked directly. If you intend to make slides or study the soft parts, keep the specimens isolated in small plastic bags or other containers provided with water. Do not permit them to overheat. Place sufficient nonmucous-secreting algae around rocks and other objects to prevent damage from encrusting forms. Where only the shell structures are required, specimens may be allowed to dry, provided they are protected from abrasion.

"Trapping" Bryozoans. During the summer months, when bryozoan larvae are especially numerous, good specimens of many species may be obtained directly on microscope slides. These specimens may later be dried or narcotized and preserved in liquid according to your need. The equipment for "trapping" consists of a sturdy wooden slide box of 25 capacity, 10 or 12 glass slides, enough plastic window screen to completely surround the box, some string, and a few pieces of stout wire. Drill a small hole through the bottom of the box in the center of either end (Fig. 12–2A). Place ten or so slides in the slots, alternately skipping a space. Next, wrap the plastic window screen around the box tightly and sew it or tie it in place. Doing a good job in this aspect will be well worth the effort. Finally, pass a piece of heavy wire through the hole at each end and twist it in such a manner that it may be attached to a rope, as shown in Fig. 12–2B. Nylon string is superior to cotton string or wire for this job, in that sea water will quickly attack the latter two.

This trap may be set in a number of ways. For example, suspend an 8-foot weighted rope from a floating dock and attach one or more traps at 2-foot intervals, as shown in Fig. 12–2C. This set has always proved very successful. Another method in offshore kelp beds that are not visited by too many swimmers is to suspend the trap on an anchored rope with a float at or below the surface, as shown in Fig. 12–2D. In protected and slightly unprotected rocky coast zones the trap may be tied to the shoreward side of a stout rock (Fig. 12–2E).

Such traps should be left out about 1 month during the warm season, or

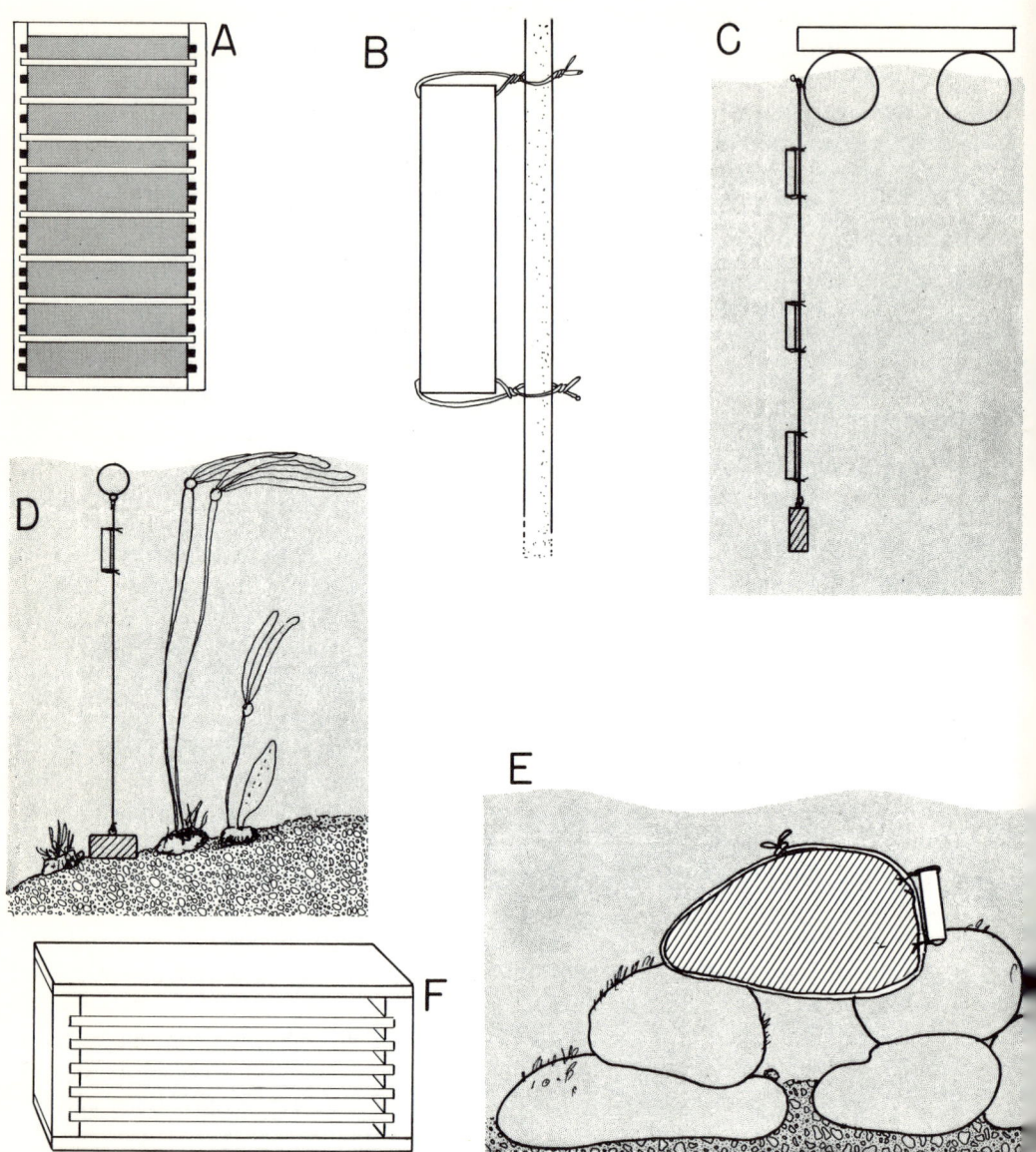

FIG. 12–2. Bryozoan "traps." A. Slide box with glass slides. B. Detail of slide box attached to rope by wire. C. A string of bryozoan traps suspended from a floating dock. D. Bryozoan trap among kelp. E. A bryozoan trap tied to a rock in the surf zone. F. A slide storage box used for maintaining specimens in liquid.

longer in the cooler part of the year. Transport the traps to the laboratory in buckets of sea water and there examine each slide, one at a time. Select the best or most desirable growth of Bryozoa on a given slide and then scrape all of the remaining debris from the slide with a razor blade. If the specimens are to be dried simply label the slides and store as directed below. If expanded animals are desired, narcotize and kill the specimens, as directed below. Wet-mount slides should be given a field number with a diamond pencil, placed in a plastic box secured with two pieces of string or two rubber bands (Fig. 12–2F), and placed in a container of 70-percent alcohol. These slides may be displayed in the classroom in Petri dishes containing alcohol.

Narcotizing, Preserving, and Cleaning. When the soft parts are important, get specimens to the laboratory as quickly as possible and transfer them to finger bowls two thirds full of cool sea water placed away from bright light. The specimens will expand quickly (ascertain with a hand lens), at which time the narcotizing agent should be introduced. Menthol is widely recommended in the literature for narcotizing bryozoans, but the author has found that Chloretone, epsom salts and clove oil, in addition to menthol, produce excellent, well-expanded specimens. If your surroundings are relatively cool so that the culture water will not quickly heat, use a Petri dish or similar small dish, as the volume of water has a bearing on the speed and effectiveness of the narcotizing agent. A small pinch of Chloretone crystals should be added when the specimens are expanded and again in about an hour and a half. The specimens may be preserved at about 6 hours, or checked intermittently after about 3 hours by gently probing the tentacles with a fine insect pin. Observe the specimens under the microscope and, when they no longer react to mechanical stimulation, add sufficient formaldehyde to make a 5-percent solution.

Menthol is used in the same manner as Chloretone except that three small pinches are added over a period of 6 hours. With clove oil, add a single drop to the water initially and again in about 2 hours. Observe periodically or preserve at 6 hours, as above. With epsom salts, make a saturated solution in sea water and add several drops to the culture jar (more if finger bowls are used). Repeat this at half-hour intervals for 3 to 4 hours, and preserve the specimen when it is insensible (about 6 hours). Note well that the times given here are based on Petri dishes and will vary considerably for other vessels. An almost foolproof method when the animals are vigorous is to add one or two small amounts of the narcotizing agent and let the specimens stand overnight, after which time they may be preserved.

Any bryozoan containing calcium carbonate is soon damaged in preservatives containing acid. Neutral formalin should be used as a killing agent for such forms, and they should be stored in 70-percent alcohol. Some kinds of alcohol may become acidic and should be checked periodically with litmus paper. Kill narcotized or other bryozoans intended for liquid preservation by adding sufficient neutral formaldehyde to the surrounding water to make a 5- to 10-percent solution. Check the specimens under a microscope; if they appear covered by an excessive amount of debris or other organisms, clean them with a soft, pointed water-color paintbrush. Ctenostome bryozoans with no calcium carbonate may be stored in 5-percent formalin. Move all others up through 50- to 70-percent alcohol.

Dry-Slide Mounts. Although specimens containing calcium carbonate may be stored in 70-percent alcohol, there is always some danger that the alcohol may develop an acidic quality and ruin the specimens. Thus, a portion of each collection should be preserved dry for taxonomic work inasmuch as the zooecia are of prime importance. Specimens may be dried on blotters. Those adhering to mollusk shells should be removed with some sort of rotary saw or hacksaw by sawing through the shell and cutting out a square around a part of the colony. This section is in turn glued to a standard microscope slide with glass cement or other glues such as Elmer's Glue or china cement. It may help to roughen the slide surface with a piece of sandpaper, though this is not usually necessary. Attach slide labels immediately and record collecting and taxonomic data. Store the specimens in standard slide boxes, adequately spacing them to avoid damage.

Colonies attached to kelp may occasionally be removed with a pocketknife and then glued directly to a slide. Rather than risking damage, however, cut a small square of kelp containing a colony and partially dry this on a blotter. Avoid any curling of the algae during the drying process, as this will destroy the colony. While it is still somewhat damp, glue the algae to the slide, label, and store as directed above.

Those animals or colonies adhering to rocks must be treated in one of two ways (besides being preserved intact in alcohol). First, with a hammer and chisel carefully break off a portion of the rock containing part of a colony. Glue these fragments onto slides for further study, and store as above. It would be to the investigator's advantage to mount the specimens as flat as possible so that the focusing of the microscope will be kept to a minimum. If the rock material is too hard, the alternate procedure is to record field data on the rock with India ink (or white ink) and store the rock intact in a small box with a lid.

Standard Slide Mounts. Slide-mounting techniques begin after the process of narcotizing and killing bryozoan specimens in formalin, as directed above. The techniques involve the following: posturing the specimen, hardening the very soft tissues, staining, dehydration, and mounting. Specimens that have been relaxed, killed, and stored in alcohol are usually too three-dimensional to fit under coverslips without distortion. Thus, specimens directed in a two-dimensional plane should be selected, stained, and processed.

Select small portions of freshly killed colonies with well-expanded zooids and place these on a sheet of glass. Next, put a coverslip support on both sides of the specimen. These supports must be as thin as possible, but thick enough to prevent crushing of the specimen. Place a regular microscope slide on top of the specimen and weight this to prevent it from floating. Pipette 50-percent alcohol under the slide and gradually work this up to 95-percent alcohol by adding a fresh solution on one side of the slide and drawing the old solution off with a blotter on the other side. Harden the tissues for 8 hours or more in 95-percent alcohol, but guard against drying out. When you remove the slide, do so carefully. Loosen the specimen if it adheres to the glass plate by directing a stream of alcohol against it from a pipette. Gray (1952) recommends flattening bryozoans between two glass slides with supports, binding the slides together with string (or rubber bands), and placing the slide "sandwich" into the alcohol solution.

Transfer flattened specimens to a proper container (see Appendix B) for washing and staining, and run them down through the alcohols to 70 percent. Stain with borax carmine for 1 hour. Destain the specimens in 70-percent acid alcohol until they are pink. Finally, dehydrate in up to 95-percent alcohol.

At this point any one of the following methods may be followed: (1) Wash in two changes of absolute alcohol, 10 to 15 minutes each, clear in xylene, move to a mixture of 1 part xylene, 1 part Permount, or other similar media, transfer to a slide, and mount in Permount; (2) transfer from alcohol to Euparal Essense to dehydrate and clear, mount in natural Euparal; or (3) mount directly in Euparal from alcohol. None of these techniques requires ringing the coverslip with a waterproof cement. Label slides with taxonomic and field data when they are dry.

Collection and Treatment of Burrowing Bryozoa. Some of the ctenostome bryozoans burrow into the shells of marine organisms, such as snails and clams, and can be detected because they make exceedingly fine holes all over the surface of the shell for the purpose of extending the lophophore. Therefore, shelled invertebrates must be examined in the field for such perforations. Bryozoan burrows may be confused with burrows of other invertebrates which are generally much larger. The host and the bryozoan should be dropped directly into Bouin's fluid, in which they may be stored. Bouin's will do some damage to the shell, but will not render it unrecognizable. The host animal must be identified for completion of the field record. Finally, remove the shell from around the specimen by decalcification. Following Gray (1952), treat the specimens in Haug's solution (Appendix C) by suspending the host on a silk thread into the solution. When the shell has been eaten away, wash the specimen in many changes of 70-percent alcohol before staining and mounting.

FRESH-WATER BRYOZOANS

Characteristics. Fresh-water bryozoans, limited in number, are more or less cosmopolitan in distribution. They possess no hard parts made of calcium carbonate but, rather, live in zooecia made of a chitonous or gelatinous material. Otherwise, the anatomy is relatively similar to that of marine bryozoans (Fig. 12–3A). Colonies may be encrusting, chain-like or creeping (Fig. 12–3B), or erect and moss-like in appearance. Taxonomically, the soft parts are important. It is essential, therefore, to narcotize specimens before preserving them or carefully noting taxonomic features. Badly contracted specimens, however, are of little value. The shape of the lophophore (U-shape or circular), the number of tentacles, and various internal structures must be seen. These animals form asexually reproductive structures (statoblasts) in response to unfavorable environmental conditions. Stato-

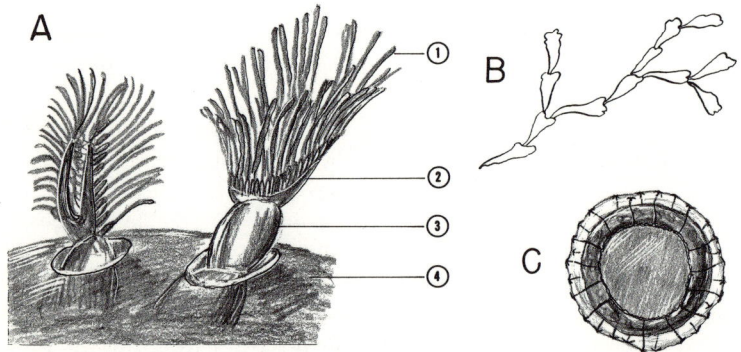

FIG. 12–3. Fresh-water bryozoans. A. Anatomy of a colony. B. An outline of a chain-like colony. C. A statoblast. (1) Tentacles, (2) lophophore, (3) trunk, (4) body of colony.

blasts (Fig. 12–3C) may be equipped to float (floatoblasts), to attach to other organisms by spines (spinoblasts), or to remain with the colony (sessoblasts). Other bryozoans that do not produce statoblasts may develop protective buds called hybernaculae. All of these structures are taxonomically important and, therefore anatomical notations and careful collections should be made.

Habitat and Collecting. Fresh-water bryozoans are common in lakes, ponds, and slow-moving streams and rivers. Rogick (in Ward and Whipple, 1959) notes that "Bryozoa live in various types of fresh waters of pH ranges 5.3 to 8.0 and up to elevations of 3950 meters. They occur at various depths from shore-line level to 214 meters." Look in areas of considerable plant growth on rocky substrates, but generally avoid sandy substrates. Pull sticks and branches from the water and look carefully for minute, gray-green gelatinous-like material which looks rather like water-soaked bread crumbs, or for fine chain-like branching growth closely applied to the substrate. The shaded or undersides of rocks or other debris are especially productive. Growing plants should also be harvested and examined on the undersides of leaves and stems. The biological dredge (Chapter 1) or the dredge net (Chapter 13) will be useful for collecting bottom debris in deeper waters. Examine your collections with a hand lens while in the field, as bryozoans may be confused with fresh-water sponges. Bryozoans like *Pectinatella* may grow to massive gelatinous colonies measuring 1 foot or more in diameter. Bryozoans are seasonal, late spring and midsummer being most satisfactory for collecting. Some species last quite some time into the fall, whereas others disintegrate as soon as the water temperature drops.

When collecting, cut out sections of branches or plant material containing colonies, chip off pieces of rock by means of a hammer and chisel (or take the entire rock), or carefully remove colonies by sliding a dull pocketknife blade between the substrate and the colony. If you are unsure of the appearance of bryozoans, consult some reliable text such as Ward and Whipple (1959) or Pennak (1953). After a little practice, recognition will become relatively easy. Keep specimens in cool water until ready for processing.

Narcotizing, Preserving, and Cleaning. In the evening after collecting, place the specimens in finger bowls of water. Menthol is highly recommended in the literature for bryozoans. Simply sprinkle a large pinch of menthol crystals on the water surface and let stand overnight. Test the narcotization by teasing individual tentacles with an insect pin (see "Marine Bryozoans"). If results are satisfactory, add enough formaldehyde to make a 4-percent solution. The author has gotten consistently good results with menthol and with benzocain used in the same manner. Chloretone, used likewise, is good but has not always been consistent; thus, specimens treated with it should be checked with a probe before preservation.

Because calcium carbonate is not present in fresh-water bryozoans, specimens may be preserved in either 70-percent alcohol or 5-percent formalin. For large gelatinous colonies of *Pectinatella* use 10-percent formalin. If the jar is relatively crowded with animals, change the formalin solution after 24 hours. Beyond this, follow the techniques used for marine bryozoans.

Slide Mounts. Narcotized and preserved specimens should be positioned, stained, and mounted as directed for marine bryozoans.

Intact-Colony Slides or Cultures. The author has long planned to obtain microscope slides with intact colonies of fresh-water Bryozoa adhering directly to them. Though this has not yet been tried, the procedure will be as follows: Collect statoblasts where they accumulate during the autumn in windrows around ponds and lakes, or directly from colonies in the autumn. Pennak (1953) gives many records of statoblast viability of 2 years or more when specimens are kept dry in the laboratory; he suggests they live even longer when kept wet in the refrigerator. In the spring roughen standard microscope slides with sandpaper and attach a few statoblasts to the center by means of glycerin jelly or some other material. Place these slides in finger bowls of previously boiled and cooled pond water; keep them in a cool room with reduced lighting. When the colonies begin to grow, exchange the water for fresh, unboiled pond water to provide nutrients. After a few days transfer the slides to a slide box, prepared as directed under " 'Trapping' Bryozoa," above, and submerge in a quiet pond or lake. Watch periodically and when the colonies are sufficiently large, narcotize, preserve, and store in liquid as directed above.

REFERENCES

Barnes, Robert D., 1963, *Invertebrate Zoology*, W. B. Saunders, Philadelphia.

Gray, Peter, 1952, *Handbook of Basic Microtechnique*, Blakiston, Philadelphia.

Edmondson, W. T., ed., 1959, *Ward and Whipple's Fresh-Water Biology*, Second Edition, John Wiley & Sons, N.Y.

Hyman, Libbie H., 1951 and 1959, *The Invertebrates*, Vols. III and V, McGraw-Hill, N.Y.

Osburn, R. C., 1950, Bryozoa of the Pacific Coast of America: Part 1, Cheilostomata-Anasca, *Allan Hancock Pacific Expeditions*, 14(1):1–269.

Osburn, R. C., 1952, Bryozoa of the Pacific Coast of America: Part 2, Cheilostomata-Ascophora, *Allan Hancock Pacific Expeditions*, 14(2):270–611.

Osburn, R. C., 1953, Bryozoa of the Pacific Coast of America: Part 3, Cyclostomata, Ctenostomata, Entoprocta and Addenda, *Allan Hancock Pacific Expeditions,* 14(3):612–841.

Pennak, Robert W., 1953, *Fresh-Water Invertebrates of the United States,* Ronald Press, N.Y.

(See Chapter 1 for other References.)

CHAPTER 13

The Insects

Perhaps the sheer number of known species of insects is sufficient to explain the uncommon interest man has always had in this group of animals. Estimates of the number of species of insects range on the conservative side from 750,000 to over 900,000, and to considerably over 1,000,000 species by workers who attempt to estimate those insects still undescribed. These estimates demonstrate very vividly that the insects dominate all other natural populations by sheer numbers. In addition, they are highly advanced organisms and can be used to demonstrate almost all of the important zoological principles. They are almost always available, abundant, and seemingly in little danger from overcollecting. For the pure delight of field study, for the study of structure or reproductive activity, for studies of color or ecological adaptations, the insects offer perhaps more opportunity and challenge than any other animal group. In addition, insects are man's chief competitors for food, lodging, field crops, and belongings; thus economic entomology also stirs the imagination.

Insects may be studied for many reasons, and in each field of interest special techniques are likely to be involved. Because of the tremendous variety in insect morphology, in habitat and behavioral patterns, and in modes of nutrition and reproduction, insect study requires more techniques and a greater variety of equipment than are employed for study of most groups of organisms. Many tools and techniques may be borrowed from other disciplines, but most of the equipment used in entomology has been developed exclusively for that study. Such equipment is easily made, however, and is inexpensive.

Building an insect collection includes the steps of field collecting, mounting or preserving, identifying, and storing and caring for specimens. This chapter will detail the techniques for building and using the equipment needed and will provide information for all levels of insect study.

COLLECTING METHODS AND EQUIPMENT

Basic Net-Making Techniques

A simple net consists of a wire ring or hoop, a cloth bag of some sort, and a suitable handle. Usually, instructions for the construction of a net begin with exact measurements for the hoop diameter, the length of the cloth bag, the size of the handle, and so on. Too frequently students give up attempts to build a net because of the rigid measurements imposed in matching the bag diameter to that of the hoop. Nets can be constructed very cheaply and very easily without resort to a complex formula of measurement. In brief, all one has to do is to sew the bag to any rough dimension that he chooses, then fit the wire hoop into the bag, and finally, attach the hoop onto some suitable handle.

Many kinds of bag fabrics are available locally or through biological supply houses (see the General Biological Supply House catalog for fabric illustrations). The fabrics most frequently used for entomological work are cotton, nylon or dacron marquisette (this is always obtainable in dry goods or catalog stores), nylon or cotton bobbinet, or Brussels netting of a cotton or nylon grade. The specifications and uses of these various materials will be discussed below under specific kinds of nets.

After the netting is selected, determine the approximate hoop diameter you want. This may be 8 or 10 inches for dip nets or ground nets, 12 to 15 inches for aerial nets, and so on. In cutting and sewing of the fabric, the circumference of the net bag should be approximately three times the diameter of the hoop; the depth should be from one to two times the hoop diameter, depending on the particular kind of net being made. For example, if the hoop diameter is to be approximately 10 inches the bag circumference

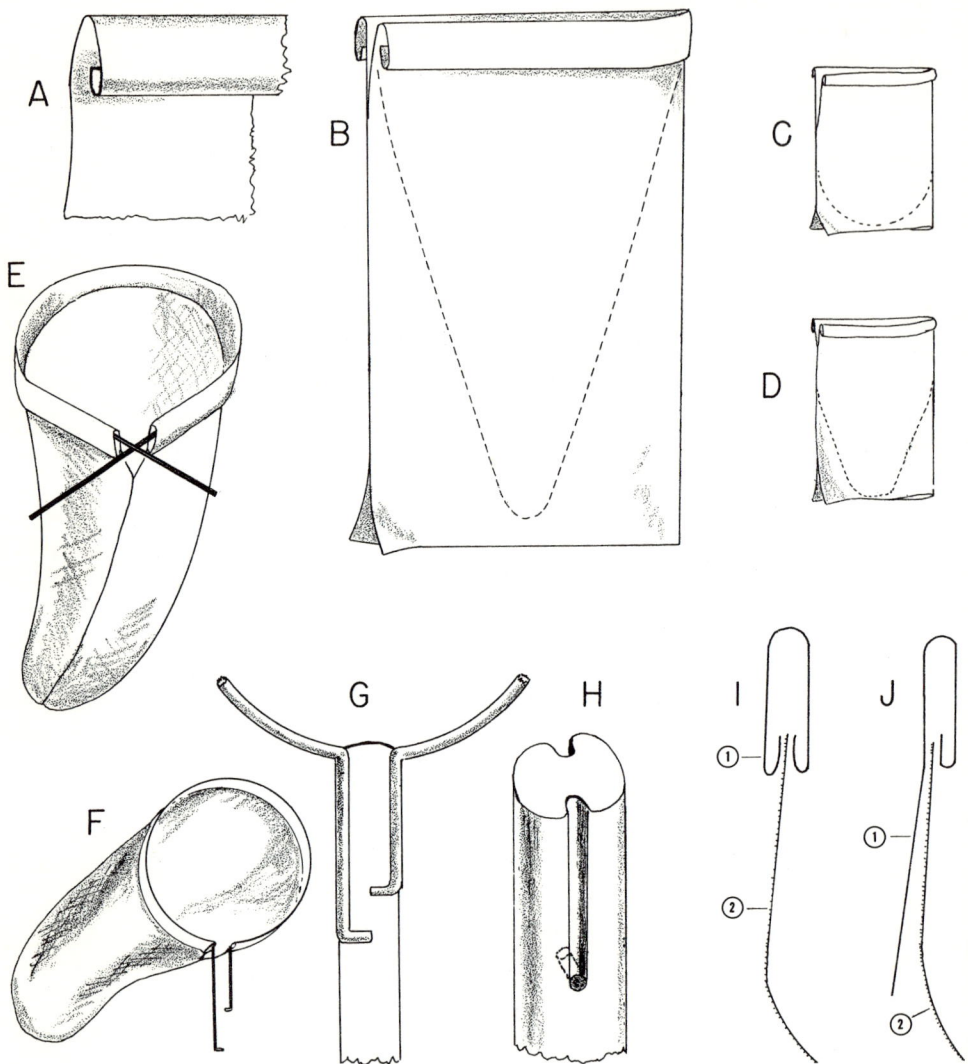

FIG. 13–1. Net-making technique. A. Folding the hem. B. Stitching the bag. C. Aerial net pattern. D. sweep-net pattern. E. Forming the hoop. F. Finished hoop. G–H. Attaching hoop to handle. I. Reinforced hem. J. Reinforced hem with apron. (1) Canvas reinforcing, (2) net material.

is 30 inches and the depth 15 to 20 inches. After the fabric has been cut, fold a 1-inch hem along the top edge of the bag, as illustrated in Fig 13–1A. This hem will receive the hoop and thus should be stitched at least twice on the sewing machine to ensure that it will hold. If the net is to be used around brush or in water it is a good idea to make the hem of a stout fabric such as muslin, white duck, or canvas. Figure 13–1I shows a simple hem made of canvas and attached to the netting fabric. Figure 13–1J shows a hem made of canvas with an additional apron attached along the outside of the net to protect the net from weeds or other obstructions. Once the top hem is complete, fold the cloth double and stitch the outline of the net bag as shown in Fig. 13–1B. Particular attention should be paid to stitching the bag around the bottom curvature, for frequently the meshes will tear loose at this point. Stitch the bag two or three times to ensure holding the fabric. Figure 13–1C shows the outline of the bag used for aerial nets; Fig. 13–1D shows the outline used for sweep nets. Trim away the surplus cloth around the bottom of the net bag, leaving at least ¼ inch of fabric to prevent raveling.

Once the net bag is completed, a stout wire is introduced into the hem at the top of the bag. Number 8- or 9-gauge galvanized wire, available in hardware stores, proves very suitable for most kinds of nets. Round off the end of the wire with a file to prevent snagging. Once the wire is part way into the hem it may be necessary to bend the wire slightly, but smoothly, so that it will round the curvature of the hem, as shown in Fig. 13–1E. Cut the wire off, leaving at least 6 inches projecting from each side of the hem. Next, clamp the wire into a vise and bend it, as shown in Fig. 13–1, F and G. Finally, bend the ends of the wires inward, at a 90° angle, to form "ears" which will project into the net handle about ⅜ inch.

Although any suitable stick (willow, etc.) may be used for a handle, a ⅝-inch hardwood dowel (¾ inch at the maximum) proves to be the most suitable for general aerial nets. The exact length and diameter of the handle will be determined by the intended use of the net, however. For example, a long, stout handle is required for a dip net. The handle must be as straight as possible in order to give the net the proper balance. If your net handle has a bend, place the bend at right angles to the plane of the hoop to prevent the net from twisting in your hand. Place the hoop alongside the handle and mark the location of the wire "ears." Next, drill holes for the ears and cut a partial or complete groove along the side of the handle from the hole to the tip (Fig. 13–1H). Finally, attach the hoop to the handle by placing the wires into the grooves and taping the hoop in place. Electricians' plastic tape (not rubber tape or friction tape) should be used to tape the hoop in place. This completes the net, except for adjustment of the handle to its proper length. If your net is an aerial net the handle should be fairly long, but not so long that balance and maneuverability are impaired. Therefore, test the net out of doors and, if it seems too heavy and clumsy, reduce the length of the handle somewhat. This will ensure a greater catch of insects in the field, in spite of the reduced length.

Specialized Nets

Aerial Net. Most winged insects that are active during the warm summer months must be captured with an aerial net. Build this net as described above, using a hoop diameter of from 12 to 15 inches. Any of the marquisette fabrics are very useful, though the nylon fabric will stand up under wet conditions somewhat better than the cotton. The nylon, however, is more difficult to stitch and often very difficult to dye if special colors are desired. For collecting butterflies the lightweight nylon fabrics are desirable in that they do the least amount of damage to the wings. The depth of the bag should always be from $1\frac{1}{2}$ to 2 times the diameter to ensure that when bag is closed it will still provide ample space for the specimen at the tip. The bottom of the bag should have a rounded curve rather than coming to a point, in order to prevent butterflies or other fragile insects from becoming lodged and damaged at the tip. The length of the net handle should depend on maneuverability rather than on the over-all reach attained. However, in the tropics and elsewhere extremely long handles are often necessary to reach butterflies and other insects that are high above the trail. Sections of aluminum tubing which can be joined together, or a collapsible handle may be made from the leg of an old camera tripod.

Sweep Net. The sweep net is designed to be used in heavy vegetation and, therefore, should be constructed of muslin or white duck or light canvas. The bag hoop and handle are constructed as described above, the bag being more pointed, however, as illustrated in Fig. 13–1D. The sweep net is fashioned with a fairly short, stout handle. The collector sweeps it back and forth through the grass or bushes as he walks (Fig. 13–2K). This net collects a tremendous number and variety of small insects in a very short time, and therefore the contents of the net should be checked frequently and the specimens sorted and placed in proper killing vials. Sweep netting is usually very good on the third or fourth warm sunny day in spring and will continue to be quite good during most of the summer period.

Umbrella, or Beat, Net. The umbrella, or beat, net has proved to be an invaluable collecting tool which is too little used by most collectors. It is

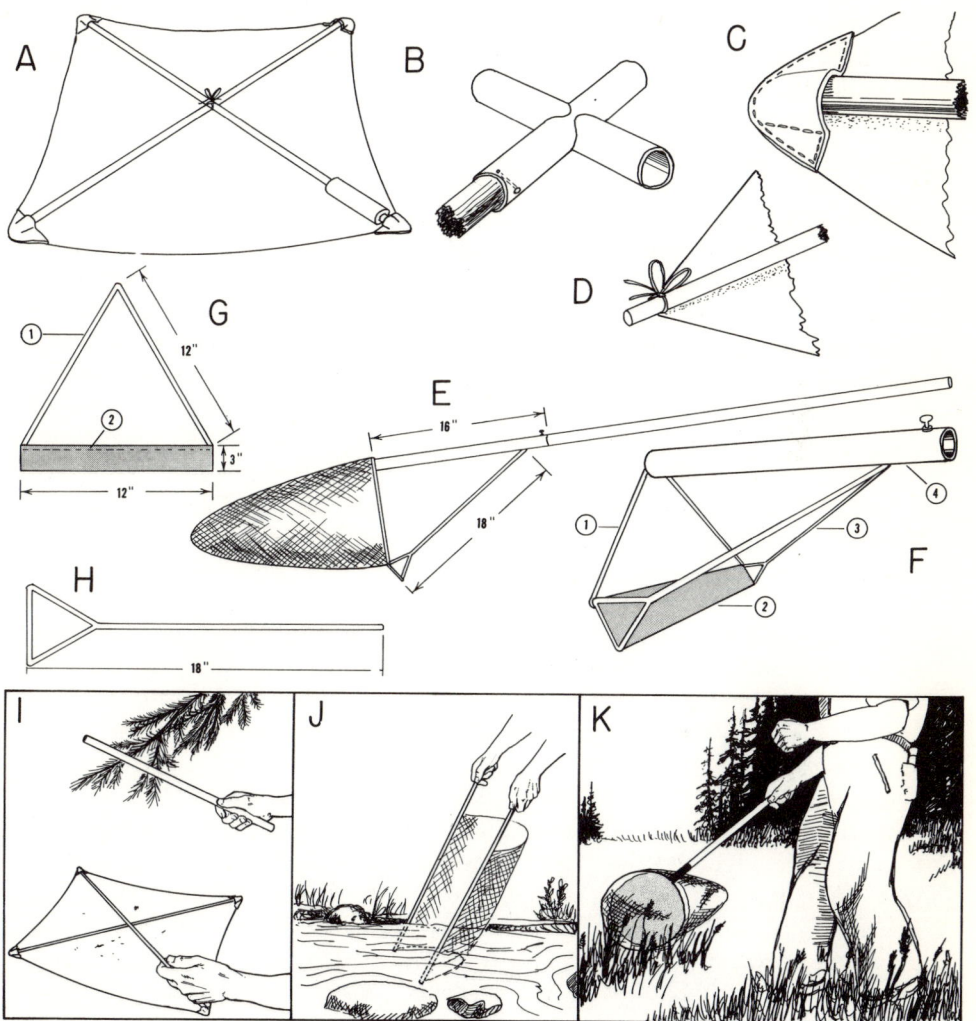

FIG. 13–2. Some additional nets. A. Finished beat net. B. Central cross. C–D. Corner construction. E. Completed dredge net. F. Frame. G. Triangular hoop. H. Side-brace. I. Using beat net. J. Using aquatic-dip net. K. Using sweep net. (1) Triangular hoop, (2) rake, (3) brace, (4) metal tube with thumbscrew.

especially useful for collecting the many species of insects that move up onto higher bushes and trees during the spring and summer months. To make the beat net (Fig. 13–2, A and I) some stout cloth such as muslin or very light canvas should be used. Cut a piece of cloth 42 by 42 inches and sew a double 2½-inch hem on all margins. When this is completed, a square 32 by 32 inches will be obtained. The corners may be reinforced with leather in order to receive the cross sticks (Fig. 13–2C) or provided with tie strings as shown in Fig. 13–2D. Next, cut two sticks or dowels long enough to go from corner to corner. These may be tied together in the center or bolted together with a ⅛-inch bolt and wing nut. One of these sticks may be wrapped with tape to provide a stouter grip and thus make the net more maneuverable. It is often desirable to reduce the length of the sticks so that they may be packed into knapsacks. This may be done by making a cross of ½-inch metal tubing (not pipe) of the variety used in plumbing water faucets. Cut the tubing about 2½ inches long and weld or braze these together in the center. Cut the crossbars to half the original length and fit them into the cross. If one of the sticks is wrapped to produce a handle, this should be permanently riveted into the metal cross to prevent it from rotating. With this kind of construction the net will be fairly light and compact and easily transported into high mountain country or other difficult terrain.

The beat net is used by holding it beneath protruding branches of trees and bushes and beating upon the bushes with some stout stick, as shown in Fig. 13–2I. The handle of a sweep net or aerial net makes an excellent beating stick and will thus provide an alternate net should a butterfly come into view.

Ground Net, or "Tiger" Net. Many insects sun themselves on the ground but will jump or fly very rapidly when approached by the collector. For example, tiger beetles, flies, grasshoppers, and other insects are often very difficult to obtain unless a special net is used. A ground net is constructed very much like the general net, described above, but has a hoop diameter of about 10 inches. The handle, however, is relatively long and may measure from 42 to 48 inches over all. The net bag may be constructed of marquisette or bobbinet, but should be provided with a stout canvas reinforcement around the hem to prevent chafing when the net is "slammed" down to the ground.

When approaching a specimen such as a tiger beetle the collector moves in very cautiously, with the net held directly up in the air. He then crouches down when within striking distance and swings the net at its full length so that the hoop goes over the insect. Highly active insects of this nature will often escape by crawling out from underneath the net, so it is necessary to get hold of the specimen as quickly as possible. If the net material is dyed dark green or blue or black the collector can usually look through the net and see the specimen, even in brilliant sunshine. When the collector becomes accustomed to the length of the handle he will soon learn to swat the net down with considerable speed and accuracy. Using this method he will obtain a much greater catch than would be possible by simply sweeping the area with an aerial or sweep net.

Aquatic-Dip Net. The aquatic-dip net is extremely useful for collecting many kinds of plant and animal life. A net fabric should be selected to provide the best mesh for the job desired. For example, the dip net is very difficult to move rapidly through the water, and speed is often very important. Therefore, the largest mesh obtainable should be used, provided it will still hold the organisms which are being captured. One-eighth inch or one-quarter minnow netting is therefore much more desirable for catching amphibians and reptiles than would be marquisette or bobbinet which are best for insect and general collecting. The depth of the net should be approximately equal to the diameter of the hoop. Slightly deeper nets may be used if extremely active and agile specimens such as fishes or amphibians are being captured. The hem at the top of the net bag should be reinforced with canvas and it is often a good idea to provide a canvas apron around the outside of the net bag, as illustrated in Fig. 13–1J. Usually number 9 galvanized wire is stout enough to make an adequate hoop, but for vigorous collecting especially among weeds and rocks a hoop of $\frac{1}{4}$-inch iron dowel is better than the lighter wire. The length and diameter of the handle again must be determined by the distance at which one must work with the net (water depth and other factors should be considered here) and the amount of pressure that will be placed upon the net as it is swung through the water. Otherwise, the general instructions given on page 205 should be followed.

Dredge Net. The dredge net has proved to be a very popular and valuable net for collectors working in streams, ponds, lakes, and even in marine habitats. Many designs are available on the market. The dredge net shown in Fig. 13–2E is a net designed by the author for most kinds of aquatic collecting. The special feature of any dredge net is that it has a lip which can be dragged along the bottom, in the manner of a larger biological dredge, to pick up both attached and motile plants and animals.

The net (Fig. 13–2) is constructed by making a triangular hoop of $\frac{1}{4}$-inch metal rod. This is welded to a 3-inch piece of bar iron to serve as the metal rake. Two side braces are constructed of $\frac{1}{4}$-inch metal rod, as shown in Fig. 13–2H, and then attached to the dredge, as shown in Fig. 13–2, E and

F. Cut a 16-inch piece of metal tubing or light pipe, of a diameter suitable to receive the net handle, and weld this to the top of the triangular hoop and to the two side braces. Provide the tubing with a wing bolt at one end so that the handle may be removed when the net is not in use. The net bag may be fabricated from metal screen (hardware cloth), plastic window screen, or any suitable cloth netting such as bobbinet or Brussels netting. The size of the mesh in the net will have to be determined by the specimens being sought. The more flexible cloth or plastic netting is much more desirable than the rigid screen netting, even though it may be necessary to protect the cloth netting in some way to prevent snagging. This net or any other net used in the marine environment must be carefully washed with fresh water and hung up to dry to prevent rusting or rotting.

Stream Net. Figure 13-2J shows a very simple but useful net that may be used in streams in place of the dredge net. This is constructed by attaching either plastic window screen or cloth netting to two handles. These handles may either be held as illustrated, while another collector dislodges organisms upstream and allows them to float into the net, or the handles may be sharpened so that they can be driven into the substratum, thus permitting one collector to operate the net.

Other Collecting Devices

Aspirators. The aspirator is an invaluable piece of collecting apparatus when small, fragile insects or other arthropods are being sought. Figure 13-3, A and B, shows two popular designs. Plastic (preferably) or glass vials or tubing should be used for the main body of the aspirator. In Fig. 13-3A a two-hole rubber stopper is provided with two metal, plastic, or glass tubes. One tube is used to receive the small insect; the other is provided with a piece of surgical tubing about 20 to 24 inches long with a piece of gauze tied or taped onto the opposite end of the tubing. The second model (Fig. 13-3B) is essentially the same, except that two one-hole rubber stoppers are used in a piece of plastic or glass tubing. The collector holds one end of the rubber tubing between his teeth and sucks in quickly when the collecting tube is placed near some small insect. The insect is automatically pulled back into the collecting chamber and is held there until transferred to a killing bottle. The gauze prevents these insects from being inhaled by the collector.

Dippers. In some forms of aquatic collecting the dipper proves quite handy for obtaining small specimens such as mosquito larvae and pupae. Anything from a kitchen strainer to a regular dipper provided with a piece of window screen on one side will suffice. The dipper is convenient in that specimens may be carefully worked into the collecting chamber without unduly disturbing the bottom debris in which they hide.

Berlese Funnel. The Berlese funnel provides the best method for removing very small insects or other animals from litter and debris, such as leaf litter, coarse humus, nesting material such as bird nests or rat nests which might harbor mites and fleas, and so on. There are dozens of modifications for the Berlese funnel, but all of them consist of the same general parts. A fairly large metal funnel (illustrated in Fig. 13–3D) is equipped with a collecting bottle at the bottom, a shelf for holding the debris inside, and a source of heat above. The principle upon which this funnel functions is that the heat slowly drives all of the motile organisms down through the debris, either by overheating or by drying out the litter, and thus forces the specimens to fall through the screen and into the collecting bottle at the

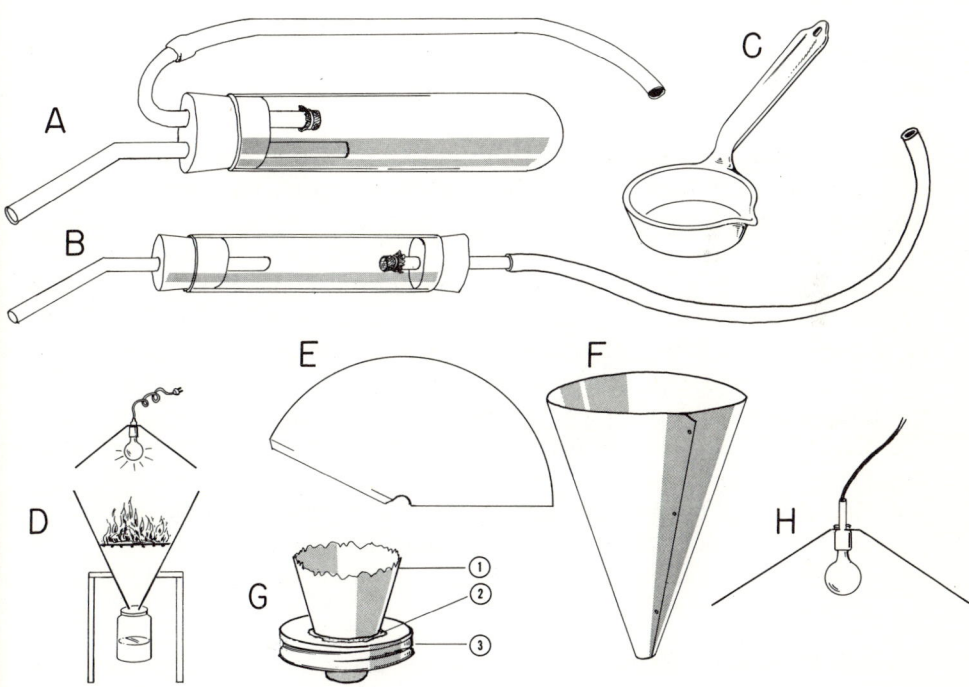

FIG. 13–3. Additional collecting equipment, A–B, Aspirators. C. Dipper. D. Berlese funnel in operation. E. Pattern for funnel. F. Completed funnel. G. Attachment of jar lid. H. Reflector assembly. (1) Base of funnel, (2) solder, (3) jar lid.

bottom. The writer has used this device to process hundreds of rat nests in order to obtain fleas, mites, and pseudoscorpions.

The diameter (14 to 20 inches) and the height (16 to 30 inches) of the funnel must be determined by the nature of the material to be processed and the approximate volume that must be handled at one time. Twenty-eight-gauge galvanized iron or other suitable metal should be cut into a half circle (Fig. 13–3E) and, after trimming, should be rolled and the sides either riveted or soldered together (Fig. 13–3F). Next, solder the outer metal ring from a fruit-jar lid to the lower side of the funnel (Fig. 13–3G). Construct some sort of stand for the funnel to prevent tipping. Finally, attach a light socket to a commercial light reflector or a shallow funnel (made in the same manner as described above). Suspend this above the funnel or rest it directly on top of the funnel. Experimentation may be necessary to select the proper size of light bulb. Factors to consider are the over-all dimensions of the funnel, the volume of litter being processed, and the amount of speed required in processing. Light bulbs that are too large are unsatisfactory in that specimens are killed before they can move down through the litter. It is better to take several extra hours, or even days, in processing than to attempt to rush the job. The collecting bottle may be dry or provided with some killing agent such as alcohol or FAA if the specimens are to be preserved.

The collector goes into the field and picks the kind of debris to be examined and brings it back to the laboratory or campsite for processing with the Berlese funnel. Plastic bags or paper bags are very suitable for picking up litter such as leaves, rat nests, and so on. Tightly seal the bag at the top by tying plastic bags with string or by rolling the tops of paper bags and stapling them shut. A slip of paper with all the necessary field data should be included in each bag; this should indicate the locality, date, the time of collecting, a description of the habitat in which the collection was made, the temperature and other environmental conditions, and any other data that may prove of value in research. Process the bags of litter as soon as possible to prevent specimens from dying inside. If there is some delay, be sure that the bags of litter are not subjected to overheating, dehydration, or freezing. Occasionally, as long as three days may be required in the total processing of material, and specimens will continue to fall into the collecting jar over that period of time.

Insect Trapping. Many kinds of insects are conveniently collected in large numbers with some form of insect trap. Any material that is of some peculiar significance to the insect may be used as bait. For example, many kinds of food materials or food plants used by the larvae of certain insects will lure the adults and thus may be used as bait. Chemicals which produce odors resembling those of mating insects may also be used. Lights, and even mechanically produced sounds (e.g., to attract mosquitoes), have been used very successfully. In order to devise a suitable trap and a bait the habits and requirements of the particular species must be fairly well studied by the student. For example, fresh or rotten meat or vegetable matter will lure many kinds of beetles, flies, bees and wasps, and other forms. Cow dung or other forms of dung will lure many flies, beetles, and (in the tropics) butterflies, as well as other insects. Pine pitch and fir pitch, xylene, turpentine, or other aromatic compounds have been painted on fence posts, log stumps, and so on, to attract beetles, flies, wasps, and many other insects that are attracted to freshly injured trees. Sugar solutions, molasses, stale beer, and even perfume scents have been used to attract moths and butterflies, as well as other insects. Rotting fruit or alcohol will attract a whole host of species. Students interested in constructing some of the many kinds of specific traps not described herein should see *Entomological Techniques* (Alvah Peterson, 1959).

Tin-Can Traps. When camping and collecting in a given locality entomologists frequently bury tin cans around their campsite in order to trap and collect beetles and other insects. Figure 13–4A shows a typical setting of a tin-can trap. A tin can is buried and baited with some suitable bait such as meat scrap, fat, or peanut butter. A board or piece of bark or even a flat stone should be placed over the top of the tin can to keep rodents or dogs out of the trap and yet provide enough room for the insect to enter. Traps of this nature may work for weeks at a time and can be visited daily by the collector.

Funnel Trap. The author has devised a funnel trap for catching carrion beetles which are attracted to dead and decaying animals. Figure 13–4B shows a cross-section through the trap. A large hole is drilled on either side of a wooden box and a screen funnel is tacked into this hole. One of the boards on the roof of the box is removable and serves as a door. The trap is partially buried, as illustrated, and is provided with 2 inches of dirt on the bottom. A suitable carrion bait such as a dead mouse is placed in the trap and a few drops of scent are also placed in the trap. The scent is prepared by rotting some meat or fish in an airtight jar which is half filled with water. After this has rotted for about two weeks it makes an excellent scent to be used if the bait provided is too fresh. This trap is much more elaborate than is needed to collect carrion beetles locally, for by placing a bait under a board or a stone one can obtain large numbers of carrion beetles. The

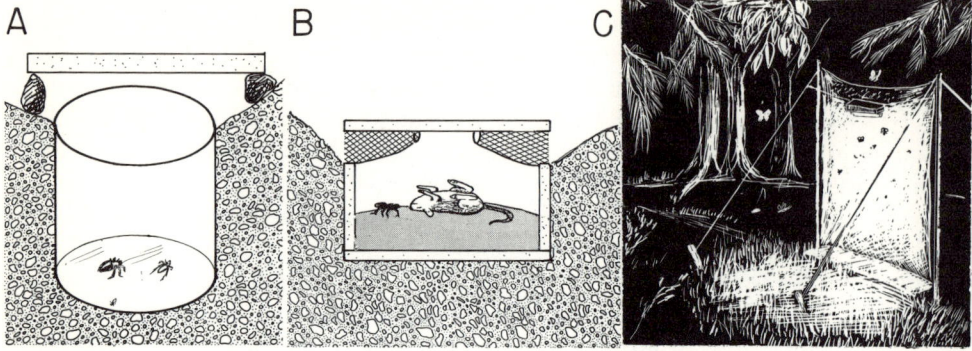

FIG. 13–4. Insect traps. A. Tin-can trap. B. Funnel trap with bait. C. Night lighting for moths.

trap is designed for long-range field trips where large numbers of traps are set out along the side of the road and retrieved three or four days later.

Light Traps. Many kinds of light traps can be devised which will attract nocturnal insects. Some of these are provided with holding chambers and even with killing chambers so that they are essentially self-running. The moth sheet (Fig. 13–4C) is perhaps the most interesting of the light traps, but it must be attended by the collector. In principle a moth trap can be made by using any white surface (e.g., the side of a building) and a suitable light. However, in the field some sturdy fabric such as muslin, white duck, or even strong bed sheeting will do very nicely. The dimensions of the sheet should be approximately 6 by 8 feet. The sheet is suspended between two poles 6 feet high and 6 feet apart. This allows about 2 feet of the sheet to be stretched out as a ground flap.

A suitable light source is required to attract the insects. Pressure gas lanterns have often been used with success. These may be improved somewhat by staining the lens of the lantern to provide a blue or almost "black light" quality. By far the best light source is the 15-watt black-light neon tube. The light fixture and black-light tube can be obtained for only a few dollars and may be run from either alternating current or direct current. Generators, automobile batteries, or long extension cords will provide adequate power. The light should be suspended slightly in front of the upper margin of the sheet, thus illuminating the sheet and ground flap.

Collectors using light traps in the late spring and summer months will notice that the species composition differs from hour to hour during the entire night. Unless the temperature becomes unfavorable, one may collect the whole night through and get new and different species every few hours. The location of the moth sheet makes a lot of difference, for the species will vary tremendously according to whether you are collecting near water, in woods, out on an open meadow, or on a desert hillside. The collector will sometimes need five or six cyanide bottles so that no one bottle becomes too full of living specimens. When a reasonable number of moths or other insects have been placed in a given bottle, the bottle should be set aside until the specimens are dead. Specimens may be transferred to holding boxes, prior to field pinning or papering, thus freeing the killing bottles for new specimens. Do not mix beetles or other robust insects with moths, but instead kill them separately. On field trips the chlorocresol method of field storage is an extremely time-saving technique.

General Collecting Sites

Insects are available almost all year long, except when winter conditions are extremely severe. Large numbers of insects overwinter under stones, logs, leaf litter, or down in the ground, and may be found in the early spring in any likely situation which will have afforded winter protection. In the late spring and early summer many species metamorphose from juvenile stages and emerge as adults for the first time. In the summer, breaking old logs and looking under boards is less profitable than during the early spring or late fall, since only nocturnal specimens will be found. Toward the end of summer another new wave of adult insects metamorphoses from the larval stage and appears in time for fall hibernation. When the collector learns the general life histories of insects, along with their daily behavioral patterns, he can usually predict where the best collecting will take place.

Daily weather conditions have a lot to do with the availability of insects. Cold weather will inhibit activity and will prevent insects from leaving their hiding places or will keep them inactive once they have left their hiding places. Rain will usually drive insects down out of the vegetation and cause them to hide. The first warm, sunny day after a cool spell will produce a flurry of flying insects. However, several days may be required before those specimens in well-protected hiding places are warmed and activated.

The most productive collecting area is the ecotone. The ecotone is defined as that distinct area where two habitats overlap. For example, collecting is always better where the habitats of a stream and a meadow overlap, or where the meadow and the forest habitats overlap, since species from both habitats will be present. The seasonal appearance of many species is geared to the growth and development of plant populations, and thus any given habitat will have a year-long succession of plant and insect populations.

When collecting in the woods, turn stones and logs or remove loose bark from trees to find overwintering and nocturnal specimens. Tree trunks should be carefully examined and sweep or beat nets should be applied to bushes and low trees. Any sunny glade within the forest should be given particular attention, as this represents an ecotone. Mushrooms or other fungi will harbor both adult and juvenile insects. On warm days sap flowing from injured trees will attract a continuous succession of beetles and other insects.

In meadows the scarcity of hiding places makes turning any object on the ground worthwhile for the collector. The sweep net should be applied to low grass and bushes. If flowers are available, the aerial net will be of extreme value. Fungus, dead animals, or cow dung should always be investigated for the dozens of probable species that will be found there.

In streams and ponds collectors will find a continuous parade of insects, changing with the seasons. The dip net and dredge net will prove valuable in both day and night collecting. Bottom debris should be collected and examined on the shore for small insects; larger insects may be dipped directly from the water. Many insects associated with water, such as dragonflies and damsel flies, will be available either in the juvenile stages in the water or as adults flying over the water.

In desert and semidesert situations where sagebrush and sand dunes prevail, both day and night collecting prove very profitable in season. One should use the sweep net or beat net on the standing brushy vegetation. Pay particular attention to any group of flowering plants, since these will be short-lived in the desert and will attract many insects. The moth sheet is an indispensable tool in night work for capturing a wide variety of insects. In the desert the collector will notice that populations shift, so that the daytime species often disappear in the late afternoon and are replaced by a crepuscular population (animals living in the gray hours), and that these in turn are replaced by the nocturnal segment which may appear in several distinct groups during the night.

Collecting Immature or Parasitic Insects

Immature Insects. Immature insects will be encountered continuously in the process of general collecting. Species with incomplete metamorphosis which resemble the adult may often be preserved in the same manner as the adult. However, the collector who is interested in immature forms should provide himself with a number of vials and small boxes or plastic bags so that larval forms can be safely transported to the laboratory and there preserved or placed in rearing chambers. Special techniques for the liquid preservation or the rearing of immature insects will be discussed below. Students interested in immature insects should see Chu (1949) and Peterson (1959).

Parasitic Insects. Although some parasitic insects are free-living (mosquitoes, flies, true bugs, and so on), chewing lice, biting lice, and fleas remain with their hosts, and thus will necessitate the examination of mammals or birds. Methods of obtaining birds and mammals are discussed in Chapters 21 and 22.

KILLING, FIELD STORAGE, AND PRESERVATION

Construction and Use of Poison Bottles

General Remarks. Unfortunately, the most useful killing bottles are also the most poisonous and, hence, the most dangerous. Anyone working in this area must recognize that the cyanide poisons are deadly. It cannot be emphasized too strongly that parents or teachers take considerable risk in permitting young collectors to use such devices unless they are thoroughly reliable and trustworthy individuals.

Potassium Cyanide Bottle. To make this killing bottle, select a strong glass bottle with a tight-fitting, one-piece lid equipped with a wax sealer or a strong, heavy-walled glass vial equipped with a rubber stopper. Unless large insects are being collected the bottle should be 5 or 6 inches tall and not much more than 2 inches in diameter, so that it will readily fit into the pocket. Put from $\frac{1}{4}$ to $\frac{1}{2}$ inch of potassium cyanide in the bottom of the bottle and add $\frac{1}{2}$ inch of sawdust over this (Fig. 13–5, A or B). After this is tamped in place, pour in about $\frac{3}{8}$ inch of plaster of Paris which has been freshly mixed. If the bottle is tapped down on a table top the plaster of Paris will flow evenly to all parts of the bottle. Put the bottle in a hood or some other place where it can safely dry out. The sawdust will absorb some of the moisture from the plaster and in turn wet the cyanide. After the lid has been placed on the bottle, a day or two may be required before the cyanide permeates the plaster and "charges" the bottle. Crumpled paper or some dry leaves should be placed in the bottle for insects to hide in, and the word "poison" must be boldly printed across the outside of the bottle.

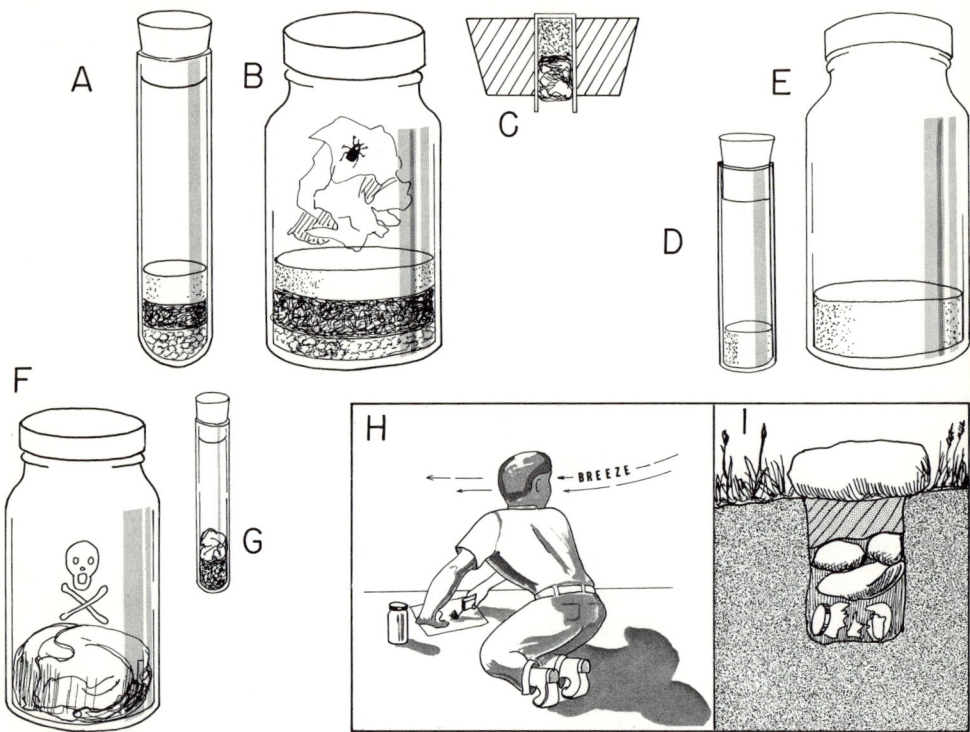

FIG. 13–5. Killing bottles. A–B. Potassium cyanide method. C. Cork stopper with cyanide vial. D–E. Ethyl acetate method. F–G. Cyan-o-gas method. H. Constructing a Cyan-o-gas bottle. I. Disposal of broken cyanide bottle.

The portion of the bottle from the plaster down to the bottom should be taped with adhesive tape to prevent the loss of cyanide in the event that the bottle is broken. Figure 13–5I shows how to dispose of a broken bottle in the field by digging a hole and placing rocks and dirt over the top of the bottle. One general drawback to the potassium cyanide bottle is that the bottle often sweats and becomes wet enough inside to damage and discolor some insects. Changes in temperature or humidity may start this sweating. Thus, the bottle must be moist enough to activate the cyanide and dry enough to prevent excessive sweating. The use of paper toweling in the bottle will help greatly.

Liquid Poisons. Liquid poisons are generally safer than cyanide when used in killing bottles, but not as effective in killing insects. Ethyl acetate, carbon tetrachloride, or ether may be used. Select a stout bottle as described above and pour into this ¾ inch of plaster of Paris (Fig. 13–5, D and E). After this has dried, saturate the plaster with the liquid poison. Crumpled paper or dried leaves should be added as a baffle for the insects. This will prevent them from fighting while they are in the process of dying. Specimens may be left for many days in a poison bottle made with ethyl acetate, provided the bottle is tightly sealed. If the ethyl acetate evaporates the specimens will become dried, and they must be relaxed before pinning. The liquid poison bottle is not too suitable for butterflies and moths as the wings become wet and thus are ruined.

Cyan-o-gas. Perhaps one of the best killing agents, for butterflies and moths in particular, and other insects in general, is Cyan-o-gas, a commercial ant poison available at most hardware or nursery stores. This comes in ¼-pound and 1-pound tins and is absolutely dry. Strong bottles or vials with tight sealing lids should be selected for this killing bottle as described above. A bottle of this type (Fig. 13–5F) should be made under a hood or out of doors where a breeze will dissipate the fumes (Fig. 13–5H). Spread three large paper table napkins on the ground, one on top of another. Pour approximately one rounded teaspoon of Cyan-o-gas on top of the first napkin and loosely roll this up in such a fashion that the chemical will not leak out. The first towel is rolled into the next and this into the third paper napkin so that a round ball of loosely crushed paper is formed. Force the finished ball, which is almost too large to go into the top of the bottle, down into the bottle and crush it into place with a blunt stick. This will hold the paper and Cyan-o-gas in place. Face the breeze while the bottle is being made so that the fumes will be blown away from you. When the lid is placed back on the bottle the Cyan-o-gas is ready to be used. A minor drawback in the Cyan-o-gas bottle is that specimens will be dehydrated somewhat more quickly than in other kinds of killing bottles and should not remain inside for long periods of time. Also, any cyanide preparation may affect colors of some bees and wasps if they are left exposed to it too long.

Small Cyan-o-gas vials may be made by simply pouring some Cyan-o-gas into the bottom of a vial and tamping enough paper toweling on top to keep the powder from escaping. A third technique that may be used is to fit a large cork (Fig. 13–5C) with a vial. The vial is filled with Cyan-o-gas and a paper toweling plug. The cork then is put back in the killing bottle which is then ready to use.

The Use of Poison Bottles. It is unwise to mix robust insects, such as large beetles, with delicate insects, as the latter will almost always be partially destroyed. Butterflies and moths usually contribute so many scales to other insect specimens that mutual damage is done by mixing them together. Make it a habit to keep the killing bottle dry inside by changing the toweling frequently. Learn automatically to hold your breath while the top of the cyanide bottle is open.

Most insects can be tossed directly into the bottle while the lid is briefly removed. Bees, wasps, butterflies, and similar insects cannot be handled freely, and thus a special technique is needed for putting them in the bottle. In handling such insects, loosen the lid of the poison bottle and place the bottle on the ground. Place the net over the poison bottle and hold the tip up while the insect instinctively goes to the top of the bag. Take hold of the lid through the net and remove it while the specimen is worked into the bottle. Then replace the lid and screw it tight.

Large butterflies are most conveniently demobilized by pinching the thorax through the butterfly net. This is done by allowing the wings to fold up over the thorax in a normal position and then grasping the butterfly through the net and firmly pinching the thorax without crushing. Then the butterfly is removed from the net and dropped into the killing bottle without wing damage.

Collecting Juveniles and Adults in Liquids

Many adult species may be collected into 70- to 80-percent isopropyl (rubbing alcohol) or ethyl alcohol when it is inconvenient to pin or layer in the field. Insects with furry or scaly bodies or with certain kinds of colors may be damaged in liquid preservatives. However, a wide variety of specimens can be placed in alcohol and pinned later with no damage whatever to the specimen. Juvenile specimens such as larvae and pupae are collected and maintained in liquids. If these are heavily chitinized and are to be studied externally they may be killed and preserved in alcohol. However, with soft-bodied specimens particular care should be taken in killing.

Specimens should be killed so that important taxonomic characteristics can be studied. Among these characteristics are various head structures, mouth parts, antennae, appendages, hooks, spines, bristles, hairs, spinnerets, and the like. Color may be of importance and thus color notes should be taken (see Index).

A number of methods and fluids for killing which will either harden or expand the specimen for study have been devised. If internal tissues or structures are to be studied, FAA or Bouin's fixative may be used (see Appendix C). Soft-bodied larvae such as mosquito wigglers, butterfly caterpillars, and so on, may be killed by dropping them in near-boiling water for a few moments and then transferring them to a preservative, usually 80-percent ethyl alcohol. Peterson (*Larvae of Insects*, 1959) recommends several killing solutions.

The Peterson XA solution will distend and harden soft-bodied butterfly, moth, beetle, and other larvae. Specimens should be removed after a day or so and transferred to alcohol. The XA solution consists of one part xylene and one part 95-percent ethyl alcohol.

Peterson's XAAD solution is used for beetle or butterfly larvae; it consists of four parts xylene, six parts isopropyl alcohol, five parts glacial acetic acid, and four parts dioxane. (*Note,* dioxane is poisonous.)

Peterson suggests that the KAAD solution is perhaps the best for killing most kinds of larvae. He cautions that bright colors may be lost; but nevertheless finds that this solution will distend and harden butterfly, fly, beetle, wasp, bee, and other larvae except for those with thick exoskeletons. This solution is made of one part kerosene, seven to nine parts 95-percent ethyl alcohol, one part glacial acetic acid, and one part of dioxane. Students interested in working with immature insects should see Peterson (1959*b*) and Chu (1949).

Field Storage of Insects

On longer field trips it is often necessary to store insect specimens temporarily so that they will not be damaged as a result of drying or mishandling before they can be properly prepared in the museum. Many different techniques must be used, depending on the kind of material that has been collected and the conditions under which the collecting has occurred. These will be discussed below.

Field Pinning and Pinning Boxes. Unless the collector is going to be in the field for many weeks, so that the number of specimens taken is prohibitive, one of the best methods for dealing with specimens is to pin them into temporary field boxes, using mounting methods described below. Roaches, grasshoppers, earwigs, larger true bugs, flies, beetles, bees and wasps, moths, etc., are often handled in this manner. Serviceable field-pinning boxes may be made out of stationery boxes or similar cardboard boxes with tight-fitting lids, or of old cigar boxes. Fine-grain fiber wallboard (CeloteX) or plastic styrofoam makes excellent pinning surfaces. These should be cut to the inside dimensions of the box and securely fastened, either by gluing or by pushing pins in through the outside of the box. If styrofoam is used as a pinning base, paradichlorobenzene (P.D.B.) should be avoided as a fumigant, as it will destroy the plastic. In its place moth crystals (naphthalene flakes) can safely be substituted.

After the insects are pinned they may be arranged in rows across the top of the box starting from left to right. A complete set of field data should be placed on the last pin of a series from a particular collecting station. This then will separate one series from the next so long as the insects are placed in rows. The alternative, of course, would be to put complete data or field numbers on each insect pin.

Field Storage in Liquids. Any of the liquids described above may be used for temporary or, in most cases, permanent storage. Ethyl alcohol in concentrations of 75 to 80 percent is by far the best. One must be careful not to dilute the alcohol too much by placing too many insects into it. To avoid this danger the collector may change the alcohol on a full jar of specimens after a day or two has elapsed. Insects that are very hairy or covered with scales, such as bees and butterflies, should not be preserved in liquids. Alcohol and other liquids may affect insect colors drastically, and thus insects which have pale yellow or light green colors should not be placed in alcohol. Most insects with metallic colors or with black or dark brown colors will not be affected by alcohol.

Ethyl acetate serves as an excellent temporary storage medium for many insects, but insects whose color is readily affected by alcohol should not be stored in ethyl acetate. To store specimens such as beetles, equip vials with a small amount of paper toweling tamped firmly into the bottom. This, in turn, is wetted by ethyl acetate, the insects and field data are placed in the vial, and the vial is tightly capped or corked. Once sealed, the vials may be taped with plastic tape around the cap and stored in larger jars. This technique proves very useful for storing as long as six months or more, and will render specimens in excellent condition for genitalia dissections or routine pinning. The only danger with this technique is that the ethyl acetate may evaporate, leaving dried specimens, or it may pull grease from the bodies of insects which will necessitate a washing in ethyl acetate before pinning can be completed.

Papering. Large winged insects such as dragonflies, damsel flies, butterflies, and the like may be stored in paper envelopes. These are folded from rectangular pieces of paper of suitable size, as illustrated in Fig. 13–6, A through E. One of the best methods for making envelopes is to purchase several sizes of small note pads of a good quality of white paper and to fold the envelopes from the pads as they are needed. The data should be recorded on each envelope before the specimen is placed inside. The data should include the name of the species or kind of insect if this is known. Bend the tips of the envelopes over (Fig. 13–6D) to ensure that the specimen will not slip out. Papered insects are then placed in any convenient

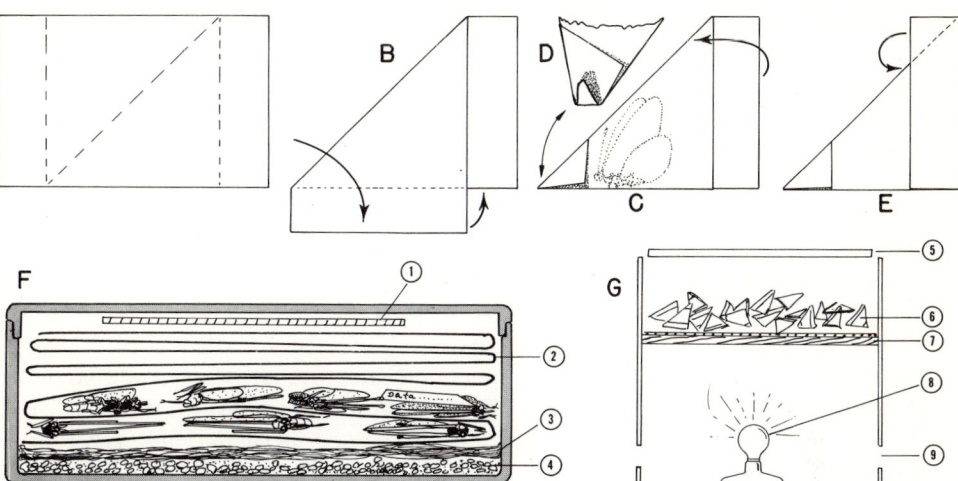

FIG. 13–6. Field storage techniques. A–E. Folding butterfly envelope. F. Sandwich box with chlorocresol and specimens. G. Field drying box. (1) Blotter with water, (2) folded toilet paper, (3) cello-cotton or tissue, (4) chlorocresol, (5) vent, (6) papered specimen, (7) shelf, (8) heat source, (9) air intake.

box, along with a few moth crystals, and taken to the laboratory for pinning at any later date.

Layering. One very common field method for storing large quantities of small insects is layering. Select a box with a tight-fitting lid for this technique. After a large number of specimens have been caught they are placed upon a layer of facial tissue or cello-cotton in the bottom of the box. Regular cotton or any material on which the insects will snag their legs should be avoided. Write the field data on a slip of paper and place this with each layer of insects. Additional facial tissue or cello-cotton is added and subsequent layers of insects and tissues may be placed in this box until the box is full. A few moth crystals should be added before the box is taped closed. In the laboratory the now-dried insects will have to be carefully removed with a forceps and placed in a humidity chamber for relaxing. It is possible to relax the entire contents of the box by placing it in a relaxing jar, but the problem of molding and improper relaxation may develop.

Chlorocresol Method. A new method which removes the necessity of field pinning or papering all kinds of insects has been described by Tindale

(1962) as the chlorocresol method. This method has tremendous advantages over field pinning and papering, not only in reducing the large amount of meticulous work necessary but in the fact that the specimens are already relaxed when they are taken out of the collecting boxes 6 months or more later. For this technique plastic sandwich boxes (Fig. 13–6F) are used, although other containers which are waterproof and can be sealed will be satisfactory. About a teaspoonful of chlorocresol crystals is placed in the bottom of the sandwich box. This, in turn, is covered by a tight-fitting layer of cello-cotton or other material which will keep the crystals from leaving the bottom of the box. Toilet tissue is then folded back and forth into the box to form an accordion, between the layers of which the insects are ultimately placed. A strip of blotter paper ½ inch by 3 inches is placed in the top of the box. In the field the specimens are placed between the layers of toilet tissue, beginning at the bottom and working toward the top. A sheet containing the field data should be placed in with each layer of specimens or at the top of the specimens from one particular locality. Between collections the boxes are sealed with masking tape to prevent loss of moisture. When the box is filled the blotter is saturated with water, placed inside, and the lid tightly taped in place. Upon returning from the field it is desirable to store the chlorocresol boxes in the refrigerator (not the freezer) until the insects are to be pinned. Upon opening the box you will find the specimens perfectly relaxed and usually needing no moisturizing before being pinned. Do not take too many specimens out of the container at one time, however, as some may dry out prior to pinning.

The chlorocresol method has a few drawbacks which are readily overcome. This compound will melt and vaporize at 156° F. and may subsequently recrystallize upon the stored specimens. Therefore, extreme heat should be avoided after the specimens are in the storage boxes. The boxes should be stored flat, not on end, and they should always be sealed tightly with masking tape. Specimens may be shipped in these sandwich boxes by placing them in a larger box with a small amount of packing.

Chlorocresol functions essentially as an insecticide and fungicide and prevents molding and decay while the specimens are in storage. This product will become generally available on the market, although currently it is being imported from British pharmacy houses. It is available through Bio Metal Associates (equipment for entomology), Box 61, Santa Monica, California.

Field Drying in the Tropics. In any damp, humid climate it is often essential to dry specimens so that molding will not occur, unless the chlorocresol method is employed. Butterflies that are papered or insects that are field pinned or layered must be sufficiently dry to overcome destruction by moisture. Figure 13–6G shows a cross-section through a simple drying box which may be used in the drying of both plant and animal material. This may be constructed of a cardboard box which is provided with a screen shelf, air holes cut at the base to permit a steady flow of air, and a lid placed over the top to prevent rapid loss of heat. A heating unit such as an electric light bulb will send a draft of warm dry air up through the pinned or papered specimens until they are dehydrated. They are then sealed into containers with some moth crystals. If a gasoline lantern is used as a source of heat, the box must be set high enough so that the heat will not injure the specimens nor burn the box. A sheet may be draped around the box and down around the lantern to funnel the heat up through the specimens.

Field Data. Specimens are of little value without adequate field data. One must recognize the danger of losing field data or failing to remember data that seemed so obvious at the time of collection. A field notebook should be kept in addition to the data that are placed with the specimens to provide a second source of information. Many workers simply put a field number with stored specimens, but they run the risk of losing the original data. The best method is to place a brief copy of the collecting data along with each group of insects preserved. This should include the locality and data along with the station number (as discussed in Chapter 1, p. 22). The field notebook should include in addition to the date and exact locality of collection, information concerning the vegetation associated with the specimens at the collection site, temperature and other weather conditions, and a description of the microhabitat. Using two sources of data in the field will ensure the value of the specimens that are brought back to the laboratory.

MOUNTING
INSECT SPECIMENS

The Construction of Mounting Equipment

In the preparation of insect specimens a few simple pieces of equipment are essential and others are greatly beneficial. Almost every common piece of equipment can be purchased through biological supply houses, but they can also be made at home at a fraction of the cost. Some of the more important pieces of equipment are discussed below.

Butterfly Boards. The butterfly boards that are the easiest and cheapest to build are made from fiberboard, wallboard, or CeloteX. A very soft fine-textured board should be selected, rather than the coarse fibrous variety. Some companies make fine-grained wallboard with a glazed white paper surface which is excellent for pinning boards. Many sizes of pinning boards will ultimately be needed. Some of the common sizes are as follows: 3-inch width with a $1/16$-inch groove, 3-inch width with a $1/8$-inch groove, 4-inch width with a $1/4$-inch groove, 4-inch width with a $3/8$-inch groove, $6 5/8$-inch width with a $5/8$-inch groove. The length of the board is optional, but 16 inches proves to be very economical.

With a very fine-tooth blade (preferably on a table saw) the wallboard should be first cut to the length of the pinning board and then ripped to the desired width. Figure 13–7, A, B, and C, shows the more simple plans of construction. All of the pieces are ripped to the proper width and a groove of the appropriate size is ripped down the middle (Fig. 13–7A). A series of saw cuts rather than a dado blade should be used for making the groove. Two identical pieces are then glued and pinned together to form a single finished board. The advantage of this style is that the butterfly pin is thrust through two $1/4$-inch thicknesses of wallboard rather than a solid $1/2$-inch piece, as in Figure 13–7B.

For the second design (Fig. 13–7B), rip a baseboard to the proper width and two pinning boards narrow enough to provide a groove when the three pieces are assembled. Wood glue and pins should also be used for assembling this board.

Many workers prefer to have tapered pinning boards so that the butterfly wings will be slightly elevated. Even with proper drying butterfly wings sometimes snap down a little when the specimen is removed from the pinning board, or the wings may droop if the specimens are stored under moist conditions. Figure 13–7D shows a simple tapered pinning board made of strips of CeloteX that have been elevated with small wooden spacers (balsa wood or other). In the construction of this pinning board, adequate amounts of glue should be placed where wood surfaces contact, and pins should be used to secure the parts. The pins should not be driven from the pinning surface, but rather from the ends or sides of the pinning board.

Many workers prefer a medium-hard pinning surface made of some soft, fine-grained wood. Figure 13–7, E and F, shows the end and side views of a tapered pinning board made with wood. This model is constructed by first cutting a baseboard of the appropriate length and width and then nailing to it two end blocks. The end blocks should be thick enough so that the finished pinning board measures 1 inch from the pinning surface down to the top of the baseboard. The boards making up the pinning surface

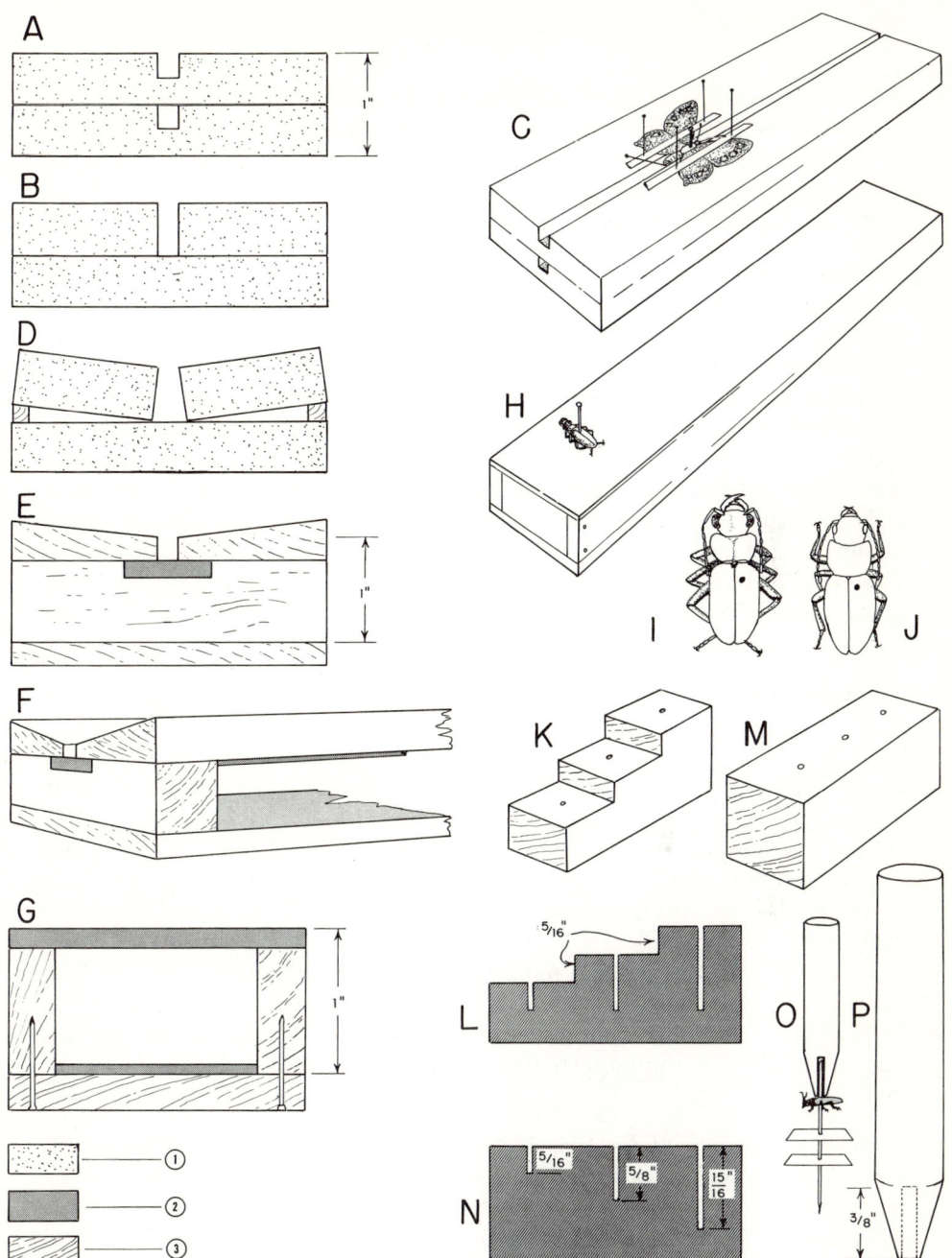

FIG. 13–7. Pinning boards and equipment. A–D. Butterfly boards, CeloteX construction. E–F. Conventional butterfly board. G–J. Insect mounting boards with specimens. K–L. Three-step pinning block. M–N. Three-hole pinning block. O–P. Pinning dowel. (1) CeloteX, (2) balsa wood, (3) soft wood.

should be of a fine-grained soft wood and may be either tapered or flat. A $\frac{1}{4}$-inch thick piece of balsa wood should be fitted into a receiving groove in the end blocks and the pinning surfaces nailed or glued in place. In mounting butterflies, the pins are thrust through the balsa wood down to the top of the baseboard.

Insect Boards. Insects should be mounted as neatly as possible; this often requires propping up the legs or body while the specimen dries (as discussed below) Fig. 13–7, I and J. To make this task easier collectors who are specializing in certain families of insects often use insect boards for the pinning and leg arrangement of specimens. Figure 13–7, G and H, shows the construction of this simple board. It consists of a baseboard, two side boards, and two end boards. These may be made of plywood or any suitable soft wood which is cut to the proper size, and assembled by means of nails and glue. A $\frac{1}{16}$-inch piece of balsa wood is placed in the bottom of the pinning board and a $\frac{1}{8}$-inch piece of balsa wood is placed over the top as a pinning surface. The finished board should measure 1 inch from the top of the pinning surface down to the baseboard, as shown in Fig. 13–7G. The use of this board is discussed below.

Pinning Blocks. Some form of pinning block is absolutely essential for placing the insect and its labels at the proper height on the pin. The pinning block is made of any fine-grained softwood in the three-step pattern (Fig. 13–7K) or level pattern (Fig. 13–7M). The three-step pinning block is cut so that each step has a surface 1 inch wide and a rise of $\frac{5}{16}$ inch. After cutting and sanding the block put a $\frac{1}{16}$-inch metal drill in a drill press and set the gauge to drill $\frac{15}{16}$ inch from the top of the upper step. All three holes are drilled with this same setting, thus giving the required depth for each step (Fig. 13–7L).

The level pinning block is more easily constructed and is made of fine-grained soft wood. Three $\frac{1}{16}$-inch holes are drilled to the depths of $\frac{5}{16}$ inch, $\frac{5}{8}$ inch, and $\frac{15}{16}$ inch; a hand drill or drill press may be used for this operation. The holes are appropriately numbered 1, 2, and 3, to indicate the depth.

Pinning Dowel. At least $\frac{3}{8}$ inch of pin must protrude above the specimen to ensure safe handling. When thick-bodied insects are pinned on the pinning blocks they often stick up too high on the pin and do not leave enough room for handling. The pinning dowel is made of $\frac{3}{8}$-inch hardwood doweling; a $\frac{1}{16}$-inch hole is drilled $\frac{3}{8}$ inch into one end of the dowel. The shoulders of the dowel are then cut away in a pencil sharpener (Fig. 13–7, O and P). The dowel is used to push specimens down on the pin to a uniform height. This necessitates, however, the placing of the labels at a slightly different level, because specimens are often lower on the pin.

Pins and Pinning Forceps. About ten sizes of insect pins, based on diameter, are usually available; they are scaled from fine diameter to large as 000, 00, 0, 1, 2, 3, 4, 5, 6, and 7. The first three sizes are good for direct pinning of very small insects, but they are difficult to thrust into a pinning bottom without the aid of a pinning forceps. Sizes 1, 2, and 3 are the most useful; 4 and 5 may be of some value with larger specimens. All of the sizes from 000 to 5 are approximately $1\frac{1}{2}$ inches in length. Size 6 is $1\frac{3}{4}$ inches and size 7 measures $2\frac{1}{16}$ inches in length. The larger sizes are sometimes useful for pinning giant tropical specimens. Pins may be obtained in the stainless steel or the black japanned finish. The japanned finish is less expensive and much more satisfactory than the stainless steel.

A pinning forceps should be available for transferring specimens into permanent storage cases. Special forceps are available, or a pair of long-nose pliers which has the jaws set at a 90° angle (this can be ordered through any hardware store) will serve just as well. Extremely lightweight forceps listed as pinning forceps are of little value. In using the forceps hold the pin right near the point and thrust it straight down into the pinning bottom. Be careful to line the pin up in the grooves on the jaws of the forceps, and not to snap the pin, for this will damage the specimen.

Insect Labels. Insect labels are commonly printed by hand with a crow quill pen and waterproof India ink. The labels should be quite small (about $\frac{3}{8}$ by $\frac{1}{2}$ inch) so that they do not take up too much space and yet are large enough to protect the specimens. Labels may be used to give the collecting locality or the collector's name or even taxonomic data. It is not essential to outline the labels. If a large number of insects have been collected at one locality, hand printing is too time-consuming and tedious. Two convenient alternatives exist: the photographic reduction method and the printing press.

In the photographic method labels are typed on a white sheet of paper with a clean new ribbon. The individual labels are spaced in columns so that they can easily be cut with a paper cutter after reduction and are provided with room for the date of collection and the collector's name. The entire sheet is marked for reduction to 3 inches wide, and then turned over to a photographer. Figure 13–8A shows a part of a typed page at normal size and Fig. 13–8B shows the entire page reduced and printed. The photographer should have a camera with a ground-glass viewer so that he can measure the reduced size of the plate. An $8\frac{1}{2}$- by 11-inch sheet of paper reduced to a width of 3 inches will produce the proper size of label. In this

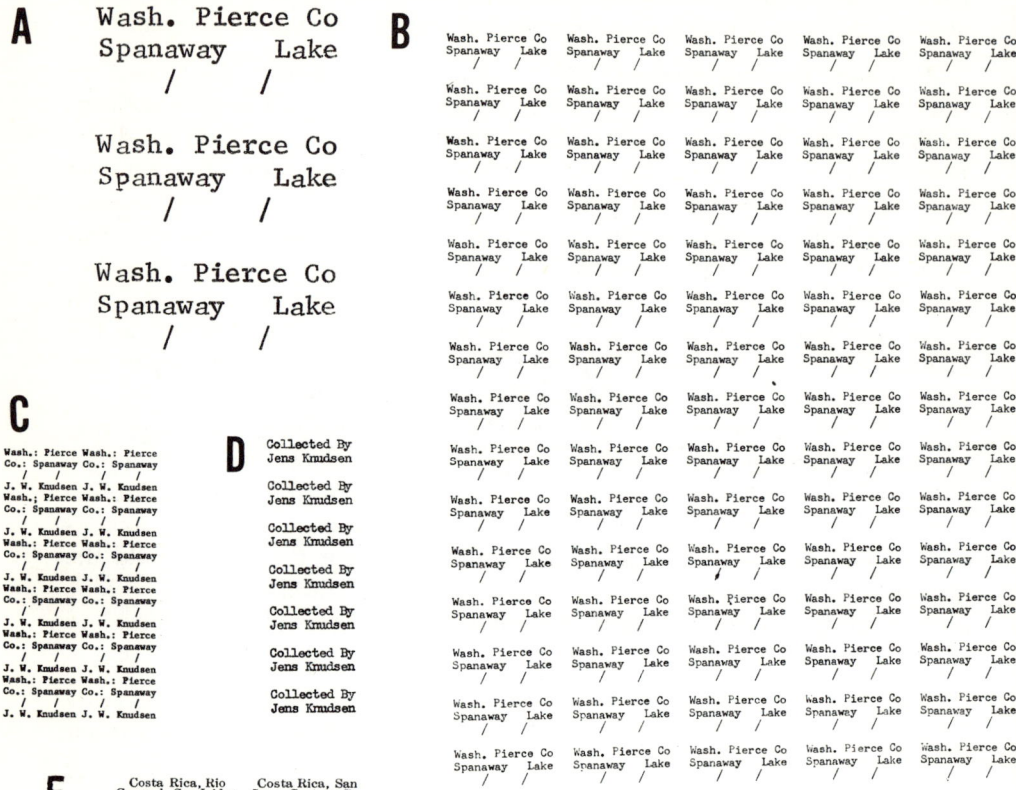

FIG. 13–8. Insect labels. A. Typed labels actual size. B. Photographically reduced labels finished size. C–D. Specimens of photographic labels. E. Printed labels of 4-point type.

procedure no enlarging process is needed for printing the finished labels; rather, contact prints are ordered. Any time new labels are needed the negative is sent to the drugstore and contact prints made. One should request dull or mat finish on the prints, as printing with India ink is somewhat easier on such surfaces. Figure 13–8, C and D, shows other labels which may be made by this method.

Figure 13–8E gives an example of insect labels printed on a museum press. The labels shown are actual size and have been set in number 4 type. At the time of writing of this book, a small 3- by 5-inch hand press equipped with ink and a galley of type costs only about $45 (Kelsey Co., Meriden, Conn.). Four or five labels are set at one time and printed over and over again on 3- by 5-inch cards until the desired number is obtained.

Labels should always be turned on the pin so that they are read from the rear of the insect or the left side of the insect. One should be consistent in label position, but may use whichever method requires the least space.

Methods and Problems of Mounting Insects

Techniques for Insect Orders. Techniques differ greatly between adult and juvenile specimens, so that one seldom uses the same technique all the way through even a single order of insects. Techniques will depend on the stage of the life cycle of the specimen, its habitat, its size, the strength of its construction, and so on. The table on "Techniques for Insect Orders" (p. 234) presents the more common methods of collecting, killing, and preserving adult and juvenile insects. The reader should check each order in the index for special information presented elsewhere.

Where to Pin Insects. Insects must be pinned when they are fresh and pliable. Figure 13–9 shows where some of the more common insects should be pinned. Most insects are pinned eccentrically so that important taxonomic characteristics which may occur along the median line will not be destroyed. Beetles (Fig. 13–9A) are pinned through the right elytron, or wing cover, close enough to the front so that the pin passes between the second and third walking legs. True bugs (Hemiptera and Homoptera, Fig. 13–9B) are pinned eccentrically through the scutellum. Grasshoppers and their allies are pinned eccentrically either with one wing spread (Fig. 13–9C) or

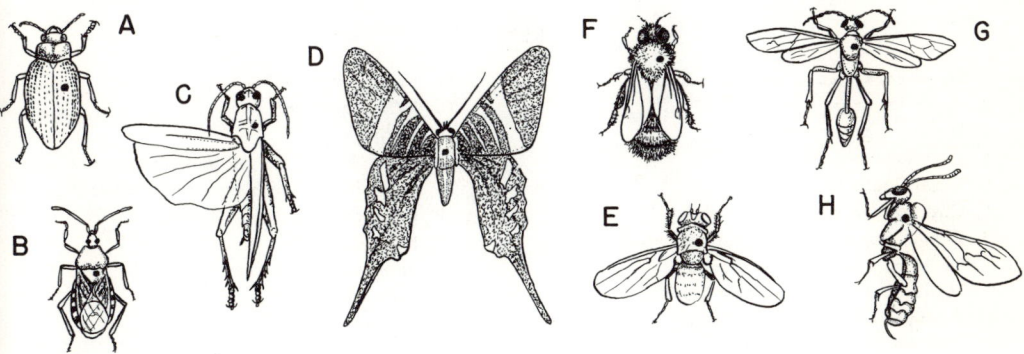

FIG. 13–9. Pinning insects. See text for instructions.

TECHNIQUES FOR INSECT ORDERS

Order	Collecting Techniques for Adults	Killing Adults	Preserving Adults	Killing Juvenile Nymphs, Naiads Larvae, Pupae	Preserving Juvenile Nymphs, Naiads Larvae, Pupae
Protura	1,7	8	17,20,21	8	17,20,21
Collembola	1,7	8	17,21,21	8	17,20,21
Diplura	1,7	8	17,21	8	17,20,21
Thysanura	1,7	10,8	19	8,10	17,19
Orthoptera	1,3	10	18,23	10,11	18,19,17
Dermaptera	1,2	10,8,9	18	10,8,11	19,18,17
Plecoptera	1,3,5	10	18	8,9,11	17
Isoptera	1,2	9,10,11	19,18,17	9,10,11	19,17
Embioptera	1,2	10,8,9,11,14	18,17	8,9,11	17
Odonata	3	10	11,18	8,10	17,19
Ephemeroptera	3	8,10	17,18	8	17
Mallophaga	6,2	8,11	20	8	17
Anoplura	6,2	8,11	20	8	17
Corrodentia	2,7	8,9	17,20,21	8,9	17,21
Hemiptera	1,2,3,5	10,8	18,19	10,8,13	19,17
Homoptera	1,2,3,5	10,8	18,19	10,8,13	19,17
Thysanoptera	1,2	8	17,21	8,9	17
Neuroptera	1,3	10,8	19,17	8,9	17
Mecoptera	1,3	10,8	19,17	8,9	17
Trichoptera	1,3	10,8	19,17	8	17
Lepidoptera	3,4,5	10	18	8,12,13,16,15	17,22
Diptera	1,2,3,4	10,8,9	18,19,21	8,13,15,16	17,21
Siphonaptera	6,7	8	20,17	8,13,15	17
Coleoptera	1,2,3,4,5,7	10,8,9	18,19,17	8,12,13,15,16	17,19,18
Hymenoptera	3,1,4	10,8	18,19,17	8,11,13	17

KEY

1. Hand collecting or with forceps.
2. Aspirator or paint brush.
3. Use an appropriate net.
4. Trapping or baiting.
5. Night lighting.
6. Capture and process host.
7. Berlese funnel, litter or nest material.
8. 75- to 80-percent ethyl alcohol.
9. 75- to 80-percent isopropyl alcohol.
10. Killing bottle (cyanide or other).
11. FAA for tissue studies.*
12. XA mixture.*
13. XAAD mixture.*
14. Ketone mixture.*
15. KAAD mixture.*
16. Boiling water.*
17. Store in 75- to 80-percent alcohol or other liquid (with or without glycerin).
18. Pin specimens.
19. Point on card.
20. Permanent slide (use KOH first).
21. Permanent slide (without KOH).
22. Inflate and dry.
23. Remove viscera of large fleshy specimens; pin. See text.

* Not suitable for all species

with both wings back in the normal position. The legs should be propped up and the antennae brought back along the sides of the body while the specimens are drying. Butterflies, moths, dragonflies, and their allies may be pinned (Fig. 13–9D) centrally and mounted on a butterfly board, as discussed below. Flies, bees, wasps, and their allies may be pinned eccentrically through the prothorax, as illustrated in Fig. 13–9, E, F, and G. Sometimes wasps are pinned on their sides (Fig. 13–9H) when the abdomen is greatly curved, so that important taxonomic characteristics may be observed. The specimens should be propped up on a card until dried in this position. Most other orders of insects are pinned eccentrically or placed upon points or minute needles (*Minuten Nadeln*).

Use of Pinning Block. Provide yourself with some suitable base upon which to start the process of pinning. This may be made by pressing clay into a bottlecap or by cutting a piece of ¼-inch balsa wood. Hold the specimen down between the thumb and first finger of the left hand and move your fingers back just far enough to expose the area that is to be pinned. Thrust the pin straight down through the body of the insect while firmly holding the specimen with the left hand (Fig. 13–10A). This technique will prevent the disarrangement of wings or other parts during pinning. Next thrust the pin down into the top hole of the three-step pinning block (Fig. 13–10B). Following this the locality label is placed on the second step of the pinning block; the collector's label or taxonomic label is placed on the third step. When these have been placed on the pin the specimen may be put away in a storage box. Figure 13–10, C and D, shows the correct way of pinning where the specimen is at the proper level on the pin and the body is level in both the longitudinal and transverse planes. Do not tilt the insect forward or backward or to one side or the other, and do not place it too high or too low on the pin, as shown in Fig. 13–10, E through H. If specimens have droopy bodies or long gangly legs (Fig. 13–10, I and J) they should be propped up on cards while the specimens dry. Drying may require a week or two for larger specimens. After drying, the card is carefully bent away from the insect and pulled down the pin, then the proper labels are added.

Evisceration and Stuffing. You may experience difficulty with large crickets or long-horned grasshoppers in that the soft abdomen will partially rot and discolor before thorough drying takes place. To overcome this the specimen may be opened on the mid-ventral side of the abdomen with a sharp-pointed dissecting scissors. Next, the viscera are removed with a forceps and a small amount of cotton is added to take the place of the viscera. The specimen is then pinned in the typical manner, as described

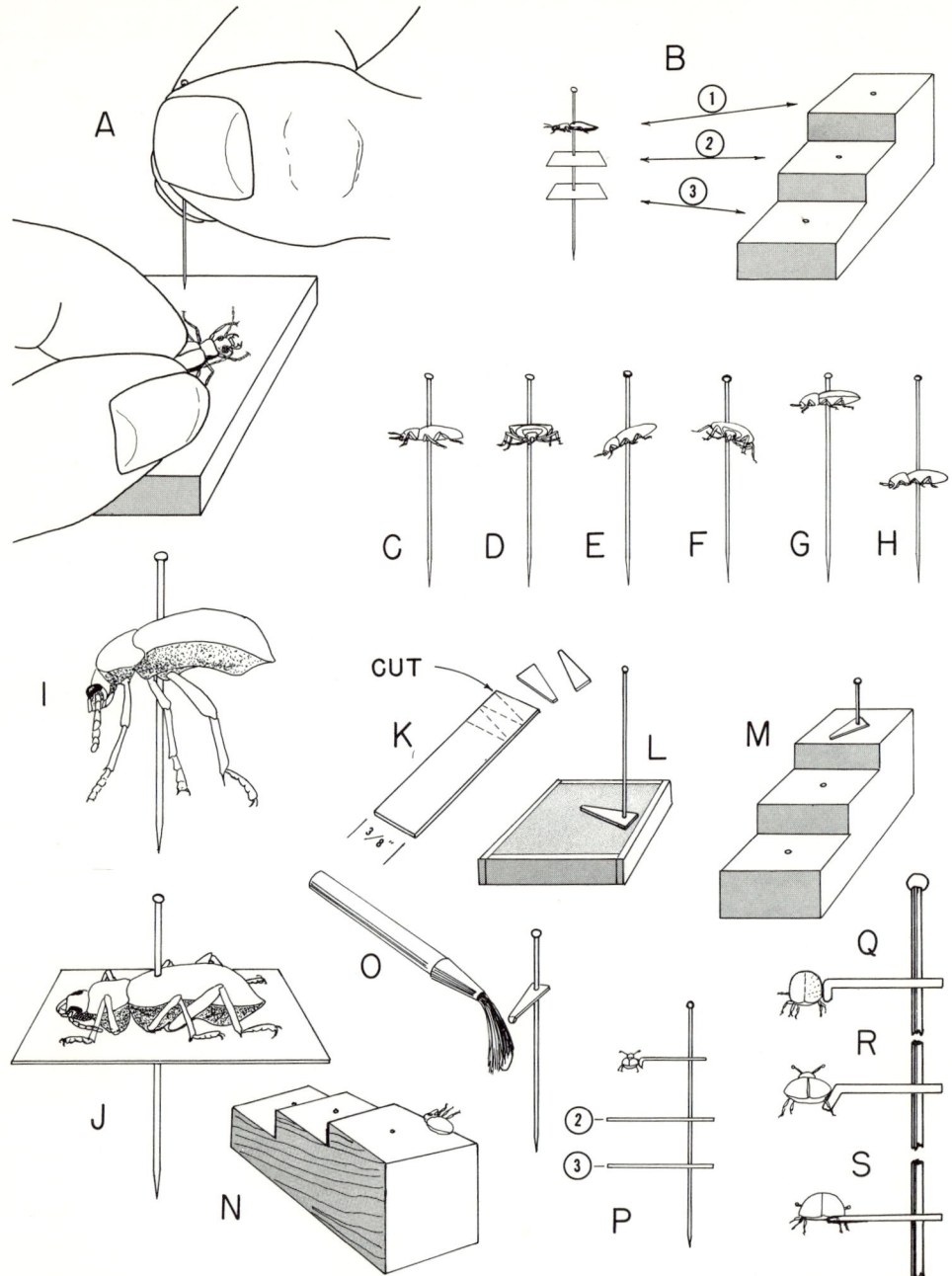

FIG. 13-10. Pinning techniques. A. Starting the pin with a pinning base. B. Use of pinning block. C–D. Correct method of pinning. E–H. Improper method of pinning. I–J. Propping drooping legs. K. Cutting points. L–M. Mounting point. N. Positioning specimen. O. Applying cement. P. Finished specimen with labels. Q–S. Bending point to match specimen. (1) Specimen label, (2) locality label, (3) collector's label.

above. Good results are sometimes obtained without evisceration by injecting the soft parts of the body with 10-percent formalin prior to pinning. However, considerable shrinkage will still occur.

Damsel flies and dragonflies almost always lose some of their color shortly after being pinned, owing to organic changes during the drying process. Moore (1951) suggests several techniques for preserving color perfectly. One technique consists of eviscerating the dragonfly in the following manner: Thread a needle, double the thread, and tie a knot in the end of the two pieces. Run the needle into the ventral side of the abdomen at the base of the thorax and pass this out the anal opening. Draw the knot up to the abdomen and then twist the thread many times. Draw the thread through the abdomen, removing with it much of the viscera. Next soak the specimen overnight in ethyl acetate or ethyl alcohol which will remove much of the grease from the body. After degreasing, pin and dry the specimen.

Relaxing Specimens. Specimens that have been dried and stored, or specimens that have been in cyanide or Cyan-o-gas bottles too long, will be brittle and should be relaxed before pinning. Any jar or can that can be tightly sealed may serve as a relaxing chamber. This should be provided with some wet paper toweling, wet sand, or similar material, as a source of moisture, and with a shelf to support the specimens and keep them off the wet substrate. The shelf may be made of a piece of cardboard, screen wire, or a bottle lid. Specimens are stored in the relaxing chamber until the limbs are movable; then they are pinned. One must watch the relaxing chamber to prevent the growing of mold while the specimens are inside. A few drops of phenol or formaldehyde or a few chlorocresol crystals will retard molding. The size of the specimen and the degree of drying, plus other factors such as fat content, will determine the length of relaxation necessary. Frequently, butterflies and moths will have to be relaxed from two or three days to as long as a week before the wings can be moved without damaging the specimen. Not all specimens relax at the same rate of speed; thus, some will be ready before others. Do not attempt to pin a partially relaxed specimen; rather, put it back until all of the body parts can be positioned easily.

Specimens may be relaxed quickly by using live steam. One method is to place a piece of screen wire over a ring stand, put the specimens on the screen, and cover them with an inverted tin can or beaker. Position the ring stand so that the screen is above the spout of a steam kettle (preferably of small size to reduce the volume of steam) and allow the live steam to penetrate the specimens until they are thoroughly relaxed. Only a few specimens at a time should be so treated. A second method is to place specimens in a shallow pan and then put a steam iron over the top so that live steam penetrates the bodies of the insects, relaxing them in just a few moments. Too much live steam may cause wing curling and distortion in butterflies.

Pointing Insects. Insects that are too small to be pinned with regular insect pins should be mounted on insect points or *Minuten.* There is no rule which tells the collector when the insect is too small for a standard pin and should be placed upon an insect point. Of chief concern is the fact that a pin may destroy or damage a small insect. By definition, pointing consists of placing a triangular piece of paper on an insect pin and gluing the small insect to the tip. It is essential that insects be mounted in a uniform way with the point glued to the right side of the specimen's body (Fig. 13–10, P through S). Although many adhesives have been used in the past, by far the best is clear fingernail polish. This dries much faster than shellac, but slower than model airplane glue, and can readily be dissolved with fingernail polish remover, should the need arise.

Make insect points of a good grade of paper, such as manilla folders or Bristol board, which will not fray after cutting. Cut points from strips of paper $3/8$ inch wide, as illustrated in Fig. 13–10K. The base of the triangle should be about $1/8$ inch wide and the tip $1/16$ inch or slightly less. The width and shape of the tip will vary, depending on the kinds of insects being mounted. Put the point on a number 2 or number 3 insect pin (never smaller), and position the point on the pin with a three-step pinning block (Fig. 13–10, L and M).

Next, position the insect on its left side on the pinning block so that the point can be applied to the right side of the body. If necessary, bend the tip of the point to conform to the insect's anatomy (Fig. 13–10, Q through S), add fingernail polish to the point (Fig. 13–10O) and attach the point to the side of the insect. Allow a few seconds for the fingernail polish to adhere to the insect's body and then lift the insect, pin and all, in a smooth easy motion and stick the pin into a pinning base. If the insect tends to roll or slip from the point you may correct this by pinning the specimen next to the three-step pinning block and resting the specimen on the step until the glue has dried. Using too much fingernail polish will only require a longer period of drying and partially obliterate important taxonomic features. Finally, place a label on the pin to complete the job. Do not neglect this important technique; it can easily be mastered with a little practice. Too frequently important smaller insects are omitted in collections because of their size. Most biological supply houses sell point punches which cut uniform points, and are well worth the price.

Minuten Nadeln (Minute Needles). *Minuten Nadeln* are pointed wire needles, about ½ inch in length, which are fine enough to pierce small insects and are attached to a larger insect pin by some suitable method. Figure 13–11, A and B, shows two methods of attaching *Minuten* to cork, balsa wood, or pith blocks. The technique in Fig. 13–11C is one in which the *Minuten* is mounted on a piece of file card attached to a larger insect pin. In all cases the pin should be of a number 2 or number 3 size and labels should be added as illustrated. Some collectors like to use special killing techniques so that specimens intended for *Minuten Nadeln* will be positioned naturally. For example, ether is often used for mosquitoes so that the wings and legs will be in a more or less normal position.

Carding Insects. This method may be used for smaller insects in place of pointing or *Minuten,* or for larger soft-bodied specimens that become very fragile when pinned and dried. The technique consists of cutting small squares of index card about the size of an insect label or of a size appropriate to the specimen to be mounted. The card is placed at the proper height on the insect pin and coated with a thin layer of clear fingernail polish. The specimen is then placed upon the card and becomes firmly attached as the cement dries. There are several drawbacks to this technique in that one side of the insect is hidden from view and overly large cards are space-consuming. When carding small specimens, mount them in different body positions so that all taxonomic characteristics can be viewed through the dissecting microscope. Perhaps this technique has its greatest value when very long series of small specimens are to be mounted for comparative studies, for dozens may be placed side by side on a single card.

Coverslip Mount. When small insects are being prepared on permanent

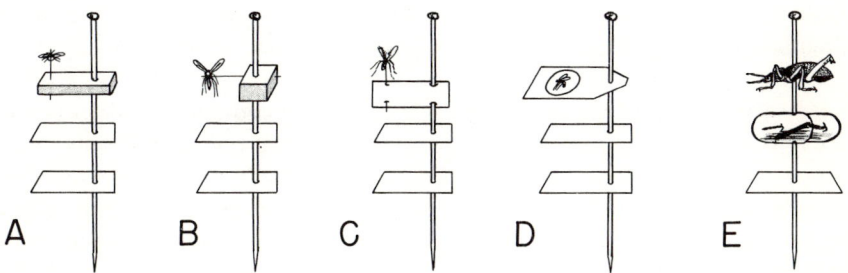

FIG. 13–11. Additional mounting techniques. A–C. Three methods of mounting minute insect pins. D. Coverslip mount. E. Using a medicine capsule for broken appendages.

glass slides it is often desirable to place a few of these specimens in the pinned collection. In order to do this, prepare the insect in the standard way for slide mounting (discussed below and in Appendix B). A celluloid coverslip is cut (Fig. 13–11D) large enough to receive a 15-millimeter circular glass coverslip. The specimen is then transferred from the clearing medium onto the plastic coverslip as it would be onto a permanent glass slide. Add sufficient mounting medium to cover the specimen and lower the glass coverslip into place by means of a dissecting needle. Next, mount the plastic coverslip on a number 3 insect pin and attach the label. This technique will not replace the permanent glass slide, as specimens are difficult to study with a compound microscope when mounted by the coverslip method. However, specimens which might not otherwise have been included in pinned synoptic collections may be mounted by this method.

Mounting Winged Insects. Special techniques must be used and care taken in the mounting of some winged insects such as dragonflies and damsel flies, butterflies, moths, and so on. Dragonflies and damsel flies are mounted on somewhat modified butterfly boards. These are constructed as described above (p. 228), but have a central groove measuring ¾ to 1 inch in width. This board not only allows room for the arrangement of the wings, but will permit the legs to be positioned as well. Wings are strapped down with paper strips as described below.

The mounting of butterflies and moths is a relatively simple procedure provided a few tricks of the trade are learned (see Fig. 13–12). The big difficulty, of course, arises from the fact that the bodies and wings are covered with scales which are easily dislodged and lost. Specimens must be fully relaxed before pinning.

After selecting the proper size of pinning board, carefully thrust a pin straight down through the center of the groove to provide a hole for the specimen pin; the pin used should be the same size or slightly larger than the one to be used for the specimen. This practice may prevent specimen damage that could otherwise result. Lift the specimen by the antennae and hold it by the legs, using the thumb and first finger of the left hand. The body itself should not be held; rather, it should rest firmly in the V formed between the thumb and first finger (Fig. 13–12A). Open the wings outward by blowing between them, maintaining a constant stream of air while the pin is lowered into position. If the wings are slightly stiff they may be opened first by inserting a closed forceps between them. When pressure is removed from the forceps the two blades will naturally open and cause the wings to separate. While blowing between the wings thrust the pin

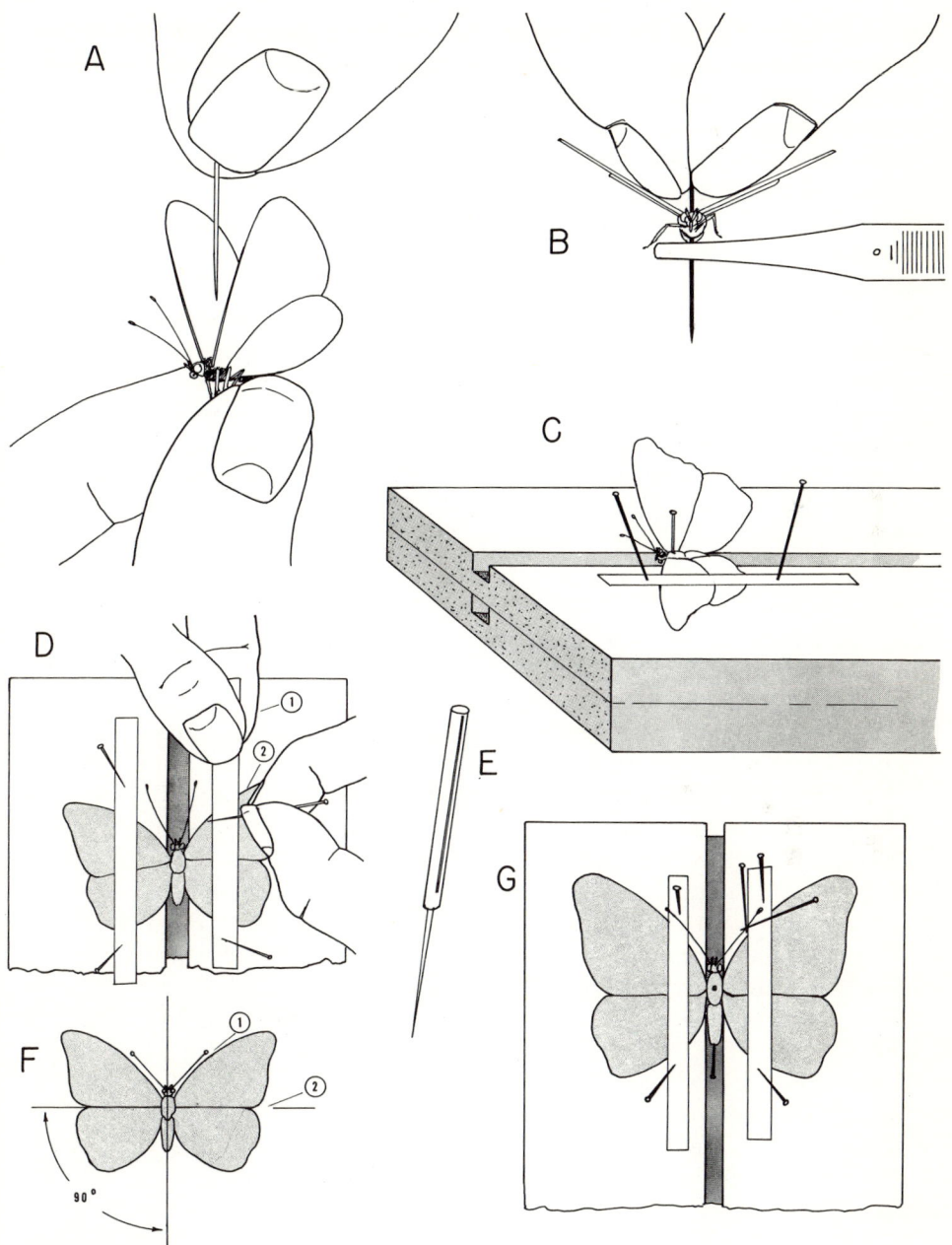

FIG. 13-12. Mounting butterflies and other winged insects. See text for instructions. (1) Proper spacing for antenna, (2) proper spacing of wings.

through the thorax. Be sure that the pin goes through the center and straight down so that the body is not tilted from side to side or from front to back. This is extremely important.

Next, grasp the head of the pin by the thumb and first finger so that when the specimen is pushed up on the pin, by forceps or fingers, the wings will strike upon the fingernails and thus will not lose their valuable scales (Fig. 13–12B). Always push the specimen higher up on the pin than is necessary. After the specimen pin is thrust into the mounting board, lower the specimen by means of forceps to the proper level. Care must be taken to have the wings exactly in line with the surface of the pinning board so that they will not be bent upward or downward when strapped in place.

Thread, string, glass, paper, and other materials have been used to strap down butterfly wings, but the author prefers to cut strips of slick-surfaced magazine covers and pages or 20-pound bond paper for this purpose. These should be 6 or 7 inches long and cut in $\frac{1}{8}$-, $\frac{1}{4}$-, $\frac{3}{8}$-, and $\frac{1}{2}$-inch widths, as needed. Many workers like glassine paper because of its smooth transparent surface, but it is not very strong or durable. A strip of paper is inserted between the butterfly's wings and one set of wings is pinned down to the pinning board (Fig. 13–12C). Following this, the other set of wings is pinned down and then the wings are moved into the proper position (Fig. 13–12D). Move the wings with a fine 0 or 00 insect pin or a small probe made by thrusting a headless 00 pin into a matchstick (Fig. 13–12E).

The wings should be positioned so that the posterior margin of the front wing is at right angles to the midline of the body as shown in Fig. 13–12F (1 and 2). The hind wings are brought up in position so that only a small gap exists between them and the front wings. Pin the antennae at a uniform distance from the front wing, as shown in Fig. 13–12G. The abdomen should finally be propped up on a pin and carefully centered.

When moving the wings into position, pin the paper strap behind the wings and pull forward, with the left hand, to exert pressure upon the wing, as shown in Fig. 13–12D. Do not move the forewing all the way forward in one operation, as this often allows the hind wing to "jump" out and get above the forewing. While holding tension on the wing, (1) lift the paper strap slightly and (2) slide the forewing anteriorly by hooking a pin just behind the costa or radius vein. Move the hind wing up to meet the forewing, alternately exerting and releasing pressure with the paper strap, and continue moving the two wings in this manner until they are in position. With larger specimens it is sometimes necessary, temporarily or permanently, to hold the forewing by pinning it just behind the costa vein with a fine insect pin. The wings on both sides of the body are positioned in this manner and the paper straps repinned close to the margin of the wing. If it is desirable, glassine or plain paper may be pinned over the tips of the wings. The antennae and abdomen are adjusted at this time.

A great deal of damage is often done during removal of pins from the wings or removal of the specimen from the pinning board. Never pull the insect pins straight out of the wings or paper straps. Rather, twist the pin between your fingers until it is loose and then pull it free. First remove the pins from the antennae and the abdomen and then remove the paper straps. In the latter operation pull the strap forward and twist the forward pin loose, slide the strap out to the side until it clears the antenna, and remove the strap. Keep labels low down on the shank of the pin to make them visible from above.

Insect Boards. Insect boards should be used when the greatest degree of pinning perfection is desired or when large numbers of "droopy" insects are to be pinned. The boards (described above) have a pinning surface of $\frac{1}{8}$-inch balsa wood. After thrusting the pin through the insect's body, push the pin all the way down into the insect board (Fig. 13–7H), and position the appendages with a pair of fine forceps or a dissecting needle (Fig. 13–7, I and J). Pins required to hold the legs are thrust into the balsa wood and left until the specimen is dry.

Riker Mounts. Riker mounts are very popular for displaying all kinds of plant and animal material, as well as insects. The Riker mount consists essentially of a cotton-filled, glass-topped box in which specimens are placed. These mounts are excellent for teaching displays and limited synoptic insect collections, but are quite unsatisfactory for growing taxonomic collections. The Riker mount will give good protection to specimens that are to be handled by students, but once they are sealed it is a lot of work to reopen them in order to add or rearrange specimens.

Small Riker mounts for one or two specimens are easily made (Fig. 13–13A) by cutting strips of wood $\frac{1}{4}$, $\frac{1}{2}$, or $\frac{3}{4}$ inch square. These are then glued or tacked around the margins of a piece of cardboard of appropriate size. This forms a simple box into which cotton and the desired specimens are placed. A piece of glass, cut to fit the top of the box, is taped in place with black photographer's masking tape.

Excellent Riker mounts can be made if boxes of the appropriate size, about 1 inch deep, are available (e.g., ditto stencil boxes). Remove the top of the box and draw a line $\frac{1}{2}$ inch in on all sides. Place the top over a wooden block and cut along the lines with a scalpel, as shown in Fig. 13–13C. Spray the cover with a dull-finish black enamel paint. Next, cut some $\frac{1}{4}$- by 1-inch wooden strips to fit inside the bottom of the box (Fig.

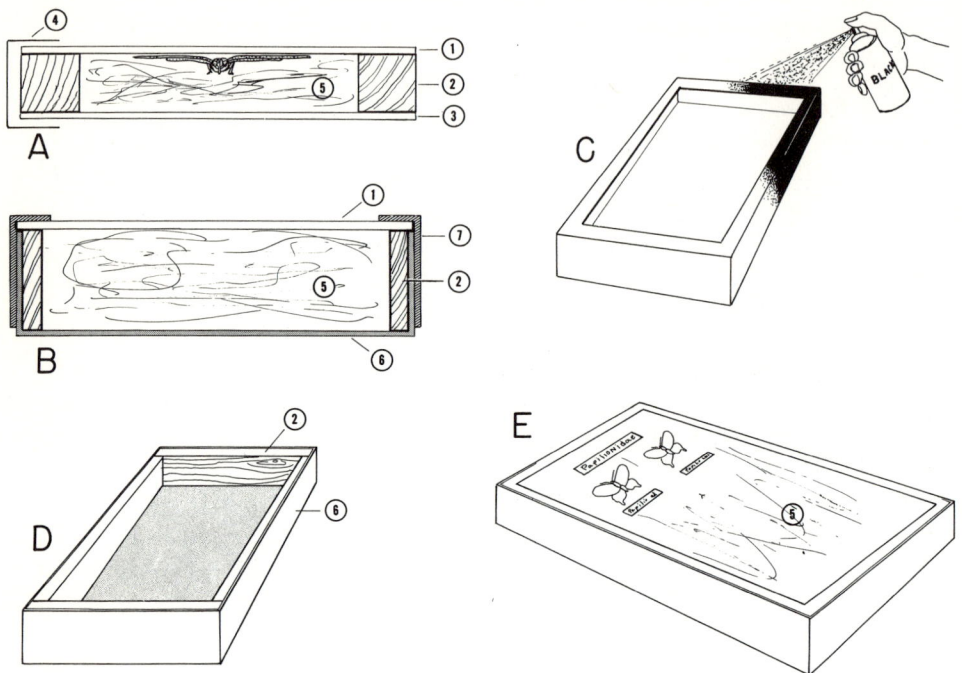

FIG. 13–13. Riker mounts. A. Simple construction. B. Box construction. C. The lid. D. Bottom with wood strips. E. Placing specimens on cotton. (1) Glass, (2) wood strip, (3) cardboard back, (4) black tape, (5) cotton, (6) box bottom, (7) box top.

13–13J), and glue these in place. Place a thick smooth layer of cotton in the box, followed by appropriate specimens and specimen labels (Fig. 13–13E). Finally, place a sheet of glass over the specimens and put the painted lid back on the box. The lid may be held in place with straight pins that have been cut off and thrust into the box or by black masking tape.

When Riker mounts are used for insect display a special technique should be used in preparing the specimens. It is essential that the specimens (butterflies, bees, dragonflies, and so on) have their wings and legs properly arranged before they are placed in the Riker mount. Sometimes specimens have a good natural position immediately after coming out of a killing bottle, but usually this is not the case. Therefore, specimens should be mounted in the conventional way until the wings and legs are set in the desired position. This poses a problem, however, since the body juices of fresh insects will weld the specimen to the insect pin. You may overcome this problem by papering butterflies and layering insects until they are dry. Then relax them and mount them in the conventional manner. After this has been done the insect is slid down the pin and placed in the Riker mount.

Pin Removal. Occasionally it is necessary to remove the pin from a specimen which tightly adheres to its pin. Holbrook (1927) suggests a simple technique for pin removal which will be slightly modified herein. Attach a fine insulated copper wire to each pole of a 5-volt storage battery. Next, scrape a little of the enamel finish from the shank of the insect pin above and below the specimen. Pin the specimen firmly into any kind of support and attach one wire to the pin beneath the specimen. Touch the other wire to the pin above the specimen just long enough to heat the body tissue surrounding the pin. As soon as contact is made by the second wire, press down on the insect gently with a pair of forceps. Within three or more seconds the specimen will become loosened and will readily slide off the end of the pin. Be careful not to heat the specimen any longer than necessary, as permanent damage can be done. The specimen should be relaxed in a humidity chamber if there is danger of breaking the brittle appendages while remounting. When large numbers of specimens are to be removed from their pins this process can be further simplified by attaching a piece of aluminum foil or tinfoil to the end of one wire. After the enamel has been scraped from the bottom of the pin, the pin is thrust through the foil. Then the wire is touched to the top of the pin and the insect removed as described above.

Larval Inflation. The technique of larval inflation is an old and useful one which every serious student should experiment with and learn. Students working with butterflies and moths find it especially useful, but the technique may be used with other orders as well. Larval inflation essentially consists of a dry preservation of the specimen by means of evisceration, drying, and mounting. This technique has the advantage that colors usually remain unaltered, the hairs and bristles are more fully displayed, and the specimens can be kept along with pinned adults. This technique is also useful when making natural history displays where lifelike specimens are desired. One drawback is that the soft parts are totally lost, necessitating preservation of some specimens in liquid for anatomical studies.

The equipment used (Fig. 13–14, A through D) consists of a metal drying oven with a glass top, an alcohol lamp for heat (Fig. 13–14A) and a holding tube equipped with an atomizer bulb and an air reservoir (Fig. 13–14, B and C). This equipment can be obtained from biological supply houses but is rather expensive compared with the very modest cost of homemade equipment. Any sheet metal (preferably 28-gauge galvanized iron) may be

246 Biological Techniques

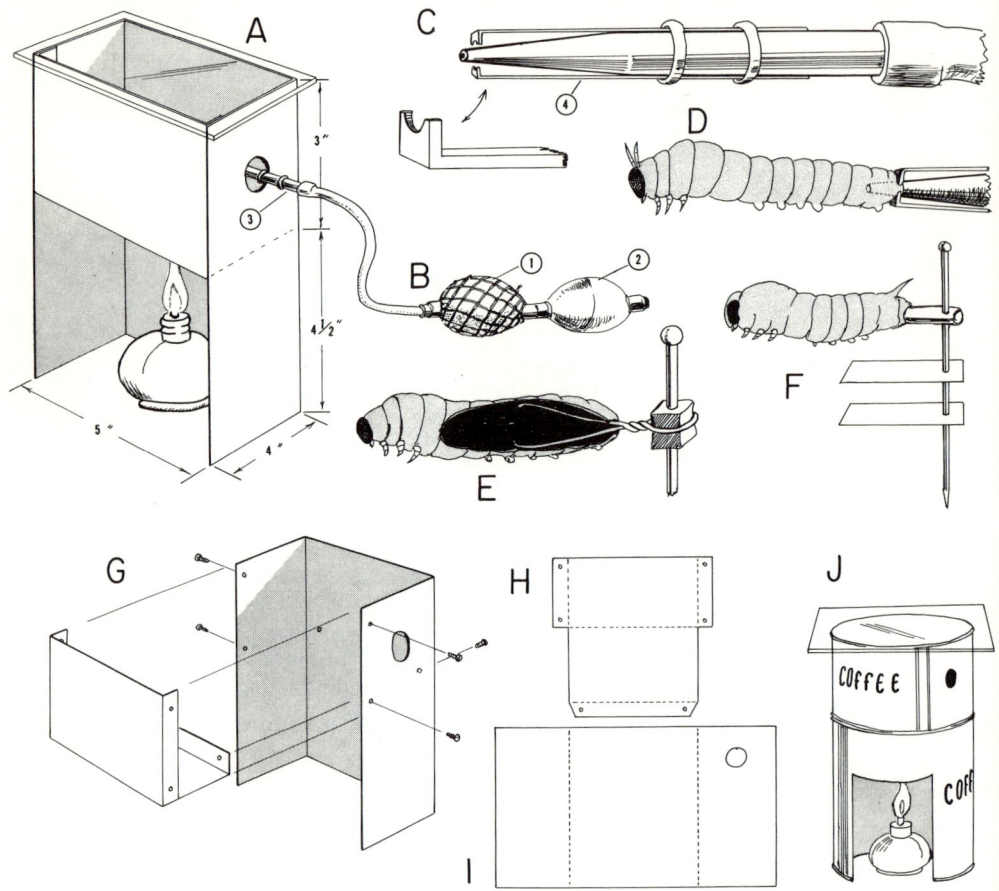

FIG. 13–14. Larval inflation equipment. A. Drying oven with alcohol lamp. B. Inflating apparatus. C. Mounting tube with clips. D. Caterpillar ready for inflation. E–F. Mounting techniques. G–I. Pattern and construction of drying oven. J. Economical drying oven. (1) Air reservoir, (2) aspirator bulb, (3) mounting tube, (4) clip.

used in the construction and any dimensions will do, though somewhat standard dimensions are given in Fig. 13–14A. Cut two pieces of metal as shown in Fig. 13–14, H and I, bend these as indicated by the dotted lines, and assemble with rivets, sheet metal screws, or nuts and bolts, as illustrated in Fig. 13–14G. A much simpler drying oven may be constructed from 1-pound and 2-pound coffee cans bolted together, bottom to bottom, as shown in Fig. 13–14J. The 1-pound can has a hole cut in the side and is provided with a glass plate; the 2-pound can has a door cut large enough to admit the alcohol lamp.

The tube on which the specimen is mounted is made of $3/16$- or $1/4$-inch (O.D.) glass tubing drawn out to a fine point of about 1 millimeter outside diameter. One or two metal clips made from pieces of "bobby pin" or other spring metal are attached to the tubing by means of rubber bands. One clip is satisfactory; two clips may make mounting somewhat difficult.

An atomizer bulb with valves attached to a rubber air reservoir and expansion net (Fig. 13–14B) is connected to the glass tube by a 1-foot piece of rubber surgical tubing. The atomizer bulb–reservoir combination may be purchased for around three dollars. An air reservoir is essential to provide a steady flow of air. Harrison and Usinger (1934) suggest using a toy rubber balloon as an air reservoir. Inflate the balloon several times to make the rubber very pliable before experimenting with larvae. Attach the balloon to a small, two-holed rubber stopper by means of a piece of string or rubber band. Run a tube from the atomizer bulb into one side of the rubber stopper and a tube going to the larval holder from the other side of the rubber stopper. When the atomizer bulb is pumped vigorously the balloon will inflate and provide a steady flow of air.

To use the larval inflation equipment, kill the specimen in fumes of ethyl acetate or cyanide and make a fine incision around the anal opening. With a piece of glass tubing or a pencil roll the contents out of the larva, starting just behind the head and applying pressure posteriorly until the soft parts are forced out of the body. Do not apply so much pressure that the skin is bruised or the spines broken. Insert the glass nozzle into the anal opening and attach the skin by means of one or two metal clips. Preheat the oven with an alcohol lamp and then inflate the larva and place it inside the oven. While watching through the glass plate, keep the larva inflated until the skin begins to dry, rotating the specimen to prevent overheating and uneven drying. If a problem with scorching develops, trim the wick on the alcohol lamp to reduce the flame and adjust the glass so that air may circulate through the oven more quickly, thus reducing the heat. When dry the specimen is mounted by pinning a wooden matchstick of the appropriate length and gluing the larva to the top of the matchstick. Or, a pinned matchstick may be sharpened and glued into the anal opening (Fig. 13–14F). Some workers pin a small block of cork or balsa wood and twist a fine, soft wire around this. The wire is then trimmed and inserted into the larval specimen and glue is applied at the anal opening (Fig. 13–14E).

Microscope Slide Mounts. On many occasions permanent slides are required in entomology. For example, studies of wings, mouth parts, entire head structures, genitalia, or small adult or larval insects such as fleas, lice, flies, mosquitoes, aphids, and the like, may all require slide-making technique.

Several kinds of mounting media are commonly used in entomology and should be selected for the material to be mounted. Complete instructions for the media mentioned below will be found in Appendix B or Chapter 14. Balsam, Permount, and Euparol require complete dehydration of the specimens prior to mounting. Other media such as Hoyer's, Turtox CMC-10, PVA, lactophenol, glycerine jelly, and others may be used with specimens taken directly from water, alcohol, or other preservatives without intermediate dehydration. The lactophenol medium, the Turtox medium, and Hoyer's medium have added advantage in that they will quickly kill larvae right on the slide and thus do the job of killing, fixing, clearing, and mounting. Some specimens will have to be macerated in KOH to render the exoskeletons clear. Stain usually is not required unless specimens are small and transparent and have a refractive index similar to that of the mounting medium. When thick specimens are mounted the coverslip may be propped up with broken pieces of glass or the specimens may be placed in a depression slide. Thick mounts usually require the addition of more mounting medium around the coverslip where the evaporation of solvents has started to form air pockets. Some instructions to be used in conjunction with Appendix B are mentioned below.

1. Whole mounts. Fleas, lice, darkly pigmented insects, or large structures such as biting-fly heads and the like are first macerated in KOH and then mounted in balsam or Euparol. Small flies, aphids, and other small insects are not usually macerated in KOH but are cleared in clove oil or some other medium prior to mounting in balsam or Euparol.

2. Mouth parts and genitalia. The head or abdomen of the specimen is removed and macerated in KOH. When working with small insects macerate the entire animal. Following this, complete the procedure required for the mounting medium selected (balsam, Euparol, Hoyer's, Turtox, glycerine jelly; see Index B or Chapter 14) and transfer the head or abdomen to a slide with a few drops of mounting medium. Dissect out the mouth parts or genitalia on the slide and throw away the excess parts; add a few more drops of mounting medium and a coverslip. Glycerine jelly mounts are often desirable in that the specimens remain quite pliable during the period of dissection.

3. Wings and legs. Dried wings or legs may be wetted in 95-percent alcohol and mounted in Euparol or Permount. Unless the elytra of beetles or large, dark, and opaque legs of insects are being mounted, no maceration in KOH is necessary.

4. Small larvae and eggs. Small larvae usually are not macerated in KOH nor stained unless they are perfectly transparent. They may be killed on the slide and mounted in Hoyer's medium, lactophenol, or Turtox CMC-10 (according to Peterson, 1959, the Turtox medium does not kill quickly). Specimens to be mounted in glycerine jelly, balsam, or Permount should be killed in hot water (see "Collecting Juveniles," p. 218) and then mounted according to Appendix B.

5. Dry mounts. Bristles, hairs, scales, and similar structures may be too transparent for mounting in a standard medium and may reject stain; they should be mounted dry. In mounting, these structures are transferred to a slide and covered with a dry, clean coverslip. A wringing medium such as asphaltum or clear fingernail polish is used to seal the coverslip in place.

DISSECTION TECHNIQUES

Specimens for Dissection. Specimens are often dissected for studies of internal anatomy, genitalia, or mouth parts. Specimens killed just before dissection (or frozen specimens) are much better than those which have been preserved. Alcohol usually will not properly preserve internal soft structures well enough, and a suitable fixative such as FAA must be used if specimens are to be preserved. The tissues of fresh specimens are variously colored and quite distinctive, whereas those of preserved specimens are uniformly colored and difficult to identify.

Mounting Dissection Specimens. Frequently the genitalia can be pulled from a specimen with a pair of watchmaker's forceps; the specimen then may be pinned and stored in the usual manner. Where internal structures are to be studied the animal must be firmly affixed to some substrate. One suitable technique for medium-sized specimens is to embed them in paraffin. Pour about ¼ inch of paraffin inside a jar lid and allow this to harden. Next cut out a piece of paraffin large enough to admit the insect. Fill this depression with hot melted paraffin and embed the specimen in it. When the paraffin has set the dissection may begin. The lid may be flooded with water which will reduce light reflection from the specimen and float the internal structures.

For larger specimens such as beetles, butterflies, and so on, the wings and legs may be held down with modeling clay. Fill the lid of a microscope slide box with clay, place the insect on the clay and fasten the legs down with additional pieces of modeling clay.

Very small insects may be glued to a glass slide by first roughing the glass surface with sandpaper, then applying a coat of adhesive such as fingernail polish, and placing the specimen in this adhesive until it is dry. The glass is then submerged in water to reduce surface reflections.

Dissection Instruments. A number of instruments described herein may prove useful for dissection: (1) Mounted needles made by forcing fine sewing needles or insect pins into wooden dowels. Some of these may be sharpened to serve as scalpels. (2) Hooked glass rods made by heating a piece of glass rod until it is cherry red and then whipping one piece off with a vigorous snap of the wrist. This produces a fine-pointed hook of glass which can be useful for probing and tearing tissue. A second method is to press the point of a fine insect pin against a piece of glass until it forms a hook. Other instruments—(3) iris scissors or other fine-pointed dissection scissors; (4) watchmaker's forceps; (5) dissection microscope, preferably with a zoom lens system; (6) a strong, cool light source such as a neon microscope light which will not overheat the specimen at close range.

Dissection Methods. If one is interested solely in the chitinous portions of the exoskeleton, the soft parts may be removed by treating the specimen in 10-percent KOH (potassium hydroxide) for several hours to a day or by boiling for a few moments. Boiling is dangerous, and thus the test tube must be aimed away from the operator. Wash treated specimens in several changes of tap water before dissecting.

When soft parts are studied, carefully open the body, beginning in the middle of the abdomen and working both anteriorly and posteriorly. Use hooked glass rods or dissecting needles, dissecting scissors, or two pairs of forceps to tear through the exoskeleton. Very carefully open the body with a forceps, starting with the abdomen and working toward the thorax, removing tissues attached to the body wall as you do so. Do not expect to locate all of the structures of any one system on a single specimen. Be prepared to study from five to ten specimens for a single system. However, separate drawings and studies can be made for many systems simultaneously, as the dissection progresses. These are added to, with each new dissection, until completed.

Freeze-drying Technique. Freeze-drying is a relatively new technique. It has some merit in preparation of soft-bodied insects for natural history displays but has little value for making study collections. The technique has been used successfully not only for insects, but for such large things as reptiles, birds, and even monkeys. However, the technique is extremely costly (in equipment) and time-consuming. A number of references are available for students interested in the full details (Davies, 1964; Merryman, 1960, 1961; Blum and Woodring, 1963; and Woodring and Blum, 1963).

The Los Angeles County Museum recently used the freeze-dry method to prepare an excellent display of termites within their native habitat. Specimens were dropped into acetone which was chilled to –65° C. by adding dry ice, transferred to a Bell jar equipped with a drying agent, and hooked up to a desiccating pump. Dehydration takes twenty-four hours or longer for small insects up to many weeks for vertebrates. The advantages of this technique are that color and form are not altered. The frozen state of the tissue prevents any shrinkage while the drying process is going on.

MUSEUM AND STORAGE TECHNIQUES

Construction and Use of Equipment

Storage Cases. Insect specimens must be stored in well-constructed cases with tight-fitting lids in order to protect them against insect pests and dust. Biological supply houses sell a wide range of insect cases which differ in size, quality, and price. However, insect cases of good quality may be constructed in the home or school workshop. The size of the case is optional, but should be small enough for convenient storage. If glass is being used, the case dimensions should be based on stock glass sizes to avoid the necessity of cutting and fitting glass. Glass is sized in graduations of 2 inches. For example, a common insect case sold commercially measures 17 by 19 inches, using a glass size of 16 by 18 inches.

When selecting materials for case construction, use a fine-grained soft wood, free of knots, for the sides. If plywood is used for tops or bottoms, $3/16$-inch or $1/8$-inch mahogany plywood is suitable and least expensive. A fine-grained fiber wallboard (such as CeloteX) makes an excellent pinning bottom although balsa wood, cork, or special bulletin-board-cork linoleum are very satisfactory.

In planning insect cases, size is the first consideration. Extra-large sizes should be avoided, and the size of storage cabinet in which the cases will be housed must be considered. The style of construction depends somewhat on the means of storage. For example, glass-topped boxes are excellent for permanent storage in light-proof cabinets. On the other hand, a solid light-proof construction must be used if storage cabinets are not used. Figure 13-15 shows several plans for constructing glass-top and solid cases. Note

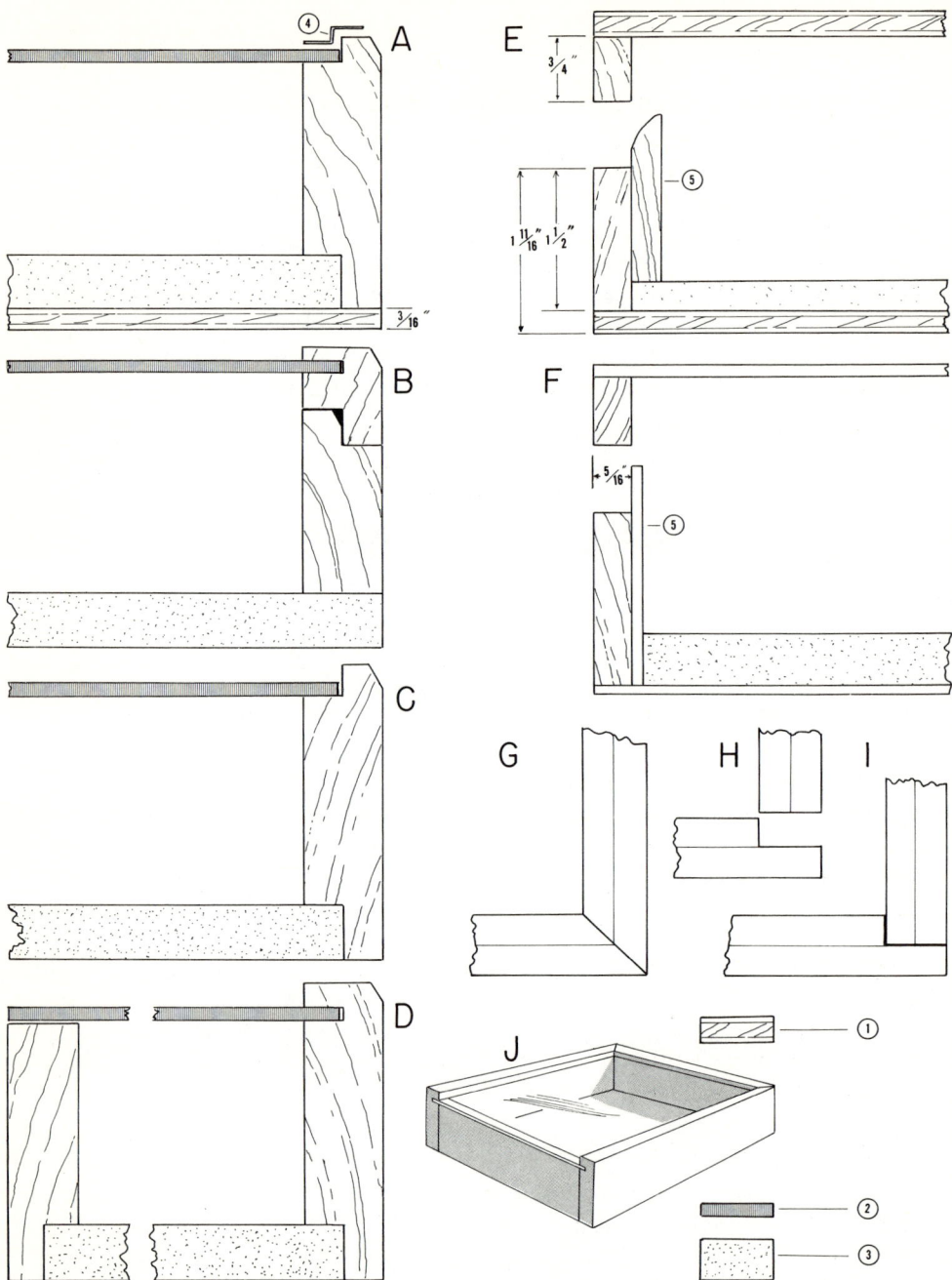

FIG. 13–15. Pinning cases. A–D. Top and bottom patterns for glass-topped cases. E. Solid case, plywood construction. F. Solid case, cardboard construction. G–I. Corner construction. J. Finished case as shown in D. (1) Plywood, (2) glass, (3) CeloteX, (4) black tape, (5) inner lip of box.

that Fig. 13–15, A, B, and D show three styles for the glass-top construction. Of these, Fig. 13–15A proves most satisfactory. Figure 13–15B shows the glass set into a wooden lid and requires extra craftsmanship. Figure 13–15, D and J, shows the construction of a sliding glass lid which is somewhat less satisfactory than that shown in Fig. 13–15A. Plans for the bottom construction are given in Fig. 13–15, A, B, and C, the latter being preferable. The bottom shown in Fig. 13–15A provides a better finish with the addition of plywood, but is more expensive; that shown in Fig. 13–15B is subject to chipping and abrasion.

When constructing a case like that in Fig. 13–15C, first select the glass size and then cut the ½-inch CeloteX bottom to the same dimensions as the glass. In figuring the height of the side, plan on no less than 1½ inches for the inside of the box, plus ½ inch for the CeloteX and ⅜ inch for the glass recess. If the inside dimension of the box is to be 2 inches deep, for example, the side boards should be ripped at 2⅞ inches. Next, set the table saw for ⅜ inch and cut out the recess for the glass. Do the same thing for the CeloteX bottom. Figure 13–15G shows the typical mitered joint of a corner (which is difficult to cut and nail). Preferably, construct the corner like that of Fig. 13–15, H and I. When cutting the side boards to fit the glass, remember it is the outside of the glass recess, not the outside of the box itself, that must be measured. Always add about $1/16$ inch extra to the size of the glass to prevent it from sticking. Glue and nail all corners in assembling the sides. Use finish nails which can be set and puttied. Nail the CeloteX in place with headed nails before the glue used in the side joints has dried. The bottom CeloteX will square the box and insure a good fit for the glass. Finally, sand the entire box, apply dull black paint or varnish to the outside of the box, and paste white paper or glazed cardboard on the inner sides of the box. Set the glass in place, using a small piece of tape at one corner so that it may easily be lifted out again. When the case is filled with insects and is to be permanently stored, the glass may be taped in place with black photographer's masking tape (Fig. 13–15A).

Figure 13–15, E and F, shows cross-sectional views of the construction of solid-topped insect cases made of plywood or ⅛-inch cardboard. In both cases the lid fits tightly over an inside shoulder, resulting in an airtight insect-proof case. When using the plywood construction, first cut the top and bottom sheets of plywood to the exact finish and size of the box. Using $5/16$-inch clear, fine-grained wood, cut side boards 2⅜ inches wide (1½ inches bottom, ¾ inch top, plus ⅛ inch for a saw cut). Cut the sides and ends to the proper length, making a square or mitered corner. Glue and nail the sides together and then glue and nail the top and bottom onto the sides, thus making a solid box. When this is dry, set up the ripping gauge on the table saw and cut the top off from the bottom of the box. In doing this, set the saw blade so that it will cut only $5/16$ inch. Next, cut some suitable pinning bottom such as $3/8$-inch CeloteX or $1/4$-inch cork and, if necessary, cover this with white paper. Place the pinning bottom into the box and then glue in the mitered inner shoulders as diagrammed. Carefully remove all excess glue and place the lid on the bottom of the box while the shoulders set and dry. This will insure a tight fit. Finish the box with light sanding and a coat of shellac and wax or a coat or two of varnish.

Figure 13–15F shows a solid insect case constructed with $1/8$-inch cardboard. The cardboard and side boards are cut as described above. The side boards are glued and nailed and the top and bottom cardboard are glued into place. When they are dry, saw the lid and bottom apart, add a shoulder piece of cardboard, and finally press a tight-fitting pinning bottom of $3/8$- or $1/2$-inch CeloteX into the bottom of the box. The pinning bottom is designed to hold the shoulder boards in place, and thus glue should be added to give additional strength to the box. This box may be unfinished or varnished or painted, as desired.

Insect-Case Layout. One of the most frustrating jobs in entomology is to determine a policy for collecting, for this will govern the way the insect cases should be laid out. To begin with, a taxonomic layout is always used where insects are grouped by orders, families, genera, and species. When adequate literature is not available, beginning collectors key their specimens down to the family level and then group all members of that family under one heading. The problem that develops is knowing beforehand how much space to allow for each family or genus or species, for the job of moving thousands of pinned insects to make room for an additional group is tremendous.

Some of the decisions that must be made ahead of time are as follows: it must be decided whether specimens will be identified beyond the order, family, or genus level; it must also be decided whether a general representational collection of each family, genus, or species is to be obtained or whether, ultimately, every species of a group will be sought. Also, one should decide whether he will collect and pin one specimen of each species or up to several hundred of each species. This will depend on the use of the specimens, whether they are for display and teaching or for taxonomic statistics. Some consideration in allotting space must also be given to the different sizes of insects. Many beginning collectors attempt to set aside room for orders and families depending upon their relative local numbers and general size. When insect cases with solid pinning bottoms are used, preplanning of space allotment is essential. This becomes much less serious, however, when unit pinning trays are used.

When using solid pinning bottoms, rule the bottom of the pinning case into columns 2 to 3 inches or more wide (the columns should be some fraction of the inside dimension of the box), as shown in Fig. 13–16C. Next, print or type labels for orders and families and affix these with $3/8$-inch straight pins (sold as sequin pins in dime stores). These labels may be plain or ruled around the margins, according to preference (Fig. 13–16D). All of the insects belonging to the particular group, as labeled, are pinned in place. Leave room for additional specimens, then attach the next group label along with its insects, and so on.

Unit Pinning Trays. Almost all large museums and serious collectors have adopted the use of unit pinning trays as opposed to the solid pinning bottom. Unit pinning trays consist of cardboard boxes of uniform width and depth, but varying in length. The sizes are designed to receive any group of insects, small or large. Usually one tray is reserved for each species, genus, or family, depending on policy. By this method, trays of insects can be rearranged and moved from case to case without one's having to repin the specimens. Figure 13–16A shows an insect case with pinning trays of various sizes inside. Each tray is provided with its own pinning bottom. Figure 13–16B gives some of the common dimensions for the pinning trays. Only the lengths are given, since the depth of the box is always $1 5/8$ inches and the width is usually $4 3/8$ inches (the width is a fraction of the insect-case width).

Unless a very large quantity of pinning trays is to be used, it is perhaps cheaper to buy these readymade (see Wards, Monterey, Calif.; Bio Metal Associates, Santa Monica, Calif.; etc.). Museums that make their own pinning trays order the boxes readymade at 1 or 2 cents each, and glue in a piexe of $3/8$-inch CeloteX. One can make his own boxes by cutting cardboard to form the bottom and sides, folding the sides upward and taping the corners. This is tedious work and uniformity is often lacking in the finished products. (However, see p. 191 for technique.)

Liquid Storage. Museums that keep collections of larvae, nymphs, or adults for study and dissection find it necessary to store specimens in liquid preservatives. Perhaps the best method is to keep specimens with their data in vials of preservative, stoppered with cotton (or a cork or screw cap). These, in turn, should be kept in wide-mouth museum bottles, along with enough additional preservative to cover the vials. A label placed in the

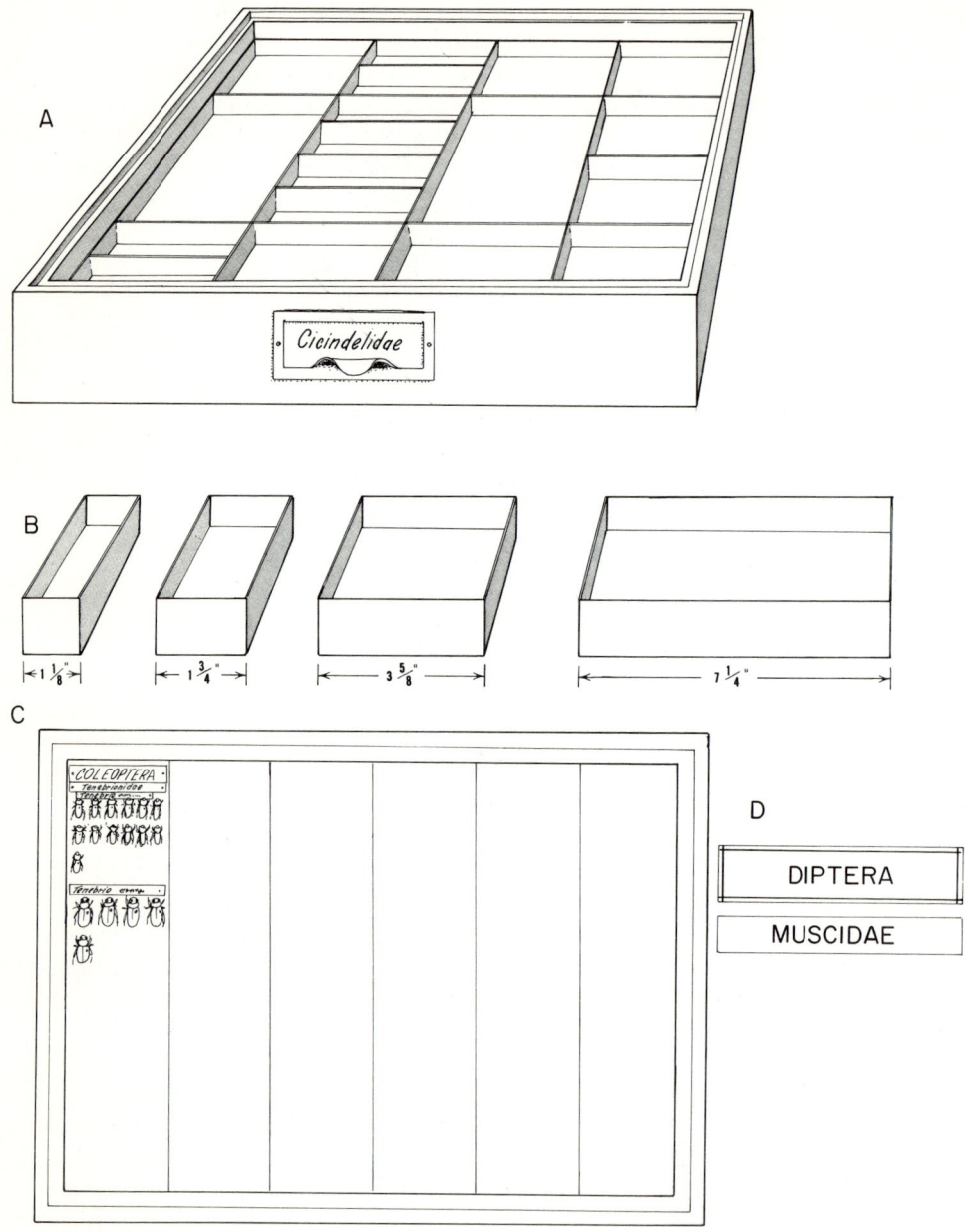

FIG. 13–16. Pinning cases. A. Case with unit pinning trays. B. Sizes of pinning trays. C. Layout of case with solid pinning bottom. D. Case labels.

bottle denotes the vial contents at a glance. This method reduces evaporation from the vials and requires little maintenance.

Storage Cabinets. Storage cabinets which hold cases of pinned insects are not essential, provided the cases are of solid construction with tight-fitting lids that will exclude light, dust, and insect pests. However, glass-topped pinning cases must be protected from undue exposure to light. Any tightly built cabinet with shelves may be adapted for this purpose by adding a foam rubber or foam plastic strip around the door. Special storage cabinets are easily built and prove most economical in cost and the use of space. Figure 13–17A shows a single cabinet design for twelve large insect cases or drawers. No dimensions are given on the drawings, since the measurements will depend on the size of case to be stored. Figure 13–17, B, C, and F, deals with a double and quadruple unit which will house 24 to 48 pinning cases.

The first step in constructing the single cabinet (Fig. 13–17A) is the determination of the exact measurements—first, of the insect cases to be stored; then, of the total cabinet. The inside height of the cabinet should measure twelve times the height of the insect case, plus $\frac{1}{8}$ inch for clearance for each drawer and $\frac{1}{2}$ inch for each runner that is to be used (Fig. 13–17, A, D, and E). The inside width of the cabinet is the same as that of the insect case, plus $\frac{1}{8}$ inch for clearance. The inside depth of the cabinet should be great enough for the insect case, plus any hardware (such as drawer pulls) included on the drawers. Select $\frac{3}{4}$-inch plywood of an A-A or A-B grade for construction.

First cut the side boards long enough to include $\frac{3}{4}$ inch each for top and bottom boards, plus 3 inches for the toe board, if it is used. Cut the top and bottom boards the width of the drawer, plus $\frac{1}{8}$ inch for clearance. Determine the size of the back board, which may be made of $\frac{1}{4}$-inch plywood, and cut this, being sure that it is absolutely square. The top and bottom and sides are glued and nailed together, and then the back is glued and nailed on, thus squaring the box. Drawer runners are best attached as shown in Fig. 13–17E where a nail-and-glue joint is used. Some people may wish to dado and inset the drawer runner (Fig. 13–17D), but this is not necessary. To space the drawer runners cut a spacer from a piece of wood the height of the drawer plus $\frac{1}{8}$ inch. Place the cabinet on its side and lay the spacer in place against the bottom board. Glue and nail the first runner in place; then move the spacer board up in position and nail the next runner in place, and so on. The front door is made of $\frac{3}{4}$-inch plywood and provided with three hinges (Fig. 13–17G) and two window catches (Fig. 13–17H). Apply $\frac{1}{8}$- by $\frac{1}{2}$-inch foam rubber or foam plastic all the way around the door so that it

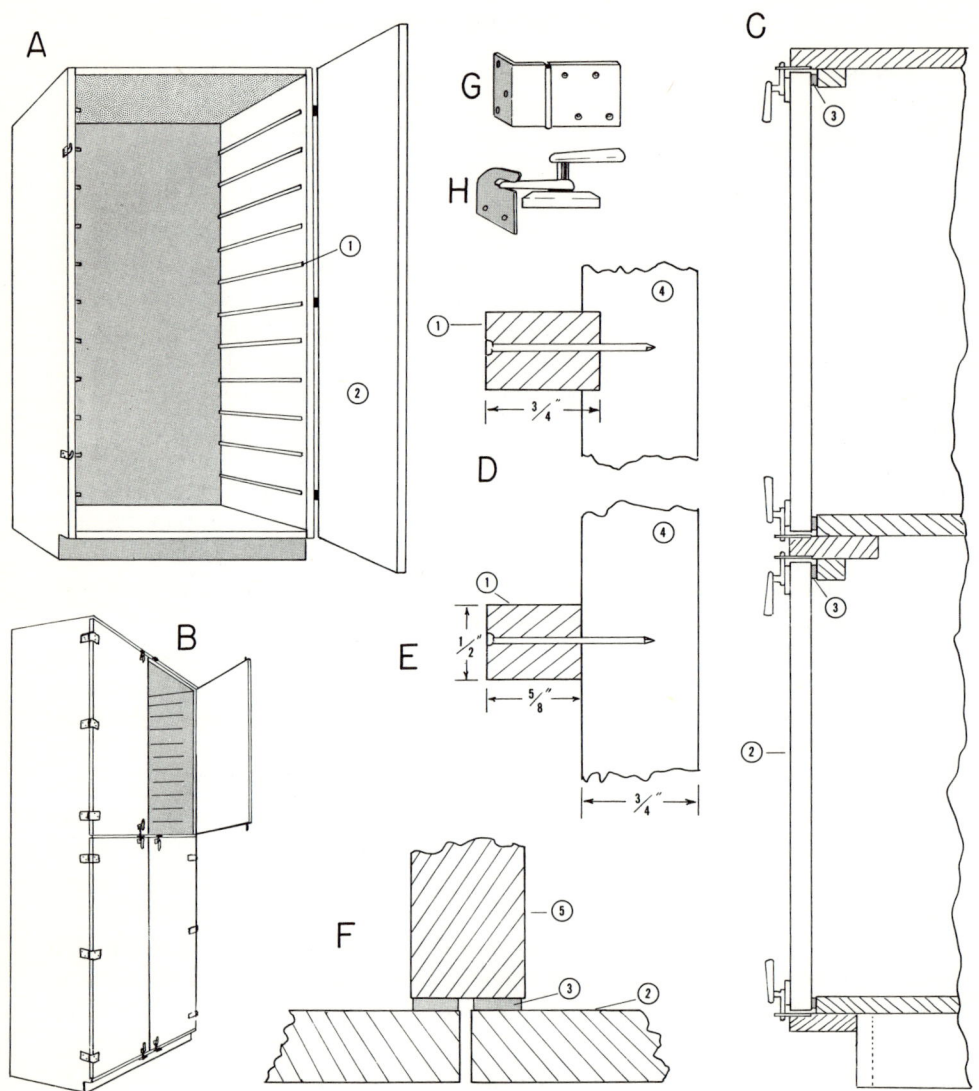

FIG. 13–17. Storage cabinets. A. Single cabinet for twelve cases. B. Cabinet for forty-eight cases. C. Longitudinal section, Cabinet B. D–E. Drawer-runner construction. F. Middle partition, Cabinet B. G. Suitable hinge. H. Window lock. (1) Drawer runners, (2) doors, (3) foam sealer, (4) side of cabinet, (5) central partition as seen from above.

will seal against the side, top, and bottom boards. Make sure that all joints are tight and fill any small openings with plastic wood or wood putty. Finish the cabinet by sanding and varnishing.

When making a multiple unit for 48 drawers, determine the height, width, and depth in the manner described above. The construction is entirely of ¾-inch plywood, with the exception that the vertical partition in the center of the cabinet should be of 1-inch plywood (Fig. 13–17F). Figure 13–17C shows a longitudinal cut down through the cabinet and gives the detail of the construction of the top, middle, and bottom partitions and the placement of hardware and foam-rubber sealers. Notice that a lip is provided to receive the door-catch hardware and to receive the foam rubber. The hinges are placed on the sides of the doors (Fig. 13–17B), but the catch hardware is placed at the top and bottom of the doors. The runners are attached and spaced in the same manner as that described above. Finish this cabinet by sanding and varnishing, taking care not to get varnish into the foam-rubber sealer.

Special Storage Problems

Pests and Fumigation. So-called "clothes moths" and dermestid beetles will quickly render hundreds of hours of work useless by destroying insect specimens. These pests must be kept out of insect cases by means of tight-fitting lids, tight storage cabinets, and fumigation. Fumigation should be a constant process in any museum, rather than one used only when an infestation has started. Paradichlorobenzene (PDB) or naphthalene flakes (moth crystals) should be added to each pinning case in small quantities about twice each year. The fumigant may be placed directly on the bottom of the case, but may leave a dust residue. Large pieces of PDB may roll around and damage delicate insect specimens. Workers usually make a small cardboard box for the fumigant and pin this in one corner of the case. PDB should not be used with a styrofoam pinning bottom, as PDB will make the styrofoam evaporate; use naphthalene flakes instead.

When looking for infections of dermestid beetles or clothes moths, look for what appear to be little piles of sawdust under the insect specimens. These may be on the labels or on the floor of the pinning box. If you observe them, remove the specimen and look carefully for a small hole where the larval pest has entered the specimen. Destroy the entire specimen unless it is valuable, in which case it may be especially fumigated with carbon bisulfide, strong PDB, or some killing agent. Next, fumigate the entire case of apparently uninfected insects with a large quantity of PDB or carbon bisulfide. Carbon bisulfide is very dangerous, and one should follow the precautions given on the container. Children should not be allowed to use or have access to these fumigants.

Light and Color. Insect colors are produced either by pigments or by structural devices. Some common pigments are carotene, which produces yellow and orange colors; melanin, which produces light tan to black; metabolic waste products (uric acid derivatives), which produce white, light gray, and metallic silver; and plant pigments that have been ingested and deposited under the cuticle. Structural colorations are produced over some common pigment, such as melanin, by irregularities on the cuticle of the insect which refract light to produce the brilliant metallic colors. Structural colors are the most stable, colors deposited within the cuticle are next; pigments deposited just beneath the cuticle are likely to be altered by organic changes that occur after the insect dies. Light will affect any of these colors in time, some more rapidly than others. Change of color is extremely slow and hard to detect and damage may occur before one is aware of it. The only sound advice is to keep insects in a light-proof case when they are not in use. Short periods of exposure will do no harm, but longer exposures of many weeks or years should be avoided.

Dust and Cleaning. Dust and lint will rapidly collect on insects that are not covered and will greatly alter their appearance and usefulness. Specimens that are greasy will have to be cleaned, as described below; those which are dry, but dusty, may be cleaned by gently brushing them with a water-color brush and then blowing the dust away with a fine stream of air.

Greasing and Degreasing. Some insects, especially beetles, flies, bees, and the like, contain large quantities of stored fat which may in time work through the cuticle and discolor the specimens, both by saturating natural pigments and collecting dust and lint. Wings of butterflies may become so greased that the scales will be matted and discolored. When this occurs curators usually degrease specimens by placing them in a solvent such as carbon tetrachloride, ethyl acetate, or others. Probably the most effective solvent is clean xylene or xylol. Occasionally, some of these degreasing compounds will affect insect labels, but usually the specimen, label and all, can be pinned to the cork of a large vial and then placed in a vial filled with the degreasing solution. Degreasing may take from one to several days, but is well worth the effort.

Moisture, Drooping, and Mold. Excessive moisture will penetrate the joints of appendages and eventually cause legs and wings to droop. With prolonged exposure to moisture, mold may begin to grow on certain insects which not only destroys them but ensures that spores will be distributed throughout the entire pinning case. The author once lost a box of extremely valuable butterfly specimens which were left by mistake in a damp basement in southern California. The storage room should be heated sufficiently to reduce excessive moisture.

Packing Insects for Shipment

Dried insects are shipped as pinned, layered, papered, or chlorocresol specimens, as described above. When shipping pinned specimens, outfit a cigar box or similar box with a soft pinning bottom, such as styrofoam, and pin the specimens deeply and securely into this. Cut a piece of cardboard just large enough to fit over the pins inside the box when the sides and ends of the cardboard have been folded upward. The sides and ends should be flush with the top of the cigar box, so that the box lid will hold the cardboard firmly in place. Tape all cracks and openings to the box and label the box as to its contents. Next, pack the specimen box inside a much larger pasteboard box and surround it on all sides with several inches of crumpled newspaper. Finally, tie and tape the pasteboard box (use paper gummed tape) and clearly label it on all sides with "Fragile" stickers. When declaring specimens for custom purposes, list them as museum specimens of no commercial value. If the box is to be shipped or hand-carried across a state or national border where agricultural checks are routine, the inclusion of paradichlorobenzene in copious quantities in the outer box seems to soothe the nerves of agricultural agents and also ensures against the transportation of some pest organism.

Living material is more difficult to ship in that problems of high or low temperatures, dehydration, feeding, agricultural regulations and laws, and the like, all have to be dealt with. Many states and countries have laws forbidding the shipment of living material and require that a prior permit be obtained. Air freight is the most suitable commercial method of transportation for live specimens. The container must be appropriately marked as having living material for research purposes and warnings to keep the specimens from extreme heat or cold.

SOME METHODS AND EQUIPMENT FOR REARING INSECTS

Juvenile insects are reared to obtain adult stages, to study life histories, to work with physiological or ecological factors which may affect growth and development, and so on. The best place to study insect development is, of course, where the insects occur normally and under natural conditions. However, it is often necessary to move specimens to a laboratory for closer observations. When this is done, one should duplicate as many of the natural environmental factors (pH, humidity, temperature, light, and so on) as possible. A constant source of food must also be provided.

pH, Humidity, and Temperature

Students attempting to do serious life history studies should consult Peterson (1959) for the excellent discussions of temperature, light, humidity, and other factors. In dealings with any specimen that lives within a liquid medium (water, mud, dung), the pH should be controlled—with buffers, if necessary. Humidity can easily be controlled in culture jars with various chemical solutions. Temperature should closely approximate outdoor conditions. Temperatures may be raised inside small breeding cages housed in a cool room, but reducing the temperature becomes a problem without refrigeration equipment. Light in the laboratory may be artificial or natural. With artificial lighting the problems of creating additional heat and providing the proper color (wave length) of light are important. The breeding habits of many insects are controlled by light—more particularly, by certain colors of light within the total daylight spectrum. Very inexpensive time-clocks can be purchased (Sears and elsewhere) which will turn lights off and on at desired times.

Rearing Cages

Plaster Cages. Insects that live in soil, dung, or similar media must have the moisture content of the rearing chamber perfectly controlled. Figure 13–18A shows a plaster of Paris cage that can be made very inexpensively. The large chamber holds soil and is covered by a piece of glass or screen; the small chamber provides water which will permeate the soil, thus maintaining the proper level of moisture. Figure 13–18B shows a cut-away view of the mold used to make this cage. After determining the size and depth of the cage that you wish to use, cut out a tapered block of wood the exact size of the water chamber and one the exact size of the rearing chamber. Make sure the sides are well tapered so that the plaster will easily slip away from the mold. Next, cut a piece of cardboard the size of the glass you intend to use. Nail the blocks and cardboard onto a baseboard and then paint these thoroughly with boiling paraffin so that all of the sharp corners are rounded with the paraffin. Finally, cut side and end boards (which are held in place by dowels or masking tape) and impregnate these with hot paraffin. Paint the entire mold with tincture of green soap (a parting compound) and pour in fresh plaster of Paris. Allow the plaster to become thoroughly hardened before removing from the mold. While removing, take off the side boards and then gently tap the plaster away from the bottom mold.

Screened Cages. Figure 13–18, C through F, shows several commonly used screen cages. Figure 13–18C consists of a gallon bottle with a flower pot and a food plant inside. Gauze or plastic window screen is tied over the top. Figure 13–18D shows a wooden screen cage built to the proper size of a plant flat. Food plants are reared or transplanted in wooden flats and the cage is set over the top. The cage and the specimens may be moved from one flat to another as the food is consumed. Figure 13–18E demonstrates the old technique of using a glass lamp chimney with a piece of gauze tied to the top; Fig. 13–18F shows an outdoor cage, essentially a screened box.

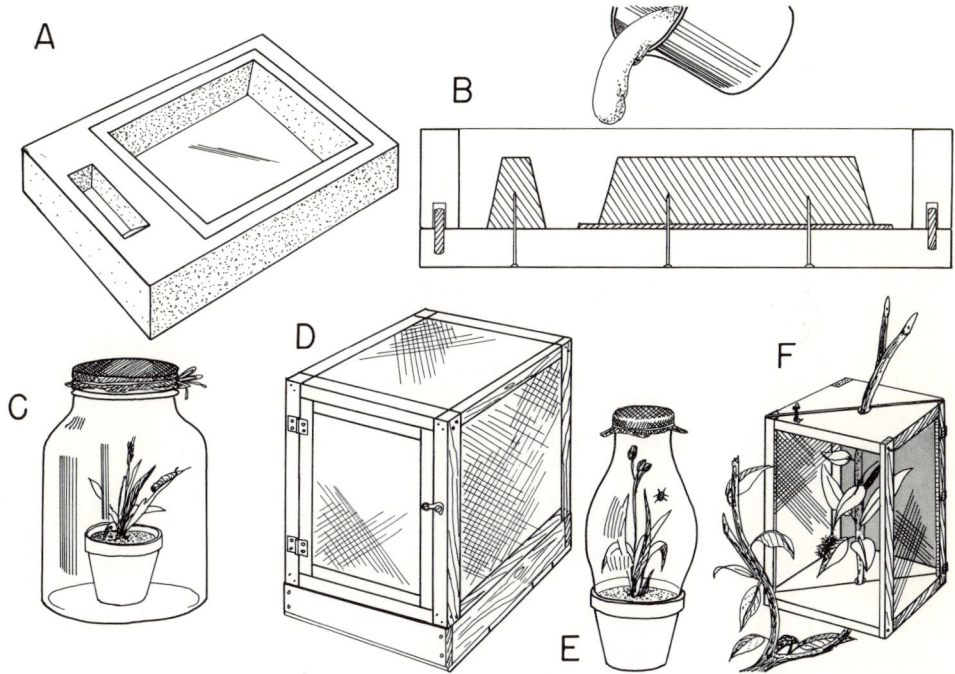

FIG. 13–18. Rearing devices. A–B. Construction of plaster of Paris rearing case. C. Rearing jar. D. Screen cage. E. Lamp-chimney construction. F. Movable outdoor rearing cage.

This box is split diagonally and is hinged so that it may be moved from one site to another on a tree as the leaves are consumed inside. This kind of a cage is also handy for catching newly emerged insects that have been parasitic inside the branch itself.

REFERENCES

Belkin, John N., 1954, *Laboratory Manual for Medical Entomology,* printed by Dept. of Entomology, Univ. Calif. at Los Angeles.

Blum, Murray S., and J. Porter Woodring, 1963, Preservation of Insect Larvae by Vacuum Dehydration, *J. of Kansas Ento. Soc.,* 36(2):96–101.

Chu, H. F., 1949, *How to Know the Immature Insects,* William C. Brown, Dubuque, Iowa.

Comstock, J. H., 1940, *Introduction to Entomology,* Comstock Publishing Assoc., Ithaca, N.Y.

Davies, D. A. L., 1954, On the Preservation of Insects by Drying in Vacuo at Low Temperature, *The Entomologist,* 87:34–35.

Ford, E. B., 1955, *Moths,* Macmillan, N.Y.

Gaul, A., 1953, *The Wonderful World of Insects,* Rinehart, N.Y.

Harman, I., 1950, *Collecting Butterflies and Moths,* John de Graff, N.Y.

Harrison, A. S., and R. L. Usinger, 1934, Methods and Techniques, *Bull. of the Brooklyn Ent. Soc.,* 39:168–170.

Holbrook, J. E. R., 1927, Apparatus and Method Used to Remove Pins from Insect Specimens, *J. Econ. Ent.,* 20:642–643.

Holland, W. J., 1934, *The Moth Book,* Doubleday, N.Y.

Holland, W. J., 1949, *The Butterfly Book,* Doubleday, N.Y.

Hussey, L., and C. Pessino, 1953, *Collecting Cocoons,* Crowell, N.Y.

Imms, A. D., 1951, *Insect Natural History,* Blakiston (McGraw-Hill), N.Y.

Jaques, H. E., 1947, *How to Know the Insects,* William C. Brown, Dubuque, Iowa.

Klots, A., 1951, *Field Guide to the Butterflies,* Houghton Mifflin, Boston.

Lutz, Frank, 1935, *Field Book of Insects,* Putnam, N.Y.

Merryman, H. T., 1960, The Preparation of Biological Museum Specimens by Freeze-Drying, *Curator,* 3(1):5–19.

Merryman, H. T., 1961, The Preparation of Biological Museum Specimens by Freeze-Drying: II. Instrumentation, *Curator,* 4(2):153–174.

Moore, B. P., 1951, On Preserving the Colours of Dragonflies and Other Insects, *Proc. S. Lond. Ent. Nat. Hist. Soc.,* Vol. 1949–50, 179–186.

Needham, J., and M. Westfall, 1954, *A Manual of Dragonflies of North America,* Univ. of California Press, Berkeley, Calif.

Peterson, Alvah, 1959a, *Entomological Techniques,* Edwards Bros., Ann Arbor, Mich.

Peterson, Alvah, 1959b, *Larvae of Insects,* Edwards Bros., Ann Arbor, Mich.
Ross, Edward S., 1953, *Insects Close Up,* Univ. of Calif. Press, Berkeley, Calif.
Ross, H., 1948, *A Textbook of Entomology,* John Wiley, N.Y.
Swain, Ralph, 1948, *The Insect Guide,* Doubleday, N.Y.
Tindale, Norman B., 1962, The Chlorocresol Method for Field Collecting, *J. of the Lepidopterists' Society,* 15(3):195–197.
Woodring, J. P., and M. S. Blum, 1963, Freeze-drying of Spiders and Immature Insects, *Annals Ento. Soc. Amer.,* 56:138–141.

CHAPTER 14

The Crustaceans

The crustaceans make up a class of arthropods (animals bearing jointed legs) which includes the so-called water fleas, fairy shrimp, barnacles, sow bugs, crabs, shrimp, and others. With the exceptions of some terrestrial and semiterrestrial crabs and isopods, etc., the crustaceans are aquatic and marine animals. They tend to be the "insects" of the aquatic world and occupy, to a large degree, all of the niches occupied on land by true insects. Because of the exoskeleton, the jointed appendages, and the well-developed sensory system, members of the Class Crustacea are capable of a great degree of activity, may be fast-moving, and will require special techniques for capture.

REMARKS CONCERNING PRESERVATIVES

The only practical method for maintaining crustaceans is in liquid preservatives such as alcohol or formalin. (Slide mounts and other techniques may be used for limited numbers of specimens.) However, three problems must be considered: (1) the stiffening of muscle tissues in jointed appendages as a result of preservation, (2) the loss of pigments in larger crustaceans, (3) the destruction of those exoskeletons containing calcium carbonate. A solution of 70-percent alcohol, properly used, is the best general preservative in that jointed appendages do not stiffen and the solution has no effect on calcium carbonate. It does, on the other hand, quickly leach out the pigments of crabs and other such animals. If neutral formalin is used (see Appendix C) color retention is much better, although a disagreeable odor and the stiffening of muscle tissue are results.

REHYDRATION OF DRIED SPECIMENS

Occasionally, museum jars are improperly sealed, causing the preservative to evaporate totally. The resulting dried specimens are fragile and do not respond well to represervation. However, if specimens are placed in a ½-percent solution of trisodium phosphate they will become thoroughly moistened and relaxed so as to permit the movement of appendages. After such a treatment they may be washed briefly in fresh water and returned to their original preservative.

SMALL CRUSTACEA

Included in this nontaxonomic assemblage (see Fig. 14–1) are orders Anostraca (fairy shrimp), Notostraca (tadpole shrimp), Conchostraca (clam shrimp), Cladocera (water fleas); the copepods such as *Cyclops;* the parasitic copepods; the isopods (aquatic sowbugs); the amphipods (side-swimmers); and the larval forms of all crustaceans, partially illustrated in Fig. 14–1, I, J, and K. These animals are lumped together because the techniques for collecting, preserving, and mounting are applicable to all. Therefore, unless other instructions are given, the general methods presented below may be used for any of the individuals.

General Collecting Methods

Processing Protective Plants. Many small crustaceans hide among freshwater and marine plants and other debris, some even taking on characteristic colors of green, red, or brown to match the plant host. Such plants may, of course, be placed in pans and examined with the aid of a magnifying lens. The simplest procedure for processing crustaceans is to suspend

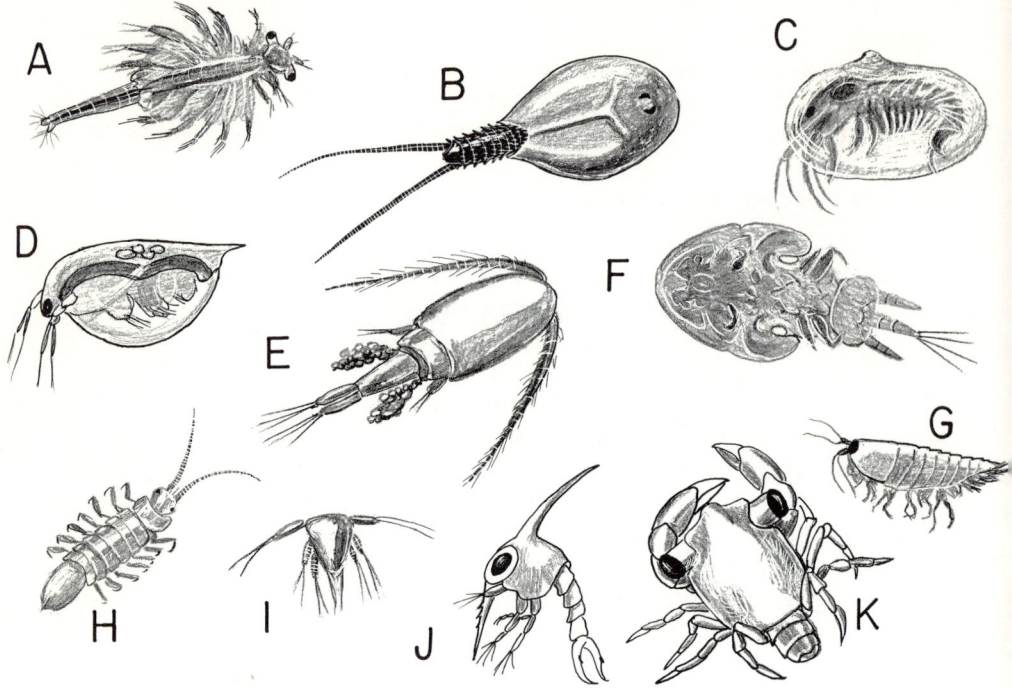

FIG. 14–1. Representative small crustaceans. A. Fairy shrimp. B. Tadpole shrimp. C. Clam shrimp. D. *Daphnia,* a water flea. E. Copepod. F. Parasitic copepod. G. Amphipod. H. Isopod. I. Nauplius larva. J. Zoea. K. Megalops.

quantities of such weed in a bucket of habitat water with a very small quantity of formalin added. The formalin is irritating and, sooner or later, causes the animal to abandon the host plant, swim about in the water for a brief time, sink to the bottom, and die. The plants may be shaken in the water once or twice before removal to dislodge any additional specimens.

Birge Cone-Net Technique. The Birge cone net is designed to work through vegetation of either shallow or deep water, and collects small organisms without including large quantities of weeds. The net (Fig. 14–2) is either towed behind a skiff or thrown out into weedy areas and retrieved.

A simple, modified Birge net consists of a forward wire-screen cone, a muslin net, and a collecting bottle at the tip of the net (Fig. 14–2A). To construct this net, make a hoop of copper, brass, or galvanized iron $\frac{1}{8}$ inch by 2 inches and about 8 to 10 inches in diameter, as desired. The metal should be overlapped and riveted or soldered to form the hoop. Next, make a smaller hoop of $1/16$-inch by 1-inch metal which is approximately $1/8$ inch smaller in diameter than the larger hoop, as seen in the cutaway diagrams (Fig. 14–2, B and C). Make a towing wire from 9-gauge galvanized wire,

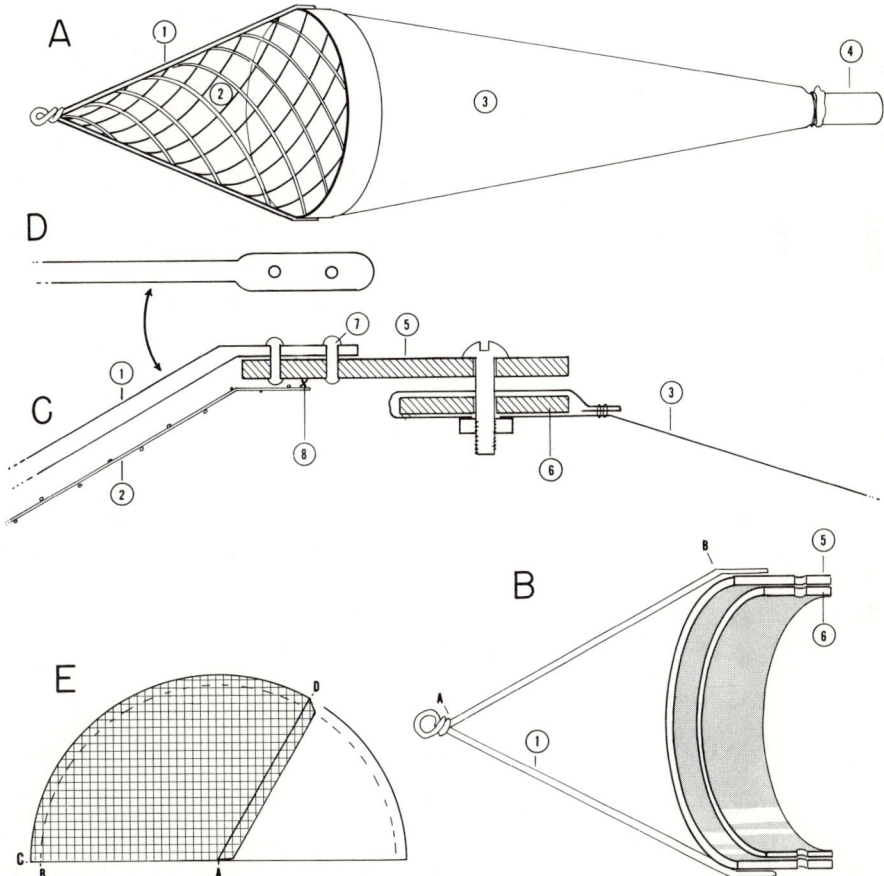

FIG. 14–2. Birge cone net. A. The assembled net. B. A cutaway diagram of the net rings. C. A cross-sectional diagram through the net. D. A detailed drawing of the wire leader. E. Cutting the screen for the cone. (1) Wire bridle, (2) screen wire cone, (3) cloth net, (4) receiving vial, (5) outer ring for wire cone, (6) inner ring for cloth net, (7) rivet, (8) solder screen at this point.

twist to form a forward loop, and bend to fit over the screen cone (Fig. 14–2B), flatten and attach to the large hoop by means of rivets (Fig. 14–2, C and D). The screen cone is made of ½- or ¼-inch mesh hardware cloth. Cut a trial pattern out of newspaper to determine the exact shape of the wire. Figure 14–2E shows the half-circular layout of a cone where A and B give the measurement corresponding to A and B on Fig. 14–2B. Note that an extra ½ inch is added beyond the length (A to B) which is bent inside the hoop and is soldered in place, as shown in Fig. 14–2C. The measurement C to D on Fig. 14–2E represents the circumference of the large hoop. Leave enough extra wire beyond point D to overlap and strengthen the cone. Roll the cone, wire it together, and solder it in place in the large hoop.

Next, place the small hoop inside the large hoop and drill a $3/16$-inch hole through both hoops, as shown in Fig. 14–2B. Solder a $3/16$-inch nut inside the small hoop over the hole, as shown in Fig. 14–2C. The large and small hoops may be joined with short $3/16$-inch bolts. Almost any stout cloth, such as muslin, may be used for the net cone. Make your measurements of length (A to B) and diameter (B to D), as shown in Fig. 14–2E. Be sure to leave a surplus of cloth to go around the smaller hoop and form a hem (Fig. 14–2C). The net bag may be blind or may be open to receive a collecting vial, as desired.

This net is very durable, but should be washed promptly with fresh water and dried immediately after use. The towing wire should be fastened to the apex of the cone to give greater strength.

Other Common Methods. The dip net, or water net (Chapter 13), is valuable for collecting single specimens from the water and for obtaining protective plants; the dredge net (Chapter 13) is also valuable. The pelagic forms of crustaceans, including larvae, are taken primarily with the coarse plankton net in open water. The night light (see Chapter 1, p. 14) used in marine waters is especially productive for small crustaceans.

Many species of small, fresh-water crustaceans produce thick-shelled resting eggs which withstand low winter temperatures or complete drying during droughts or summer periods. Therefore, the dry bottom mud from temporary ponds and pools may be rich with these so-called "resting" eggs (ephippia). To study the probable crustacean population (and that of other invertebrates) of such temporary ponds, scrape the upper ⅛ to ¼ inch of dry mud into a container, place this in an aquarium in the laboratory, and add sterile pond water (boil and cool) or distilled water. Within a few days to five weeks, many species of invertebrates will appear in such cultures. Small quantities of yeast, malted milk, or other suitable foods may be used if feeding is desired.

Specific Collecting Methods

Anostraca: Fairy Shrimps. These animals (Fig. 14–1A) are found primarily in fresh-water and inland salt-water ponds and lakes of either temporary or permanent nature. The dip net and Birge cone net, working near and on the bottom, are suitable for all but the youngest stages which are taken with a plankton net in more open water.

Notostraca: Tadpole Shrimps. These animals (Fig 14–1B) are found in temporary pools, springs, and ponds, especially in more arid regions, and occupy the substratum. Generally shallow waters permit collecting by means of a dip net, small aquarium net, or a Birge cone net.

Conchostraca: Clam Shrimps. These small organisms (Fig. 14–1C) are found in fresh-water lakes and in ponds of a temporary or permanent nature. They occupy areas of dense vegetation, often in shallow water, and may be collected by means of a Birge net or a dip net, or by removing vegetation and other debris and processing these protective plants. These organisms are quite seasonal in appearance.

Cladocera: Water Fleas. Water fleas such as *Daphnia* (Fig. 14–1D) are common in fresh-water lakes and ponds. They tend to be seasonal, population peaks occurring in the late spring and again in the late summer. They may be collected individually with the dip net among weeds, or in large numbers in open water and among vegetation by means of the plankton net and Birge cone net, respectively. Some marine Cladocera may also be collected in this manner.

Free-living Copepods. These organisms (Fig. 14–1E) inhabit both marine and fresh water. Pelagic species are taken with the tow net or plankton net in open water, and with the Birge cone net in vegetated areas. Protective plants should be processed as above. Some species may be obtained only by collecting mud and debris, such as rotting plants, and carefully examining this material once it has settled in finger bowls.

Parasitic Copepods. Go to any fishing dock or commercial or public fishing boats and examine freshly caught fish on the exterior surfaces, around the gills and gill chambers, and within the pharyngeal chamber. Parasitic copepods (Fig. 14–1F) are flattened and often exhibit a pair of egg sacs attached to the abdomen.

Isopods. Isopods (Fig. 14–1H) are found in fresh-water, terrestrial, and

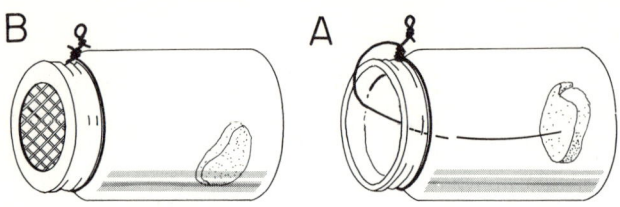

FIG. 14-3.
Two simple traps used for small crustaceans.

marine habitats. Sowbugs are found in moist, garden-like situations under rocks or boards in the daytime; they are active at night. They may be baited with pieces of potato placed under rocks or boards. Strictly aquatic forms occur in intertidal and subintertidal situations, and may be obtained with a Birge net, dip net, or by processing protective plants. Simple traps, made by including a meat bait in an empty jar or a screen-top jar (Fig. 14-3, A and B), are extremely effective in the marine habitat. Such traps should be lowered to or near the bottom and be left to "fish" overnight. High populations of marine isopods are not uniformly distributed, but may be found especially in sandy bays which support marine grass below the intertidal level. The sea slaters are obtained along rocky coasts, hiding under rocks or darting across open spaces. Slaters are usually dark-green or gray in color and are relatively large. Otherwise, they resemble terrestrial sowbugs.

Amphipods. These animals (Fig. 14-1G) occur in fresh-water and marine habitats. They are collected, like the isopods, by means of the plankton net, the Birge cone net, the simple bottle trap, and by processing protective vegetation.

Larval Crustaceans. Most early larval forms (Fig. 14-1, I and J) are pelagic, photopositive animals which are taken in the upper layers of open water by means of the plankton net. The night light is effective for luring specimens. Any drifting material, such as kelp, found some miles offshore may possess latter larval stages which are ready to metamorphose into the adult form. The metamorphic individuals usually work in toward shore and settle to the bottom, where they may be collected among plants or other debris. The megalops (Fig. 14-1K), which is the last larval stage of the crab, could be caught in large numbers by processing protective weeds or simply turning large rocks and looking for transparent organisms clinging to the undersides.

Narcotizing or Quieting Small Crustaceans

Small crustaceans may be examined directly under the microscope in depression slides. They may be slowed mechanically by adding methyl cellulose to the water, as described for the protozoans. Guyer (1929) suggests adding 2 parts of 1-percent Chloretone for every 5 parts of water in the culture. Specimens left briefly in this mixture may be revived by washing in clean, habitat water.

Killing and Preservation

Isolate specimens in a vial or small bottle and add 4 or 5 drops of 10-percent formalin to poison the animals. When all specimens have settled to the bottom, carefully remove the water and replace first with 35-percent and then with 50-percent alcohol, for 30 minutes each. Store in vials of 70-percent alcohol, with the collecting data included. Vials should be of screw-cap construction. Place vials in a larger, airtight bottle with the same preservative, for permanent storage.

The formalin-habitat water solution may be replaced with 3- to 4-percent formalin for both temporary and permanent storage, if desired. Some workers like to kill specimens by pipetting them directly into 70- to 80-percent alcohol for killing and preservation. This may cause some distortion of the fragile forms. Pennak (1953) recommends killing copepods directly in 95-percent alcohol and storing them in 70-percent alcohol.

Slide Mounts

Stains are not always essential for small crustaceans. Should stains be desired, acid fuchsin may be used at the water level (before specimens are placed in alcohol), or borax carmine may be used at the 70-percent alcohol level, as directed below. These stains are especially good when animals are treated in bulk. If appendages are to be dissected from larval and adult forms, especially from copepods, mount them in glycerin jelly, Monk's mounting medium in Permount, or similar media, as directed below. Pennak (1953) suggests that all copepod appendages may be seen if from five to ten whole specimens are mounted, ventral side up, on the same slide, for the appendages become positioned in such manner that all may be seen satisfactorily by examining the entire series.

Resinous Mounts Without Stain. Although the author prefers to use stain, the present basic method will be presented, as it is satisfactory in itself and will serve as a basis for alternate mounting methods and staining techniques. Freshly killed specimens, or those preserved in formalin, should be slowly moved up through 35-, 50-, 70-, 85-, and 95-percent alcohol, and two washes of absolute alcohol, with 15 to 30 minutes per wash. Clearing may be carried out in clove oil, cedarwood oil, or xylene. Equal parts of absolute alcohol and the clearing agent should be used as an intermediate step into the clearing agent itself. At this point, check specimens under the microscope to determine if they are too transparent; if so, stain as directed below. Depression slides, slides with a spun cell or a plastic slide cell, or slides with supported coverslips may be required for thicker specimens (see Appendix B). Transfer a specimen to an appropriate slide and blot away the excess clearing agent, leaving only enough to prevent air bubbles from forming under the specimen. Add enough mounting medium (balsam, Permount, or similar mounting media) to fill the space under the coverslip, tease any bubbles away from the specimen with a fine needle, and add a coverslip by dropping it straight down on the mounting medium. Keep the slide flat until thoroughly dry. No ringing is necessary. Add labels with taxonomic and collecting data.

Staining Overcleared Specimens. In a modified technique from French (1942), specimens which are too transparent may be stained while in the clearing medium by adding a few drops of light green, eosin, chlorazol black, or orange G stain to the clearing agent. To make these stains add excessive quantities of powdered stain to absolute alcohol, filter, and use. Overstaining is not likely to occur. This method may be used as a standard practice for all slide mounts, as well as an emergency treatment.

Alternate Mounting Methods. Specimens may be mounted in Turtox CMC-10 or CMC-S. Both of these media are nonresinous and will mount, clear, and, in the latter case, stain specimens transferred from water or any of the alcohol solutions (see Appendix B). The author prefers to mount specimens from 50- or 70-percent alcohol.

When specimens are being prepared for mounting in resinous media, they may be transferred from 90-percent alcohol into Terpineol, which will dehydrate and clear them for mounting in balsam or Permount, thus eliminating the need for absolute alcohol. Alternately, specimens may be transferred from 90-percent alcohol into Euparal Essence until cleared. They are then mounted in Euparal (see Appendix B).

Acid Fuchsin Stain Technique. Acid fuchsin (Appendix C) is an aqueous stain; thus, specimens are stained before they enter alcohol or are rehydrated (Appendix B) down to the water level and stained. This stain is soluble in water but becomes less soluble in increasing percentages of alcohol. It may be intensified in its action by acidifying the stain. Should overstaining occur, tap water or alkaline alcohol will destain the specimens to the desired density. Some experimentation may be required to determine the time required for staining. Place specimens in the stain from a few minutes to an hour or more, removing one to water occasionally to check its intensity. Remember that the stain becomes intensified and apparently darkened in the clearing process; thus, specimens should not be too dark. From this point, dehydrate through the alcohols and mount in a resinous or alternate medium, as described above.

Borax Carmine Stain Technique. Dehydrate specimens to the level of 70-percent alcohol. Transfer to borax carmine (Appendix C) from 1 to 24 hours. Destain in acid alcohol until the specimens are light pink. Dehydrate up through absolute alcohol, clear, and mount in a resinous medium or by one of the alternate methods mentioned above.

Fluid Mounts. Permanent fluid mounts for small crustaceans such as the cladocerans (French, 1942) are simply made with depression slides, spun cells, or plastic ring cells (Appendix B). Make a solution of 30 parts glycerin, 3 parts formaldehyde, and 67 parts water. Place specimens to be mounted in this solution. Select coverslips that are slightly larger than the inside of the cell. Paint the edge of the depression or cell, all the way around, with a waterproof and alcohol-proof cement such as Murrayite. Place the specimens in the cell with an excessive amount of the fluid medium, position the specimens, and drop the coverslip directly down on the mount, pressing gently so as to force out excess fluid. The Murrayite will cement the coverslip in position, but should be augmented with two or three additional coats applied to the outside of the coverslip after 2 hours.

Morrison (1943) notes that two common defects in fluid mounts are (1) the eventual development of bubbles which ultimately ruin the slides and (2) the cracking of the ring seal around the coverslip. The latter problem may be overcome by using a good sealing cement such as Murrayite or Turtox Special Sealing Cement. The former problem, however, develops because the fluid medium, especially formalin, contains large quantities of dissolved oxygen which becomes liberated. This can be overcome by (1) boiling the formalin under a hood in order to drive off the oxygen or (2) subjecting the formalin to a vacuum pump for a few minutes. After the oxygen has been removed from the formalin, protect it from absorbing new oxygen by placing it in a siphon bottle and covering it with a layer of mineral oil. Small crustaceans may, if such proves necessary, be subjected to a moderate vacuum for a few minutes, along with the formalin.

Glycerin Jelly Mounts. Glycerin jelly mounts have long been used for permanent and semipermanent slides of small crustaceans. The material need not be stained, but is rather transferred to a solution of 70-percent alcohol containing 5- to 10-percent glycerin. Place specimens in this fluid for five days to a week, until most of the alcohol and water has evaporated. During this period the glycerin impregnates the specimens slowly enough to prevent the collapsing of tissues. Next, select a suitable depression slide (or create a spun cell or mount a plastic ring cell, see Appendix B) so as to accommodate the thickness of the specimens. Place a piece of glycerin jelly large enough to fill the cell to a slight excess, and heat (do not boil) the slide gently over a flame until the jelly is liquefied. With a fine forceps or bristle transfer a specimen to the glycerin, work it down into this medium, and position it in the desired way. The yet-warm glycerin jelly should flow evenly over the specimen, creating a slightly convex surface. Put a ring of Murrayite around the cell and lower the coverslip directly, squeezing out the excess glycerin jelly with a soft cloth. If a flat slide, with or without coverslip supports, is used, cut the excess glycerin jelly from around the margin of the coverslip after it has hardened. With either type of slide, seal the coverslip all the way around with two or more coats of Murrayite or some other suitable cement. When the labeling is completed the slides may be stored, preferably flat, in a slide box.

Monk's Mounting Medium. Monk (1938) developed a medium for mounting small appendages and mouth parts of crustaceans (copepods) in a fixed position. This medium, made primarily of clear corn syrup (see Appendix C) and pectin, sets up within a few moments after exposure to air. Specimens are dissected in glycerin, as described below. When the appendages are removed, a thin layer of Monk's medium is spread on a slide and the specimens are immediately arranged on it. Within 2 minutes the medium begins to set and will continue to dry with or without gentle heat. There are two methods for placing coverslips on such slides. In the first, simply add another drop of the medium and a coverslip. In the second, the slide is premitted to dry thoroughly in a dust-free place, at which time a coverslip is cemented on with Permount, Euparal, or similar media.

Dissection Techniques

Small crustaceans may be dissected directly in water or alcohol, but are best handled in glycerin. Transfer specimens to 70-percent alcohol which has 5- to 10-percent glycerin added, and allow them to stand in a dust-free place for 5 to 8 days; this permits the glycerin to impregnate the body while water and alcohol evaporate. Specimens are dissected in depression slides under the dissecting microscope in an adequate quantity of fresh glycerin. Many types of dissecting tools, hooks, and needles may be made. These are discussed in Chapter 13 under dissecting techniques.

Dissecting will require considerable skill, especially when three or four pairs of appendages must be removed from specimens measuring 1 to 2 millimeters in length. The author's preference for dissecting crab larvae (which are always present in large numbers when cultured) is to hold the entire animal down with his fingernail and tease away the desired appendage. One specimen is used for each appendage desired, but the tedium of delicate dissection is removed in this manner.

Rearing Larval Crabs and Other Small Crustaceans. Many methods for rearing small crustaceans in the laboratory are to be found in the literature. As an example, the author's method for rearing crab larvae (Knudsen, 1958) is given. Select female specimens whose eggs, which are attached to the swimming legs, show well-developed eyes and are grayish in color. Transport females back to the laboratory and isolate them, one per aquarium, in several gallons of sea water. Running sea water cannot be used, as larvae will be lost; thus, aeration is essential. If temperature is difficult to control, place the aquaria in a large water bath with running sea water or tap water. Hatching generally occurs at night, almost all of the larvae emerging within an hour or two. Isolate between 20 to 50 larvae in a wide-mouth, squat pint jar. Prepare twenty or thirty such jars per species. Place these jars in a water bath with circulating water, to maintain reduced temperatures. Ideal temperatures should approximate those of the natural habitat.

Fresh brine shrimp eggs should be hatched daily and the first larvae to emerge may be used for food for the larval crabs. These are pipetted into the culture jars and allowed to remain for several hours. Following feeding, the crab larvae are removed into a fresh jar of sea water. Each jar is, in turn, washed and replenished with sea water. It is essential to get rid of the brine shrimp larvae, as they grow quickly and soon will attack the crab larvae. Feeding may be required daily or, at least, every other day. As specimens die, remove them from the culture jars.

Crab larvae will molt four or five times during their cycle. Molting may be detected by the presence of the shed "skin" on the bottom of the culture jar. Newly molted individuals are evident in that they are much larger than the first zoeal stages. An adequate number of specimens of each stage should be preserved, as mentioned above. Molting is the most dangerous stage for crab larvae and the highest mortality will occur at such times. With the use of a strong light, crab larvae or brine shrimp can be made to gather along the side or top of the culture jar. Very weak individuals usually sink to the bottom and will not respond to light. These should be transferred or removed. By preserving representatives of all larval stages the observer may study the life history over 5 to 10 weeks, when the last larval stage (megalops) will metamorphose into the first crab stage. For other methods of rearing crab larvae and other crustaceans consult Galtsoff, and others (1959).

CIRRIPEDIA:
THE BARNACLES

Habitat and Collecting

Barnacles occur in brackish and marine water, in and below the intertidal zone. Adult barnacles are sessile and thus will be found only on some solid substrate such as rocks or pilings. Many of the gooseneck barnacles are found attached to rocks only in the unprotected rocky coast zones. Others are pelagic (Fig. 14-4) and are attached to any solid object that has long been afloat in the ocean, or may actually produce their own floats (*Lepas fascicularis*).

Sessile barnacles, such as *Balanus*, cement themselves directly to their substrate and produce, in addition to the conical shell that surrounds the body, a basement shell between the soft parts and the substrate. Because of the basement shell, portions of the substrate are usually collected. Barnacles can generally be loosened from wood and the like with a pocketknife, but a hammer and chisel are essential on rocky substrates.

Narcotizing and/or Preservation

Preservation. Direct preservation is useful unless open, expanded specimens are desired, in which case narcotizing is essential. The immediate preservation of live specimens causes contractions and exhibits only the outer shell. If the shell alone is desired, kill specimens in 5-percent formalin, dry after 12 hours, and store in small boxes provided with tissue paper and the field data. Specimens for taxonomic purposes should be killed, as above, but then washed in fresh water, transferred through 35- and 50-percent alcohol, and stored in 70-percent alcohol.

Narcotizing. Specimens may be narcotized in the following way: Place barnacles in finger bowls with sufficient sea water to permit complete expan-

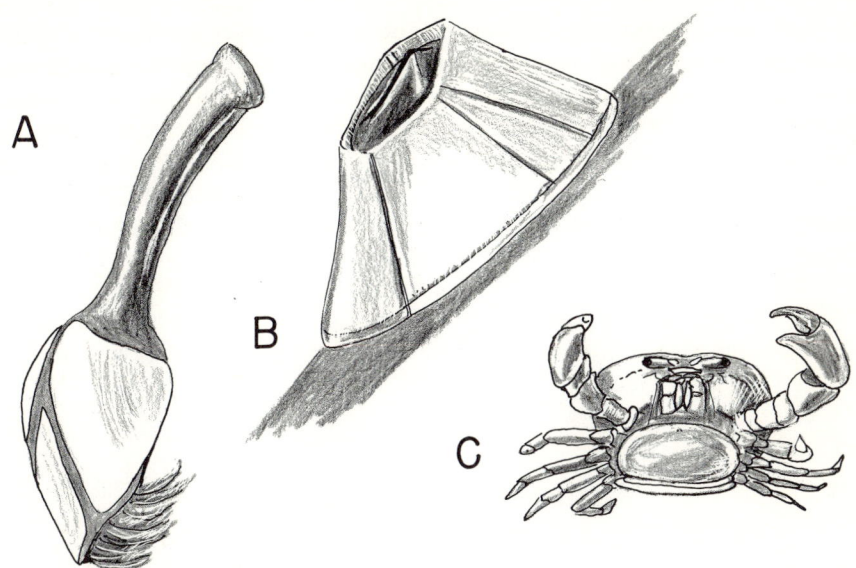

FIG. 14–4. Barnacle types. A. Gooseneck barnacle, showing the stalk for attachment, calcareous plates, and extended appendages. B. Sessile acorn barnacle. C. *Sacculina*, a parasitic barnacle attached to the abdomen of a crab.

sion. When they are expanded and active, add a large pinch of Chloretone or menthol crystals and let stand undisturbed for 6 hours. Specimens are usually fully expanded and insensible to gentle probing at this time. Introduce a small drop of formalin with a pipette directly on the appendages of one individual. If there is no contraction, preserve by adding enough formaldehyde to make a 5-percent solution; then wash and transfer to alcohol, as described above. The author finds that epsom salts are usable but must be added in increasing amounts every 10 to 20 minutes, require longer than 6 hours as a rule, and often fail to properly anesthetize all specimens. Barnacles may also be overcome in collecting containers or sealed jars which have some organic material (such as the visceral mass of a sea cucumber) which will putrefy the water, use up the oxygen, and kill the barnacles in an expanded condition.

Parasitic Cirripedia: Rhizocephala

Rhizocephalans such as the familiar *Sacculina* are parasites of crabs, shrimps, and the like. These generally occur in a low percentage of the host animals, tend to be somewhat host-specific (infecting only one species), and thus are not directly collected, but rather are taken incidentally in general field work. Look for irregular swellings, hair-like or sac-like growths extending from the appendages, or large protuberances underneath the abdomen (Fig. 14-4C). Specimens intended for display only should be killed in the manner of larger crustaceans (described below) and preserved in 70-percent alcohol. Specimens intended for taxonomic work should be plunged into Bouin's fixative in which they may be stored. Bouin's fixative will attack the shell of the host, but this is of no consequence. Eventually the entire host must be stained and then sectioned in order to reveal the root-like growth of the barnacle which penetrates all the internal tissues and appendages. Consult some textbook on microtechniques for staining and sectioning methods.

LARGER CRUSTACEANS: LOBSTERS, CRAYFISH, CRABS, AND OTHERS

Collecting

General Methods. The large crustaceans are more active at night than during the day. Fresh-water species are most easily collected by trapping, but may be taken with a dip net in shallow weedy areas in lakes, streams, and rivers. Marine forms are generally trapped, dredged, or collected by hand in the intertidal zone. These specimens hide under and around any kind of debris or rubble, live within burrows, hide within the sand, and in general occupy almost all possible microhabitats. The dip net (Chapter 13) is useful for securing free-moving specimens. Crabs and the like are easily secured by hand, though often they must be grasped by the back edge of the carapace or in such a manner as to pin both pinching hands against the carapace. On small and medium-sized crabs and crayfish you can position your fingers immediately behind the pinching arms, which will prevent the animal from reaching back far enough to pinch during collection. Always collect specimens into plastic buckets provided with large quantities of nonmucous-secreting seaweed to provide hiding places. Without such a baffle specimens will fight and tend to shed their appendages. Small pinnotherid crabs may be screened from gravel near the low-water mark in bays and estuaries, or secured from inside the mantle cavity of clams, mussels, and the like. Hermit crabs remain motionless when alarmed, but quickly give their identity away if the collector remains quiet and watchful. Pieces of bait (fish or meat) lowered into tide pools will lure shrimp from their hiding places in 5 to 10 minutes, at which time they may be netted. Larger crabs and lobsters are readily obtainable by skin diving and looking carefully in crevices along reefs, and similar places. Spider crabs and decorator crabs occupy kelp beds, pilings, weed-covered rocks, and the like, but usually betray their presence by movement after a few moments of observation.

Traps. Crustaceans are easily lured to traps containing fish, meat, liver, or other baits. Solid baits may be tied directly inside the trap, while small pieces of bait may be placed in a plastic bottle with holes drilled through the side (Fig. 14–5C). The building materials for traps must be considered. Fresh-water traps may be made of most convenient materials, whereas salt-water traps are quickly destroyed (in 1 to 2 weeks) when in constant use. Heavy-gauge, galvanized hardware cloth or screen will make excellent traps for occasional use, but is attacked by electrolysis very quickly. Untreated cotton netting will also be destroyed in a matter of days in the marine situation, but treated cotton lasts a long time. Nylon netting stretched over a metal frame proves very durable and quite satisfactory. Traps should be left in the salt water no longer than necessary, and last longer if thoroughly washed in fresh water and dried after using.

Most traps work on the funnel principle. One such trap is made of ½-inch hardware cloth 36 inches wide and 58 inches long (Fig. 14–5, A through E). Cut the hardware cloth as indicated (Fig. 14–5D) and bend along the dotted lines, using a board to achieve a sharp bend. First make the two bends along the length of the screen. After bending, flatten the screen again and make the four bends across the screen so that the box is formed. Next, fold the side pieces up and down, so that they overlap, and fold the two end flaps in. Wire the whole trap together with 18-gauge galvanized wire. Cut a 3- by 4-inch opening in the top of the trap for a door. Make the door 4 by 5 inches and attach it to the trap by three loops of wire (Fig. 14–5B). The door may be tied shut or held closed with a rubber band and hook, as shown. Finally, make a bridle and a trapline to lower the trap to the bottom (Fig. 14–5E). The line may be secured to a float or tied to a piling.

Another simple, cylindrical funnel trap (Fig. 14–6) with one or two funnels may also be made from hardware cloth. A trap diameter of 10 to 12 inches with a funnel opening of 1½ to 2 inches is adequate. The bait is suspended between the funnels and a door is cut and attached as described above (Fig. 14–6A). After deciding the dimensions (length and diameter), roll the newly cut piece of hardware cloth and wire it so as to form the outside drum. Next, lay out a half-circle according to the dimensions of

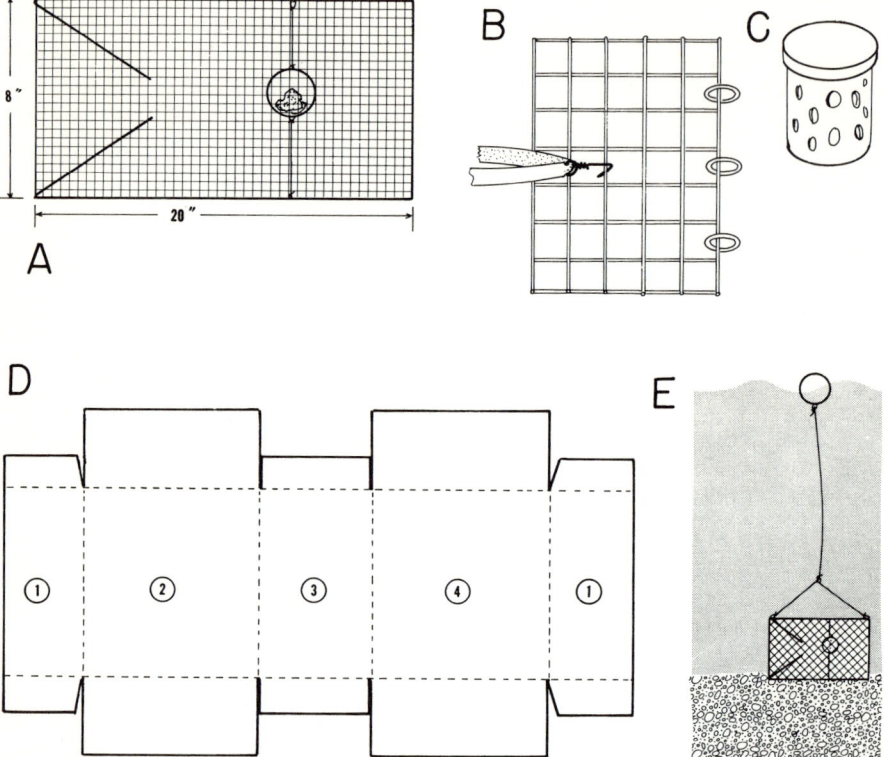

FIG. 14–5. A funnel trap used for small or large crustaceans. A. Side view of trap. B. Detail of door and rubber-band lock. C. Bait container. D. Pattern for cutting screen. E. Using a float marker for locating the trap. (1) Funnel, (2) bottom, (3) end, (4) top.

Fig. 14–6B, where A to B equals the length of the side of the funnel (note Fig. 14–6A) and A to C represents the over-all diameter of the trap. Leave a little surplus of wire, as shown, cut and roll the funnel, wire, and install by wiring around the entire end of the trap. When both funnels are in place, cut a door and add a bait container, as shown in Fig. 14–5C. Finally, make a sling along the top for quiet water settings, or on one end for streams and river use (Fig. 14–6C).

Another useful trap which the collector tends constantly is made from a square of hardware cloth or fish net. The corners of the net are attached to a crisscross frame made of Number-9 galvanized wire, which is formed in the center as shown in Fig. 14–6E. Attach a bait to the center of the

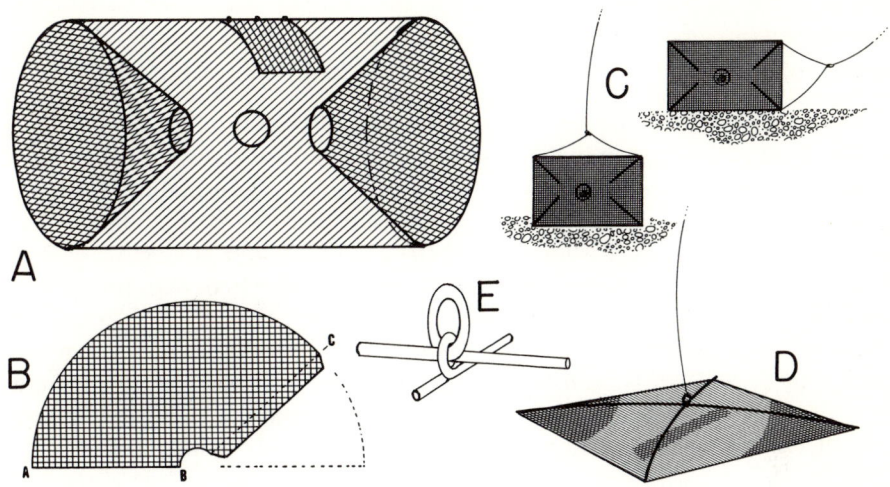

FIG. 14–6. A cylindrical funnel trap for crustaceans. A. Side view of trap. B. Cutting the funnel cone. C. Setting the trap in quiet water or moving stream water. D. A manual crab trap. E. Detail of the cross-wires for the manual crab trap.

netting, lower the net in a likely collecting spot, and retrieve the net rapidly every 5 to 10 minutes. Crabs and shrimps will move onto the net for the bait, and are usually pinned to the net when it is swiftly pulled to the surface.

Preservation

Some Problems. Almost all of the large crustaceans (decapods) can autotomize or break off any of their five walking legs at will. Since shedding the legs is related to a defense reaction, these animals frequently lose one or all of their major appendages if placed directly in formalin or alcohol. In addition, muscle tissues will become brittle if improperly preserved. (The reader should consult the section on preservatives at the beginning of this chapter.)

Killing. All marine specimens may be killed by adding fresh water (50 to 75 percent) to the collecting containers. Either provide sufficient seaweed so that these crabs may hide, or else separate them into small containers. Unseparated animals will fight and automatically shed some of their appendages. Crabs from tropical waters may be killed in iced sea water; fresh-water specimens, and others for that matter, may be placed out of water where they will quickly die. Putrid sea water (containing rotting organic matter) will asphyxiate crabs, generally with no loss of appendages.

Hermit Crabs. Hermit crabs are prone to withdraw into their borrowed shells during preservation. They may be evicted by placing them in a mixture of fresh water and sea water. A few drops of formalin, insufficient to kill them, may hasten this process. If undisturbed by other individuals, hermit crabs will first withdraw into their shells under the stress of fresh water or formalin, but soon will leave their shells when withdrawal does not better their condition. Attempting to pull the crab free from its shell usually results in breaking off the abdomen. When the animal voluntarily leaves its shell, quickly transfer it to a mixture of salt and fresh water until it is dead. Some workers chip a hole into the mollusk shell and insert a curved wire, which generally causes the crab to move out of the shell.

Dried, Intact Specimens. Although the molted exoskeleton of the decapod is desirable for dried display specimens, entire animals may be dried very much like insects. Place small individuals in a solution of 5 parts formaldehyde, 5 parts glycerin, 30 parts alcohol, and 60 parts water. After 12 hours, remove the specimens and posture their appendages, crisscrossing pins over their legs if necessary, and dry in a warm place out of direct sunlight. Probably the most useful method for drying decapods is the technique developed by Walter Windsor of Los Angeles. This method is suitable for specimens of any size, including very large king crabs. Kill large crabs or similar decapods as directed above, and remove the carapace by carefully cutting the membrane around the posterolateral margins. Remove all of the internal tissues of the body with a scalpel, scoop, and bent wire. Next, soak the specimen in plain water for 24 hours or more, to loosen the meat. Following this, insert a large hypodermic needle through the membrane between the last and the next-to-last leg segments. Inject enough water to force the meat out of the leg. Repeat this process for all legs and the pinching hands, by inserting the needle through the membrane around the thumb of the hand. When the meat is removed, position the crab for drying, and glue the carapace back on. Finally, when the specimen is dry, paint the entire animal with boat resin, mixing the resin with its catalyst just before using. This material will penetrate the exoskeleton, will strengthen it greatly, but will not give it an artificial varnished look. Specimens prepared by Windsor in this manner have retained almost perfect color for years.

Skeletal Studies. For study of the complex maze of internal skeletal structures and calcified tendons, the soft parts of the tissue may quickly be reduced by injecting a 20-percent solution of potassium hydroxide (KOH) through the thin membranes of the body. As this reduces the internal tissues the carapace can easily be removed, at which time the entire animal may be submerged in a solution of KOH. As soon as most of the tissues are removed, wash the animal thoroughly in several changes of fresh water. The legs present a definite problem here and may be punctured at the membranes and submerged for several hours in each wash. Finally, rinse, dry, and store the specimen in a small box with its label.

REFERENCES

Darwin, C., 1851, *A Monograph on the Sub-class Cirripedia,* Ray Soc., London, Vol. 1.

Darwin, C., 1853, *A Monograph on the Sub-class Cirripedia,* Ray Soc., London, Vol. 3.

Edney, E. B., 1954, Woodlice and the Land Habitat, *Biol. Rev.,* 29:185–219.

French, A. J., 1942, Notes on the Preservation, Mounting and Staining of Entomostraca, *The Microscope,* 5:20–22.

Galtsoff, P., et al., 1959, *Culture Methods for Invertebrate Animals,* Dover Publications, N.Y.

Guyer, M. F., 1929, *Animal Micrology,* Univ. of Chicago Press, Chicago.

Kiser, R. W., 1950, *A Revision of North American Species of the Cladorceran Genus* Daphnia, Edward Bros., Seattle, Wash.

Knudsen, Jens W., 1958, Life Cycle Studies of the Brachyura of Western North America, I. General Culture Methods and the Life Cycle of *Lophopanopeus leucomanus leucomanus* (Lockington), *Bull. So. Cal. Acad. Sci.,* 57(1):51–59.

Monk, C. R., 1938, An Aqueous Media for Mounting Small Objects, *Science,* 88:174.

Morrison, R., 1943, Aqueous Media for Microscope Slides, *Turtox News,* 21(1):23–24.

Pennak, Robert W., 1953, *Fresh-Water Invertebrates of the United States,* Ronald Press, N.Y.

Richardson, H., 1905, A Monograph on the Isopods of North America, Washington, *U.S. Nat. Mus. Bull.,* 54:i–liii + 727.

Van Name, W. G., 1936, The American Land and Fresh-Water Isopod Crustacea, New York, *Bull. Amer. Mus. Nat. Hist.,* 71:iii–vii + 535.

(See Chapter 1 for other References.)

CHAPTER 15

Other Arthropods: Spiders, Scorpions and Their Allies

The remainder of the arthropods are grouped here. Members of the class Arachnida include the scorpions, whip scorpions, false scorpions, true spiders, sun spiders, harvestmen, mites, and ticks, some of which are shown in Fig. 15–1. The class Merostomata is represented by the so-called horseshoe or eastern king crab *Xyiphosura*, Fig. 15–1G. Finally, the two subclasses, Chilopoda and Diplopoda, which include the centipedes and millipedes, are depicted in Fig. 15–1, H and I.

ARANEAE: THE TRUE SPIDERS

Habitat and Collecting

Spiders occupy almost all terrestrial habitats, and a few have invaded the aquatic habitat. They are mostly nocturnal in behavior. Not all spiders build large nets between bushes; some hunt on the ground or hide in ambush on flowers, others live in the deep grasses and make irregular nets on the surface, and still others make cone-like webs extending back into

Other Arthropods: Spiders, Scorpions and Their Allies

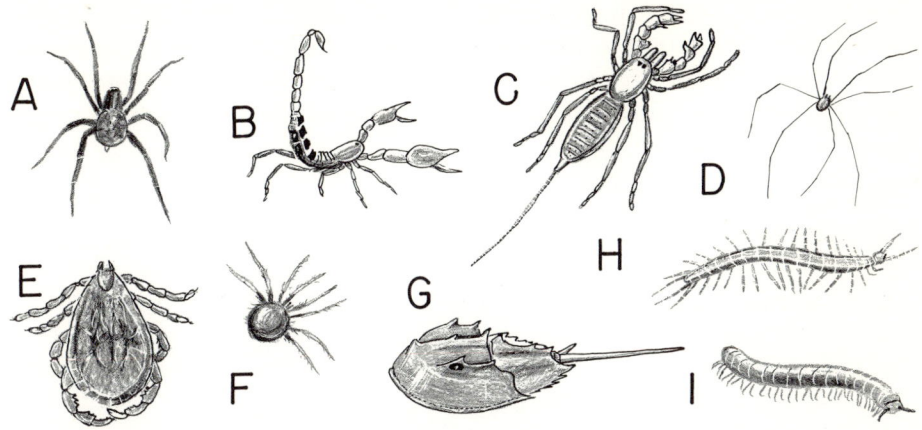

FIG. 15–1. Representative types of other arthropods. A. Spider. B. Scorpion. C. Whip scorpion. D. Harvestman, or long-leg spider. E. Tick. F. Water mite. G. Horseshoe crab. H. Centipede. I. Millipede.

burrows in the ground. Thus, the many species are extremely varied in structure, habitat, and especially in their behavioral patterns of feeding and self-protection. The number of people studying spiders is small compared to those interested in insects. This may be due to the fact that insects can be displayed more easily and in a much more attractive manner. Nevertheless, the study of spiders and other arachnids can hold a lifetime of interest.

Some Collecting Materials. The collector should carry a number of small jars and one or two large, wide-mouth (preferably plastic) jars with 70-percent alcohol. A small aquarium net is useful for snatching spiders from their webs before they have a chance to escape. Specimens may be picked up directly from the ground and placed into alcohol, caused to drop into alcohol from their webs and hiding places or quickly netted and then transferred to the alcohol solution. Field notes should be taken at the time of each collection, including the locality data (Chapter 1) and notes describing the nature of the web, the hiding place, and other critical factors concerning natural history. Killed specimens may be transferred in the field to smaller vials or bottles, with a field number which corresponds with the notebook description. The author uses very small paper bags for collecting live specimens. Bottles will do just as well, but they are quite heavy. Two spiders should not be kept alive in the same container. A specimen is netted and allowed to drop into the bag. The bag should then be closed, the top rolled down two or three turns and stapled shut. When large numbers of living spiders are to be collected, drop the bags into a cardboard box attached to a pack board. This will prevent crushing which might occur in a knapsack.

The Tuning-Fork Technique. A tuning fork or similar vibrating object is extremely useful, in addition to a small aquarium net, for capturing all kinds of web-building spiders. If the tuning fork is twanged and then placed against the web, spiders will be lured to the point of the vibration, just as to a vibrating fly. Many spiders are extremely quick to detect that the tuning fork is not a fly and will race right back to their hiding place. Nevertheless, if the aquarium net is held in readiness before the tuning fork touches the web the spider may be captured before it again reaches the safety of a crack or burrow.

Trap-Door Spiders and Tarantulas. In the country where trap-door spiders occur they may be located by walking slowly up on the slopes and shoulders of the brushy foothills. One must develop an "eye" to see the faint ring denoting the outline of the door. The door can usually be opened by inserting a needle or pocketknife blade opposite the hinge. Occasionally a green oat straw or other thin reed may be inserted down the length of the spider tube and then be carefully withdrawn with the spider in pursuit. Usually, however, the spider must be dug from the ground, tube and all. If most of the dirt is cut from around the tube, as the excavation is made, most of the tube can be placed in a tall, wide-mouth museum jar. Additional soil may be tamped in around the tube to hold it in position, and the spider placed inside.

Tarantulas are quite active in late spring in the Southwestern foothills and other parts of their range. They are frequently seen walking in the late afternoon and around lights in the evening. These spiders may be collected in a butterfly net and extracted in the manner of a bee (see Chapter 13). They are very excitable and tend to jump when first approached. After a short time in captivity, however, they become very tame and can easily be handled. These animals are poisonous and must be treated with respect. When agitated, tarantulas tend to "kick" hairs from their abdomen. These poison-bearing hairs may cause considerable pain if they are inhaled or enter the eyes.

Sweep Net. The sweep net is a most useful tool for working through tall grass and bushes where spiders hide. The construction and use of this net are described in Chapter 13. Basically, this heavy net is beat into bushes from which hiding spiders fall into the net bag and are captured.

Sorting Pans. Immature and small species of spiders often hide in leaf litter beneath trees, or in similar debris. A shallow, plastic sorting pan, at least 8 by 12 inches, is excellent for obtaining these animals. When quantities of such debris are sorted, small spiders can be seen running on the white surface of the pan. These may be picked up by forceps or, more readily, by a small water-color paintbrush freshly dipped in alcohol.

Collecting Egg Sacs. Egg sacs should be collected when available. These are either stored in 70-percent alcohol or placed on insect pins with labels like those used in insect collections (Chapter 13). Preserved specimens are kept with the adult spider. Pinned specimens should first be killed in alcohol to prevent the eggs from hatching.

Preservation

Spiders are usually killed and preserved in 70- or 80-percent alcohol. If large or numerous specimens are placed in one jar, the alcohol should be changed after 24 hours. Five-percent formalin may be used in place of alcohol, though it is not preferred. Specimens intended for critical internal dissection should be preserved in FAA (Appendix C). Color does not preserve well and ultimately will be altered. Color notes and records may be made, however, as described in Chapter 1 (p. 23). Occasionally, spiders are placed in Riker mounts (see Chapter 13) where they are kept in a dry state. Owing to the extreme amount of shrinkage that occurs in the abdomen, however, it is best to draw off some of the abdominal fluid with a hypodermic needle and replace it with latex rubber (which is used to inject the veins and arteries of vertebrates). Alternately, the abdomen may be opened from the ventral surface and the visceral mass removed and replaced with cotton. Specimens are then placed in the Riker mounts and permitted to dry normally.

Collecting Webs

The collection of spiderwebs has long fascinated specialists. There are some rather awkward and unsatisfactory techniques for mounting two-dimensional spider webs between sheets of glass or on paper treated by ink. The author has devised a method which is extremely simple, quick, and very satisfactory. The materials needed are an aerosol spray-paint can containing a dull black, oil-base paint or lacquer, a 1-inch paintbrush with soft bristles, a jar of 3 parts of Elmer's Glue and two parts water (thoroughly mixed), and some heavy white-surfaced Bristol board or posterboard cut to appropriate sizes. Thinner cardboard and heavy paper are less satisfactory, as will become obvious. Locate a well-formed, two-dimensional spider web, preferably before it is destroyed by insects. Chase the spider from the web by blowing on it, but note carefully where it hides. Now spray the net from a distance of about 18 inches, using circular strokes. The spray must be a fine mist rather than a splatter and should be directed in such a manner that all of the web is evenly blackened, but none of it becomes loaded with heavy drops of paint. The result of such a spraying is that very fine particles of wet paint cling to the spider web. Now, select a suitable piece of Bristol board large enough to contain the central portion of the web. Paint the entire surface of the board with a moist coat of glue (avoid an excess of glue). The surface should be tacky, but not dripping wet, as the spider web will slip on the surface. Next, carefully maneuver the board behind the web and then bring it forward in such a manner that the entire web is pressed into the wet glue. With your fingers, or with a pocketknife, carefully break each of the supporting guide "wires" where they leave the cardboard. Be careful not to move the cardboard up and down, or from side to side, after the cutting has begun, as the web will be pulled out of its true symmetry. Place the collected net in the shade where it may dry, and then record the field number or other data at the bottom. Figure 15–2 is made of an actual spider web which was collected in this manner. Such webs are permanent and may be stored in filing boxes. If they are to be subjected to heavy usage, a coat or two of artist's spray fixative, used as directed on the can, will render them very durable.

Photographing Webs

Webs are difficult to photograph in the field, owing to improper lighting. If possible, select webs which receive considerable natural light and which will permit the insertion of a black cloth (preferably velvet) behind the web. The object is to photograph a well-illuminated web against a dull black background. Various kinds of paper, cloth, or metal reflectors may be used to enhance the illumination by directing additional light to the web. The tripod, telephoto, and extension-tube combination are best adapted to this work.

Spiders may be removed to the laboratory by breaking their existing web (but leaving the place of hiding intact) and cutting the supporting vegetation in such a manner that the spider can easily respin the web. When

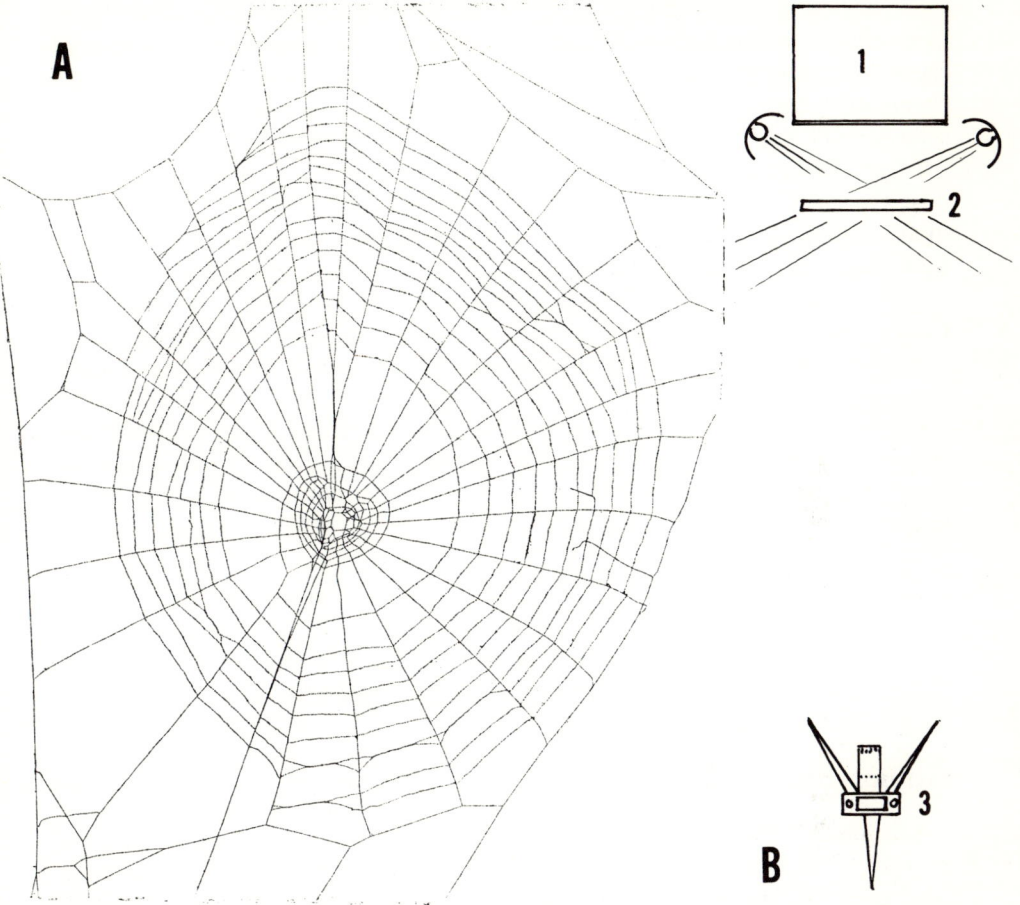

FIG. 15–2. Spider-web techniques (see text). A. The center of an actual spider web which was collected on a sheet of cardboard and is here reproduced at its original size. B. Looking down on a setup for photographing spider webs in the laboratory. (1) A blackened box or velvet cloth, (2) the spider web supported on a frame or piece of vegetation, (3) the camera supported on a tripod.

the vegetation is firmly anchored, the spider will usually spin a new web on its first opportunity. The author transferred his spiders to 10- by 14-inch wooden frames equipped with a small cone of paper, thumbtacked to the upper corner, in which the spider was allowed to hide. These frames were then hung from wires in the laboratory where the spider would spin a new web nightly. When making such a transfer, collect spiders in paper bags, as described above, and then place them in a paper cone in the laboratory. Thumbtack this cone to the frame and plug the cone with a piece of crumpled paper to inhibit the spider's escape. After a few hours remove the crumpled paper. About 90 percent of the spiders will remain in their new hiding place and will perform nicely.

In the laboratory the simplest photographic setup includes a large black box. Obtain a large cardboard packing box roughly 30 by 30 by 36 or 48 inches deep. (Any dimensions will do.) Spray the inside of this box with dull black paint. The box provides a dull-black background for photography. Figure 15–2B gives a view down onto the camera setup. The black box extends out of the top of the picture, the web is placed just in front of the box with one or two lights angled from behind and to the side, as shown. The camera is placed on its tripod with the telephoto lens and extension-tube combination. Usually, the light is sufficient to make excellent photographs. A light coat of fast-drying, white, aerosol spray paint may be applied to the web in order to make it reflect more light and thus photograph more fully.

SCORPIONS PSEUDOSCORPIONS, SUN SPIDERS, AND OTHERS

Habitat and Collecting

Scorpions. These animals (Fig. 15–1B) are found in the semitropical southeastern United States and the arid southwestern United States extending north through eastern Washingon into British Columbia, and south into Mexico and elsewhere. They are generally thought of as desert animals and occupy various habitats, from brushy foothills to sparsely vegetated sand dunes. Like all nocturnal animals, scorpions must locate some suitable hiding place for daytime dormancy. Many of them occupy self-made burrows; others crawl under any debris such as bark, rocks, fallen vegetation, and the like. Until you are familiar with scorpions, gloves or forceps should be used for picking these animals up. There are a few dangerously poisonous species in the Southwest. These are quite small over all, with pinching hands which are smaller in diameter than the wrists.

Among the sagebrush hills in the Southwest scorpions make burrows which are somewhat oval at the mouth (Fig. 15–3A). The burrow extends

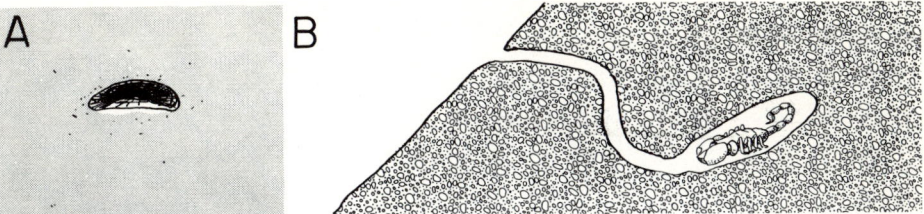

FIG. 15–3. Scorpion burrow. A. Mouth of the burrow. B. Cross-section of the burrow.

back into the hillside and then curves, as diagrammed in Fig. 15–3B, the scorpion hiding in the chamber at the end. When digging burrows, push a piece of soft grass down through the tunnel so that it may be located should it fill with dirt. Begin digging by cutting a hole in front of and below the burrow level. Then with a pocketknife chip away the dirt of the burrow so that it falls into the depression below. This way the burrow will remain clean and open and there will be no danger of encountering the scorpion before you are ready to. As you chip back through the burrow the scorpion will eventually drop down into the depression and may be captured.

Sun Spiders. Sun spiders, whip scorpions, and the like, occupy much of the same range as the true scorpions. Many of these, like scorpions, are located hiding under temporary shelters or may be picked up under street lights or by walking the sand dunes with a lantern. These animals may be handled by forceps or with a gloved hand, although, like scorpions, they may be picked up with the bare hands once the collector is thoroughly familiar with their behavior. Since it is very easy to become lost on desert sand dunes at night, have one member in your party (never work alone) drag a heavy stick to make a uniform mark which will serve as a trail back to your automobile.

Pseudoscorpions. Pseudoscorpions (Fig. 15–1C) are common the world over, from the tropics on up into the higher latitudes. They are small and secretive and frequent leaf litter, grassy fields, sea shores, nest cups in rodent burrows, or any place where food (notably soft-bodied insects and mites) is available. In some parts of the country they are even house guests. Leaf litter and other debris may be sorted in pans to isolate the specimens or may be treated with the Berlese funnel (see Chapter 13, p. 213). Knudsen (1956) noted a significant population of pseudoscorpions in wood-rat nests in southern California.

Preservation

These animals should be preserved in 70-percent alcohol (see the discussion under "Preservatives" in Chapter 14). They may be killed by dropping directly into this or some lesser solution of alcohol. In the latter case they should be transferred to 70-percent alcohol. If numerous animals or large individuals are preserved, change the alcohol solution after 24 hours. Pseudoscorpions may be mounted on slides intact or dissected. Follow the dissecting techniques given for small crustaceans (Chapter 14) when such techniques are needed. Specimens do not necessarily require maceration in KOH to destroy the soft tissues prior to mounting. However, follow the instructions given for ticks when making slide mounts.

TICKS

Habitat and Collecting

Juvenile and adult ticks will either be found on host animals (mammals, reptiles, and occasionally birds) or waiting on vegetation to attach to some host animal. (Hosts must be secured by methods given in Chapters 20–22.) Rodent runways and deer trails are sometimes alarmingly alive with adult ticks. The sweep net (Chapter 13) is excellent for securing these specimens, or the tick drag net (Baker and Wharton, 1952) will also give good results. The drag is made of flannel cloth 1 yard square tacked on one side to a stick 36 inches long. A short piece of rope is tied to the stick, on either end, to form a loop which the collector pulls along bushes and grass near animal trails. The drag should be inspected periodically for ticks which cling to it.

Preservation

Ticks may be collected into, and preserved in, 70-percent alcohol. They should be kept in vials along with their field data which, in turn, are enclosed in large, airtight jars filled with preservative.

Slide Mounts

It is not always essential to make slide mounts of ticks for taxonomic work. Also, ticks need not be treated in potassium hydroxide (KOH) unless such treatment is desired. This treatment removes all of the soft tissues, permits dorsal and ventral details to be observed simultaneously, and allows the specimen to be flattened somewhat more easily. Otherwise, ticks should be killed in a flat position between two slides.

To macerate, treat in a 10-percent KOH solution overnight and check periodically until the soft tissues are reduced. The alternative method is to use a stronger solution in the same manner, or to gently heat the specimens for a matter of minutes in 10-percent KOH. When the tissue is removed, wash specimens in two changes of 35-percent alcohol, the second being acid alcohol. Dehydrate up through absolute alcohol, clear in xylene, and mount in balsam or Permount. Specimens may be mounted directly from water or alcohol, whether macerated or not, in Turtox CMC-10. Stain is generally not needed for ticks. Larval ticks should be treated like mites (see below).

MITES AND WATER MITES

Habitat and Collecting

Mites are small and fragile animals which require special care. They are best collected and preserved in small vials of 70-percent alcohol. The simplest way to capture and transfer specimens to alcohol is by dipping a small, pointed, water-color brush in alcohol, touching the specimen, which will then adhere to the brush, and washing it off in the alcohol.

Parasitic mites may be collected from host animals (see Chapters 21 and 22) or from nesting materials occupied by these animals. Place fresh-killed hosts in white enamel pans and comb the hair or feathers in search of mites. Mites will be seen walking across the pan and can be captured with a paintbrush. Nesting material should be processed in the Berlese funnel (see Chapter 13, p. 213). The alternate method is to place such material in large white enamel pans and carefully pick through it. If the debris is in small quantities, mites may be seen walking on the debris or on the pans. White paper plates may be placed on grass and observed periodically for small, pinkish-colored mites which are present. Finally, a Birge's cone net (Chapter 14) is excellent for obtaining water mites from ponds and lakes where they occur in heavy vegetation. A small aquarium dip net is useful for this same purpose.

Preservation

Mites are best preserved in 70-percent alcohol in small vials containing the field data. Screw-cap or corked vials are preferable to cotton-stoppered vials because of the small size of the animals. Vials should be placed in an airtight jar filled with preservative. Slide mounts are useful and are often made for critical taxonomic study.

Slide Mounts

Gray (1952) recommends two types of mounting media for mites: "(1) A high refractive-index medium like Berlese's for the very heavy-walled forms, such as the Oribatid mites and the pseudoscorpions and (2) a low refractive-index mountant like Gray and Wess', for the thin-walled forms, such as the Tyroglyphid and Gamasid mites." Place a suitable amount of the proper medium (Appendix C) on a blank slide and transfer a live or preserved mite with as little water or alcohol as possible. Press the specimen down in the mounting medium, expel air bubbles, and add a coverslip. Label and dry in a flat position. When thoroughly dry, ring the slide with a waterproof cement such as Murrayite, ringing varnish, or other.

Baker and Wharton (1952) in their excellent text on ticks and mites observe that "workers at the University of California have developed a methocellulose formula which has proved to be excellent for many mites. They found the best procedure was to clear thoroughly in lactophenol before mounting, although some of the more delicate mites needed no special preparation as the lactic acid in the medium cleared the specimens sufficiently." See Appendix C for the methocellulose medium.

In addition to these media mites may be mounted in Permount or balsam by treating the heavy-bodied forms with KOH as described for the ticks, moving all forms up through the alcohols, using 35-, 50-, 70-, 85-, and 95-percent, and absolute alcohol, two changes at 10 to 15 minutes each. Transfer to a mixture of half absolute alcohol and half clove oil and, finally, to pure clove oil for clearing. When the specimen is sufficiently clear transfer to a clean slide, add the desired resinous medium, lower a coverslip into position, and dry. Add labels with both the locality and host data in the taxonomic data.

MEROSTOMATA

These strang-appearing animals (Fig. 15–1G), known as horseshoe crabs or eastern king crabs, are found intertidally and in pools above sandy beaches on the eastern seaboard. Specimens are readily captured by hand and picked up directly by the long, sword-like telson (tail). Although they are not related to crustaceans, the body is constructed of similar materials, permitting these organisms to be killed and preserved exactly as the large crustaceans (see Chapter 14).

CENTIPEDES AND MILLIPEDES

Habitat and Collecting

Centipedes and millipedes (Fig. 15–1, H and I) are broadly distributed; they are common in the tropics, desert and arid regions, grasslands, and forests. Centipedes are almost exclusively nocturnal, whereas millipedes may be active either during the day or at night. They retreat into hiding places such as under stones, wood, downed vegetation, leaves, or the burrows of other animals. Some establish home sites, whereas others find new hiding places at random. In higher latitudes these animals are commonly found overwintering in rotten logs.

Collecting should be conducted by searching likely hiding places. The gloved hand or a pair of forceps suffices to capture these animals. Many species are mildly poisonous; few are really dangerous. Nevertheless, treat all as if they were poisonous, for safety's sake.

These animals may be killed in an insect-killing bottle (Chapter 13) or by dropping into alcohol. Place fresh-killed animals in pans, arrange the legs and straighten the body, and add 70-percent alcohol. After a few days transfer specimens to vials or museum jars with their field data; keep them in 70-percent alcohol. Specimens intended for dissection should be preserved directly in FAA.

For study collections, specimens may also be pinned in the manner of insects, but with less satisfactory results. They may be arranged and glued to cardboard which is in turn pinned, or they may be pinned directly with a temporary cardboard support, as directed in Chapter 13.

REFERENCES

Baker, Edward W., and G. W. Wharton, 1952, *An Introduction to Acarology*, Macmillan, N.Y.

Baker, W. W., *et al.*, 1958, "Guide to the Families of Mites," Contribution No. 3, Institute of Acarology, Univ. of Maryland, College Park.

Comstock J. H., 1940, *The Spider Book*, Doubleday, N.Y.

Crompton, J., 1954, *The Life of the Spider*, Mentor (New American Library), N.Y.

Gertsch, W. J., 1949, *American Spiders*, Van Nostrand, Princeton, N.J.

Gray, Peter, 1952, *Handbook of Basic Microtechnique*, Blakiston, N.Y.

Kaston, B. J., and E. Kaston, 1953, *How to Know the Spiders*, William C. Brown, Dubuque, Iowa.

Knudsen, Jens W., 1956, Pseudoscorpions, A Natural Control of Siphonaptera in Neotoma Nests, *Bull. So. Calif. Acad. of Sci.*, **55**(1):10.

Petrunkevitch, A., 1933, The Natural Classification of Spiders Based on a Study of their Internal Anatomy, *Conn. Acad. Sci. Trans.*, **31**:299–389.

Savory, T. H., 1935, *The Arachnida*, Ed. Arnold, London.

(See Chapter 1 for other References.)

CHAPTER 16

The Echinoderms: Sea Stars, Sea Cucumbers, and Others

The five classes of echinoderms commonly overlap in their distribution; thus, collecting techniques will be presented for the phylum as a whole. A number of problems arise in preservation. (1) Because of the calcium carbonate skeleton, echinoderms cannot be left in any acid preservative. On the other hand, (2) many echinoderms can be dried and stored more economically than by preservation in liquid. However, some of the echinoderms, notably the starfishes, are identified on the basis of muscle attachments for which dried material is unsatisfactory. Many of the echinoderms have (3) the tendency to break off "arms" when subjected to harsh preservatives; (4) others, notably the sea cucumbers, are highly contractile, and at least some representative specimens should be preserved in a relaxed state, with the tentacles expanded.

COLLECTING ECHINODERMS

Brittle stars, serpent stars, starfishes, sea urchins, sand dollars, sea cucumbers, and sea lilies occupy intertidal and deep-water habitats with both rocky and sandy substrates. The feather stars (crinoids) are exceptions;

they spend part of their time as swimming animals. Hand-collecting and dredging (or dragging) are the two chief methods of collection.

Drags and Dredges

In deeper water with sand, mud, or rock-rubble bottoms, the starfish drag or dredge proves excellent for collecting. The drag (Fig. 16–1A) utilizes string mops which entangle starfishes and brittle stars. String mops may be attached to any suitable object, but the pattern described here is recommended. To make a starfish drag, make a 90° bend in the center of a solid 6-foot bar of metal or heavy pipe. Weld a ring at the bend to receive a tow rope, and four additional rings as shown. Attach a 4-foot piece of heavy, galvanized chain, by means of a split ring, to each of the four rings. Finally, attach two or three string mops to each chain, as shown. To operate, simply attach a rope to the tow ring, pay out sufficient line for the depth of the water, so as to allow the drag to work flatly along the bottom, and tow from a skiff. In suitable habitats collectors will be amazed at the volume of organisms obtained.

The biological dredge (see Chapter 1, p. 9, for construction and use) will retrieve a fantastic number of echinoderms. Sort dredge hauls or drag hauls immediately, and place specimens in buckets of clean sea water. Avoid overcrowding, and exchange the sea water frequently if the specimens must remain some time before preservation.

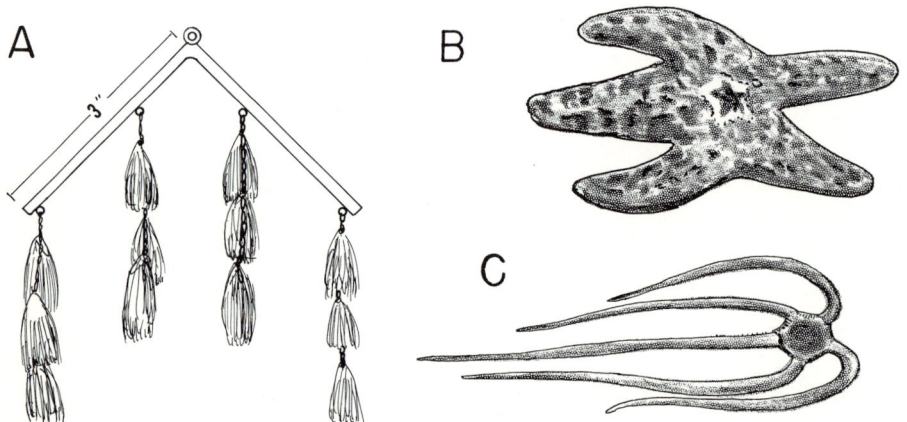

FIG. 16–1. Echinoderm techniques. A. A starfish tangle containing string mops. B. The desired position of large starfish specimens intended for drying. C. The position for drying brittle stars.

Hand-Collecting

Skin diving, wading in the water, and collecting during periods of low tides are all satisfactory for echinoderms. On sandy bottoms the tentacles of burrowing sea cucumbers denote where to dig; starfishes and brittle stars are exposed on the bottom; sand dollars bury themselves, but leave telltale rings in the sand where they are buried. In the Puget Sound area, tremendous concentrations of sand dollars are found on sandy beaches near where some source of fresh water enters the sea water.

When collecting on rock and sand bottoms, or on pure rock substrates, look for the tentacles of sea cucumbers projecting into the water from between rocks or from the sand just underneath protective rocks. Others that may be exposed by turning rocks are small starfishes, sea urchins, and brittle stars. Many of the large starfishes seek cover from the receding tides in order to avoid drying. Thus, they are usually found in the more shaded and protected crevices. Tidepools along rocky shores provide many echinoderms, especially sea cucumbers and brittle stars.

Specimens should be placed in buckets of moist, nonmucous-secreting seaweed. Periodically, fill the bucket with fresh sea water and drain again. Sea cucumbers should not be crowded or mixed with other animals, in that they tend to eviscerate or slough off the outer skin, and thus quickly decompose. Sea cucumbers should be preserved as quickly as possible.

One should not overlook pilings and floating docks, for starfishes and sea cucumbers are especially numerous there, hiding among other organisms such as mussels and barnacles.

Always bear in mind the method of preservation and the size of containers available. There is no value in collecting extremely large specimens for general use, especially if the collector has no container which will accomodate large specimens. Although size is less a factor for dried specimens, it should, nevertheless, be considered at the time of collecting to avoid wasted specimens.

PRESERVATION TECHNIQUES

Classes Asteroidea and Ophiuroidea: The Sea Stars and Others

Museum specimens are generally dried or preserved in liquid; specimens for classroom dissection must be preserved in liquid. Rapid-killing techniques may be tried, but narcotizing is recommended for starfishes, and is essential for brittle stars and serpent stars since these animals will break off their appendages.

Narcotization and Liquid Preservation. Both epsom salts and fresh water may be used to narcotize and kill these animals, but epsom salts is greatly preferred. Here, crude epsom salts seem to work much better than the chemically pure form of magnesium sulphate. Place specimens in flat-bottomed pans with just enough sea water to cover them. Add a small handful of epsom salts (depending somewhat on the volume of water) every hour during the day, and let stand overnight. After this, test the specimens, first by gently probing them, and second by introducing a drop of formalin near the specimens. This will cause contraction or squirming if the specimens are not ready for preservation. When they are ready, transfer the specimens to a fresh pan of water and arrange the arms so that they take up the least amount of room (Fig. 16–1B). Brittle stars and serpent stars are always positioned as follows: Place the specimen in a pan of water, put your finger on the disc and move the animal in a direction away from the longest arm, so as to position it as shown in Fig. 16–1C. The longest arm should be perfectly straight for the measurements and plate counts essential to taxonomy. When the specimens are arranged, add enough formalin (preferably neutralized) to make a 5- or 10-percent solution, and let stand for 24 hours.

From this point specimens are either dried or preserved in alcohol. For liquid preservation, wash the specimens to remove the formalin, transfer to 50-percent alcohol for 1 hour, and store in 70-percent alcohol. Isopropyl alcohol (70 percent) may be used in place of ethyl alcohol for echinoderms.

The fresh-water technique is essentially the same as that described above, except that specimens are placed in pans of fresh water from 6 to 12 hours or more. After this they are arranged and preserved as above.

Drying Methods. Echinoderms make excellent dried specimens when fixed in formalin, as described above. Untreated specimens may be dried directly, but with less success for larger specimens. Dried specimens tend to retain their color quite well if stored in a dark container. Air movement, rather than heat, is the key factor in drying. Thus, moderate heat with sufficient air circulation is preferable to hot ovens or direct sunlight. Arrange specimens on screen wire to permit air circulation above and below. The author finds the attic of the science hall an excellent, warm place for drying. Specimens are placed on screen near the discharge of the air ducts. One to three weeks may be required for complete drying of large specimens. However, this process can be hastened with a drying oven, described below.

Drying Oven. A very convenient dryer can be cheaply built which will meet all of the requirements for drying echinoderms and many other types of biological specimens (Fig. 16–2). The heating element may consist of a 200-watt bulb, a heating coil, or a hot plate, as desired. The dryer may be of any dimension, so long as the basic principle is adhered to.

The materials needed are: two side pieces of ¾-inch plywood, 18 by 42 inches; one top piece of ¾-inch plywood, 18 by 24 inches; one back piece of ¼-inch plywood, 24 by 40¾ inches; two (door) pieces of ¼-inch plywood, 12 by 40¾ inches; fourteen drawer runners, 1 inch by 1 inch by 18 inches (note, these will measure ¾ inch by ¾ inch); fourteen tray ends, 1 inch by 1 inch by 17⅜ inches; fourteen tray sides, 1 inch by 1 inch by 22⅜ inches; one lower front brace, 1 inch by 2 inches by 24 inches; seven pieces of ¼-inch hardware cloth, 17¾ by 22¼ inches; two pairs of butt hinges and one small hook and eye.

First, attach the drawer runners to the inside of the side pieces, 5 inches on center, measuring from the top, as shown in Fig. 16–2A, using glue and nails, as shown in Fig. 16–2D. Now, nail the top piece to the two side boards, as shown in Fig. 16–2C. Drill four ¾-inch holes about equally spaced in the upper part of the back plywood piece (Fig. 16–2A, fourth hole hidden on left), and two holes in the top of each door piece (Fig. 16–2B). Glue and nail the back piece to the sides and top, making sure that the cabinet is square. Mark and cut the lower front edges of the side boards to accommodate the front brace (Fig. 16–2E). Glue and nail the front brace in position. Attach the two doors so that they overlap the top piece (Fig. 16–2B). Finally, install the hook and eye, or other door catch, as desired. Cut the ends of the drawer pieces so that they will dovetail, as shown in Fig. 16–2F. Glue and nail the trays, square the assembled pieces, and install the hardware cloth on the bottoms (preferably with cleats, or simply with large-headed tacks).

If a light bulb is used as a heat source, obtain a porcelain light receptacle, a 4- by 4-inch electrical outlet box, an appropriate length of lamp cord, and a male wall plug. It may be necessary to shield the light bulb from dripping preservatives by placing a sheet of tin directly above it on the bottom screen. If a light is the heat source, the seventh or bottom screen will probably be omitted.

To use this dryer simply narcotize and fix specimens as directed above, permit them to drip off excess preservatives for a few moments, and then place them on the trays. When the doors are closed air will circulate in

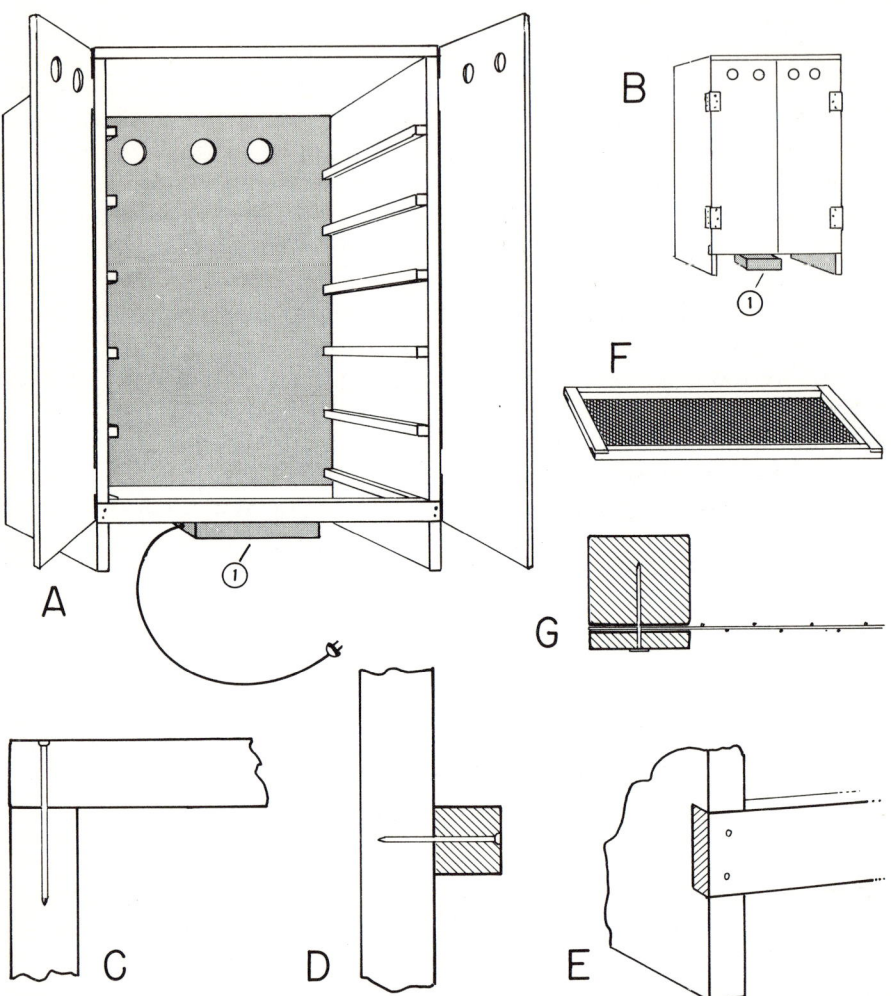

FIG. 16–2. A simple drying box useful for echinoderms or other biological specimens such as mushrooms. A. A front view of the drying box. B. Front view with the doors closed. C. Detail of top and side assembly. D. Adding the shelf cleats. E. Detail on installing the lower front brace. F. The finished screen shelf. G. Detail of attaching the screen to its frame by means of a cleat. (1) Heating element.

through the bottom, move up and around the specimens, and will finally leave through the eight holes in the top. Small specimens will dry overnight, whereas larger ones will require a longer time.

Tube-Feet Preparation. When it is necessary to preserve specimens in liquid with the tube feet extended, the following techniques may be used. Narcotize the animal as directed above, except that the specimen is positioned with the tube feet upward, and place enough water in the container for the tube feet to expand. In the morning, test the tube feet with formalin, as directed above; when the specimen is ready, fix in formalin and then transfer to alcohol; or, preserve directly in alcohol.

An alternate method is to narcotize the specimen, tube feet upward, by gradually adding enough alcohol to make a 20-percent solution. Following narcotization, the specimen is placed in 70-percent alcohol.

When the tube feet are not well extended, the following procedure may be tried. Pierce a hole in the madreporite plate, insert a flat-ended hypodermic needle into this opening, and inject large quantities of 95-percent alcohol. If the animal has been properly narcotized the tube feet should expand and remain expanded. Some workers prefer to inject the tip of each radial arm by inserting the hypodermic into the radial canal. Specimens should be placed in 70-percent alcohol or neutral formalin for final preservation.

Preservation for Histological Study. Select only very small specimens, approximately 1 inch in diameter over-all. Fix in corrosive sublimate and treat with iodine (see Appendix C), decalcify with a weak solution of hydrochloric acid, stain with alum hematoxylin and eosin (Galigher, 1934). Consult some handbook of microtechnique (such as Gray, 1952) for methods of embedding in wax, sectioning on the microtome, and subsequent mounting. Bouin's fixative may be substituted for the corrosive sublimate.

Pedicellaria. Most echinoderms have pedicellaria which are small, two- or three-jawed devices used for cleaning and protection. Dissect away portions of the dorsal epidermis containing the pedicellaria and fix these in formalin. Tease away the excessive tissue and mount on microscope slides, with coverslip supports (Appendix B) if needed. Some workers remove all excess tissues by treating overnight in 10-percent potassium hydroxide (KOH). If KOH is used, wash thoroughly before mounting. Specimens may be mounted directly in Turtox CMC-S mounting medium which will mount, clear, and stain in one operation. Specimens may be transferred directly from water or alcohol into this medium. The alternate procedure is to stain, dehydrate, clear in clove oil, and mount in balsam or Permount (see Appendix B).

Skeletal Techniques. The soft epidermal tissue may be removed from starfishes to display the skeletal structure. Select small specimens (3 to 4 inches in diameter) or portions of larger specimens. Macerate the tissues by placing the entire specimen in straight Clorox, Purex, or a 10-percent solution of sodium hydroxide. Remove the specimen from time to time, wash, and then scrub vigorously with a toothbrush to determine if the tissue is ready for removal. It is preferable not to let maceration go to a point where all tissues are removed, as the skeletal plates may become disassociated. When the tissue will readily slough off, clean externally and flush the internal contents from the body by removing the oral membrane and the stomach. Thoroughly wash in two changes of fresh water, for ½ hour each, and in one change of acid 30-percent alcohol (Appendix C). Dry and store in a small cardboard box with a lid.

Techniques for Larvae. The larvae of all echinoderms must be narcotized to avoid contraction and distortion. Bianco (1899) recommends using chloral hydrate, but Chloretone works just as well. Place specimens in a small container of sea water, add some crystals of Chloretone, and let stand from 4 to 6 hours. Test the specimens with a dissecting needle to make sure they will not react to stimuli. They may then be moved up through 35-, 50-, 75-, and 95-percent alcohol, where they are hardened for 24 hours, and then returned to 70-percent alcohol. (Bianco recommends killing all advanced stages by placing them in corrosive sublimate for a few moments. See Appendix C for the use of this fixative.) Specimens may be stained in borax carmine (p. 134) and mounted by one of the methods given for the Bryozoa (Chapter 12, p. 199).

Storage of Specimens. Over long periods of time, echinoderm colors will remain more stable if the specimens are stored in the dark. Dried echinoderms will attract dermestid beetles and other insect pests if they are improperly stored. A storage cabinet with a rubber or plastic seal to prevent the invasion of insects is recommended. The cabinet design presented for the mollusks (Chapter 11) is well suited for echinoderms if provided with a seal as described for insect cabinets (Chapter 13).

Class Echinoidea: Sea and Heart Urchins, Sand Dollars

Liquid Preservation. Although urchins can be suitably dried, some should be preserved in liquid, as the structure of the mouth parts, the so-called "lantern," and the associated muscle attachments are of taxonomic importance. Specimens may be killed and preserved in 70-percent alcohol or killed in 5-percent formalin, washed, and preserved in 70-percent alcohol. The alcohol solution should be changed after 24 hours. Isopropyl alcohol is suitable in place of ethyl alcohol. As soon as larger specimens are placed in the preservative, arrange the spines in such manner that they will more easily fit into a jar. Inject specimens intended for dissection through the oral membrane, to ensure rapid internal preservation.

Tube-Feet Preservation. Bianco's method (1899) for preserving urchins with the ambulacral feet well extended is as follows: Place specimens in a small container with just enough water to cover them. When they are well expanded pour in a quantity of chrom-acetic fixative equal to the amount of water. After a few moments, transfer the specimens to 35-percent alcohol, 50-percent alcohol, and store in 75-percent alcohol.

Positioning Spines for Preservation. Frequently, urchins are desired for wet or dry specimens with the spines positioned in a lifelike attitude. Usual methods of preservation often cause the drooping of spines. You can obtain lifelike specimens by placing the urchin in a container (a coffee can will do) with ample sea water. When the animal has fully erected the spines in the desired position, quickly sift fine sand all around the animal until it is completely buried (Fig. 16–3, A and B). With your hand or some flat object hold the sand in place and pour off the excess water. Replace this with 10-percent formalin, let stand for 24 hours until the muscles holding the spines harden, and then gently wash away the sand. The specimens should then be washed thoroughly in fresh water and preserved in alcohol, as described above, or dried directly without any washing. If drooping persists in large specimens during the drying process, simply pour off all the fixative and dry the specimen in the container, sand and all. After drying is completed, the sand can easily be removed.

Drying Methods. Dried specimens are easily maintained, require no expensive preservatives, and retain their color quite well. However, the spines will become broken if the specimen is mistreated. It has long been the practice to remove the visceral mass by cutting around the oral membrane (Fig. 16–3C), removing the lantern and scooping out the viscera with a wire. This procedure is unnecessary, however, as the viscera contain a high percentage of water and dry very well if they have been previously preserved. Likewise, it is desirable to keep the lantern intact for future reference. The procedure, therefore, should be preservation followed by drying, as described for the starfishes above.

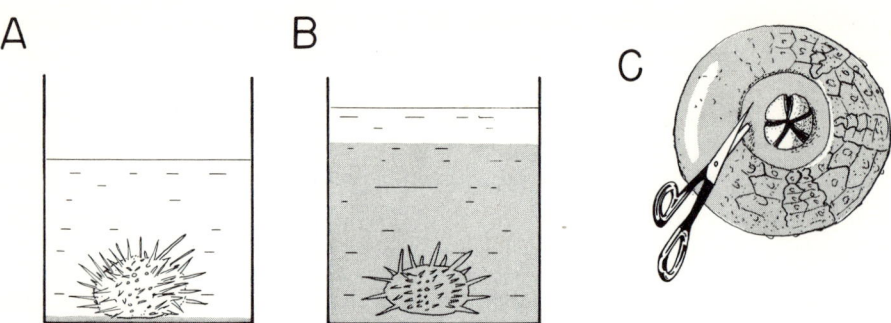

FIG. 16–3. Techniques for working with sea urchins. See the text for a detailed explanation.

Larvae. Treat larvae as described for the starfishes (p. 306).

Class Crinoidea: Sea Lilies and Feather Stars

These fragile echinoderms live below the low-tide line, on down to the deep-sea habitat. They are therefore seldom sought, but are taken incidentally during standard dredging. The inclusion of one or two sea lilies is always the high point of a dredge haul.

Liquid Preservation. Crinoids are prone to break up during preservation. Bianco (1899) recommends killing specimens by plunging them directly into 90-percent alcohol, after which they should be stored in 70-percent alcohol. The author has not preserved crinoids, but would assume that narcotizing in epsom salts, as described for the starfishes, would be desirable prior to preservation. Some specialists relax crinoids in the dark (they contend that light causes crinoids to contract) and hold the specimens flat to prevent movement until they are killed by 95-percent alcohol. Quick freezing with CO_2 or slow freezing in the refrigerator may also be attempted.

Drying Methods. Place killed crinoids in a mixture of 2 parts glycerin and 3 parts 50-percent alcohol for 24 hours, remove, dry, and store in a small cardboard box, as described above.

Class Holothuroidea: Sea Cucumbers

Sea cucumbers require special treatment from the time they are collected until the time preservation is completed. They tend to slough off the outer epidermis and eviscerate as soon as they are vaguely disturbed. Therefore, preserve specimens as quickly as possible.

Narcotizing and Preserving. One method for relaxing cucumbers is to place them in sea water until they are expanded, and then to add small quantities of epsom salts every 30 minutes over a period of many hours, increasing the amount each time. It may be essential to let them stand overnight until they are completely insensible to touch. At this point, kill in 10-percent formalin, wash, and preserve in 70-percent alcohol, or kill them directly in 70-percent alcohol. Inject the body cavity with preservative. Replace the alcohol after 24 hours.

An alternate method that may be tried is to permit the animal to expand completely in sea water. At this point grasp the animal with a pair of wooden forceps where the tentacles join the body, lift the body with the opposite hand, and submerge the tentacles in concentrated acetic acid. Simultaneously, an assistant should inject large quantities of 95-percent alcohol through the anal opening by means of a syringe. The anal opening is then plugged and the animal is placed in 70-percent alcohol. Variations of this method are injection through the body wall or through the oral aperture.

Specimens left in collecting containers are often found beautifully expanded in the water that becomes putrid. These are usually not the best for dissection, but are perfectly adequate for classroom demonstration. The use of putrid sea water as a killing agent is very successful for other animals and should be tried for sea cucumbers if other procedures fail.

When adequate time is not available for the careful relaxation of specimens, simply preserve directly in 10-percent formalin or 70-percent alcohol. Such specimens are the least desirable, but their use is preferable to wasting the animal. Sea cucumbers must never be left in formalin unless intended solely for classroom observation. Formalin destroys the dermal ossicles which are found in the skin and are used taxonomically.

Dermal Ossicles. Remove a piece of cucumber skin and treat it in 10-percent KOH or full-strength Clorox or Purex. This reduces the dermal tissue, leaving the ossicles. These should be thoroughly washed in fresh water and stored in a screw-cap vial of alcohol, along with the field data and specimen identification. Such vials are, in turn, sealed in an airtight jar filled with 70-percent alcohol.

Dermal ossicles are conveniently mounted on slides by the methods described for pedicellaria (p. 305).

REFERENCES

Bianco, S. L., 1899, The Methods Employed at the Naples Zoological Station for the Preservation of Marine Animals, *Bull. U.S. Nat. Mus.*, 39(M):1–37.

Clark, A. H., 1921, Sea-lilies and Feather Stars, *Smithsonian Misc. Coll.*, 72(7):1–43.

Fisher, W. K., 1911–1930, Asteroidea of the North Pacific and Adjacent Waters, *Bull. U.S. Nat. Mus.*, 76(1–3):1–1020.

Galigher, A. E., 1934, "The Essentials of Practical Microtechnique," published privately.

Gray, Peter, 1952, *Handbook of Basic Microtechnique*, Blakiston, Philadelphia.

Hyman, L. H., 1955, *The Invertebrates: Vol. 4, Echinodermata*, McGraw-Hill, N.Y.

Mortensen, T., 1927, *Handbook of Echinoderms of the British Isles*, Oxford Univ. Press, London.

(See Chapter 1 for other References.)

CHAPTER 17

The Lower Chordates

HEMICHORDATA: ACORN, OR COLLAR, WORMS

The taxonomy of the Hemichordata is probably not yet fully settled. This group has long been listed as a subphylum of the Chordata but more recently has been recognized as a distinct phylum. The hemichordates include three distinct "subgroups": the Enteropneusta, popularly called the acorn worms, and two small groups restricted to rather deep ocean bottoms (50 to 8000 meters), the Pterobranchia and Pogonophora.

Habitat and Collecting

The acorn worms (Fig. 17–1A) live in burrows in mud or mud and sand bottoms from just below the high-tide mark on down to relatively deep water. Burrows are often simple and U-shaped, although some species have burrows with more than one anterior opening. The burrows can be detected by the funnel-shaped anterior openings and the telltale, coiled fecal casting of the posterior opening (Fig. 17–1B). For hand-collecting dig specimens with a shovel. The best procedure is to transfer a large shovelful of mud which surrounds the worm to a bucket or screen. The mud should then be

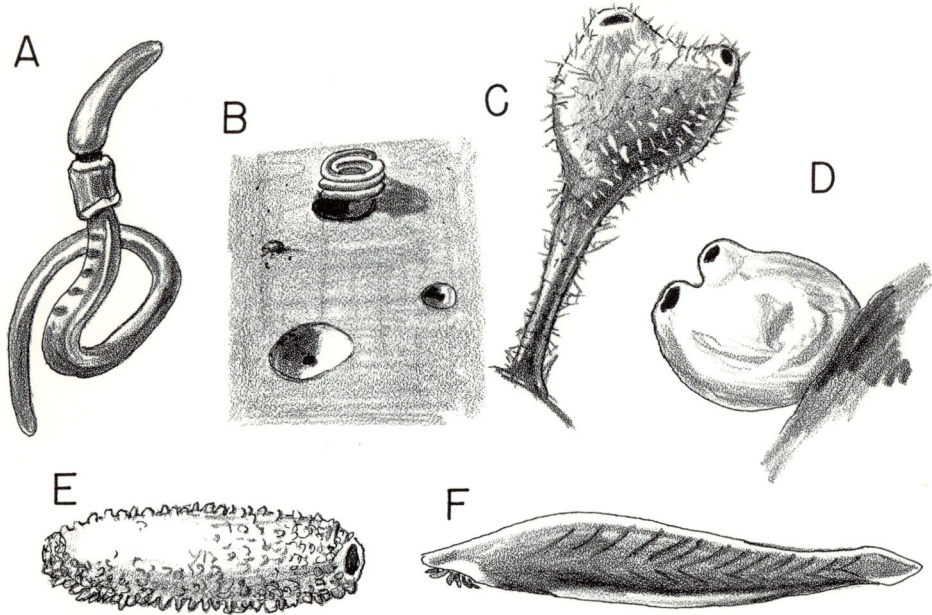

FIG. 17–1. Representative lower chordates. A. An acorn or collar worm. B. The burrow openings and a casting of an acorn worm as they would appear on a muddy substrate. C. A stalked tunicate. D. A sessile tunicate such as *Molgula*. E. A colonial, pelagic tunicate such as *Pyrosoma*. F. *Amphioxus*.

submerged in the water and carefully picked away from the specimen in order to avoid breaking the animal. Specimens should be isolated from other invertebrate groups in separate containers of sea water. The shores of bays and inlets in the intertidal zone are excellent hunting places for these worms. Below the intertidal level, however, the orange-peel bucket and biological dredge become the chief means of collecting (see Chapter 1).

Narcotizing and/or Preservation

Crude Preservation. Collected specimens may be killed directly by placing them in 4-percent formalin made with sea water. This usually causes violent contraction and thus produces specimens in which the taxonomic features are hard to distinguish. If time permits, careful narcotization should be undertaken.

Narcotizing. Several methods of narcotizing are useful. (1) Place specimens in containers of sea water and permit them to expand completely. Next, add, drop by drop over a period of an hour or more, enough alcohol to make a 10-percent solution; complete relaxation usually takes from 4 to 6 hours. Test the specimen after 4 hours by probing the posterior end of the body. If no contractions result, pipette a drop or two of formalin directly onto the body and observe. When insensible the specimen may be killed in 5-percent formalin, Bouin's fixative, Kleinenberg's solution, or 50-percent alcohol (Appendix C). Specimens may be stored in 70-percent ethyl or isopropyl alcohol or 5-percent formalin.

Bianco (1899) obtained good results by killing specimens directly in Kleinenberg's solution, presumably without narcotization. Specimens placed in stale sea water which contains some organic material (such as the blood or visceral mass from some invertebrate) and allowed to warm slowly to room temperature will usually die in a nicely expanded condition. However, the specimens must be transferred to either a fixative or preservative immediately after death, as tissue breakdown will soon occur.

Storage

Smaller specimens are placed in vials, along with their preservative and field data. The vials are cotton-stoppered and put inside an airtight museum jar filled with the same preservative. Specimens should be kept in a dark place when not in classroom use. Larger specimens are placed directly in museum bottles with their field data.

UROCHORDATA: THE TUNICATES

Urochordata is a relatively large subphylum of the phylum Chordata which contains about 2000 species (Abbot, 1961). Sessile species are covered by a tunic (outer coat) which may be tough and horny, leathery, or gelatinous in consistency. Pelagic species have a gelatinous tunic and are usually transparent; some are luminescent.

Habitats and Collecting

Sessile Tunicates. Sessile tunicates require some form of solid substrate for attachment and, thus, will be found only where rocks, coarse species of algae, pilings, and similar objects are found in the intertidal and subintertidal zones. They attach firmly to the substrate by means of a broad or narrow base, and may be stalked or unstalked (Fig. 17–1, C and D). The only problem in collecting tunicates is dislodging them from their substrate. With the gelatinous forms care must be taken not to rupture the tunic. With the thumbnail or a very dull pocketknife, specimens can usually be pried away from their substrate. Never attempt to loosen specimens by pulling directly on the body. Animals attached to wood or algae are quite easily recovered, but those attached to rocks will require more care. Place specimens in a small quantity of sea water and keep them cool while transporting them to the laboratory (see Chapter 1 for methods of transportation).

Pelagic Tunicates. Tunicates are common in the offshore waters, especially in the summertime, where they drift with the current. They are frequently carried into bays and estuaries, often in large numbers during the fall. Many species practice diurnal migration and thus are most numerous in the surface waters at night and in the deeper waters by day. The transparency or very light coloration of the pelagic-tunicate body requires that the observer look in the water in that angle from the sun at which the outline of the body will be conspicuous. Specimens are captured, as they drift along the surface, with a dip net (Chapter 13) or tow net (bathypelagic net, Chapter 18). Working under a night light (Chapter 1) is very productive in the summertime. However, the luminescent species, such as *Pyrosoma* shown in Fig. 17–1E, are conspicuous even in total darkness. Pelagic tunicates should always be isolated in containers of sea water and kept as cool as possible until narcotization or preservation.

Narcotization and/or Preservation

Crude Preservation. The most desirable specimens of tunicates are those which are fully expanded and have both siphons completely dilated and open. Specimens which are mildly disturbed close the siphon and those which are greatly disturbed close and contract the siphons and the body as well. Five-percent formalin is an excellent preservative for tunicates. If fully expanded specimens are not required, kill and store tunicates in 5-percent formalin. When the specimens in one bottle are numerous and bulky, replace the solution after 24 hours.

Narcotization and Preservation. The author wishes to acknowledge that many of the methods presented below are summarized from Bianco's work (1899) on the preservation of marine animals. Several procedures are available.

(1) Place specimens in containers of sea water to which enough chloral

hydrate has been added to make a .2-percent solution. (2) Add a small pinch of chloral hydrate each hour for 2 hours after the specimen has expanded in pure sea water. Permit the tunicates to remain in the solution between 12 and 24 hours, or longer, and then kill them in chrom-acetic fixative (Appendix C), 5-percent formalin, or FAA.

Another procedure given by Bianco is: (3) expand specimens in a large volume of sea water and then add a small quantity of chrom-acetic fixative to the surface of the water. This will diffuse into the water and kill the specimens within 30 minutes to 24 hours. Formalin may be tried in place of the chrom-acetic fixative. Next, siphon off most of the water without disturbing the specimens and add large quantities of chrom-acetic fixative or 5-percent formalin. Store specimens in formalin.

Transparent pelagic species such as *Pyrosoma* may be (4) placed directly into 50-percent alcohol and then transferred to 70-percent alcohol with good results. The author has found that the direct preservation of *Pyrosoma* in 5-percent formalin also gives good results. The author has obtained slightly less satisfactory results (5 to 8) with sessile tunicates (the siphons remain extended but not widely opened) by adding small quantities of alcohol, epsom salts, menthol, or fresh water periodically over a time span of 6 to 8 hours, and then allowing the specimens to remain an additional 16 hours, at which time they were killed in 5-percent formalin. Chloretone, clove oil, and freezing gave mixed but less satisfactory results.

Storage

Follow the procedure given for hemichordates above.

CEPHALOCHORDATA: THE LANCELETS

Habitat and Collecting

The lancelet, which is called *Amphioxus*, is a small fish-like animal which lives in sandy bottoms along the ocean. Potter (1961) notes that there is a total of 28 species in this group, four occurring in the United States, and that they are limited to subtemperate and tropical parts of the world. Lancelets are small, measuring about 1½ inches on the average, although Potter records one species in China and India that measures 6 inches.

Wells (1932) collected *Amphioxus* on the Gulf Coast of Florida, taking specimens from the landward side of sand bars where they occurred in burrows at about the low-tide level. He detected the presence of specimens by holes in the sand. Ricketts and Calvin (1956) further add that *Amphioxus* is found on low-tide sand bars opposite the mouths of sheltered marine bays far enough in to be protected from wave shocks. They note that "violent stamping on the packed sand in just the right area will cause some of the animals to pop out." This author has collected *Amphioxus* on the first sandy beaches inside the channel at Newport Bay, California, which receive protection from both waves and strong channel currents.

The usual collecting method is to screen sand with a coarse, $\frac{1}{8}$-inch collecting screen (see Chapter 10). When the habitat is located, sand is washed through the screen, one shovelful at a time, and the animals are transferred to a bucket of water. The dredge net (Chapter 13) can also be adapted for collecting *Amphioxus*.

Preservation Techniques

The exact use of the specimens will determine the type of preservation. For general use specimens may be killed and stored in 4-percent formalin. Specimens intended for histological purposes should be killed in Bouin's fixative, or at least in FAA. To prepare *Amphioxus* with the oral cirri extended, Bianco (1899) placed specimens in 9 parts of sea water and 1 part of alcohol until they died. They were then transferred to 50-percent alcohol and gradually moved up to 70-percent alcohol. Alcohol is often preferred as a proper preservative, especially if specimens are to be used directly in the classroom.

Storage

Museum collections of cephalochordates are kept in vials, along with their data, and are sealed in airtight museum jars, as described for Hemichordata, above.

REFERENCES

Abbot, D. P., 1961, "Urochordata" in *The Encyclopedia of the Biological Sciences*, Reinhold, N.Y.

Berrill, N. J., 1950, *The Tunicata*, Ray Society, London.

Bianco, S. L., 1899, The Methods Employed at the Naples Zoological Station for the Preservation of Marine Animals, *Bull. U.S. Nat. Mus.*, 39(M):1–37.

Potter, G. E., 1961, "Amphioxus" in *The Encyclopedia of the Biological Sciences*, Reinhold, N.Y.

Ricketts, E. F., and J. Calvin, 1956, *Between Pacific Tides*, Stanford Univ. Press, Stanford, Calif.

Van Name, W. G., 1945, *The North and South American Ascidians*, Bull. Amer. Mus. Nat. Hist., 84.

Wells, Morris M., 1932, *The Collection and Preservation of Animal Forms*, General Biological Supply House, Chicago.

(See Chapter 1 for other References.)

CHAPTER 18

The Fishes

For simplicity, the scope of this chapter has been broadened to include the cartilaginous fishes such as the sharks and rays; the bony fishes, which include the familiar game and food fish; and the cyclostomes, referred to as lampreys and hagfish. Although some of these groups are more or less widely separated taxonomically, techniques of collection and preservation are generally similar and permit common treatment in this text.

Fishes are collected for many reasons. Academic interests range from the classic use of the dogfish shark or fresh-water perch in the study of fish anatomy to fish identification or taxonomy, structural adaptations and ecology, and natural history of fishes. Many fishes are collected because they are directly or indirectly related to economic studies dealing with sport fishing, food fish, and the like. Obviously, specimens must be well preserved if they are to yield maximum value. Specimens must be accompanied by adequate field data if they are to be of any scientific value at all.

One should be aware of the taxonomic features of any group of organisms, and specimens must be preserved in a way that yields maximum value. Figure 18–1 shows some of the more common morphological structures of the bony fishes and cartilaginous fishes. The over-all shape of the body, the form of the tail, the type of scale, the relative position of various fins, the

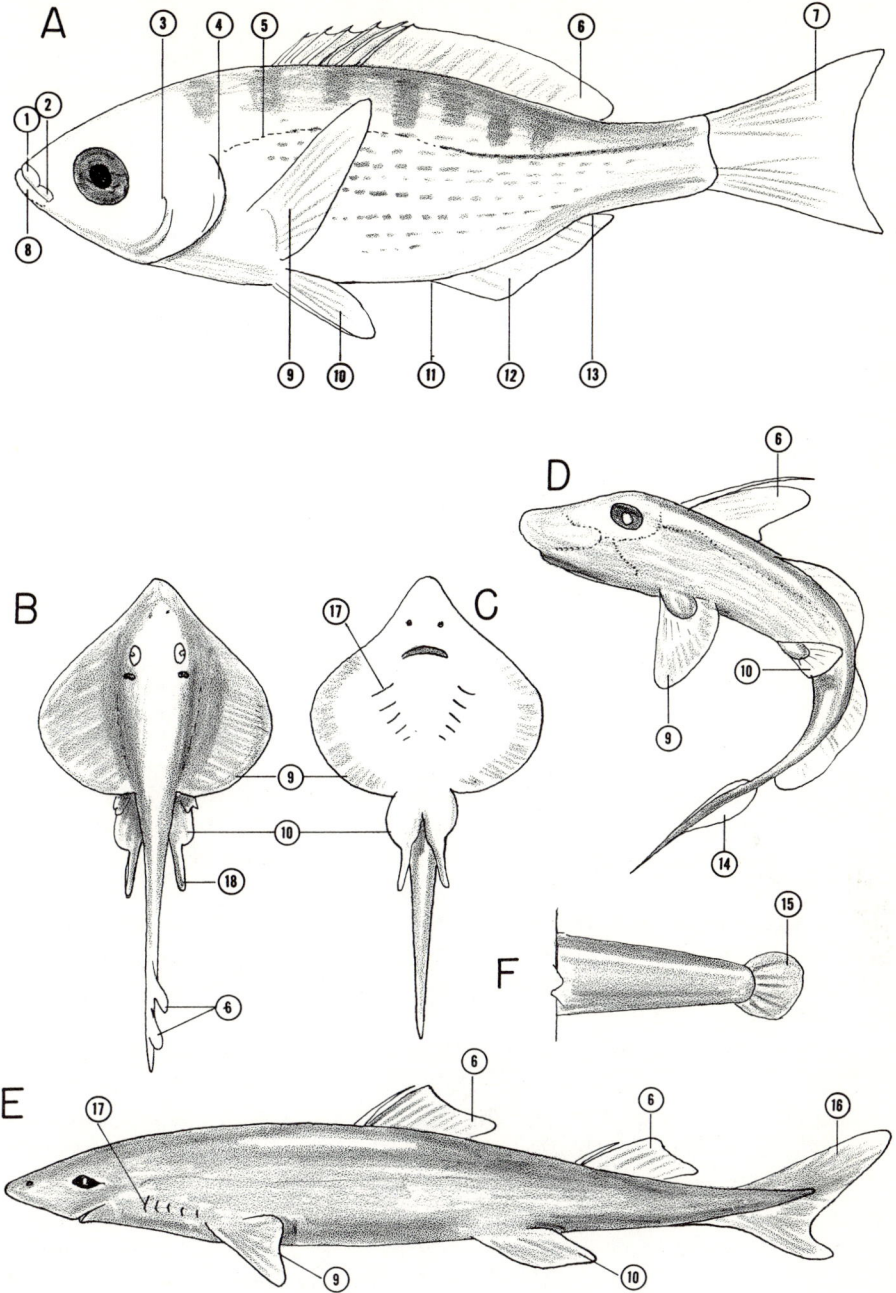

FIG. 18–1. General fish morphology. A. A bony fish. B–C. A skate or ray. D. The ratfish. E. The dogfish shark. F. Homocercal tail. (1) Premaxillary, (2) maxillary, (3) preopercle, (4) operculum, (5) lateral-line system, (6) dorsal fin, (7) caudal fin, a homocercal type, (8) mandible, (9) pectoral fin, (10) pelvic fin, (11) origin of fin, (12) anal fin, (13) insertion of fin, (14) tail, (15) homocercal tail, (16) heterocercal tail, (17) gill slits, (18) male copulatory organ.

320 Biological Techniques

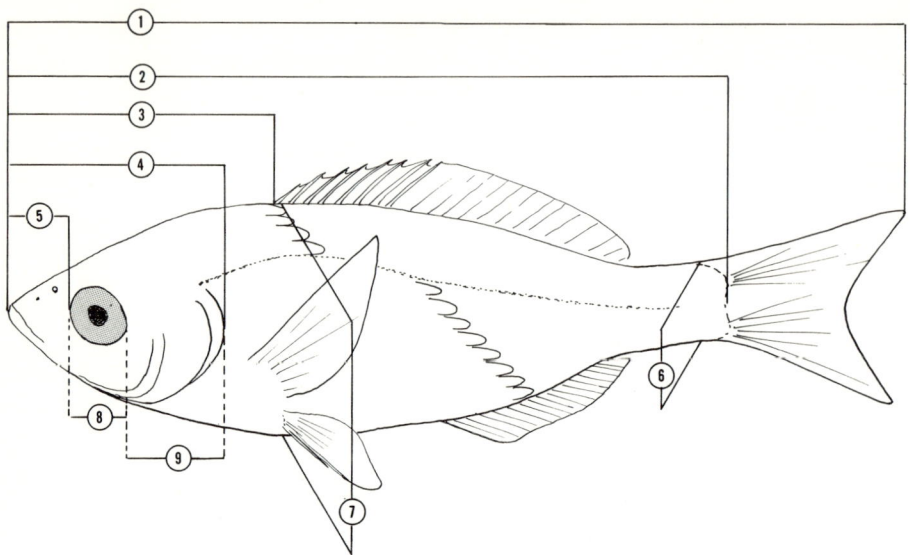

FIG. 18–2. Standard fish measurements. (1) Total length, (2) standard length, (3) predorsal length, (4) head length, (5) snout length, (6) least height of caudal peduncle, (7) greatest body height, (8) ocular length, (9) post-orbital length.

exposure or concealment of gills, the number of gills or gill slits, and the like, become important taxonomic features.

Body measurements are also critical in taxonomic and growth studies. Figure 18–2 shows nine of the more common measurements: some of these are the total length from the snout to the end of the tail, the standard length from the snout to the end of the vertebral column, the predorsal length from the snout to the first dorsal fin, the head length from the snout to the back of the operculum.

Scales and otoliths (ear bones) are used both for determining the age of specimens and for identification. Many fish are scaleless, including the cyclostomes, some catfishes, and the like. Cartilaginous fishes generally have placoid scales which possess small posteriorly orientated spines (Fig. 18–3A). Other common scales are the ctenoid which bear spines posteriorly, the cycloid which lack spines posteriorly, and the ganoid which are diamond shape and possess small spines or denticles (Fig. 18–3, B, C, and D, respectively). Both scales and otoliths show growth rings. Some workers have discovered that otoliths are morphologically different at least on the generic level. Improper preservation (see below) will immediately destroy otoliths and make them of no value for taxonomic work.

Finally, adequate field data must be taken if museum specimens are to be of all-around value. This must include not only the geographic locality of the collection and the date (see Chapter 1), but should include a description of the habitat in which the specimens were collected, color notes taken from the living specimens (see Chapter 1), and the like. Specimens that are well preserved lend themselves to internal dissection and studies of the food content of the stomach. However, food organisms should be removed and preserved in 75-percent alcohol or neutral formalin, since calcium carbonate contained by many food organisms would be altered by regular formalin. Also, parasitic organisms such as tapeworms and flukes will not be satisfactorily preserved for slide making if they are killed along with the fish host.

COLLECTING METHODS

Laws and Regulations

Write to your department of fish and game to determine the state laws concerning fish collecting and other types of collecting. Almost all states require fishing licenses and many require additional collecting permits. These permits can easily be obtained and often require little more than simply notifying the department of fish and game. Any fish specimens may be taken during the proper season, if they are covered by the state fishing license. However, although persons bearing a special permit are entitled to more freedom in collecting than the general angler, few states permit the use of rod and reel, nets, or traps in closed waters or during closed seasons unless the collector is accompanied by a game protector. Likewise, the use of rotenone or other chemical poisons for fish collecting is prohibited in most

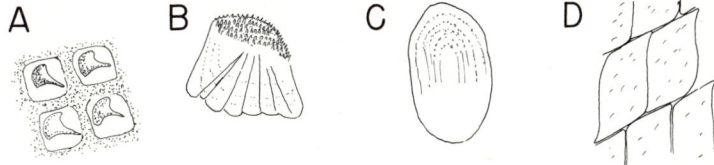

FIG 18–3. Types of fish scales. A. Placoid scales of cartilaginous fishes. B. Ctenoid scale. C. Cycloid scale. D. Ganoid scales.

states unless specific authorization is obtained from the game protector. Although these laws may be slightly cumbersome, the serious student or teacher is not in any way inhibited from conducting his scientific research. Such laws make for a greater degree of plant and animal conservation and, therefore, should be honored.

Collecting Sites and Methods

Rivers and streams which are uniformly deep may afford good collecting along the entire course. Deep holes, eddies, bends, and brushy places attract adult fishes; eddies and brushy areas attract juvenile specimens. Where the river or stream consists of a shallow, rock-strewn bed interrupted periodically by large boulders and deep pools, species groups and age groups will be more cleanly separated, according to the physical conditions of the stream. In addition to the above-named habitats, one should explore overhangs of rock or brush. Depending on the physical conditions of the water, seines, dip nets, minnow traps or other traps, hook and line, or electrical shockers can be used. The last-named are very useful in shallow streams rippling over rocks. Conversely, fish poisons are of little value in any local spot, but do widespread damage along the course of the water and therefore should be avoided completely.

Lakes and ponds vary tremendously in their physical makeup. The amount of weed, natural fish food, hiding places and nesting places, along with other physical factors mentioned in Chapter 1, will all help determine the species composition and the size of each population. One should collect in all habitats, from gravel to sand bottoms and from open water to extremely weedy areas in order to get a representative sample of adults and juveniles. Dip nets and seines work well inshore; fish traps, minnow traps, biological dredges, otter trawls, and hook and line are useful in deeper water.

Tide pools, rocky ocean shores, and coral reefs are difficult to collect from, owing to the irregular shapes of the pools, the interference of surf, the numerous hiding places, and so on. Turning rocks and examining other likely hiding places will prove productive, and the dip net is useful in smaller pools. Occasionally, fishes may be flushed from coral or rock hiding places by a geology pick and will swim into a carefully placed dip net. Seines may be placed in a stationary position in such a way as to close the entrance to a larger tide pool; the collectors move through the water and flush the fish ahead of them into the net. Fish poisons work extremely well in tide pools and similar areas, where the water exchange is not too great.

For deep ocean or lake bottoms the otter trawl or biological dredge is extremely successful. The trawl (see below) or the dredge (Chapter 1) can be made small enough to be handled from a skiff with an 18-horsepower motor. In open, surface waters large tow nets or bathypelagic nets may be put into operation, as well as the Chapoton surface trawl (see below). For collecting in deep ocean water ($\frac{1}{8}$ mile to 1 mile or more) the bathypelagic net or the Isaacs-Kidd midwater trawl are the most satisfactory.

Construction and Use of Fish Nets

Dip Nets. The construction of a light-weight dip net is described in Chapter 13. For working with fishes a larger mesh in the netting is required as well as a more sturdy hoop. Round iron rod measuring $\frac{1}{4}$ or $\frac{3}{8}$ inch in diameter proves quite suitable. The rod should be bent and fastened to the handle as explained in Chapter 13. For a dip-net bag, a so-called "smelt and minnow" replacement dip-net bag (or other replacement bag available commercially) is most satisfactory. On the other hand, $\frac{1}{8}$- or $\frac{1}{4}$-inch square minnow-seine netting may be fashioned into a dip-net bag in the manner described in Chapter 13. One problem that develops is the abrasion of the bag where it goes over the hoop. Fish dip nets are therefore usually threaded with wire, as shown in Fig. 18–4A, and the net bag is then applied to this. Stefanich suggests drilling $\frac{5}{8}$-inch steel airplane tubing, which is used for the hoop, and threading steel wire (size 14) in such a way that it occupies the inner diameter of the hoop (see Fig. 18–4B). When the dip-net bag is threaded to the wire, this type of construction allows the bag to be reversed for rapid emptying or cleaning.

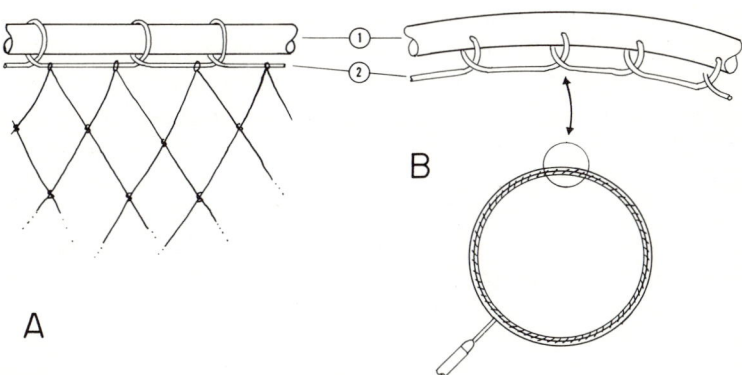

FIG. 18–4. Methods of making dip nets. See text for complete explanation. (1) Net hoop, (2) wire.

FIG. 18–5. Measuring fish netting. A. Square-hung netting. B. Diamond-hung netting, or stretch netting. C. Making a stretch measurement of diamond-hung netting.

Measuring Mesh Size. The reader should understand the difference in terminology for net meshes. For very fine netting, such as plankton netting, the meshes are called by the number per inch. For intermediate-sized netting, such as minnow netting, the meshes are square and are measured along one side of the square—¼-inch square mesh or ⅛-inch square mesh, etc. (Fig. 18–5A). In the larger commercial meshes the net may be square-hung or diamond-hung (the latter is shown in Fig. 18–5B). In either case, the meshes are stretched until closed, as shown in Fig. 18–5C, and then measured. Thus 1-inch square mesh by the former system would measure as a 2-inch stretch mesh in the commercial terminology.

Seines. The seine net is perhaps one of the oldest and most useful types of net for general collecting, especially for smaller fishes. Minnow seines usually are 4 feet deep, or more, and may measure from 6 to 20 feet or more in length. They are equipped with a lead-sinker line on the bottom and a float line on the top, and usually have a pair of sticks attached to either end to keep the seine open. The netting may be of cotton construction, although nylon is much preferred as it is practically rot-resistant. The two most popular mesh sizes are ⅛- or ¼-inch square.

A simple minnow seine is made as follows: After determining the size of the net required, select a suitable mesh. Square meshes are hung on a 1:1 basis, or 1 foot of netting applied directly to 1 foot of top or bottom line. On the other hand, diamond-hung stretch mesh is hung on a 3:2 basis where 18 inches of stretched netting are hung per foot of top or bottom line. The top and bottom lines may be made of ³⁄₁₆- or ¼-inch cotton or, preferably, nylon line. Tapered hardwood floats, 3 by 1½ inches, are spaced at 2-foot intervals along the top line. Tubular lead seine weights, 1½ inches long by ½ inch outside diameter, are distributed at 1-foot intervals along the bottom. Stout nylon or cotton twine should be used to apply the net, floats, and sinkers to the top and bottom lines. Cut the top line several feet longer than required for the net. Attach one end of the line to a tree at chest height, and then thread the required number of floats on to the line. Tie the opposite end of the line to some object also at chest height. Firmly attach a piece of cord, preferably waxed with bees wax, to the top line and begin attaching the net and floats as shown in Fig. 18–6, A, B, and E. Square mesh may be applied with a running-chain stitch rather than the clove-hitch (Fig. 18–6E). Repeat the procedure when hanging the bottom line with its leads and netting. When larger seine nets are being made, the seine weights may be replaced by a heavy galvanized chain applied firmly to the bottom line, as shown in Fig. 18–6D. This will give added strength to the seine and make it fish on the bottom along its entire length. Seines measuring longer than 20 feet are best hung in this fashion. Finally, a pair of sticks 4 feet or longer should be attached to either end, to facilitate handling the net.

Figure 18–6F shows a typical lake or pond set. One worker pulls his end of the net far out into the water while the second collector remains close to shore. The end of the net is brought around in a wide arc and, finally, both ends are brought up on the beach. As the net moves into shallow water it tends to bag out to trap the small fishes.

A typical beach set along sandy-bottomed ocean or lake shore may be made with a large seine, as shown in Fig. 18–6, G and H. Two lines 300 or 400 feet in length are attached to bridles on either end of a 100-foot seine. This seine should be equipped with a chain along the bottom, as mentioned above. The net is loaded into the skiff as follows: one side rope is payed into the skiff hand over hand, followed by the net, and in turn, by the second side rope. A collector holds the end of the first side rope on the shore while two men in a skiff row directly away from the shore, paying out rope until they reach the seine. They then move parallel to the beach until all of the seine has been payed out, and then they row in toward shore, paying out the second side line. The seine should be given adequate time to sink to the bottom before it is retrieved toward the shore. By putting a marker on the ropes every 50 feet the two teams pulling in the seine can notify each other of their progress and thus keep the seine parallel to the shore (Fig. 18–6H). The seine is brought in at only moderate speed, in order to keep fishes moving ahead of it, rather than greatly alarming the fishes and causing them to swim around the ends of the net. When the net reaches the shore the fishes are removed into buckets and other containers.

Otter Trawl. The otter trawl is an excellent piece of gear for collecting fishes and other organisms over sandy or muddy bottoms. The trawl consists of a funnel-like anterior section which tapers posteriorly to a fine-mesh cod end. The net is weighted by a heavy chain along its foot and kept

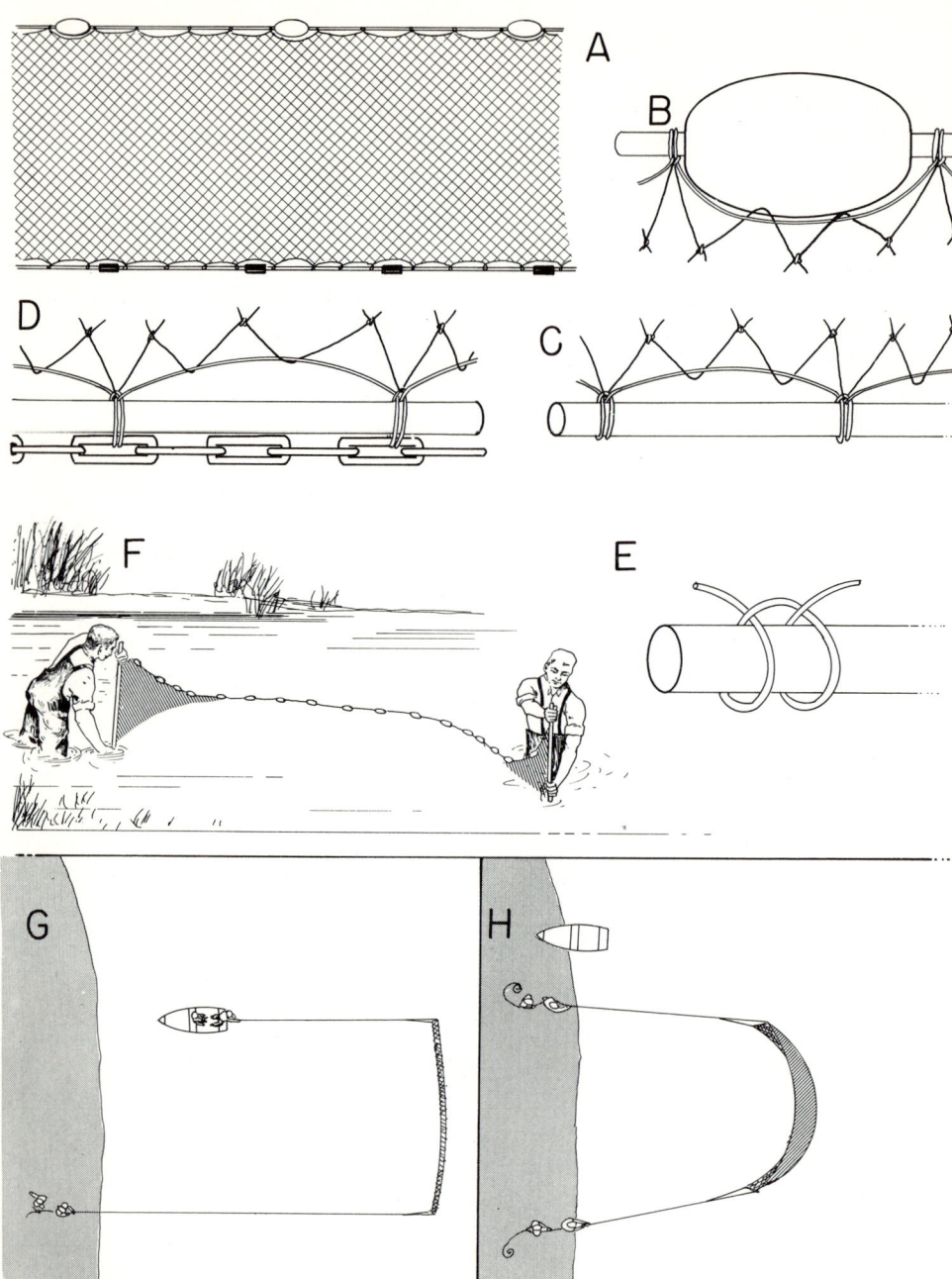

FIG. 18–6. Hanging and using fish nets. A–E. Making the minnow seine (see text for complete instructions). F. Using the minnow seine. G. Setting a large beach seine offshore by means of a skiff. H. Retrieving the beach seine in such a way that it remains parallel with the shore (see text for details).

perpetually open by two otter boards which plane outward as the net is pulled along the bottom. Four or more floats suspend the head of the net. The otter trawl described here follows the diagrams of Rupp and DeRoche (1960) who developed this net for the Maine Department of Inland Fisheries and Game.

The important pieces of equipment needed for the otter trawl are as follows: two pieces of marine plywood ½ inch by 15 inches by 24 inches, four styrofoam floats 3¼ by 3 inches, one piece of ¼-inch nylon rope 280 feet long, one tow rope 300 feet or more long, one ½-inch nylon head rope 12 feet long, one ½-inch nylon foot rope 13 feet long, one piece of galvanized chain of 3/16-inch diameter link material measuring 22 feet long, four pieces of chain (as above) each 31 links long. The netting may be cotton or, preferably, nylon as follows: netting for the anterior section 1¼-inch stretch measure, netting for the belly section 1-inch stretch measure, netting for the cod end ½-inch stretched measure. (The netting measurements shown in Fig. 18–8 are given in meshes rather than inches.) In addition, two pieces of steel bar 2 inches by ⅜ inch by 24 inches will be needed to weight the bottom of the otter boards, and two pieces of hardwood stiffener 1½ inches by ¾ inch will be needed for the top of the otter boards.

Cut the otter boards to size and trim off the lower forward corner, as shown in Fig. 18–7B. Drill 1-inch holes 2½ inches in from each corner, as shown, and then bolt a single steel bar to the bottom of each otter board and a single hardwood stiffener to the top. Attach the chains so that they crisscross, exposing 12 links in front of the cross, 1 link per chain is used in the cross, and 16 links after the cross, as shown. In addition to the 29 mentioned above, a link is used at each end for attaching the chain to the otter board by means of a staple made of ¼-inch steel (see Fig. 18–8A). Paint the assembled otter board with marine paint.

Measure, cut, and assemble the 1¼-inch stretched measure netting, as shown in Fig. 18–8C. Note that the top part "A" on the right joins to the upper side part "C" on the left, forming a funnel. The belly of the net and the cod end (Fig. 18–8, D and E) are cut and rolled as shown in Fig. 18–8F. The three sections are next tied together and the terminal of the cod end is tied shut with a piece of cord. The head rope and its floats must now be applied to the upper anterior part of the net, as shown in Fig. 18–8B. The 13-foot rope is next applied to the base of the net and the 22-foot chain is fastened to this. Finally, four 20-foot pieces of rope are used to attach the net to the otter boards. Two 100-foot pieces of rope make up the bridle which attach the otter boards to the tow rope, as shown in Fig. 18–7A.

Rupp and DeRoche (1960, p. 137) used a 14-foot boat powered with a

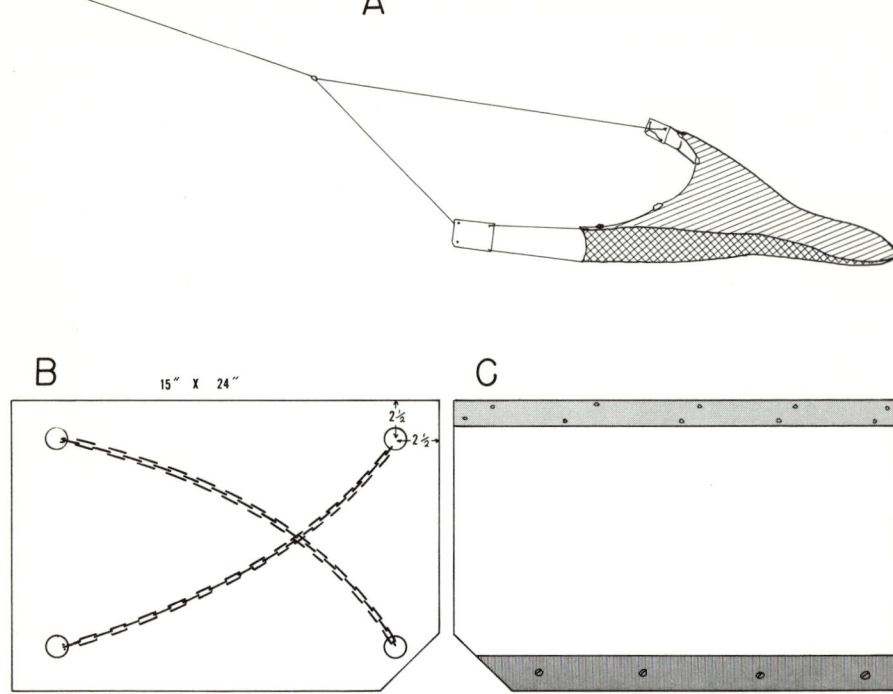

FIG. 18–7. The otter trawl. Redrawn from Rupp and DeRoche (1960). A. A diagram of the working trawl. B–C. Otter boards (see text for instructions).

10-horsepower outboard motor. They state that

> . . . the trawl can be set most easily if there is a slight breeze and the course can be arranged to trawl downwind. The boat is turned sideways to the breeze, the motor is turned off, and the trawl and the otter boards are put over the windward side. When the gear is hanging free 15' to 20' below and behind the drifting boat, the motor is started, the boat is turned downwind, and the rest of the warp is paid out with the motor running about half speed. When the warp (tow rope) is all out and secured, the motor is opened to full power. Full power with this gear gives a speed of 3 to 4 miles per hour.

They suggest that some fish may be lost after the tow during the period when the tow rope and bridle are being pulled into the boat. They suggest also that a funnel-like net placed inside the anterior part of the cod end will overcome most of this loss.

It would be well to attach the boat-end of the tow rope to a large float which can be thrown overboard should the trawl hang up on the bottom.

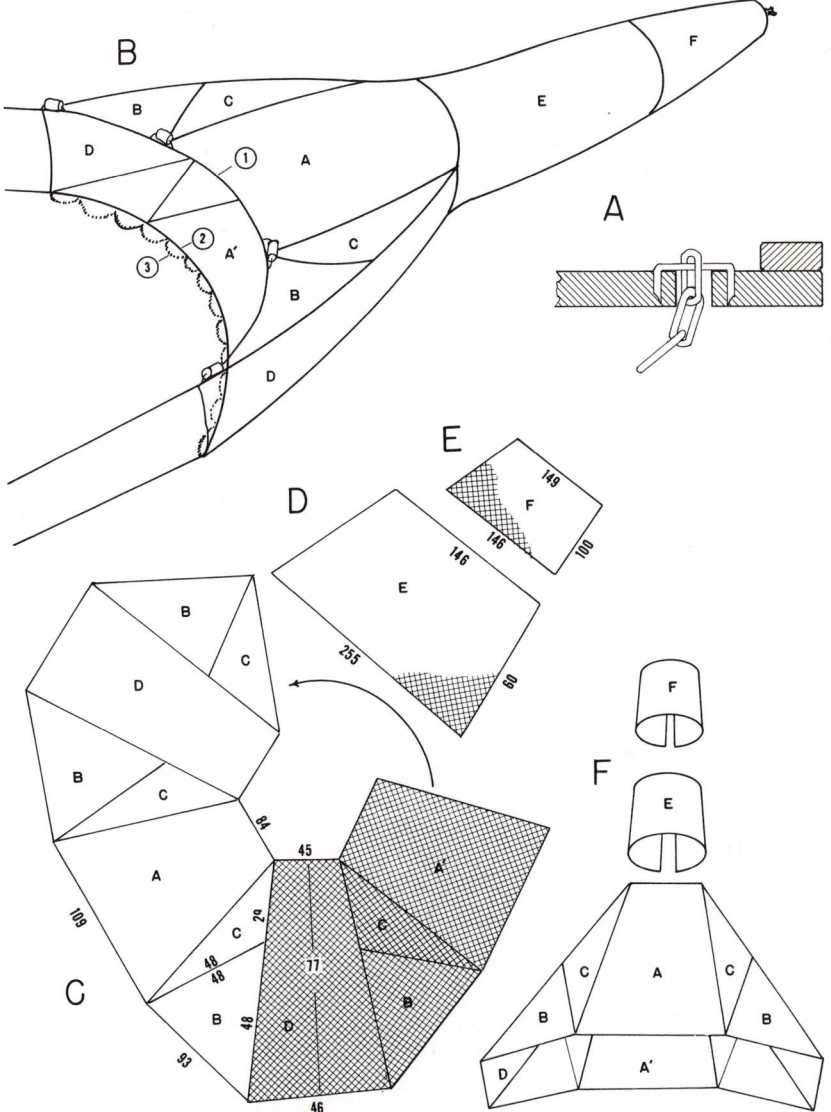

FIG. 18–8. Details of the otter trawl. Redrawn from Rupp and DeRoche (1960). A. Attaching the chain to the otter board by means of a large staple. B. The assembled otter trawl. C–F. Various aspects of the pattern and assembly of the otter trawl, showing the lay of the mesh and the number of meshes required for the different aspects of the trawl. Note that numbers here refer to meshes rather than inches. (1) Head rope, (2) base rope, (3) chain.

It would then be necessary to retrieve the float and to pull the trawl back in the way it had come in order to free it from bottom obstructions.

Surface Trawl. Chapoton (1964) developed the following surface trawl for catching juvenile American shad. The trawl net is made of ½-inch stretched mesh nylon netting; it measures 4 by 12 feet at the opening, is 15 feet long, and tapers to form the cod end as shown in Fig. 18–9. The two ends of the trawl are attached to a frame made of 2- by 2-inch pieces of wood which are strengthened at the corners with ½-inch plywood. The head rope of the trawl has the usual floats, whereas the foot rope is supplied with lead weights. A short bridle is attached to either end of the trawl frame.

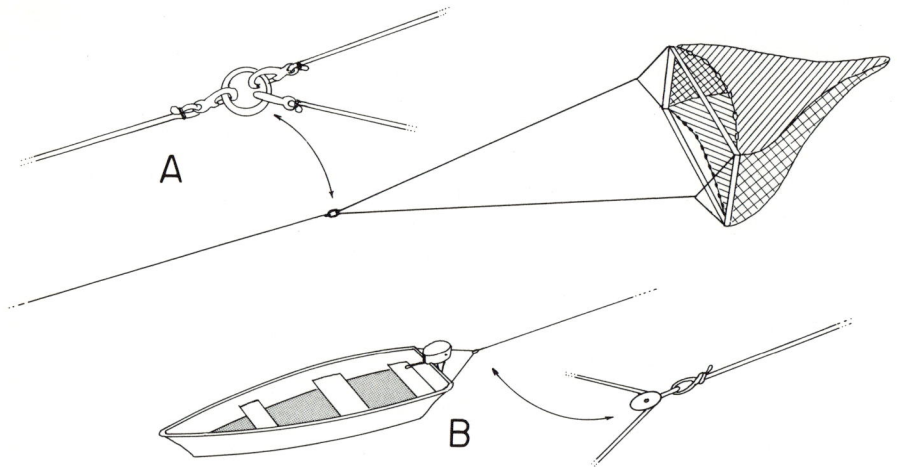

FIG. 18–9. The Chapoton surface trawl, redrawn after Chapoton (1964). A. A diagram of the surface trawl and a detail of the lead lines. B. A detail of attaching the trawl to the skiff. See text for complete instructions.

Chapoton used $5/16$-inch polyethylene rope for pulling the surface trawl. The main tow rope is 25 feet long and has a pulley, which attaches to the bridle of the boat, at one end and a snap hook at the other. Two additional pieces of polyethylene rope 50 feet long are equipped with snap hooks at either end. One is attached to either bridle on the trawl frame and both are snapped into a ring at the end of the tow rope, as shown in Fig. 18–9.

A 16-foot aluminum boat with a 10 or 18-horsepower outboard motor is used to pull the trawl. To operate, the trawl is floated away from the boat. When it is trawled, the net uprights into a "fishing" position. At the end of the trawl, the wooden frame tilts backward and the lead-line immediately buoys to the surface with the frame, thus preventing the escape of caught fish. (From Chapoton, 1964, p. 144.)

Care of Nets. Any net or line should be washed in fresh water immediately after it is used. Fish and other marine animals and plants contaminate nets with mucus which will begin to rot and will, in turn, destroy the fabric. Salt water alone is sufficient to damage cotton netting if it is not quickly washed and dried. Nylon netting tends to withstand much more abuse, can be kept wet for long periods of time, and requires less cleaning than cotton netting. Cotton netting should be periodically treated with tar, copper sulfate, or other preservatives. Consult Graumont and Wenstrom (1948) or other good texts for procedures of treating and making nets.

Fish Traps

Minnow Trap. Figure 14–6 (p. 283) illustrates a funnel trap which is appropriate for both crustaceans and fishes. The basic principle of the minnow traps is the funnel which leads toward a bait. Fishes enter the trap in an attempt to get at the bait and then find it difficult to escape. The trap may be square or round or any form one chooses. One funnel is often sufficient, but two or more may increase the catch. Figure 18–10 illustrates a collapsible minnow trap made of ½-inch stretched nylon mesh. This trap utilizes two rings of No. 9 galvanized wire, 1½ inches in diameter, and two rings of No. 9 galvanized wire, 10 inches in diameter and provided with four small eyes (Fig. 18–10, D and C, respectively). Four green willow sticks serve to hold the trap open (Fig. 18–10, A, D, and E). Each funnel is hung between its 10-inch ring and its 1½-inch rings. The funnels are then assembled and the small rings are tied together, as shown in Fig. 18–10D. Next, a piece of netting is applied between the 10-inch hoops to form the outside wall of the trap. A cut in this mesh forms a door which is closed by a piece of twine (Fig. 18–10A). If the four supporting sticks are removed, this trap will collapse and store flat. When the trap is put into operation the sticks are placed in position, the trap is baited and is lowered into a stream or pond by means of an anchor string. The bait generally consists of chopped fish tied in cheesecloth and suspended between the two funnels.

Fyke Net. The Fyke net (Fig. 18–11A) may be modified for catching fish coming up or down a stream, moving in or out with a tide, and so on. The seven stout hoops are made of ¼-inch iron rod 3 to 4 feet in diameter. In most Fyke nets the hoops become gradually smaller toward the cod end,

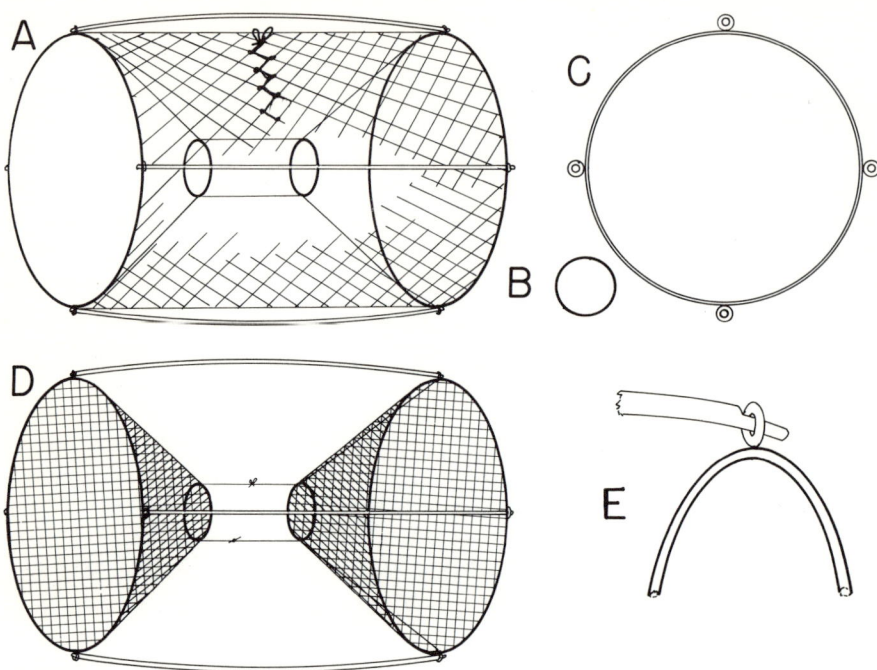

FIG. 18–10. A collapsible minnow trap made of fish netting. A. A side view of the trap. B–C. The supporting rings. D. Tying the funnels in position. E. Detail of inserting the spacing sticks.

but to simplify the cutting of webbing all the hoops may be of the same diameter. As noted, two of the hoops support funnels which are held open by means of strings directed toward the cod end. Fishes are directed into the Fyke net by a regular minnow seine which extends for some distance on either side of the opening. The net should be so designed that when it is stretched between the three anchor poles the funnels will be fully expanded. This net needs no bait, but rather takes advantage of normal fish movement.

Trap Net. The trap net also takes advantage of normal fish migration which directs them into the funnel trap. Figure 18–11B shows a simple form of this trap. This net is generally 6 to 10 feet high on the side walls and may have a leader net which extends out 100 feet or more. Generally, the use of this net would be considered beyond the scope of normal field collecting. Those interested should consult Crowe (1950) for a full description of the pattern and materials, anchor devices, and setting and tending of this net.

Electric Shockers

Electric shockers using both direct current and alternating current have been developed for fish collecting. They are especially effective in shallow streams and rivers where collecting is otherwise quite difficult. A portable power source is used to create a field between two electrodes held down in the water. Direct current units tend to attract fishes from their hiding places, but are slightly less effective in immobilizing fishes than are the

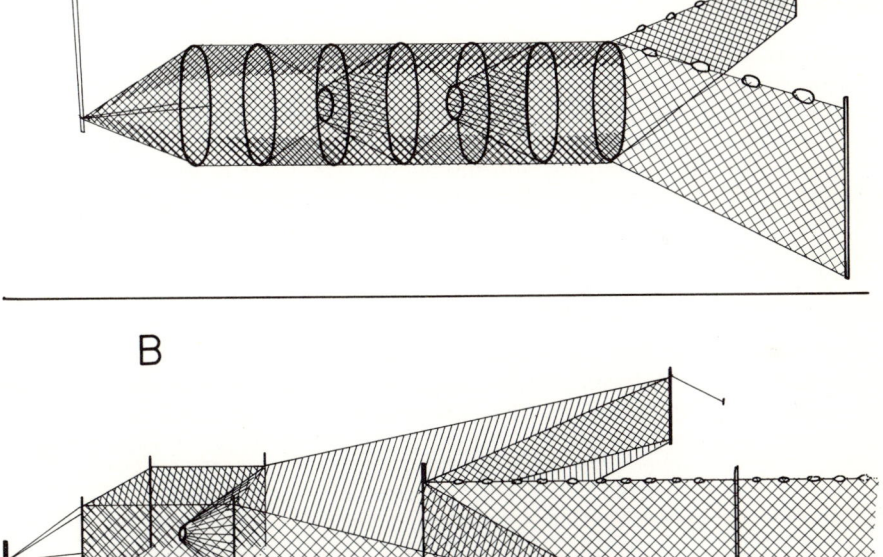

FIG. 18-11. More elaborate fish traps. A. The Fyke net. B. The trap net. See text for a discussion of these nets.

alternating current units. Alternating currents quickly paralyze fish and are very useful for gathering specimens for preservation.

McCrimmon and Berst (1963) have published a paper on a portable fish shocker which may be used as an alternating current or direct current unit. The following description and quotations are taken directly from their article. Their back-pack unit includes two 6-volt motorcycle batteries connected in series to provide a 12-volt power source. When a 117-volt alternating current 60-cycle Heathkit MP-10 power converter is mounted on top of the battery box the completed unit measures 13 by 6½ by 9½ inches and weighs approximately 15 pounds. The electrodes measure 6 by 9 inches and are mounted on 6-foot hardwood handles. One handle is provided with an "off-on" switch and an AC-DC converter switch to permit the operator to select the type of current required according to local conditions.

Figure 18–12A shows the basic circuit diagram for the alternating-current unit. This unit alone would probably be entirely satisfactory for fish collecting. Figure 18–12B, however, shows how the basic alternating-current unit can be modified to include direct current.

> The power converter is equipped with two outlets, one of which may be used as a direct-current power source for the electrodes by inserting four silicon rectifiers in a bridge configuration between the alternating-current circuit and the outlet. Selection of alternating-current or direct-current output may be effected by inserting the electrode plug into the appropriate outlet or by using a double-pole double-throw relay controlled by a toggle switch on the handle of the electrode.

McCrimmon and Berst (1963, p. 161) suggest that this unit can be further modified to utilize a secondary 12-volt activating circuit.

> A secondary circuit utilizing a relay controlled by the microswitch may be added to the basic unit in the primary low-voltage circuit to activate the current between the electrodes [Fig. 18–12B]. This modification has two advantages: first, it eliminates the idling current by switching off the converter except when the field is required between the electrodes; second, it requires that a current of only 12 volts, rather than the 110 volts in the basic unit, appear at the microswitch in the hand of the operator.

In using the fish shocker two or more individuals should be present for safety's sake. All should be provided with full-length rubber waders and elbow-length rubber gloves. The primary operator works upstream, holding the electrodes between 1 and 6 feet apart. Stunned fishes are retrieved by a secondary operator with a dip net. McCrimmon and Berst suggest adding sodium chloride upstream when extremely soft water is encountered, in order to make the shocker effective. Collectors interested in other references

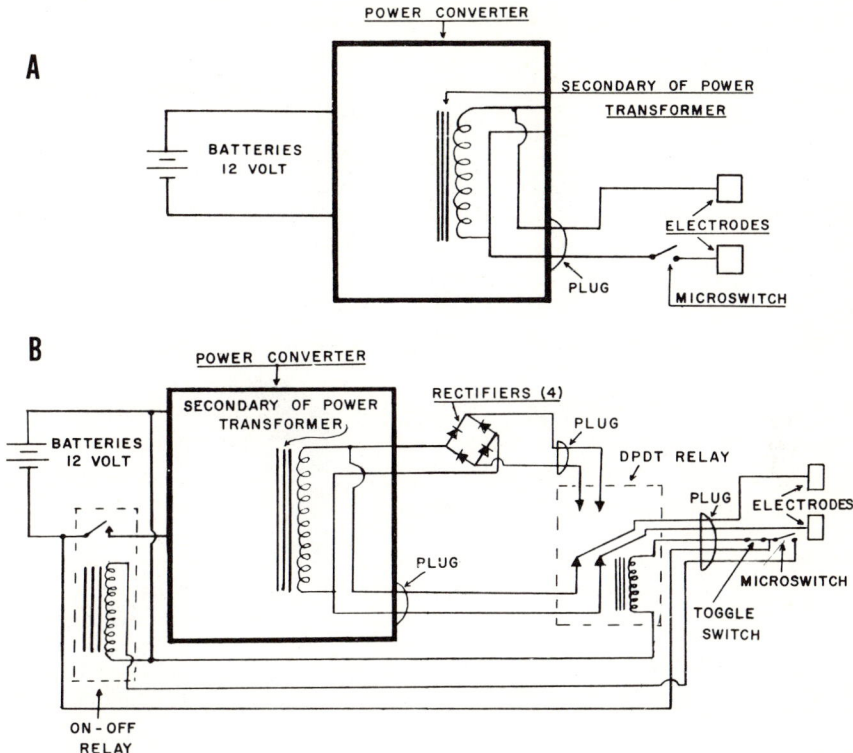

FIG. 18–12. Details of the fish shocker circuit diagrams taken from McCrimmon and Berst (1963). A. Circuit diagram of the basic A.C. unit. B. Circuit diagram of modified A.C.-D.C. unit.

to shockers should see Lowry (1964), Holton and Sullivan (1954), Morris (1950), Rayner (1949), and Hazzard et al. (1947).

Lennon (1961) describes a unique modification of electrofishing wherein a fly rod is equipped with a power unit and an electrode. Through experimentation a fly rod was developed which can be operated by one man for trout surveys in small streams.

Fish Poisons

The use of fish poison is controlled by law, as mentioned above. The most common compounds contain rotenone and come in either liquid or powdered

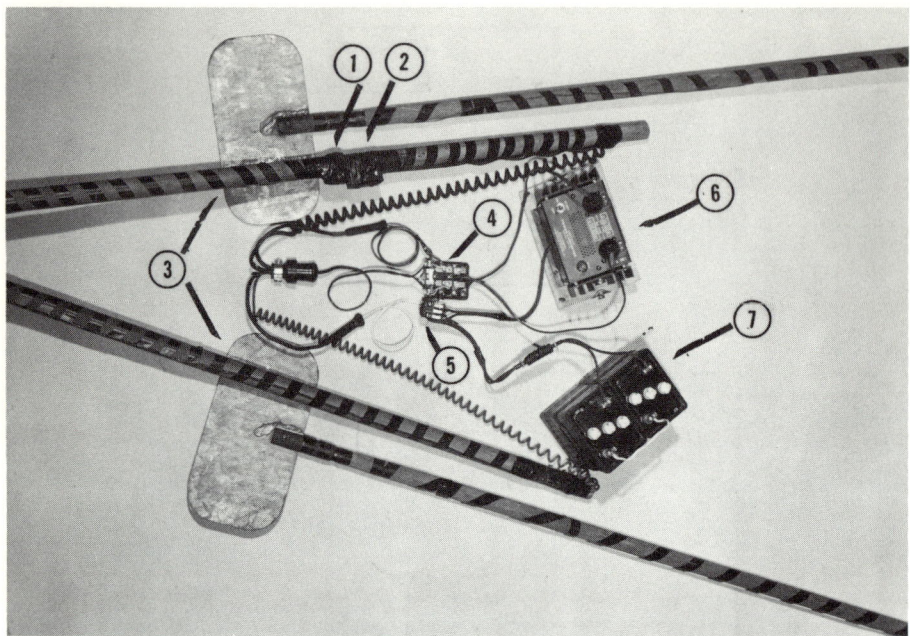

FIG. 18–13. The assembled back-pack fish shocker, after McCrimmon and Berst (1963). Photograph courtesy of McCrimmon. (1) A.C.-D.C. switch, (2) on-off switch, (3) electrodes, (4) A.C.-D.C. relay, (5) on-off relay, (6) power converter, (7) batteries.

form. These are diluted with water and are used primarily in tide pools and similar areas where the poison will eventually become diluted. Such materials should not be used in small ponds, simply because a total kill of the pond will eventually result. Rotenone and similar compounds eventually kill gill-breathing invertebrates as well as fishes. For this reason the collector must always bear in mind the need for conservation while collecting. Rotenone works better in warmer water temperatures; it becomes less effective and requires a longer period of time in cold water. A new liquid product known as "Chem Fish Collector" has been developed (Chemical Insecticide Corporation, New Jersey) which is not affected by water chemistry or temperature; the manufacturers claim that it works more quickly than simple rotenone.

Manufacturers of fish poison recommend a base dilution for their products; the poison is diluted and added to the pool to be collected. Within a few minutes fishes will begin to float to the surface and can be retrieved by means of a dip net.

Hook-and-Line Fishing

An excellent way to obtain carnivorous fishes is by hook-and-line fishing. Rod and reel, throw lines, and set lines are all appropriate. The throw line consists of a heavy cord about 100 feet long. One end is staked to the ground and the remainder is loosely coiled in front of the stake. A heavy lead sinker is provided at the opposite end of the line, along with one or two leadered hooks placed at 1-foot intervals from the sinker. The collector twirls the weight around overhead and casts the entire throw line far out into the water. Long set lines, which may measure a mile or more in length, are made of heavy $3/16$-inch marine cord. One end is weighted and baited hooks are placed along the length of this cord every 10 to 15 feet. The set line may terminate on shore or may be tied to an anchored float. The set line should be tended by a skiff every hour or so.

Skin Diving

Skin diving is a very effective way of capturing fishes and other aquatic organisms, both along ocean shores and in lakes and ponds. Swim fins, snorkel tube, and face mask are essential. Small dip nets with very large mesh netting are sometimes effective for capturing fishes. The spear gun, however, is better suited for this type of collecting. A very effective and simple spear gun may be made by using a simple fish or frog spearhead (gig) which is attached to an 8- to 10-foot piece of $1/2$-inch thin-walled electrical conduit tubing (Fig. 18–14, A and C). A wooden plug in the

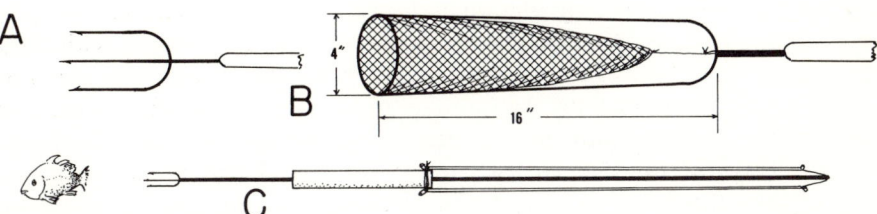

FIG. 18–14. Fish spears and spear nets. A. Head of fish spear. B. Spear-net head, used for catching small live fishes underwater. C. The spear gun, using surgical tubing to propel the spear. See text for complete details.

front end of the tube receives the spearhead, while a wooden plug in the other end is notched to receive the rubber-band propellent. A piece of ¾-inch conduit or bamboo tubing is needed to make the sling for propelling the spear. Tie to this two pieces of ¼- or ⅜-inch surgical tubing. Tie the other ends of the surgical tubing together with a 3-inch length of stout cord. To operate, run the butt of the spear through the sling tube, fit the spear into the string connecting the surgical tubing, and draw the sling tube as far forward on the spear as practical. This is held by one hand (Fig. 18–14C). When the spear is released, it lunges forward for a considerable distance. If the spear is too buoyant, one or two small holes should be drilled, in order to release the air and allow it to fill with water.

The author has used a small spear net (Fig. 18–14B) which is very successful for catching small fishes. A 4-inch net hoop is welded to a 16-inch long U-shaped piece. This, in turn, is attached to the spear handle, as shown. A cone-shaped net 4 inches in diameter is sewn to the hoop and lightly tied in place. To use the spear net, simply select a fish, wait until it is positioned most favorably, and fire the spear net so that it captures the specimen.

TRANSPORTATION OF LIVE FISHES

Some simple techniques for the transportation of live organisms and the control of temperature, metabolic waste, oxygen concentration, and the like, are given in Chapter 1. Small fishes may be transported in polyethylene bags filled with water (if the bags are kept cool), as the bags tend to permit the exchange of gases. However, the science of live-fish transport has developed far beyond the scope of methods needed by a student or teacher. Nevertheless, the reader should be aware of some of these developments. There is a large literature dealing with tank design, insulation, and handling techniques; more concerning circulation and pumping devices, aeration by spray and compressed gas liberators, etc.; still other literature dealing with filtration equipment, temperature control, and the like. Anesthetics are now widely used in transporting fishes. Some of these are quinaldine, chlorobutanol, ether, methanesulfonate, urethane, sodium amytal, and thiopental sodium. Three of the best recent papers which deal with these techniques give tables concerning the effect of anesthetics and present lengthy bibliographies of additional literature: McFarland (1960), Norris *et al.* (1960), and McFarland and Norris (1958).

PRESERVATION TECHNIQUES

Killing and Hardening

Kill fish specimens rather than letting them die in the air. Unless otoliths are required, add a small quantity of formaldehyde to a large container of water and place the fishes in this. If the container is large enough the specimens will die perfectly straight. Smaller fishes should then be transferred to bottles with 5-percent formalin. Fishes longer than 5 or 6 inches should have a cut made in the right side of the abdominal cavity and should be put in 10-percent formalin for hardening. Those fishes that will not readily fit into jars (extremely long individuals, for example) may have to be bent and tied with string during the hardening process. One must always bear in mind the size of containers ultimately available. Extremely large specimens should be rejected when more convenient smaller ones are available. Fishes may also be killed by putting them into 70-percent alcohol while they are still alive. Regardless of the preservative, if the fishes are dead the slime should be removed from the body prior to preservation. Many workers feel that fishes required for skeletons should never be placed into formalin, as softening of the bones may occur.

Shipment of Field Specimens

When a large number of field specimens has been amassed and preserved in formalin, wrap individual specimens or lots of specimens in cheesecloth, along with their field data. Pack these into metal containers or tin cans (see Chapter 1) in such a way that no movement of the specimens will be possible. Wet the specimens with the initial preservative, but pour off the excess, and seal the container. If the container is sealed properly the specimens will not become dried and will travel very nicely for periods of many weeks. The critical thing is to be sure field data are included with each lot of specimens.

Soaking Out and Final Preservation

If field specimens preserved in formalin are to be transferred to alcohol for final preservation, most workers first remove the formalin before transferring to alcohol. This is done simply so that the fish specimens will not reek of formalin when handled later on. Place specimens in bottles of clean water and change the water each day. Two or three days (or more) may be required to remove most of the formalin. Some workers watch for a telltale scum to form on the surface of the water as a sign that the soaking-out process is completed. This indicates that the formalin is now so weak that bacteria can begin to grow. At this time specimens should be placed in 70-percent ethyl alcohol or 40- or 50-percent isopropyl alcohol. If a large volume of specimens is placed in one container the alcohol should be changed after 24 hours.

Specimens are kept in tightly sealed museum jars, along with their field data. The preservatives always modify the color; light will further bleach the specimens. Thus, it is advisable to store specimens in a dark place.

Color Preservation

Probably no solution yet devised will preserve fish color (or that of other animals) perfectly. This is one notable area in which research is still required. Many methods have been attempted, including preservation in gasoline. Martin and Terrence (1962) experimented with ten different solutions for color preservation in fishes. At the time of their publication the specimens had been preserved only 7 months, which is significant but not quite conclusive. The solution that showed the most promise for them is made as follows: 20 cubic centimeters of 1-percent phenol, 12 cubic centimeters of formaldehyde, 8 cubic centimeters of glycerin, and 60 cubic centimeters of water. At the time of their writing, both the color and the animal tissue were in a good state of preservation.

Kotthaus (personal communication, 1963) comments that small numbers of fishes preserved in an arsenic-glycerin solution in 1950 were not yet rendered colorless at the time of writing. His preservative is made as follows: 30 cubic centimeters of glycerin, 6 grams of potassium acetate, 0.08 grams of sodium acetate, 0.012 grams of arsenic trioxide, and 63 cubic centimeters of water. This fluid becomes dark with time, but tends to maintain color longer than direct formalin-preserved specimens.

Scales

Fish scales should be collected during tagging operations and at canneries and similar places, because they are important for determining fish age (by the concentric growth rings) and growth and, to a lesser degree, for taxonomic studies. These may be placed into dry, cotton-stoppered vials, along with the fish's field data, or they may be preserved in vials of alcohol. These scales are later placed on a microscope slide, covered with a slip, and projected with a microscope slide projector such as the Bausch and Lomb Tri-simplex. The age is determined directly from the projected scale, or the scale is traced on a record card. Scales may be mounted permanently under coverslips as follows: Place a dry scale on a slide, put a large coverslip (No. 2 thickness) over the scale, and seal the coverslip in position by applying melted paraffin to the edges with a cotton swab.

New techniques have been devised by professional ichthyologists for making scale impressions on 1- by 3-inch cellulose acetate slides. The scale is imprinted on the plastic slide in an arbor press or roller press, with or without heat. The description for building the press necessary for the work is beyond the scope of this book; the interested reader is referred to Smith (1954), Butler and Smith (1953), Arnold (1951), or Greenbank and O'Donnell (1950).

Otoliths

Otoliths, fish earbones, are becoming increasingly important in fish studies, for they show growth rings and are morphologically different on the generic level and, thus, are useful for aging specimens and for taxonomy. Otoliths should be removed from fresh-caught fish or from those killed and preserved in alcohol. Fishes that have been in formalin, even for a few hours, are not satisfactory in that the fine characteristics on the otoliths are quickly eroded.

To remove the otoliths cut through the skin on the side of the head and remove the opercular bone. Next, shave away the thin bone making up the capsule of the ear. When exposed, the otolith can be removed with a forceps. Occasionally, the otoliths will fall down into the brain cavity. Should this occur, simply go on with the removal of the otolith from the opposite side and then flush the "lost" otolith from the brain cavity by means of a stream of water directed through the ears. Otoliths have very little tissue clinging to them and may be placed directly into small dry vials, along with their data. Screw caps rather than cotton stoppers should be used, since the fine sculpturing on the otolith may become entangled in the cotton.

Puffers

Puffers are tropical fishes which inflate themselves with water until they become almost perfectly round and lose their normal fish-like appearance. When the skin is extended hidden spines become erected, giving the fish the appearance of a pin cushion. Puffers are usually preserved in formalin, but occasionally are skinned and dried to demonstrate the spines or for use as ornaments.

The following procedure is used for skinning puffers: Make a slit on the ventral abdominal surface near the anal opening; make this slit no longer than necessary. Pull the skin to one side and work in toward the body with the objective of cutting the body in two. This achieved, draw out the tail section by cutting the body free from the skin. Leave enough of the fin rays intact to give the fin a normal appearance. When the tail has been removed repeat the process by removing the tissue from the anterior end of the body. When skinning around the eyes use sufficient care so as to leave the eyes intact and attached to the skin rather than the skull (the skull is removed). Likewise, allow the bony plates and the jaws to remain with the skin rather than with the skull. Remove all excess flesh from the skin, turn the fish right side out and sew up the incision on the ventral surface. Next, insert a round balloon through the open jaws and inflate this until the skin is completely distended. Now tie off the neck of the balloon, arrange the fins in a natural position and submerge the specimen in a bucket containing 10-percent formalin for just a few minutes. Dry the specimen in an airy but shady place. After a few days of drying, let the air out of the balloon and pull the balloon back out through the mouth.

Drying Fishes

Some of the very bony fishes, such as sea horses, cow fishes, or trunk fishes, are occasionally dried for museum and taxonomic purposes. These are generally injected and briefly preserved in formalin and then placed in a shady but well-ventilated place to dry. Older workers use corrosive sublimate in place of formalin. If specimens are intended for student use, however, corrosive sublimate should be avoided, because even dry specimens will contaminate the hands with this very poisonous material. When fishes are completely dried, store them in boxes with their field data and sufficient paradichlorobenzene to prevent insect infestation.

Skinning Large Fishes

Fishes that are too large for any museum container may nevertheless be preserved in a satisfactory condition by skinning out the nonessential parts. For this procedure, make an incision to the left or right of the midventral line, remove the visceral mass, and work up through the main body of the fish so as to cut the body (but not the skin) in two. Remove the posterior portion of the trunk by freeing muscle and fin attachments. Leave enough of each fin base and of the tail base to keep the fins in a normal position. Now remove the anterior part of the body up to the posterior end of the skull, but leave the skull intact in the skin. If the visceral mass is desired, preserve this separately. The skin can now be rolled or folded and placed in a convenient-sized container, wherein it is preserved in 10-percent formalin and treated as described above.

Gut Contents

If the gut contents are to be examined for food studies one should recognize the following: (1) Immediate preservation of the contents must be achieved by injecting preservatives into the gut or by removing the gut for special preservation. (2) Formalin may destroy the calcareous shells of food organisms and thus should be avoided if these shells are important.

SKELETAL TECHNIQUES

The study of fish skeletons is one of the keys to fish taxonomy. Smaller fishes, up to 10 or 12 inches, should be cleared and stained unless they are rather deep and thick-bodied. Elongated specimens up to this length are very suitable for this technique. Bigger fishes are usually prepared by dermestid beetles, which do a very satisfactory job. Specimens may also be boiled and hand-picked, but are less satisfactory. Because these techniques are almost identical for all vertebrates they will be discussed collectively in Chapter 23.

REFERENCES

Arnold, E. L., 1951, An Impression Method for Preparing Fish Scales for Age and Growth Analysis, *Prog. Fish-Cult.*, 13(1):11–16.

Butler, R. L., and L. L. Smith, 1953, A Method for Cellulose Acetate Impressions of Fish Scales, With a Measurement of Its Reliability, *Prog. Fish-Cult.*, 15(4): 175–178.

Chapoton, R. B., 1964, Surface Trawl for Catching Juvenile American Shad, *Prog. Fish-Cult.*, 26(3):143–144.

Clemens, W. A., and G. V. Wilby, 1946, Fishes of the Pacific Coast of Canada, *Bull. Fish Research Board of Canada*, No. 68.

Crowe, W. R., 1950, Construction and Use of Small Trap Nets, *Prog. Fish-Cult.*, 12(4):185–192.

Eddy, S., 1957, *How to Know the Freshwater Fishes,* William C. Brown, Dubuque, Iowa.

Graumont, R., and E. Wentstrom, 1948, *Fisherman's Knots and Nets,* I–XV, Cornell Maritime Press, N.Y., 1–203.

Greenbank, J., and D. J. O'Donnell, 1950, Hydraulic Presses for Making Impressions of Fish Scales, *Trans. Am. Fish. Soc.,* 78:32–37.

Hazzard, A. S., et al., 1947, Hunt Creek Fisheries Experiment Station, *Prog. Fish-Cult.,* 9(2):65–70.

Holton, G. D., and C. R. Sullivan, 1954, West Virginia's Electrical Fish-Collecting Methods, *Prog. Fish-Cult.,* 16(1):10–18.

Hubbs, Carl, and Karl Lagler, 1949, *Fishes of the Great Lakes Region,* Cranbrook Institute of Science, Bull. No. 26, Bloomfield Hills, Mich.

Lennon, R. E., 1961, A Fly-Rod Electrode System for Electrofishing, *Prog. Fish-Cult.,* 23(2):92–93.

Lewis, W. M., 1963, *Maintaining Fishes for Experimental and Instructional Purposes,* Southern Illinois Univ. Press, Carbondale, Ill.

Lowry, G. R., 1964, Some Ways of Increasing the Utility of a Backpack Fish Shocker, *Prog. Fish-Cult.,* 26(3):127–130.

Martin, M., and M. Terrence, 1962, A Solution for the Preservation of Color in Fishes, *Turtox News,* 40(4):122–123.

McCrimmon, H. R., and A. H. Berst, 1963, A Portable A.C.-D.C. Back-Pack Fish Shocker Designed for Operation in Ontario Streams, *Prog. Fish-Cult.,* 25(3): 159–162.

McFarland, W. N., 1960, The Use of Anesthetics for the Handling and the Transport of Fishes, *Calif. Fish and Game,* 46(4):407–431.

McFarland, W. N., and K. S. Norris, 1958, The Control of pH by Buffers in Fish Transport, *Calif. Fish and Game,* 44(4):291–310.

Morris, R. W., 1950, An Application of Electricity to Collection of Fish, *Prog. Fish-Cult.,* 12(1):39–43.

Norman, J. R., 1951, *A History of Fishes,* A. A. Wyn, N.Y.

Norris, K. S., et al., 1960, A Survey of Fish Transportation Methods and Equipment, *Calif. Fish and Game,* 46(1):5–33.

Rayner, H. J., 1949, Direct Current as Aid to the Fishery Worker, *Prog. Fish-Cult.,* 11(3):169–170.

Rounsefell, G. A., and W. H. Everhart, 1953, *Fishery Science: Its Methods and Applications,* John Wiley, N.Y.

Rupp, R. S., and S. E. DeRoche, 1960, Use of an Otter Trawl to Sample Deep-Water Fishes in Maine Lakes, *Prog. Fish-Cult.,* 22(3):134–137.

Schrenkheisen, Ray, 1938, *Field Book of Freshwater Fishes of North America North of Mexico,* Putnam, N.Y.

Schultz, L. P., 1936, Keys to the Fishes of Washington, Oregon, and Closely Adjoining Regions, *Univ. Washington Publ. Biol.,* 2(4).

Schultz, L. P., and E. M. Stern, 1948, *Ways of Fishes,* Van Nostrand, Princeton, N.J.

Scott, W. B., 1954, *Freshwater Fishes of Eastern Canada,* Univ. of Toronto Press, Toronto.

Smith, S. H., 1954, Method of Producing Plastic Impressions of Fish Scales Without Heat, *Prog. Fish-Cult.,* 16(2):75–78.

Stefanich, F. A., 1952, An Improved Dip Net, *Prog. Fish-Cult.,* 14(4):172.

Walford, L. A., 1937, *Marine Game Fishes of the Pacific Coast from Alaska to Ecuador,* Univ. of Calif. Press, Berkeley, Calif.

CHAPTER 19

The Amphibians

The amphibians are soft-skinned, mostly tetrapod vertebrates which are familiar to all of us. The common frog has long been used for the classical dissection and study of vertebrate anatomy in general classes. Frogs are becoming more important in physiological tests and studies in embryology. By inducing female frogs to deposit eggs and supplying these with sperm obtained from the testes of the male, one can learn much about the early stages of egg cleavage. In addition to their use in studies of muscle and heart physiology, pregnancy tests, egg development, anatomy, and the like, the amphibians have long been studied for their natural history, breeding, feeding, and other activities. Surprisingly, we still have a lot to learn about the natural history and correct taxonomy of amphibians.

CHARACTERISTICS

Before collecting and preserving amphibians the collector should have a thorough knowledge of those characteristics of taxonomic importance, so that his specimens will be preserved to yield the maximum amount of information. In addition to gross morphology, ornamentation, and the like, body

The Amphibians   347

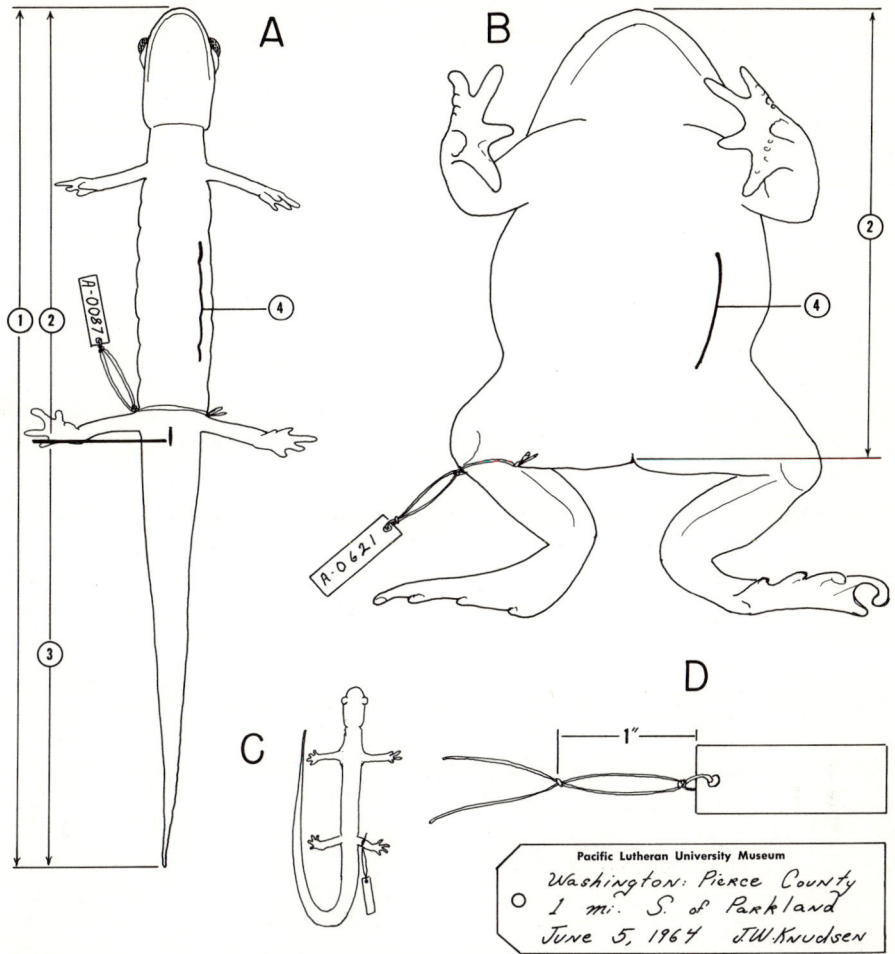

FIG. 19–1. Measuring, tagging and posturing amphibians. A–B. Measurement and tagging procedure for salamanders and frogs, respectively. C. Posture of long-tailed salamanders. D. Method of tying the tag and an enlargement of a typical leg tag. (1) Total length, (2) snout-to-vent length, (3) tail length, (4) cut here or inject with formalin. See text.

measurements, skeletal structures, color in life, and even voice are important. Figure 19–1 shows some of the common measurements made on salamanders and frogs; these are total length, snout to vent length, and tail length. It is imperative that specimens be preserved in a uniform way, with the body straight and the legs uniformly arranged, in order to facilitate measurements and study. Many mispreserved amphibians which have been dropped directly into strong formalin are so grotesquely twisted and out of "posture" that they are both unattractive and difficult to work with. Thus, if specimens are to be preserved at all, the collector should be prepared to relax and fix his specimens properly, to take color notes while the animals are still alive, to record the voice if this is essential, and so on.

FIELD DATA

Make a practice of recording complete field data in a notebook at the time of collection. This should include not only the date and geographic locality (see Chapter 1), but should also have such things as the time of day, the weather conditions, air and water temperatures, the general habitat, the microhabitat in which the collection was made, and notes concerning the activity or lack of activity of the specimens. All of these data help to create a picture of amphibian natural history. A field notation such as "collected in woods" is of little value as compared with "collected in woods, north-facing slope, heavy overgrowth with 90-percent coverage, specimens under rocks and heavy leaf litter only in areas with ground-water seepage." Many collectors record either ground or water temperature and rectal temperatures of the specimens. Any unusual observations, such as breeding activity, egg-laying, feeding, and so on, should be recorded. If possible, color notes should also be recorded, either in the field or immediately after returning to the laboratory. (See Chapter 1 for techniques of color notations.) Field notes are usually arranged by collecting station, each station having its own individual number. This same number will accompany both the specimens from that station and the logging of specimens in the museum.

NEED FOR CONSERVATION

There is a notable tendency on the part of collectors of amphibians and reptiles to overcollect in areas of low population density. For example, spade-foot toads which live in desert and semidesert situations will migrate en masse to temporary waterholes after the first heavy rain during the breeding season. Owing to the harshness of such habitats, the population is naturally quite low. Thus, overzealous collectors may destroy the total population by simply taking ten or fifteen specimens. This is especially true where the microhabitat of amphibians is discontinuous, confined to small areas which have some unique physical factors that make amphibian existence possible. Such situations are usually well enough isolated to make reoccupation of the habitat quite unlikely. Conversely, where the physical requirements of certain species are continuous, such as those of the common tree frog *Hyla regilla* which is continuous from central California up into British Columbia on the west side of the Cascade Mountain range and elsewhere, there is little chance of overcollecting by taking moderate numbers, since reoccupation and migration are always possible. Usually there are few laws pertaining to the collection of amphibians. However, some states, such as California, require a collector's permit for every kind of plant and animal. These are readily obtained from fish and game departments and usually cost about 1 dollar per year.

COLLECTING METHODS

Equipment and Field Problems

The kind of equipment used to capture amphibians will depend on the habitat. Collecting containers, which are required in large numbers, present one of the basic field problems. Specimens must be isolated first by their collecting station, and second by species or, at least, genera. Almost all amphibians secrete mucus which may be poisonous to other species. Specimens cannot be overcrowded in containers and must be kept both moist and cool. Therefore, the field car should be parked in shade and specimens should be kept in the car itself rather than in the trunk, which will quickly overheat in warm weather. In collecting from a car, cartons of quart and pint museum jars make excellent containers. Collect specimens into plastic bags, plastic bottles, or wet cloth sacks provided with some moss or leaf litter. About ½ inch of water should be placed in each jar, along with a small quantity of paper toweling for frogs. For salamanders, loosely crumple up one or two paper towels and put them in a bottle. Thoroughly saturate the towels with water before adding the specimens. On longer field trips the paper towels should be changed daily and the bottles washed out in order to remove mucus and excreta. Place a complete set of field notes and a field station number, written on a good grade of paper with a medium-soft pencil, in each bottle, along with the specimens. Specimens will require no food for long periods of time if kept cool. Do not screw the lids down so tightly that oxygen cannot enter the bottle. Keep the bottles in their pasteboard cartons in order to provide a dark place for the specimens.

Caecilians

Caecilians are tropical amphibians which are legless and superficially segmented, thus often resembling brightly colored and overgrown earthworms. These specimens are usually found in or around rain forests and require considerable moisture. They live in rodent burrows or make their own burrows in soil with a high humus content. They are frequently found in the center of thoroughly rotted logs, under bark on old stumps and logs, in heaps of rotting vegetation, in and under manure piles, and are occasionally seen out of their burrows at night during heavy rainstorms. There is more luck than prediction involved in collecting caecilians. When uncovered, they must usually be grabbed quickly to prevent their escape. They can burrow through soft soil at an amazing rate of speed if given the chance. Once they are captured they are treated like any other amphibian.

Frogs and Salamanders

Ponds and Lakes. Some species of frogs and salamanders are permanent residents in ponds and lakes and thus can be found during all favorable seasons. The presence or absence of large predaceous frogs such as the bullfrog will frequently determine the presence or absence of other species. Almost every pond and lake has its breeding population that migrates from the terrestrial habitat during its particular breeding season. Migrations may start as early as the time when the winter snows leave the ground (about January 15 in the Pacific Northwest) and will continue up into the early summer months, each species taking its turn. Salamanders, tree frogs, and toads can be collected during this migration or found hiding under boards, logs, and so on, near the breeding pond. Use a coarse-meshed dip net with a moderately long handle for netting frogs and salamanders. Fine-meshed dip nets are difficult to move through the water rapidly and are less suitable for amphibians. The minnow seine (see Chapter 18, p. 324, for construction and use) is excellent for collecting adult and larval amphibians. A small triangular dredge, such as that described in Figure 1–2, p. 10, is excellent for sampling ponds and small lakes. Fig. 19–2, A, B, and C, shows how a small dredge on 100 feet of line can be carried around the end of a small pond, released, and retrieved in order to catch larval and adult forms.

Collecting at night with a gasoline pressure lantern or flashlight is quite satisfactory. Frogs that are "singing" can be cautiously approached and netted. When approaching a singing male frog use the light only during periods when it is singing. When it stops singing, switch the light off and

FIG. 19–2. Steps in sampling a small pond with a hand dredge.

wait quietly until singing resumes. Frog eyes will glow in the beam of a flashlight directed parallel with the surface of the water. It is often necessary to swat the dip net directly down over the frog and pull the net in toward shore in one continuous motion. Techniques that may work for frogs during daylight hours are the use of a trout fly to literally hook the frog or the use of a .22-caliber pistol with bird shot for large bullfrogs.

Minnow traps (see Chapters 14 and 18) baited with fresh-cut liver will often attract aquatic salamanders. During the breeding season an unmated female salamander isolated in a small screen cage within the minnow trap will usually lure dozens of males (and also females) with the chemical secretions given off in the water. The author has had unbaited minnow traps that caught salamanders for days after an unmated female had been present in the trap. Frog, toad, and salamander eggs may be found singly or in chains or clumps on the bottom of shallow ponds or attached to vegetation. When collecting eggs, keep them isolated in plastic bags of water or jars of water. Attempt to select egg masses with different rates of development, in order to preserve a complete developmental series.

Rivers and Streams. In the tropics hundreds of species of salamanders and frogs live near or in small streams and rivers. This is also true, but to a lesser degree, as one moves away from the equator. Large aquatic salamanders, such as the northwestern *Dicamptodon,* are easily caught on hook and line or in baited minnow traps. When collecting along streams and rivers, slowly turn debris such as logs, loose stones, and the like, in search of adult animals. Night collecting will often reveal considerable amphibian activity where both adults and larval forms move to the shallow waters for feeding. Specimens that live in swift streams usually swim rapidly downstream when alarmed. Therefore, they should be approached from a downstream position to ensure their capture. Many tropical frogs live on the vegetation just above small streams and rivers. They can be heard calling and can be located by flashlight at night. Some species attach their eggs to vegetation hanging directly over such streams. There the larval forms develop and, upon hatching, fall into the stream to complete their growth. Where the terrain permits, a seine can be quite useful for catching amphibians.

Terrestrial Habitats. Many toads and frogs, newts and salamanders spend most of their time in the terrestrial habitat. One must be familiar with the distribution and habitat requirement of such animals before collecting can be successful. In forested situations where salamanders are common, specimens may be found in and under rotten logs, bark, leaf litter, underground burrows, roadside seeps, in rock slides, and so on. Collecting at night with a lantern will often reveal numerous salamanders that are up for either mating or feeding. During the very heavy spring rains and early fall rains numerous frogs and salamanders can be collected by driving the roads at night in a likely habitat. Blacktop roads with little traffic are the best for collecting. Start collecting soon after dark during the height of heavy pelting rainstorms. Drive slowly (between 10 and 15 miles an hour), using low beams from your automobile for light. You will quickly develop an "eye" that can distinguish very small salamanders from earthworms that are seen on the road. You will recognize that some areas are used as migratory paths by amphibians. Such runways should be frequented throughout the year in order to obtain a complete pattern of migration.

Some of the hole-nesting frogs of the tropics are extremely large and very quick to escape. These are usually detected by shining the flashlight parallel to the surface of the ground, whereupon the eyes of the frogs glow in the dark. A very large net (with a hoop diameter of 16 inches or more) is best for capturing such specimens. They have to be approached cautiously and netted before they have a chance to escape into their burrows. On the other hand, some frogs and toads are very easily approached and may be picked up by hand. The arrow-poison frogs of the tropics are diurnal and are easily approached. These brilliantly colored animals, however, secrete a very poisonous mucus which is not dangerous unless the collector rubs his eyes or gets the mucus into his mouth.

Recording Frog Calls. Of late there has been considerable interest in the use of frog calls, even for taxonomic purposes. While frog calls cannot be used satisfactorily in taxonomic keys, they may be important in the study of species relationship. Researchers attempt to record both the individual voices and the chorus of singing frogs. Legler (1964) suggests that frog calls can be adequately recorded on inexpensive foreign tape-recorder equipment.

He recommends that the entire chorus be taped initially before isolated individuals are approached. After the chorus is "on tape," the individual may be approached. Frequently, the collector startles the singing male and causes him to be silent. Here, Legler advocates playing back the entire chorus of singing frogs with the volume of the tape recorder turned way up. This usually will stimulate the individual specimens to begin singing, at which time their isolated voices can be taped.

TRANSPORTING LIVE SPECIMENS

It is often necessary to transport or ship live specimens over long distances. When working by automobile, prepare the specimens as described above (p. 349). The primary requirement is that the specimens be kept cool, moist, and clean. In warm, dry climates amphibians can easily be kept alive by placing their containers in the new, large, and inexpensive camping iceboxes. In extremely high temperatures small quantities of ice will keep the insides of such compartments relatively cool.

When shipping live specimens, air freight is the most desirable. Place specimens in a large stout plastic bag with wet toweling or wet moss. Have the bag almost completely inflated with air, twist the neck shut and seal the bag with rubber bands. Next, place the plastic bag in a large pasteboard box with a thick layer of loosely crumpled newspapers surrounding the bag; this will serve as a source of insulation and will keep the specimen from being damaged. Mark such containers "Living specimens for biological research," and "Keep away from extremes of high and low temperature."

PRESERVATION TECHNIQUES

Procedure and Problems

Normally the process following collection includes (1) narcotizing, (2) killing, (3) positioning, (4) fixing, and (5) transfer to the final storage solution. Many workers do not wish to be bothered with this elaborate procedure while field collecting and simply drop living specimens into a proper preservative. However, the extra time involved in preservation is well worth the trouble, since the specimens are better looking and are more easily measured and stored. One problem that must be overcome in preservation is that of keeping the field data along with the specimens. This may mean that narcotizing and fixing will have to be conducted in numerous small containers to prevent the mixing of specimens. Sooner or later, a label must be attached to each specimen to keep its identity separate.

Narcotizing and/or Killing

Amphibians are allowed to swim in solutions containing narcotizing agents. These are usually local anesthetics which cause the specimens to lose consciousness or even die. It renders the specimen well relaxed and, therefore, permits it to be positioned in a desired manner.

Chloretone. Perhaps the most common narcotizing agent for amphibians is Chloretone. Prepare this by adding about 1 level teaspoon of Chloretone to 1 gallon of water. Specimens should be placed in containers with sufficient Chloretone solution to completely cover them. The animals will swim for several minutes before sinking to the bottom of the container. If the Chloretone solution is too strong, specimens will tend to contract and develop "kinks" in the tail and body. When the solution is properly diluted, however, specimens will die in a completely relaxed manner. Store Chloretone solution in an airtight jar when it is not in use.

Clove Oil. The author finds that clove oil, which contains a phenol compound, is quite useful in narcotizing amphibians. Put 2 or 3 drops of clove oil in $2/3$ quart of water, cap, and shake well. Use clove oil in the same manner as Chloretone. Specimens will usually sink to the bottom within 5 minutes and will be totally senseless in 10 minutes or so. A longer time is required for larger specimens.

Freezing. If the collector has access to a refrigerator with a freezing compartment, amphibians may be killed under reduced temperatures. Put specimens in plastic bags, along with their field data, tie the bag, and place in the frezer overnight. Next, completely thaw all specimens and then transfer them to a fixing solution.

Positioning and Fixing

This process involves placing specimens in pans of 8- to 10-percent formalin for from 48 hours to 1 week. Specimens should be thoroughly narcotized before transfer. Position salamanders with the tail out straight, as shown in Fig. 19–1A. If specimens are so long that they will not readily fit into containers, tails should be bent up along the side of the body, as shown in Fig. 19–1C. If necessary, use wax-bottom trays for fixing and pin the specimen in the desired position. Figure 19–1B shows the frog position that can most easily be worked with.

Heavy-bodied frogs and toads and salamanders should be injected in the body cavity with formalin or cut along the right side of the body, as shown in Fig. 19–1. This permits the rapid preservation of the gut content, which otherwise would start to decay. Salamanders with a body diameter equal to or less than the diameter of a lead pencil need no incision or injection. Likewise, small frogs need no injection. If frogs tend to float in the fixing solution, the body and lungs are filled with air. Gently squeeze the abdomen, working from the posterior end up toward the anterior end, until the air is expelled from the body. Very large frogs and salamanders may require injection of formalin into the heavy muscles of the legs as well as in the abdomen.

Final Preservation

The following solutions are quite satisfactory for long-term preservation: 5-percent formalin, 70-percent ethyl alcohol, or 40- to 50-percent isopropyl alcohol. Specimens to be stored in formalin may be transferred directly from the fixing solution. Conversely, workers using alcohol generally prefer to soak out the formalin before transfer. To remove the formalin place the specimens in jars of fresh water for 24 hours or more. Change the water each day. Do not permit the specimens to become too soft, but rather transfer them to alcohol when most of the formalin has left the body. The alcohol should be changed after 24 hours to ensure that the fluids contained in the specimens have not diluted the preservative below a safe level.

Labeling and Records

Two types of labels may be used for amphibians, in addition to normal museum-bottle labels. These are attached directly to the specimen, either around the waist or around the hind appendage, as shown in Fig. 19–1, A and B. The most common label measures about ¼ inch by ¾ inch, and is made of a stout waterproof paper. It is tied to the specimen by means of a lightweight string or heavy button thread. This small label usually contains only the field collecting number or the museum acquisition number. Most museums log all new specimens in an acquisition book, along with the date and locality of collection and the collecting or field number. Every new acquisition has a new number and, therefore, specimens so numbered can easily be matched up with their field data. The second kind of label applied directly to specimens is somewhat larger, measuring approximately ½ inch by 2 inches. This label contains all of the pertinent field data, as shown. The larger label is much preferred for small collections where field notes are frequently separated from the collections.

Storage Methods

Although color cannot be preserved well in amphibians, every effort should be made to store specimens so as to retain as much color as possible. Therefore, specimens should be placed in museum jars, along with their data, tightly sealed, and stored in a darkened room or cabinet. Under ideal conditions, the temperature should not fluctuate in the storage room. With extreme temperature fluctuation the preservative expands and contracts and may eventually break the seal of the jar lid. Museum specimens should be checked at least twice a year and preservatives added when needed.

Dry Skins and Color

Kincaid (1948) has developed a useful method for preserving the color pattern of the skins of frogs. By this method the skin of a frog is removed, attached to a mounting board, and pressed in a botany press. Since the frog skin is attached only at a few major points, it may be removed in a matter of seconds. The author has mounted skins prepared by this technique which have excellent color after fourteen years of storage.

Provide the following: sheets of Bristol board or white poster board (white cardboard), some wax paper, and a plant press with blotters (Chapter 5, p. 76). Kill the frog with ether or by one of the methods mentioned above, and make incisions on the ventral side of the body (Fig. 19-3). Next, arrange the skin on the mounting board. This may be done by carefully spreading the skin on the board or by floating the skin in water and sliding the board up underneath the skin. Regardless of the method used, press all air bubbles from underneath the skin, cover with a sheet of wax paper, and place in a botany press with lots of newspaper above and below the specimen. Press for two or three days, changing the newspaper dryers daily. The skin will provide enough natural mucus on the underside to firmly adhere to the mounting board. If bits of wax paper stick to the specimens, simply moisten the wax paper with a sponge and rub it away with your finger. Finally, place all of the field and scientific data on the card. Store such specimens in lightproof cardboard boxes with paradichlorobenzene.

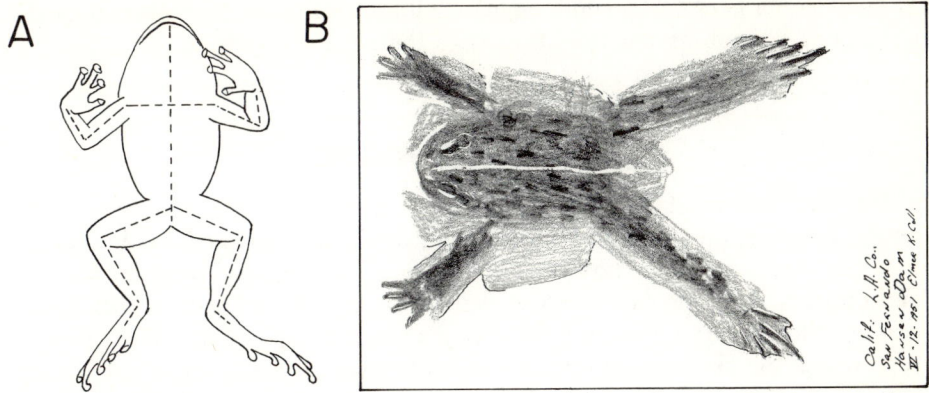

FIG. 19–3. Skinning an amphibian and pressing the skin for color preservation. See text for instructions.

SHIPPING FIELD SPECIMENS

When large numbers of specimens have been amassed on distant field trips it is simplest to ship them in bulk rather than in separate jars or other containers. Furthermore, specimens are usually held in formalin for shipment and are not soaked out and represerved in the final solution until they reach the home museum. Wrap individual specimens or groups of small specimens, together with their field data, in cheesecloth. Pour fresh preservative over the cheesecloth to keep the specimens moist. Next, place wrapped specimens in plastic bags, expel most of the air, twist the neck of the bag, and tie the bag closed with string. Numerous plastic bags of such specimens may now be packed in large metal drums which can be tightly sealed. No additional preservative is needed, so long as the plastic bag and drum are tightly sealed. If specimens do not completely fill all of the space within the drum, additional cloth or light paper toweling should be packed around the specimens to prevent any movement. Metal drums are, in turn, packed in wooden crates or in stout cardboard boxes. With the latter, pack a thick layer of crushed newspapers around the metal container.

An alternate method is that of sealing the specimens inside a series of plastic bags. Next, place the bags in a cardboard box and completely surround them with crushed newspaper. This technique will greatly reduce the weight and is more satisfactory for air freight. Specimens may remain in such containers for several months without damage. However, unpack specimens as soon as possible, to ensure their good condition.

SKELETAL TECHNIQUES

The skeleton is of extreme importance in amphibian taxonomy on all levels. Large animals may be fleshed out, dried, and cleaned by dermestid beetles. Most specimens, however, are too small for this and must be stained and cleared. Rare specimens which cannot be spared for "making skeletons" can be studied very adequately by means of stereoscopic X ray. This process reproduces the skeleton in such a way as to make it easily as valuable as a cleared and stained specimen. Because the procedures used for working with the skeletons of all vertebrates are so similar, these techniques will be discussed collectively in Chapter 23.

REFERENCES

Bishop, S., 1952, *Handbook of Salamanders,* Comstock, Ithaca, N.Y.
Cochran, D. M., 1961, *Living Amphibians of the World,* Doubleday, N.Y.
Kincaid, T., 1948, To Preserve the Color Pattern of the Skin in Frogs, *Turtox News,* 26(2):50–51.
Legler, J. M., 1964, Tape Recordings of Frog Calls, *Turtox News,* 42(2):68–69.
Ley, W., 1955, *Salamanders and Other Wonders,* Viking, N.Y.
Logier, E. B. S., 1952, *The Frogs, Toads, and Salamanders of Eastern Canada,* Clark, Irwin, Toronto.
Netting, G., and G. Orton, 1950, *A Field Guide to the Amphibia and Reptiles,* Houghton Mifflin, Boston.
Noble, G. K., 1954, *The Biology of the Amphibia,* Dover, N.Y.
Oliver, J., 1955, *Natural History of North American Amphibians and Reptiles,* Van Nostrand, Princeton, N.J.
Schmidt, K. P., 1953, *A Checklist of North American Amphibians and Reptiles,* Amer. Soc. Ichthy. and Herp., Northridge, Calif.
Smith, M. A., 1951, *The British Amphibians and Reptiles,* Collins, London.
Stebbins, Robert C., 1954, *Amphibians and Reptiles of Western North America,* McGraw-Hill, N.Y.
Wright, A. H., and A. A. Wright, 1949, *Handbook of Frogs and Toads,* Comstock, Ithaca, N.Y.
Zim, H. S., and H. M. Smith, 1953, *Reptiles and Amphibians,* Simon & Schuster, N.Y.

CHAPTER 20

The Reptiles

The reptiles constitute an old and diverse group of vertebrate animals characterized by their dry skin which is usually covered with epidermal scales. Snakes, lizards, turtles, and crocodilians make up the main groups of reptiles. These are more or less well distributed around the world, but are most numerous in the tropics. Because they produce so-called "land eggs" reptiles are not restricted to breeding in or near water, and thus make up a significant part of the fauna of arid regions. Reptiles often show amazing morphological modifications which adapt them to their environment. Among these might be cited the loss of limbs of many burrowing lizards, the modification of the feet for swimming in sea turtles, the compression of the body in sea snakes, the skin flaps used for soaring by the "flying lizards," the prehensile tail of chameleons, and the like. A few living reptiles are extremely large, for example, the crocodilians, the giant constrictors, and some lizards such as the monitors and iguanids (up to 10 feet and 7 feet, respectively). Most of the reptiles, however, are small and are generally inconspicuous in appearance.

FACTORS TO CONSIDER IN COLLECTING

A collector should advise himself of his state laws concerning the collection of reptiles. Most crocodilians and some turtles are protected by law generally throughout the United States. The Gila monster is protected throughout its range in the southwestern United States. Some states (such as California) protect all plants and animals; thus, a permit must be obtained before any collecting can be done. Because of the poor reputation of some snakes, such as the rattlers, "look-alike" species that are perfectly harmless are often killed by citizens who feel they are doing a good deed. Field collectors are often overly enthusiastic and may frequently take more specimens than a local population can afford. Teachers and collectors alike, therefore, should be highly conscious of the need for conservation and protection of reptilians and other animals to prevent them from becoming scarce or extinct.

Because reptiles are cold-blooded (poikilothermic) their metabolic rate is governed by environmental temperatures. Collectors can take advantage of this by keeping specimens relatively cool. Under such conditions snakes and lizards will require little food over long periods of time. Conversely, collectors should realize that desert reptilians are still very sensitive to extremely high temperatures. Reptiles utilize shade during the heat of the day and may suffer sunstroke if kept in the sun. When translated, these facts suggest that specimens should not be kept in closed automobiles that will overheat. Preferably, the windows should be open, the automobile should be in shade, or the specimens should be placed beneath the automobile while collectors are in the field.

Finally, body measurements are important taxonomic tools. These measurements (see Fig. 20-3, p. 368) are difficult to take on poorly preserved specimens. Therefore, collecting trips should be organized well enough so that specimens may either be properly preserved in the field or transported back to the laboratory for preservation.

FIELD DATA

The field notebook is perhaps more important than any other field tool, for unless "on the spot" records are kept, specimens will be of little value. Records should include the data and geographic locality (see Chapter 1) and as much physiographic and ecological data as possible. The plant formation, the terrain, air and soil temperatures, the microhabitat, and so on, should be described. In addition, details of any observed activity, or lack of activity, should be noted. For example, notes concerning mating, feeding, sunning, or other activities will all help to create a general description of the specimen's natural history. Color notes (see Chapter 1) should also be taken when specimens are intended for taxonomic purposes.

COLLECTING METHODS

Equipment and Field Problems

Reptiles are placed in cloth bags when captured; flour sacks or seed bags are excellent for this. Use bags made of stout canvas for poisonous snakes or large nonpoisonous specimens. Provide sacks with a tie string (not a drawstring) to secure the bag. Snakes are especially adept at working through weak spots in the seam of a bag or through the neck of the bag, unless it is properly tied. Figure 20–1 shows one method of "bagging" a specimen. The specimen is dropped into the bag, whereupon the mouth of the bag is closed and the sack is spun around to twist the neck. This prevents specimens from escaping while the neck of the bag is being tied. Note (Fig. 20–1C) the neck of the bag is doubled over and double tied.

Large lizards and nonpoisonous snakes are placed in bags as follows: While holding the specimen in your right hand reach all the way to the bottom of the bag, grip the specimen with your left hand through the bag, release your right hand while still holding the specimen with your left hand, and twist the neck of the bag closed. Tie the bag as shown in Fig. 20–1. Professional rattlesnake collectors attach a collecting sack to a stout hoop and handle. The bag then resembles a butterfly net. Specimens are transferred to this bag by means of a snake stick or hook stick. The bag is closed with a light rope tied in an overhand knot around the neck of the bag. Finally, the neck is double tied, as shown in Fig. 20–1. This technique greatly lessens the danger of being bitten.

When transporting collected specimens, do not pile one specimen bag on top of another until some of the smaller lizards and snakes are killed by crushing or suffocation. If numerous small, fragile specimens are placed in one sack, include some coarse brush which will keep the sack expanded and give the specimens room to move about.

In addition to special collecting equipment mentioned below, the geology

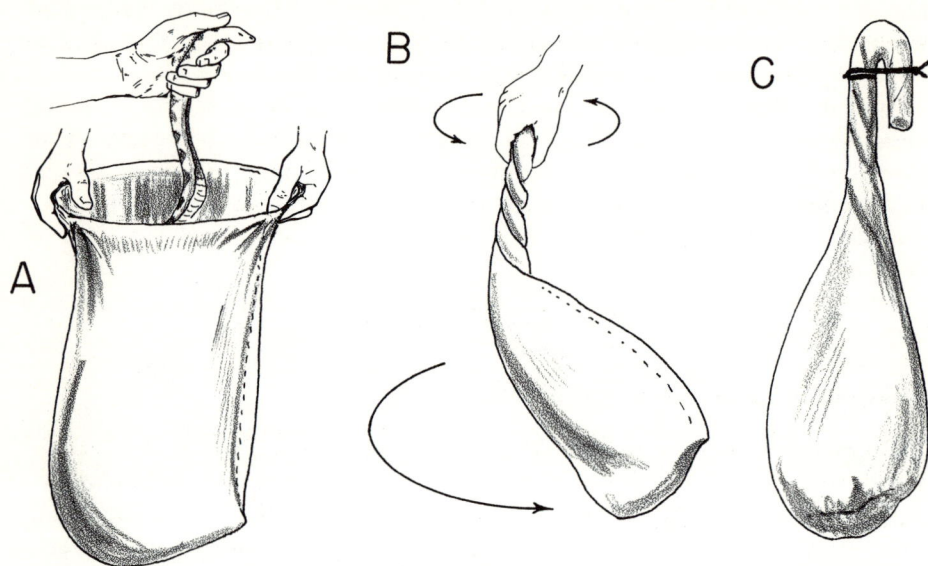

FIG. 20–1. Method of holding a snake and placing reptiles in a snake bag by first dropping the specimen in the bag, twisting the bag, and finally tying the neck.

pick is perhaps one of the most useful general field tools. This is used for turning rocks, removing bark, splitting open old logs, tearing up rodent burrows, and doing the many things too dangerous to be done by hand in rattlesnake country. High-top leather boots and heavy leather gloves are also recommended.

Reptiles, like other animals, are highly seasonal in appearance and behavior. Therefore, collectors must take advantage of periods when they are active and accessible. Not only do air and surface soil temperatures have to be favorable, but favorable temperatures must also reach down to the depth of the reptile's hiding place before the specimen will warm up and venture out for feeding.

Lizards

In the higher latitudes lizards remain in open country or brushy country, but are seldom found in heavy forests. Toward the tropics arboreal species are more numerous. Some legless lizards are found only in soft soil or sand where they burrow and hunt insect larvae for their food. Most lizards are extremely fast when sufficiently warm, and are usually alert enough to make approach difficult. Furthermore, lizards sun themselves near hiding places, into which they retreat when disturbed. Most lizards can autonomize (break off) their tails if handled roughly and, for this reason, must be captured in such a way as to prevent injury. Snares, .22-caliber bird shot shells, slingshots, pry bars, and various other tools are used for capturing lizards.

The lizard snare is one of the best all-around collecting tools. Lizards are very tolerant of a snare and will permit you to try numerous times to slip the snare around their heads. Their usual reaction, if any, is to try to grab the snare as if it were a fly. To use the snare move very cautiously and quietly toward the lizard until it can be reached by the tip of the snare. Carefully work the snare over the lizard's head and then pull it closed. In closing the snare, always move the snare stick backward along the length of the lizard's body rather than forward and in front of the lizard. This will cause the snare to tighten around the neck just ahead of the shoulders rather than slipping off the lizard's head. Use 6-pound or 8-pound woven nylon fish line, or ½-pound to 1-pound monofilament nylon leader. Tie the snare and attach it to the snare rod, as shown in Fig. 20–2, A and B. Any long stick may serve as a snare rod, but a telescopic fishing pole is the most ideal.

Professional collectors use .22-caliber pistols with bird-shot shells, approach within 10 feet of a lizard specimen, and then shoot at it. A few shot generally stun (or, sometimes, kill) the specimen, giving the collector time to capture the lizard before it can escape. Specimens collected in this manner, however, must be preserved almost immediately to prevent spoilage. In the tropics many of the arboreal (tree-dwelling) lizards and snakes can be collected in no other way than by shooting. The old-fashioned slingshot is amazingly accurate (with a little practice) and quite useful for collecting lizards.

In desert areas snakes and lizards wander about at night a great deal during the early part of their breeding season. This usually occurs on those first few nights when the air temperature remains above 80° at least until after midnight. At such times drive slowly (between 15 and 20 miles an hour) along paved back roads in the desert, using the low beams of the car for light. Specimens can be approached directly and picked up by hand under such conditions.

364 Biological Techniques

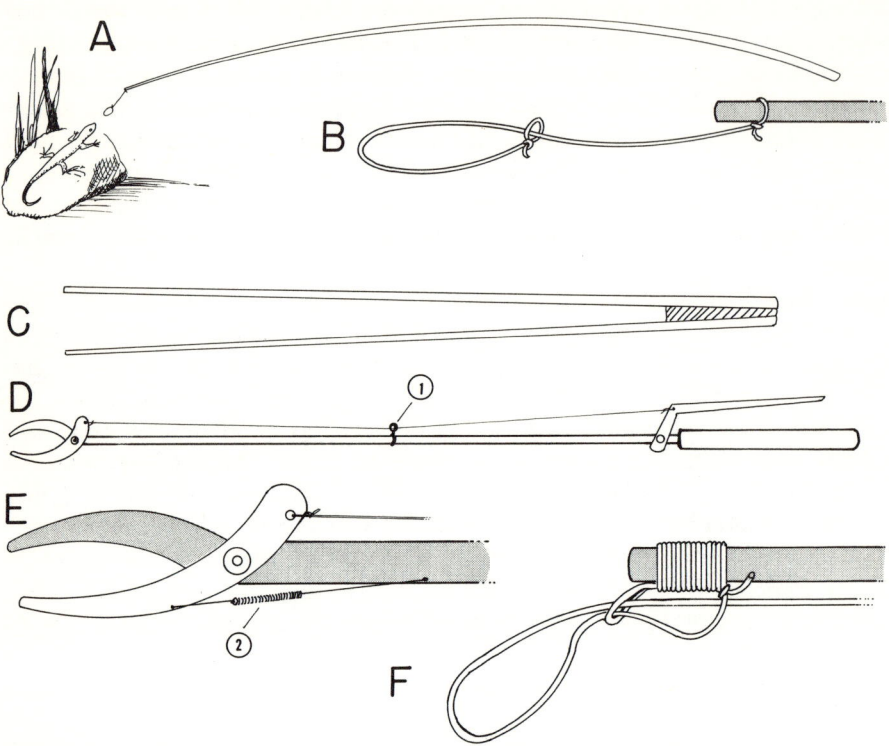

FIG. 20–2. Reptile-catching devices. A–B. Use and detail of the lizard snare. C. Wooden tongs for picking up poisonous reptiles. D–E. Details of a snake stick. F. Details of a heavy-duty snare for capturing very large snakes. (1) Guide, (2) spring.

Snakes

Many of the techniques used for lizards are directly applicable to snakes. It is more difficult to predict where snakes will be found than where lizards may occur. Snakes that are coiled and sunning frequently give the collector sufficient time to snare them or to capture them in some other manner. However, many snakes will move quickly into heavy brush or seek out rodent burrows and must be captured before they can escape. The .22-caliber pistol with bird shot is quite good, especially for poisonous specimens. When you are equipped to do so, you can stop most of the little snakes by gently stepping on them, grasping them by the neck, and transferring them to a collecting bag. Some collectors use long wooden forceps

(Fig. 20–2C) which are easily constructed in the field. A snake stick (Fig. 20–2, D and E) may be anything from a forked stick or a stick with a large screw hook in one end, to a stick with a single moveable jaw for the capture of snakes. Some of the collectors in Costa Rica used homemade snake sticks (Fig. 20–2). Note that the moveable jaw is kept open by means of a small coiled spring or a piece of surgical rubber tubing. The jaws are closed around the neck of the snake by pulling back on the string, either by means of a handle (Fig. 20–2D) or manually. Larger snakes in the open may be snared. Figure 20–2F shows how light, $\frac{1}{8}$-inch sash cord can be tied to some convenient stick to form a temporary snare.

Turtles

Turtles present a special problem in museum work because of their unusual size and dimensions. It is often impossible to find museum containers that can hold larger specimens and, therefore, the collector should be somewhat selective when this is possible. Most land turtles can be picked up and placed in cloth bags. Aquatic specimens are easily caught with skin-diving equipment if the water is clear. The fish seine, described in Chapter 18, may also be used to good advantage in shallow ponds, lakes, and slow-moving rivers.

Many aquatic turtles will go after meat baits and therefore can be trapped in large funnel traps, built somewhat in the order of minnow traps, or in half-submerged barrel traps (Fig. 20–3). Make this trap as follows: Drill some holes in a wooden barrel or metal drum and stake this in a shallow pond where turtles occur, so that the top of the drum sticks out of the water 5 or 6 inches. Hinge a piece of roughened board to the side of

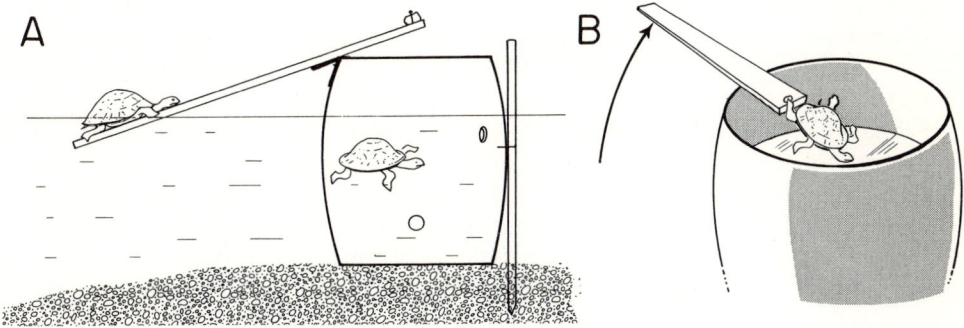

FIG. 20–3. Diagram and use of a barrel turtle trap. See text for instructions.

the drum and nail a piece of meat bait to the upper end of this board (Fig. 20–3A). Place a second piece of meat in the barrel. This will diffuse meat juices through the water and may help to attract turtles. The trap functions by gravity. Turtles attracted to the bait move up on the board until their weight causes them to fall into the trap (Fig. 20–3B).

PRESERVATION TECHNIQUES

Snakes and Lizards

Procedure. The procedure for proper preservaion of reptiles is as follows: (1) killing, (2) positioning the specimens, (3) fixing specimens, (4) labeling, (5) final preservation and storage.

Killing Methods. If the facilities are available, by far the best and simplest method for killing any reptile is freezing. Place specimens in cloth bags, along with their field data, put the specimens in the freezing compartment of the refrigerator overnight, remove, and thaw. Because reduced temperatures lower the metabolic rate of reptiles, the specimens simply go to "sleep" as they cool off and are thus killed in a very humane way. The only danger with this technique is that tails, toes, legs, or bodies may break should the specimens be dropped or mishandled before they are thawed. Once completely thawed, they are treated like any other specimen.

One widely used method of killing reptiles is that of injecting them with a 10-percent nembutal solution. Some workers use ether or chloroform in place of nembutal. The nembutal method is excellent for field work where limited facilities are available. The amount to be injected must be determined by the size of the specimen. Nembutal, being an anesthetic, renders the specimens unconscious. Because an overdose is used specimens quickly die and are ready for preservation. Be careful not to inject too much nembutal into small lizards and snakes for muscle contractions and kinking may occur.

Snakes and lizards are often drowned in warm water, but this is a slow process. It is true that hot water will rapidly kill reptiles but it also causes kinking and body contraction. One method for drowning reptiles is as follows: place the specimen in a net bag (not a cloth bag) with a weight. Drop the specimen in a bucket of water which is warm to the touch but not uncomfortably hot. Another method is as follows: place the specimen in a jar. Fill another container with about twice the amount of water required to fill the specimen jar. The water should be warm to the touch but not too hot. Quickly remove the lid and fill the jar to capacity, replace the lid and seal it tightly. If successful there will be very few or no air bubbles in the jar. The warm water accelerates the metabolic rate of reptiles and usually kills them by asphyxiation. Obviously poisonous snakes are too dangerous to handle in this manner. If kinking occurs in specimens attempt to straighten the spinal column immediately before the muscles become set.

Positioning, Fixing, and Labeling. Figure 20–4, A and B, shows the typical measurements recorded for lizards and snakes (total length, snout-to-vent length, tail length). The reptile skin inhibits preservatives from entering the body quickly enough to inhibit rotting. All small lizards should either be injected in the body cavity with formalin or should have a cut made on the left ventral side of the body. Larger lizards should also be injected in each leg segment and just underneath the skin at the base of the tail. If a hypodermic is not available, use a very sharp scalpel or razor blade and cut small slits in the limbs and tail. Caution must be used when working around the tail of a lizard since it can break off at any time. Snakes are either injected with 10-percent formalin every inch along the length of the body cavity, or else they are cut on the left ventral surface of the body. These cuts should measure 1 to 2 inches in length, should penetrate into the body cavity, and should be about 1 inch apart.

Two types of labels may be used (Fig. 20–4D). The typical label measures ¼ by ¾ inch and has either the collector's field number or the museum acquisition number. In either case, this number refers to a complete set of field data. The second label bears all of the field data on one side and the taxonomic information on the other. This label is preferred in small collections where a filing system or other method of recording data is not maintained. The label is attached around the neck of a snake or around the hind limb or waist of a lizard. Labels should be made of waterproof paper and tied to the specimens with lightweight, but strong, string.

When positioning specimens, always keep in mind the containers in which they must eventually be placed. Lizards are positioned in pans containing 10-percent formalin. Specimens that are short enough to fit in quart bottles have the tail extended; those which are extremely long have the tail brought up along the length of the body (Fig. 20–4, A and C, respectively). Snakes are coiled, belly up, in a bottle containing 10-percent formalin. This posture is essential in snakes that have been cut, to permit gases to escape. It is not essential for injected specimens, but it is, nevertheless, convenient. To coil a snake, hold the specimen by the tail, and by twisting the tail coil the specimen into the bottle. Finally, pour enough formalin to cover it by 1 or 2 inches.

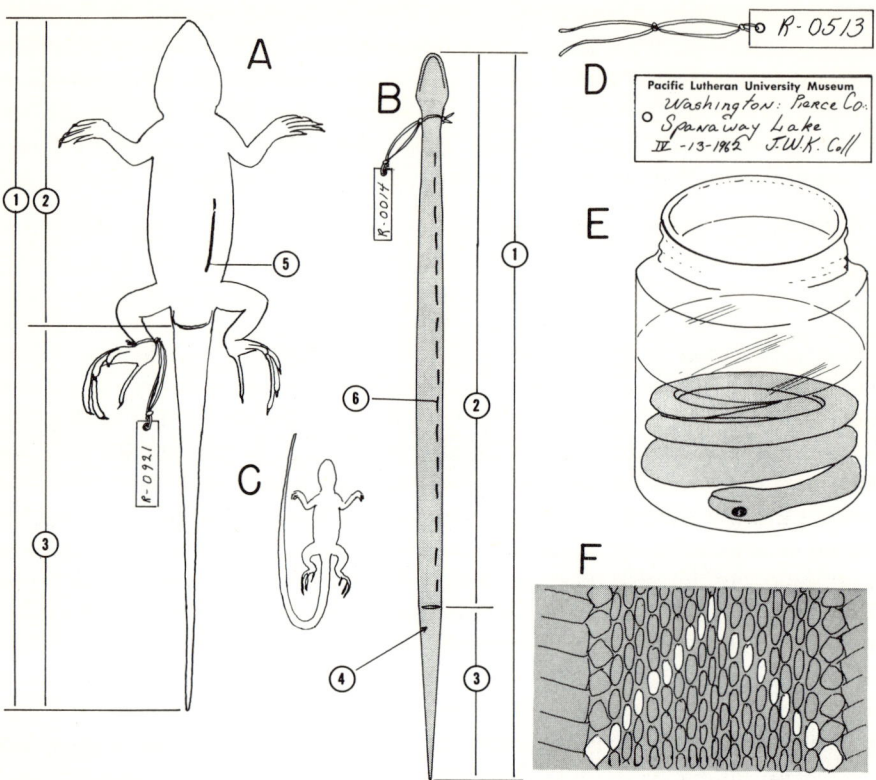

FIG. 20–4. Measuring, tagging, and positioning reptiles for preservation. A. General lizard techniques. B. General snake techniques. C. Posture for long-tailed lizards. D. Tags bearing the museum number or the complete field data. E. A snake coiled during the initial phase of preservation. F. Scale-count technique. Make the scale count following the normal scale pattern as shown by the white scales: the scale count of this specimen is 19. (1) Total length, (2) snout-to-vent length, (3) tail length, (4) inject or press here to evert the hemipenis, (5) cut or inject at this point, (6) make a series of cuts or injections along this line.

Fixation requires from 48 hours to a week. Observe the specimens carefully for any discoloring of the ventral abdominal surface. Discoloring indicates that the specimen is rotting internally. Should this happen, the rotten area should be injected with 10-percent formalin and then cut with a razor blade.

Final Preservation and Storage. Small and medium-sized reptiles may be kept in 6-percent formalin; large specimens should remain in 10-percent formalin. Because of formalin's very disagreeable nature, most workers prefer to transfer specimens from formalin to alcohol. Seventy-percent ethyl alcohol or 50-percent isopropyl alcohol is suitable. Specimens may be transferred directly from formalin to alcohol or soaked to remove the formalin, as follows: Place specimens in a jar of clear water and let stand for one or more days. Change the water daily. Watch carefully for a faint scum on the surface of the water, which indicates that the formalin level is so low that bacteria can begin to grow. Transfer the specimens to alcohol and replace the solution after 24 hours with a fresh solution. Seal specimens in airtight museum jars, along with their field data. Specimens should be kept in the dark to retard color loss. Check museum specimens at least twice a year and replace any fluid that has been lost as a result of evaporation.

Large Snakes and Lizards. When specimens captured in the field are too large for standard museum containers they should be skinned and preserved. Treat snakes as follows: Record color notes and make the three standard measurements (Fig. 20–4). Next, make a slit down the ventral surface of the body from the neck to the vent. Cut the body in two at the base of the skull and at the base of the tail, being careful not to injure the skin. Next, peel the skin away from the main part of the body; roll up the head, tail, and skin; label; and preserve in 10-percent formalin.

Large lizards are treated as follows: Record color notes and make measurements (Fig. 20–4A). Make a slit on the ventral surface of the body from the neck to the vent. Begin peeling the skin away from the body and, when possible, sever the neck, the base of the limbs, and the tail from the main trunk. Now remove the main part of the body, but leave the skull, limbs, and tail intact. Do not attempt to skin the tail, as it may autotomize at any time. Label the specimen, roll it up, and preserve it in 10-percent formalin.

Turtles

One of the chief problems in preserving turtles is finding suitable containers which can hold the large shell. As mentioned above, this problem can be overcome somewhat by selecting smaller individuals suitable for preservation. Some additional techniques will be given below.

Killing Methods. Any of the methods described for snakes and lizards may be used for turtles. Freezing is by far the easiest method; drowning in a mesh bag, injection with nembutal, and other methods may work as well.

Preserving Small Specimens. When the specimen is dead pull the head, tail, and limbs out of the shell so that they are exposed. Make slits between the neck and forelimbs and between the tail and hind limbs; inject formalin into the head, neck, and limbs. Place the specimen directly in a jar with 10-percent formalin. After several days this solution may be replaced by a 6- to 8-percent formalin solution or by alcohol, as described above for snakes and lizards. Attach a specimen label to either the neck or the hind limb.

Dried Specimens. Some workers prefer to kill, preserve, label, and then dry turtle specimens. Such specimens always have some odor and are less satisfactory than wet specimens. Two preservatives may be used: 10-percent formalin or corrosive sublimate (see Appendix C). The corrosive-sublimate technique is much older and was used because the extreme toxicity of this compound inhibits the attack of insect pests. However, corrosive sublimate must always be treated as a dangerous chemical, especially if students are handling the dried specimens. To dry a turtle inject and preserve in the same manner as that described for small specimens. When it is thoroughly preserved, place the specimen in a dark but airy room to dry slowly. When it is thoroughly dry, store in a box with ample paradichlorobenzene; this will both overcome the odor and discourage infestation of insects.

Wet-Dry Specimens. Perhaps the only suitable way to handle really large turtles, such as marine turtles, other than having them stuffed by a taxidermist, is drying the shells and preserving the head, limbs and tail. To achieve this, separate the plastron (ventral shell) from the carapace (dorsal shell) by cutting the bridge that joins these two shells. This is usually cut with a hacksaw, but may be severed with a knife in marine turtles. Next, carefully cut the skin free from the shell so that the head, neck, and two anterior limbs may be removed as a single unit. If this unit is too large to preserve in one piece, sever the forelimbs from the neck. Do the same for the tail and two hind limbs. Carefully inject the limbs or make frequent small cuts to ensure proper preservation. Preserve these portions in 10-percent formalin, replacing the solution after 3 or 4 days. Use large glass containers or metal drums lined with plastic bags for these portions.

Next, remove all of the flesh from the two shells, label each shell individually, wash it with a solution of 10-percent formalin on the inner surfaces, and dry in a dark but airy place.

Crocodilians

Killing. Whenever possible, subdue crocodilians by freezing or with nembutal. Larger specimens are usually shot in the field.

Small Specimens. Depending on the size of container available, smaller specimens should be positioned, fixed, and preserved like lizards (p. 366).

Large Specimens. Large specimens are generally skinned in the same manner as large lizards. If the specimen is of taxonomic importance, the skull, hands, and feet should be left in the skin, while the body and tail are removed. Tack the skin out upside down, and salt it heavily with sodium chloride. When it is almost dry and stiff, roll the skin into a small bundle, tie, and label. Deflesh the skull if it was separated from the skin, dry, and clean by means of dermestids, as described below.

SHIPPING FIELD SPECIMENS

All of the techniques described in Chapter 19 for the amphibians may be employed for shipping field specimens of reptiles. Such specimens are positioned, fixed, and stored in formalin until after they have been shipped.

SKELETAL TECHNIQUES

The bony structures of reptiles are perhaps as important in taxonomy as external characteristics. At any rate, taxonomists must prepare some specimens for skeletal studies. Small snakes and lizards are cleared and stained by standard, but somewhat modified, techniques. Larger specimens are skinned and then cleaned by dermestid beetles. Because of the similarity in techniques used for all vertebrates, skeletal techniques for reptiles will be discussed in Chapter 23.

REFERENCES

Carr, A., 1952, *Handbook of Turtles,* Comstock, Ithaca, N.Y.
Ditmars, R., 1936, *The Reptiles of North America,* Doubleday, Garden City, N.Y.
Ditmars, R., 1939, *Fieldbook of North American Snakes,* Doubleday, Garden City, N.Y.
Ditmars, R., 1946, *Reptiles of the World,* Macmillan, N.Y.
Pope, Clifford, 1939, *Turtles of the United States and Canada,* Alfred A. Knopf, N.Y.
Pope, Clifford, 1955, *The Reptile World,* Alfred A. Knopf, N.Y.

Schmidt, K., and D. Davis, 1941, *Fieldbook of Snakes of North America and Canada,* Putnam, N.Y.

Schmidt, K. P., and R. F. Inger, 1958, *Living Reptiles of the World,* Doubleday, N.Y.

Smith, H. M., 1946, *Handbook of Lizards,* Comstock, Ithaca, N.Y.

Wright, A. H., and A. A. Wright, 1957, *Handbook of Snakes,* Comstock, Ithaca, N.Y.

CHAPTER 21

The Birds

Happily, bird study today involves more field observation and identification than field collecting. The great popularity of bird study has supported the production of excellent field guides, inexpensive field glasses, and considerable literature dealing with the subject. In the old days the only valid records of bird distribution and other characteristics were those taken down the sights of a shotgun barrel. Today, however, field observations by competent students are adequate. Therefore, the identification of birds, the recognition of bird songs, studies of migration and growth by bird banding, and nesting studies are more important for the average student than field collecting. Perhaps there are some unique features of birds which make field studies possible. For example, birds occur in large numbers with numerous species, they are mostly diurnal and easily observed, their behavior is very conspicuous, for the most part, and can be measured and recorded by simple observations.

Freshly killed bird specimens are, however, always available to the alert student. Birds killed by striking windows or by automobiles need not be wasted. Even parts of birds so killed are valuable teaching tools. Therefore, each student should be fully aware of collecting and preservation techniques in order to conserve specimens that would otherwise be wasted.

COLLECTING TECHNIQUES

Laws and Permits for Collecting

One must assume that almost all birds are protected by federal law (if they migrate across state boundaries) and also by state law, and that a federal permit and state permit must be obtained before they can be legally possessed (this includes road kills) or collected. However, numerous migratory game birds and upland game birds may be taken in season with a state hunting license. Hunting seasons, bag limits, and the like, are set each year by the state fish and game department. In addition, certain birds listed by state as "unprotected" may be hunted at any time of the year. For example, in Washington State a recent list of unprotected birds included such species as the crow, the English sparrow, and the starling. The collector must correspond with his state's fish and game department to ascertain the legality of collecting migratory birds in season and to ascertain those species which are unprotected.

The procedure for obtaining bird permits is as follows: First, the federal permit must be obtained; then, the state permit. Contact the nearest regional office of the U.S. Fish and Wildlife Service for application blanks for the federal permit. When this permit is obtained, application blanks will be available from the state fish and game department's main headquarters. Some of the stipulations that must be met by the applicant are as follows: He must demonstrate his competence to recognize birds by species, he must be a member of a scientific institution or an academic institution or must be sponsored by such an institution, his collections must be turned over to the museum of such institutions or he must satisfactorily demonstrate that all specimens so collected will be available for use by schools and other institutions of research and that the specimens will eventually become the property of some recognized museum or institution. Letters of sponsorship and recommendation are usually called for in the initial application. Advanced students and most science teachers are directly or indirectly eligible and may qualify for collecting permits.

Permits are renewed annually, limits are sometimes set on the number of specimens that may be collected, and annual reports are required as to the number of specimens collected and the disposition of such specimens. Permits are usually not transferable to other persons, although field collectors may be assisted directly by other persons. Again, one should ascertain the exact nature of his own state's requirements in this matter. Frequently, states require persons not directly connected with institutions of higher learning or scientific research to post a bond in the sum of 100 dollars or more to ensure that collecting will be done only for scientific purposes. Obviously, then, specimens may not be used for barter or sale.

Shotguns and Shooting

Birds are most satisfactorily collected with a shotgun. Usually, no single gun serves as a good all-round collecting tool. Rather, guns of different sizes are required for different types of birds. The 12-gauge, 16-gauge, and 20-gauge shotguns are used for larger birds, with shot sizes ranging from No. 9 to No. 2 shot, depending on the size of the specimen and the distance of the shot. Whether low-base or the more powerful high-base shells are used will, again, depend on the size of the specimen and the distance of the shot. The .410 shotgun is an intermediate gun which occasionally is quite effective for large birds and may also be used for smaller specimens. Serious collectors often modify pistols, such as the .38 special or the .45, by reaming out the barrel to remove the rifling. Pistol shells are reloaded by hand with shot of various sizes. Obviously, the number of shot per load will depend on the size of the shot used and the size of the shell. The object in shooting birds is to kill the specimen with the least number of shot perforating the body. In long-range shooting the entire load should be directed toward the specimen. Conversely, when shooting specimens close at hand aim to one side of the specimen, rather than directly at it, to ensure that only a few of the shot will penetrate the bird.

Care of Specimens

Once the bird has been collected, plan to spend adequate time in temporary field preparation. With a forceps plug all shot holes with cotton to prevent bleeding. Shot holes should be plugged immediately, even if there seems to be no local bleeding, because bleeding usually commences an hour or two after the specimen has been killed. Wash any blood from the feathers with cold clear water and blot the feathers dry. If blood appears in the mouth, hang the bird head down and permit the blood to drain. Then plug the throat with cotton to prevent further bleeding. Plug the anus also, if this seems necessary. Record field notes and then tag the specimen with its field number by means of a tie-on tag. Finally, wrap the specimen in absorbent paper, such as newspaper, and pack it in a canvas field bag. Small birds are usually wrapped in a cone of paper and then placed in the field bag. Specimens so treated will remain in good condition throughout the day's hunting. In warm weather it is essential to mount the specimen soon after collecting, or to pack it in an ice chest or refrigerator.

When it is not convenient to mount specimens immediately after collecting, simply place the wrapped field specimen in two plastic bags, each sealed in turn, and place this parcel in the freezer. Make sure the field data have been included with the specimen. The double plastic bag will prevent undue loss of moisture and thus preserve the specimen until the time of mounting.

Killing Injured Birds

One encounters birds from time to time that have broken wings or have been otherwise injured seriously. Such birds are quite defensive but nevertheless should be humanely killed to prevent further suffering. Killing is accomplished by depressing the chest and causing almost instant suffocation. Grasp smaller birds beneath the wings and firmly squeeze the chest for about one minute. Grasp larger birds, such as ducks, by the two wing tips while they are lying on the ground, and kill by depressing the chest with your foot.

Window Kills and D.O.R. Specimens

More and more birds are falling victim to large picture windows. They see the reflection of the sky in the window and apparently assume they can fly directly through the window. Thus, in any neighborhood a surprising number of birds will be killed in this fashion. Collectors should let it be known, among their faculties, neighbors, scout groups, and so on, that specimens will be welcomed. If you are always ready to retrieve specimens when notified, a long and unending supply will develop. The collector should aways be prepared to care for specimens found dead on the road (D.O.R.). At certain times of the year an alarming number of hawks, owls, robins, and other birds are killed on the road. The best specimens are usually those which bounce from an automobile and land at the side of the road.

Hunters and Hunting

Another source of specimens, as mentioned above, is hunting game birds in season or obtaining the skins of specimens from hunters. One must check his own state laws concerning the preparation of specimens taken under a hunting permit. Usually, there is no problem.

Trapping and Netting

Birds are seldom trapped or netted for museum specimens. This process is slow and tedious and quite indiscriminate. Traps and nets are usually governed by law ("Bird Banding," p. 399). Bird traps are occasionally used to obtain such species as the starling or English sparrow. The Potter trap is simple to build and quite excellent for this purpose. The trap consists of three circles of $\frac{1}{2}$-inch mesh screen wire (Fig. 21-1) which are set in a cloverleaf pattern and covered by a square piece of $\frac{1}{2}$-inch screen wire. This forms three funnels and three holding chambers. Grain is sprinkled liberally inside the trap and sparingly in the funnels to attract the birds. A rock should be placed on top of the trap to hold the upper screen in position. If such a trap is used, it should be tended constantly and dismantled when not in use.

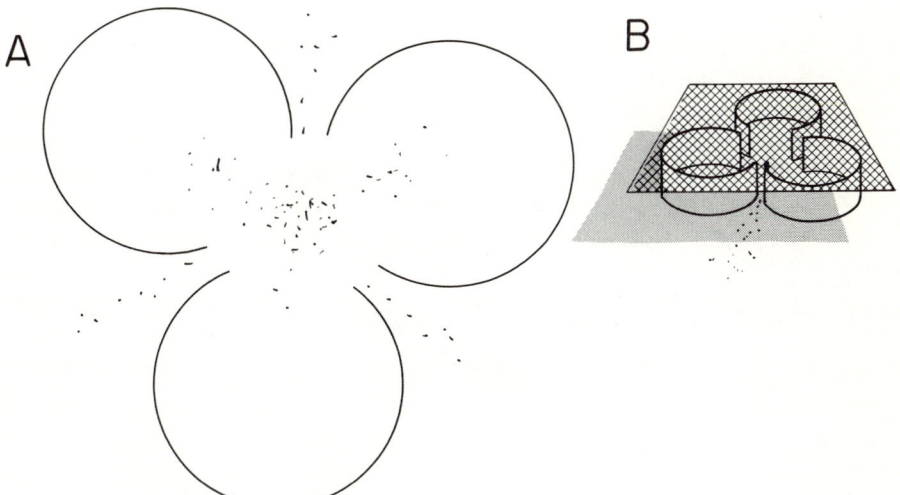

FIG. 21–1. The Potter bird trap. A. A diagrammatic view showing three partial circles of screen wire set in position to form three funnels, and bait (grain) scattered between the funnels. B. The Potter trap set and ready to work.

Skinning and Salting Field Specimens

On extensive field trips consider salting bird skins, shipping them to the laboratory, and making them up as specimens at a later date. If facilities are limited or the humidity such that skins are difficult to dry, the salting technique is often the only one for adequate field collecting.

George Hudson of Washington State University, who has prepared numerous skins in this manner, recommends the following techniques to his students: After the bird skin has been removed from the body (as directed below) carefully remove flesh and fat clinging to the inside of the skin. Turn the skin right side out and apply salt generously to all surfaces of the inside of the skin, filling the head and neck. Large birds with fleshy feet (such as hawks, ducks, and gulls) should have the tendons removed from the legs (as directed below) and salt worked down into the cavities occupied by the tendons. The present author prefers to inject the toes, feet, and tendon canals with 10-percent formalin in preference to cutting and salting. However, Hudson states that "in hot weather the skin should be loosened from the bone over the entire foot and tarsus and each toe should be cut open along the entire bottom surface. Particular care should be taken to loosen the skin at the upper end of the tarsus and to pack in much salt." Next, fill the skin full of salt, using up to 1 gallon or more for a goose, as the excess salt will soak up the moisture and help keep the feathers dry. The next day, dump the salt out of the skin and dry it (the salt) over a stove. Refill the skin with dry salt and repeat this daily until the skin is fairly dry but still flexible. Finally tie a label to the foot of the bird, shake out all excess salt, roll the skin, and seal it in a plastic bag to prevent further drying. Ship the specimen to the home laboratory by air mail, whereupon it may be stored in a freezer or cold room until it is put up as a round mount.

Field Data for Collected Specimens

The collector should keep a field notebook (see Chapter 1) wherein the collecting data are recorded. The date and geographical location should be clearly defined and notes should be taken to describe the habitat, weather conditions, behavior of the specimen, and so on. Color noting for birds is generally not essential in that feather colors are not altered if the specimen is properly cared for. However, the bills of ducks and geese and the feet of larger birds, such as sea gulls, ducks, geese, and coots, should be noted as to color, for these generally darken.

MOUNTING AND PRESERVATION

Mounting Methods

Museum specimens are not prepared like the traditional taxidermy mount in which the bird is given a lifelike pose. Rather, museum mounts or round mounts are made of bird skins. Such mounts are designed to show all of the essential areas of the plumage and other morphological characteristics while keeping the bulk of the specimen to a minimum. Dozens of round mounts may be stored in the same space required for a single taxidermy mount. Very small birds (hummingbirds and the like) lend themselves well to the wet-and-dry technique which avoids the tedious problem of skinning and stuffing. Larger birds, such as the great blue heron, are frequently prepared as flat mounts to conserve space. The collector must decide which technique is most suitable for each specimen collected. However, the standard procedure is the round mount.

The Mounting Kit

An old fishing-tackle box makes an excellent kit. The following things should be provided: a scalpel or sharp knife, a carborundum sharpening stone, a pair of fine-pointed scissors and a coarse pair of scissors, a fine-pointed forceps and a coarse pair of forceps, a package of assorted sewing needles, a spool of No. 2 cotton or linen thread and a spool of heavy cotton or linen button thread (white), a package of common pins, a 30-centimeter ruler graduated in millimeters, a 10-cubic centimeter hypodermic with an assortment of coarse and fine needles, a pound of absorbent (long-fibered) cotton, 1 pound of powdered arsenic trioxide or Boraxo, a pound of coarse corn meal or hardwood sawdust, bird labels, dip pen and waterproof ink. Optional materials that may prove useful, especially if mammals are also to be mounted, are the side-cutting pliers and a pair of dividers. The cotton is used to make up "bird bodies" for stuffing; the corn meal is used to absorb grease and blood during the skinning process. Fine hardwood sawdust is used in the tanning industry as a fur-drying compound. The liberal use of corn meal or sawdust will often make the difference between a greasy, poorly mounted skin and a clean skin. Powdered arsenic trioxide is used to poison the skin, contract the skin around the feathers, and, in general, cure it.

Boraxo (pure borax may alter color) is probably just as good as arsenic trioxide for curing skins and is more preferable for student use.

Problems to Anticipate

Blood. If there is any bleeding from the flesh of the body during the skinning process, immediately add large quantities of corn meal, soak up the blood, and remove the corn meal. Should blood get on the feathers, either from the incision or from shot holes, wash it away with cold clear water before it has a chance to dry. Wash individual feathers with a cotton swab, blot, and dry. If dried blood is found on the feathers, scrape this away from each individual feather with a pocketknife or fingernail. Brush each feather until it is clean. Usually, a feather which is too badly damaged by blood may be removed from the body without damaging the appearance of the skin.

Fat Skins and Grease. Use corn meal generously when skinning birds that contain large quantities of fat. After the skin has been removed, turn the skin inside out and scrape the fat away from the skin with a dull pocketknife. Continuously work corn meal into the fatty areas to absorb the grease. A short length of hacksaw blade is useful to break up the fatty tissue in the feather tracks of ducks and similar birds. Work this tool vigorously along the skin, but be careful not to break the skin. Next, scrape with a pocketknife and, finally, clean with several applications of corn meal.

The following technique is used by some museums when dealing with greasy skins. Most of the grease is removed as described above. The skin is then washed in warm water and detergent, particular attention being given the fatty areas. When the fat has been sufficiently removed from the skin, most of the water is squeezed from the skin and feathers. Next, a drying compound is liberally added to the skin and feathers. A product known as "Blue Cloud" (the Don Company, Gardena, California: a bath product made for drying live chinchillas) is used as a drying agent by many professional collectors. Corn meal or hardwood sawdust may also be used. When the moisture has been absorbed from the feathers, a small automobile vacuum cleaner is used to retrieve the drying compound. An air hose could be substituted to blow the compound out of the feathers. Finally, the feathers are fluffed up and the skin is then ready for stuffing.

Determining Sex. The sex of the bird must always be noted on its label, along with size and relative development of the gonad. Some birds are sexually dimorphic and present no problem, but where the sexes are identical in external appearance, one must dissect the body to locate the gonad.

As soon as the body has been removed from the skin, cut through the body wall on the left side of the rib cage. Open the body cavity and carefully move the intestines to one side. The gonads are located anteriorly to the kidneys, the male having two testes while the female has but a single ovary on the left side. Be careful not to mistake the yellowish adrenal glands, also associated with the kidneys, for the gonads. The testes are oval, whereas the ovary is irregularly lobed. Record the sex, make a diagram of the actual size of the gonad, and measure the gonad for the field notebook. Include the sex and the gonad drawing on the bird label.

Ectoparasites. Almost all species of birds have their own endemic species of lice. These are often evident on the feathers during the skinning process. Ectoparasites should be transferred to vials of alcohol and preserved, along with the identification of the host and the appropriate field data.

Crop Contents. To complete the field record, dissect out the crop as soon as the carcass has been removed from the skin. The crop contains unprocessed food which may be identified on the spot or preserved or dried for future identification. Records of the crop contents should be placed in the field notebook.

Skinning Techniques. This discussion deals with the general techniques used for all birds. In special instances additional techniques may be required for some particular species. These are: (1) techniques for large-headed birds such as ducks, geese, and woodpeckers, where the neck skin is too small in diameter to be stretched over the large head; and (2) techniques for the wings and feet of large birds where considerable muscle and other tissue must be removed. These techniques are described below.

1. The primary incision. Spread several sheets of newspaper on a table, set out your dissecting equipment, and pour out a quantity of corn meal in a wide-mouth container so that it is readily accessible. Have additional newspapers at hand to replace the original working surface should it become soiled with blood and grease. The skinning process should not be done so quickly that feathers are lost, that the skin is ripped or stretched, and so on. Nevertheless, work as quickly as possible to prevent the skin from becoming too dry once it is partially removed from the carcass. Now, place the bird on its back with its head away from you. The ventral abdominal feathers are attached to the skin in two rows or tracks, one on either side of the ventral line. With the butt of the scalpel push the feathers away from the midventral line, beginning at the middle of the sternum and moving back toward the vent. On smaller birds these feathers can be blown to one side

382 Biological Techniques

quite efficiently. With your fingers or a pair of forceps grasp the skin on the ventral line, half way between the vent and sternum, pull the skin away from the thin abdominal muscles, and snip the skin with a pair of sharp-pointed scissors. Next, insert the scissors through the cut of the skin and work forward to the middle of the sternum and then backward to the vent.

2. The knees. Grasp the skin on one side of the incision and gently loosen the skin, working toward one side. As soon as possible, add some corn meal to absorb any grease or body juices. If the primary incision cuts through the muscle wall into the body cavity, plug this with corn meal and cotton. Be constantly on guard to keep your hands as clean as possible. Now grasp the bird's leg by the heel and push the leg forward so that the knee will protrude into view (Fig. 21–2A). Separate the tibio-tarsus at the joint where it joins the femur with a scissors, and sever the remaining muscle tissue as indicated in Fig. 21–2A. Add corn meal to the cut flesh to absorb any blood. Repeat this process for the other knee.

3. The tail. With a scissors cut around the terminal end of the digestive tract, plug or replug the digestive tract with cotton, and carefully loosen

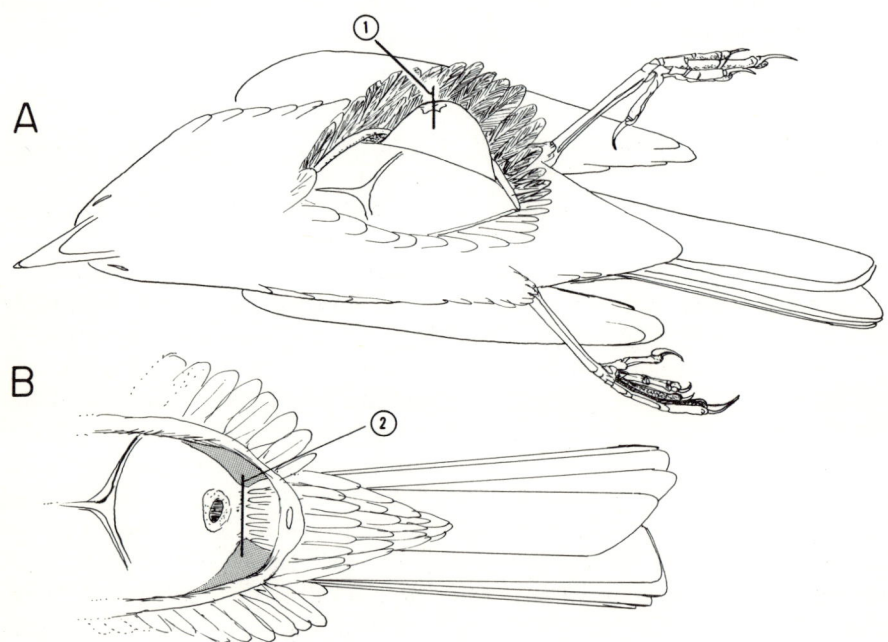

FIG. 21–2. Initial steps in mounting birds. See text for complete instructions. (1) Cut the knee at this point, (2) carefully separate the body from the tail at this point.

the skin around the body, back toward the tail. Exercise extreme care in this part of the skinning job, as the skin on the sides and back of the rump is very thin, especially in smaller birds, and is readily torn. This area of skin seems subject to extreme drying, which also facilitates tearing. In most round mounts this particular area of skin is all that holds the tail onto the body and, thus, it must be as strong as possible. In larger birds the skin will often loosen all the way around the body, even across the back of the rump. In smaller birds, however, the skin adheres very tightly to the tail. The object is to cut the tail free from the body just ahead of the tail feathers. Exercise extreme care not to cut the basal ends of the tail feathers, as this may cause them to fall out. With a fine-pointed scissors cut minute bits of flesh until the tail is severed, as indicated in Fig. 21–2B. If rips are made in the skin, these may be sewn at the time.

4. The body and neck. Now use your fingers or the butt of the scalpel to carefully push the skin away from the body, working forward toward the shoulders and neck. When you reach the base of the wings separate the humerus bones from their sockets and continue gently loosening this skin around the shoulders. The skin of smaller birds is loosely attached in this area, but is usually more firmly attached in larger birds. Again, care must be taken not to rip the skin. Continue to use corn meal to pick up grease and blood.

By this time the skin is almost turned inside out. It would seem that the feathers around the neck area would be badly damaged by this procedure, but no damage is done. Using your fingers to push the skin forward, continue to turn the skin inside out until you reach the base of the skull. As noted above, if the bird you are skinning is a large-headed specimen, it will have to be treated in a special manner (described below).

5. Skinning the skull. Using your fingers (Fig. 21–3A), work the skin up over the skull until the thin, funnel-shaped pieces of skin protruding into the ear become visible (Fig. 21–3B). The thin head skin will dry very rapidly and may need periodic moistening with clean water to prevent tearing. Now, do not cut off the "ear skin" where it joins the head skin, but rather pull it out of the auditory canal with your fingernail or a pair of blunt forceps. The skin will now slip over the top of the head and, as it is drawn forward, the eyes will be exposed. On the first dissection the student is usually amazed at the extremely large size of the eye as compared to the relatively small opening. The eyelids attach to the eyeball by means of fine connective tissue. Use extreme care in detaching the eyelids from the eye. Some workers use a very fine-pointed scissors for this procedure. The author, however, prefers to place a scalpel with the cutting edge down on the eye, as shown in Fig. 21–3C (note, the scalpel is not placed on the skin). The

384 Biological Techniques

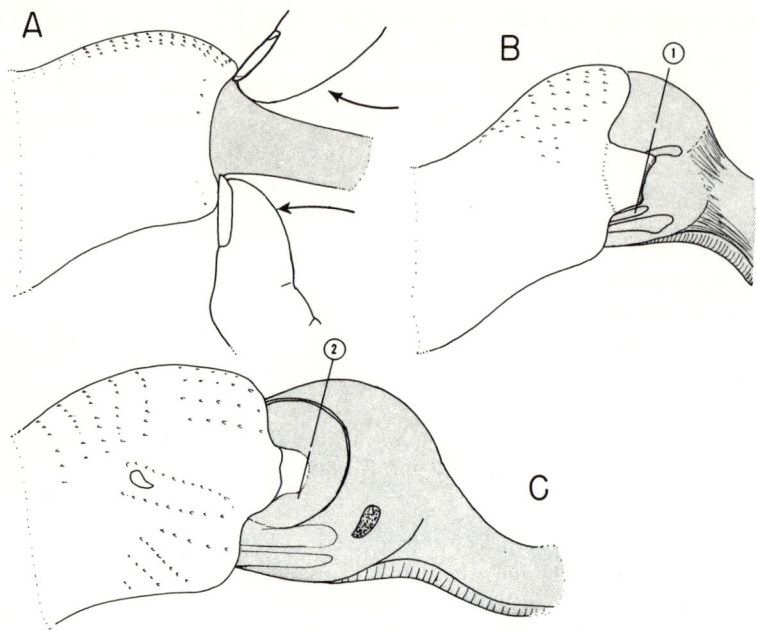

FIG. 21–3. Skinning out the bird head. A. Pushing the skin up over the skull. B. Skinning around the ear. C. Skinning around the eye. (1) Cut the ear close to the skull, (2) begin cutting the eye at this point (see text).

scalpel is drawn back and forth in a sawing motion and simultaneously is moved slowly forward directly over the pupil of the eye. During this procedure the head skin is pulled gently forward to facilitate detaching of the eyelids. After both eyelids are free, the skin needs to be worked forward only a short distance, so that it is parallel with the anterior margin of the eye socket.

6. Cleaning the skull. Loosen the eyes where they join the skull, reach in with a forceps and grasp the optic nerves, and remove the eyeballs from the skull, taking care not to puncture them as the fluid may soil the feathers. Next, pull the tongue out of the mouth by cutting where necessary. Push the points of a scissors up into the head and cut across the roof of the mouth on the ventral side of the eye sockets (Fig. 21–4B). Next, cut the floor of the brain case, starting at the base of the tongue and working posteriorly toward the base of the skull (Fig. 21–4A). Continue these cuts up the sides and across the top of the back of the skull (Fig. 21–4C). Now remove the neck and body from the skull, pick away the remaining portions of the brain, and remove any large quantities of flesh that adhere to the skull. At this point, moisten the skin which covers the skull with clear water. Dust the skin and the entire skull with arsenic or Boraxo. Keep a small piece of cotton in the arsenic jar which can be picked up by forceps and used for dusting the

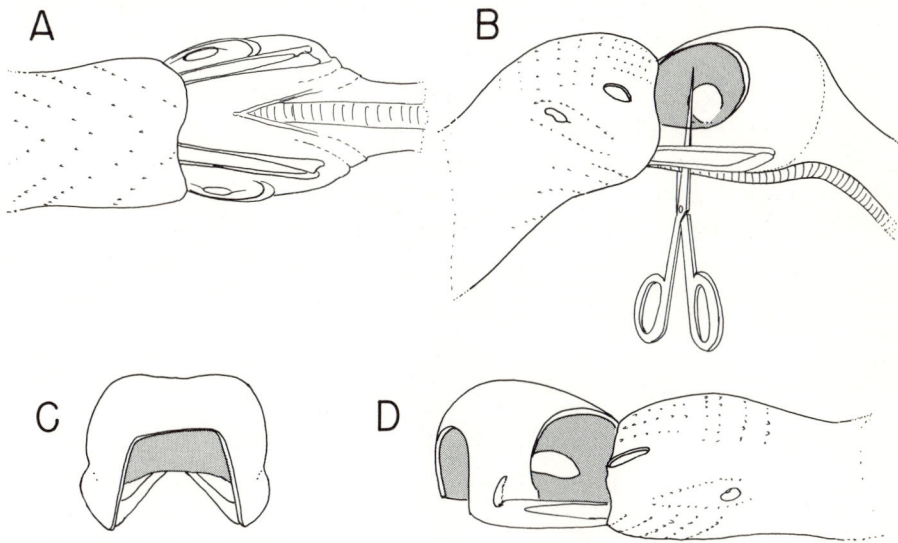

FIG. 21–4. Cleaning the bird skull. See text for complete instructions.

skin. Make two large cotton wads, each equal to the size of the eyes, and insert one in each orbit. With your fingers slowly work the skin back over the head while it is still moist. The skin must be pushed posteriorly at least until it reaches the back margin of the orbits. At this point the bill may be grasped where it protrudes through the neck skin, and the entire head and neck may be turned right side out. Never attempt this if the neck and head skin is too dry.

7. The wings and feet. Working from the inner side of the empty skin, remove the flesh from the humerus, radius, and ulna bones in the wings. Dust these bones with arsenic or Boraxo and turn the wings right side out. Next, remove the flesh from the tibio-tarsus, dust the skin and the bone with arsenic or Boraxo, and wrap the bone with enough cotton to replace the removed flesh. By pulling the feet, turn the legs right side out.

8. Treating the skin. Finally, treat the entire inner surface of the skin with arsenic or Boraxo. Next, roll the skin up and temporarily place it in a clean plastic bag to prevent it from drying out. If the skin seems exception-

386 Biological Techniques

ally dry, moisten it slightly with clean water before adding the arsenic. Set the skin aside while preparing an artificial body.

Large-Headed Birds. When dealing with large-headed birds, we deviate from the normal skinning procedure, described above, by severing the neck (not the skin) about one-third its length back from the base of the skull. Next, an incision is made on the median-dorsal part of the skin over the base of the skull (Fig. 21–5A). The cut should be just large enough to permit the head to pass through it. The head skin is now turned inside out and worked over the skull, as described above.

Don Pattie (in personal communication) recommends making the incision on the ventral part of the neck at the posterior end of the head (Fig. 21–5B). This technique requires a smaller opening, is easier to sew, and is

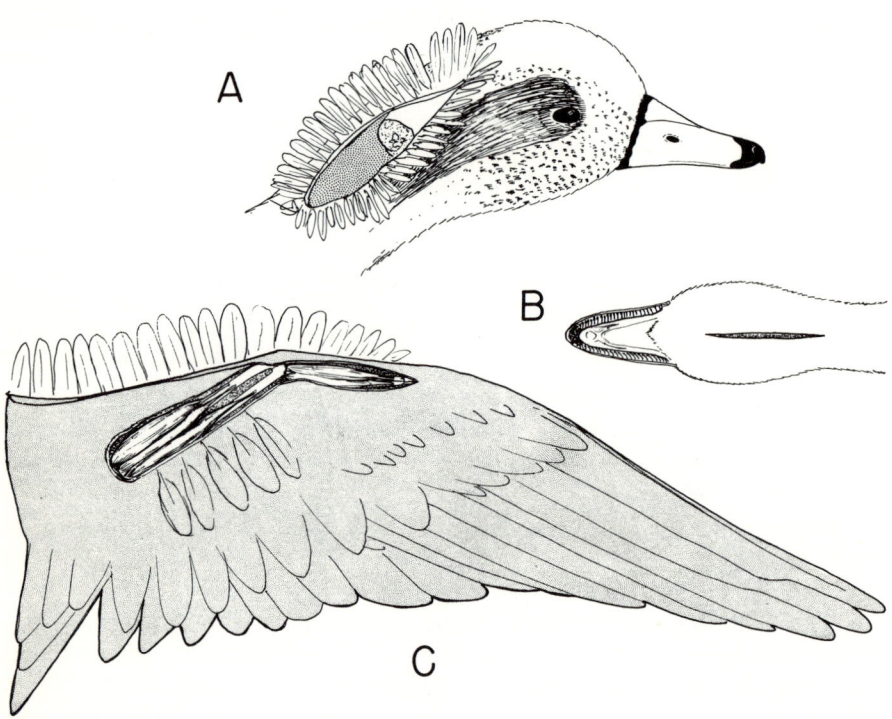

FIG. 21–5. Special bird techniques used in mounting. A. The older technique for working with large-headed birds, see text. B. The Pattie technique for working with large-headed birds, see text. C. Removing the musculature from large bird wings.

less conspicuous. You have the option with this technique of not skinning out the entire head, inasmuch as the tongue, brain, large muscles, and even the eyes can be removed by working up from the ventral surface of the skull.

Wings and Feet of Large Birds. Larger birds such as owls, hawks, sea gulls, ducks, and the like, must have the muscle tissue removed from the length of the wing. After the entire carcass has been removed from the skin, pin out one wing exposing the ventral surface. Make an incision between the radius and ulna bones, and continue this on down between the metacarpal bones (forearm and hand, Fig. 21–5C). Move the skin away and remove the major muscles and sever them at both the origin and insertion. Dust the entire area with arsenic or Boraxo and close the incision with one or two stitches. Remove the flesh from the humerus bone, working from the inside of the skin as directed above.

Many workers prefer to remove the tendons where they extend down into the feet parallel to the tarso-metatarsal bones. This is achieved by cutting the tendons at the distal end of the tibio-tarsus while this bone is being fleshed. Next, an incision is made on the underside of the foot just at the base of the toes. Through this incision the tendons may be hooked with a wire probe and pulled down out of the leg. The author prefers simply to inject the foot and tendon canal with 10-percent formalin which will preserve this tissue during the drying process and will eliminate the necessity for removing the tendons.

Stuffing. Make an artificial body which consists of a central stick wrapped with cotton (Fig. 21–6, A or B). The central stick may be made of hardwood dowel, a piece of wooden box, or any appropriate material. The length of this stick is equal to the length of the body, neck, and part of the head (Fig. 21–6C). Some workers prefer to leave an additional length protruding from the anal opening of the skin (Fig. 21–6B). The crossed feet, and even the tail, may be secured to this additional length of stick in the finished specimen, if desired. Cut a blunt point on the anterior end of the stick. Unroll a few feet of long-fibered, absorbent cotton. Such a roll is usually about 1 inch thick and 10 inches wide, with fibers running lengthwise. The trick of making the body is to pull off long, wide, but very thin layers of cotton from the roll and wrap these around the stick. Note that the bulk of the body should be equal to that of the carcass, whereas the bulk of the artificial neck is slightly greater than the real neck.

For large birds the procedure of cutting the stick and building up cotton for the neck portion is quite similar. However, excelsior or other material

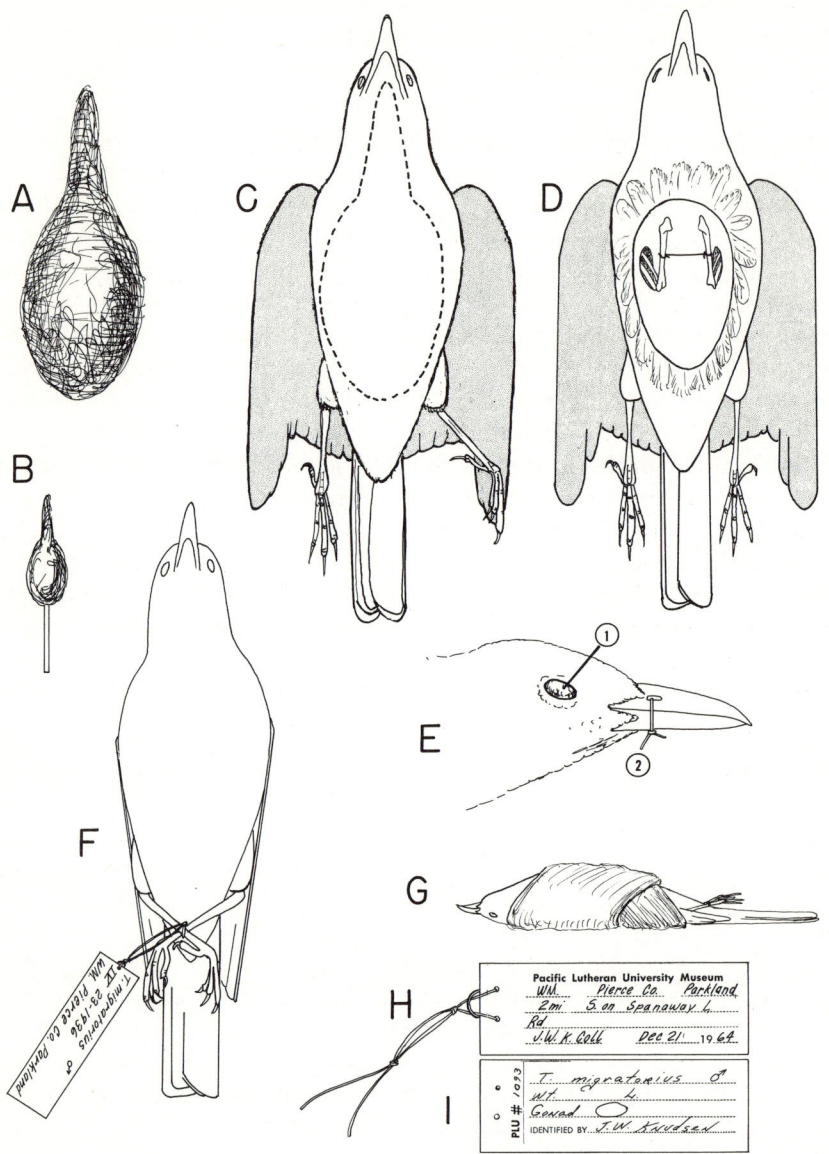

FIG. 21–6. Completing the bird mount. A–B. Two types of artificial bodies (see text). C. The position of the artificial body shown by dotted lines. D. Tying the humeral bones. E. Techniques of finishing the bird head. (1) Pull cotton part way out through the eye opening, (2) tie the bill closed. F. The finished position with the label attached to the crossed legs. G. The specimen temporarily wrapped in cotton for drying. H–I. Two sides of a standard bird tag.

may be substituted for the cotton body. Tie small quantities of excelsior to the stick and slowly build up the artificial body to the size of the actual carcass. Continuously wrap the excelsior with string which must be just tight enough to hold it in place during stuffing.

Now set the completed artificial body aside and unroll the bird skin. Attach a string to one humerus just above the elbow, and tie the other end to the opposite humerus (Fig. 21–6D). The distance between humeri varies with the size of the bird. The object of tying the humeri is to prevent the wings from drooping away from the body and to give the round mount a more natural appearance.

Next, insert the tip of the artificial body well up into the skull of the bird and work the skin down around the body. You are now ready to sew up the ventral incision. With a needle sew a thread through one side of the anterior opening. Do not tie a knot in the thread but, rather, knot the thread to the skin itself at this point. Now stitch through the opposite side a little lower down, cross back over and stitch through the skin toward the middle of the incision, and repeat the process, going from side to side and crossing back and forth four or five times. Do not draw the thread up until it has been stitched the entire length of the incision. Then, starting at the anterior end, pull the skin closed over the artificial body, knotting the thread at the last stitch. It is not serious if the skin will not completely close over the body, as the feathers will cover the incision. Tie the bill closed by passing a needle and thread through the nostrils, from side to side, and knotting the thread underneath the mandible (Fig. 21–6E). Cross the feet and tie these together with the label string (Fig. 21–6F), attaching the feet to the protruding stick, if so desired.

The bird is now ready for grooming and wrapping. While the skin is still moist the feathers can easily be arranged in a natural position. Begin at the head end and work down, stroking the feathers with your fingers or a soft brush. Arrange the wings, check the details of the rump feathers, and, in general, "put the feathers back together." Finally, the round mount is wrapped in long, wide, but thin wisps of cotton or in pieces of jewelers cotton (Fig. 21–6G). This holds the feathers in place during the drying of the skin. The cotton fibers will adhere to each other and hold firmly. Birds with large heads, crests, or other plumage should have the head turned to one side to display these features. The bird should now be placed in a dust-free, warm, and dry place for a few days to several weeks until the skin is thoroughly dry. Some workers suggest unwrapping the bird periodically during the first day of drying and rearranging the feathers if they become disarranged by the shrinking of the skin. This is usually not necessary if extreme care was taken in the initial operation.

Labeling. Each collector should adopt a standard label made of a good grade of high-rag-content paper. The information is variously arranged on such labels. One side usually has the sex of the bird, the locality and date of collection, and the collector's name. The other side of the label contains the scientific name of the bird, a description of the gonad, and may include such things as weight and length (Fig. 21–6, H and I, respectively).

Wet and Dry Mounts

Small birds, such as hummingbirds, kinglets, and creepers, are quite difficult for the beginner to skin and mount. Small birds can safely be preserved with 10-percent formalin and then dried and placed in museum trays along with other skins. If drying is done properly, such birds look as good or better than mounted specimens and are just as durable. Specimens so prepared 16 years ago for the author's museum are in perfect condition today. Some workers have prepared relatively large birds in the same manner, but this is not advocated.

Using a fine hypodermic needle inject 10-percent formalin into the body cavity and head (in the throat region). Next, prepare a small foot label and attach this to the legs, as described above. Carefully groom the feathers and position the body and then wrap the specimen in thin wisps of cotton, as described above. When it is completely wrapped, weight the body by adding a small fishing weight with some additional cotton. Submerge the specimen in 10-percent formalin for 24 hours, then remove the specimen and set it aside to dry. The weight may be removed from the specimen, but do not remove the remainer of the wrapping which holds the feathers. Remove the wrapping after the specimen is thoroughly dry. Specimens so prepared need no special care, but are kept with regular bird skins and are fumigated periodically with paradichlorobenzene.

Flat Skins. Owing to restrictions in space and other factors, larger birds are occasionally prepared as flat skins. Specimens are skinned, cleaned, and treated with arsenic or Boraxo, as described above. After this the feathers are groomed and the skin is allowed to dry in a flat position. The skin is not split and spread out as is a mammal pelt, but rather resembles the skin in Fig. 21–6D at the time it is flattened. Birds with long necks should have the head brought alongside the body and the feet directed forward if they protrude too far posteriorly. Such skins can be relaxed at any time and made up into reasonably good round mounts. To relax dried skins place them in a humidifier equipped with chlorocresol (see Chapter 13 for directions).

Bird Parts for Teaching

Badly mangled bird specimens which are of no value for whole mounts should nevertheless be saved for museum specimens. Bird wings, feet, tails, and heads are all of value to demonstrate the numerous morphological adaptations of these parts. Each individual part is treated exactly as it would be if attached to the entire skin in a round mount. For example, the eyes, tongue, and brain are removed from the bird skull and are replaced with cotton. The head is then treated with arsenic or Boraxo and the skin is drawn back over the skull. Wings are usually pinned out flat, larger ones are fleshed out, as described above. Feet are postured to their best advantage. The toes of hawks and owls, for example, may be dried in an outstretched position, whereas the web of a cormorant or duck should be fully expanded. Each individual part is given a label containing all or most of the data that would be found on a normal round-mount skin. Such parts are stored and cared for like regular museum skins.

Bird Nests

Birds nest in a fantastic number of ways, ranging from deep burrows to pits, surface nests, low bush nests, tree nests, hanging nests, tree-hole nests, and many others. Some of these may be collected for teaching purposes. A series of boxes of one or two uniform sizes which are equipped with lids will facilitate the building of a neat collection. Nests are located during the breeding season. The breeding pair should be identified while the birds are attending the nest. Do not collect any nest until it has been abandoned by the parents and young. When collecting the nest include more than enough of the supporting branches to fill the storage box. In the laboratory cut the different branches so that they will just fit inside the box, but will touch the edges of the box and support the nest in a natural position. Push tacks into the branches through the sides of the box to hold the nest in position. Label the side of the box as to the species of bird, the species of tree supporting the nest, and other ecological data. Fumigate each nest with a large quantity of paradichlorobenzene as soon as it is collected, as it may harbor small insects. Be absolutely sure that all pests are dead in the nest before such nests are placed in the museum with other specimens.

In large museums nests are placed in small open boxes along with their labels. These, in turn, are kept in museum trays inside insect-proof cabinets. Nests should be fumigated as frequently as bird skins and other specimens.

Bird Eggs

The eggs of protected birds may not be collected without a federal and state collecting permit. Excellent color plates illustrating eggs have largely done away with the need for small museum collections. Nevertheless, researchers frequently need to collect and preserve bird eggs in conjunction with life-history studies. The procedures presented herein are drawn primarily from the work of Bendire (1891).

The essential tools are: a fine-tipped blowpipe equipped with about 3 feet of surgical tubing, a needle probe, a very small countersink drill or reamer bit, a fine pair of forceps, and an embryo hook or hooked insect pin mounted in a match stick. It is preferable to collect bird eggs shortly after they have been laid. Carefully pack the eggs in a box, surrounded by cotton, and transport them to the laboratory for cleaning. With a fine needle make a small hole in one end of the egg, then enlarge the hole to the diameter of 1 millimeter, using the countersink drill. Make a second hole on the side of the egg. Next, hold the blowpipe in front of one of the holes and direct a stream of air into the egg, in order to force some of the contents out of the opposite hole. If albumen plugs the hole, attempt to remove it with the hooked needle or forceps. Sometimes the contents may be removed by forcing a jet of water from a syringe into the hole from a short distance. Never place the syringe itself into the aperture of the egg, as this will break the shell. By alternating between the two openings one can remove most of the contents. Next, fill the egg about half full of water, shake, and drain. Repeat this process until the water is clear. Place the egg on some absorbent material and dry.

If the embryo has begun to develop, remove as much of the fluid content of the egg as possible, replace it with a 10-percent solution of KOH, and allow this to stand overnight. This will reduce the embryo to a fluid state. After treatment with KOH remove the content of the egg, wash, and dry. Bendire recommends labeling the egg with a lead pencil. However, India ink may be used unless there is some reason why the label should be removed periodically. Each egg should be dated and given a museum number corresponding to that on its data card.

Skeletal Techniques

On many occasions bird skeletons are needed for taxonomic purposes or for classroom teaching. Very small birds or chicks are best cleared and stained; larger birds should be cleaned by dermestid beetles or by some other technique. Because the skeletal techniques for all vertebrates are so similar, the reader is referred to the single treatment of this topic in Chapter 23.

Display Tubes

When bird skins (round mounts) are used by beginning students they are often badly treated, wings and legs are broken, or the labels become lost. If these same specimens are displayed in plastic tubes of appropriate sizes they may be handled without danger of real damage. Such tubes are available commercially through biological supply houses or may be made quite economically. Large glass or plastic tubes can be adapted for hummingbirds, creepers, nuthatches, and other small birds.

Inexpensive bird tubes may be made as follows: Obtain a roll of 10-mil semirigid, clear plastic which is used to cover windows (the author uses a plastic sold by Sears under the trade name of "Sun-Ray"). The procedure involves (1) cutting the plastic to size, (2) presetting the plastic at the desired diameter, (3) sealing the side seam, (4) forming the cap, if one is to be used, (5) cementing the cap to the tube. Different-sized tubes will be required, depending on the sizes of birds to be displayed, but two or three standard sizes should be selected for the sake of simplicity. Presetting the plastic necessitates the use of some kind of a cylinder to wrap the plastic around. Any cylinder of a usable diameter will do. The author uses wooden closet pole (1¼ inches in diameter), wooden stair railing (1⅝ inches in diameter), 1½-inch galvanized pipe (1⅞ inches outside diameter) or 2-inch galvanized pipe (2⅜ inches outside diameter). All of these are available at lumber yards or hardware stores. If the pipe is used as a cylinder, it should be sanded smooth on the rough places, then coated with rubber cement and wrapped with a piece of smooth white paper which has also been coated with rubber cement. This will prevent the pipe from scratching the plastic. When cutting the plastic, make the sheet wide enough to go around the diameter of the cylinder plus ⅜ to ½ inch for overlap on the seam. The tube must be long enough to accommodate the bird specimens with their labels and the cotton plug at one end (Fig. 21–7A).

After cutting the plastic, wrap several sheets around the cylinder. In turn, wrap the plastic with paper and tape the paper all the way around with masking tape to secure the bundle. Next, put the cylinder in an oven and support it in such a way that the plastic does not touch the shelves of the oven. A large slide-drying oven, an incubator, or any unit that can be thermostatically controlled is suitable for this process. Heat the oven to 75° C. and subject the plastic to this temperature for 30 minutes. Remove

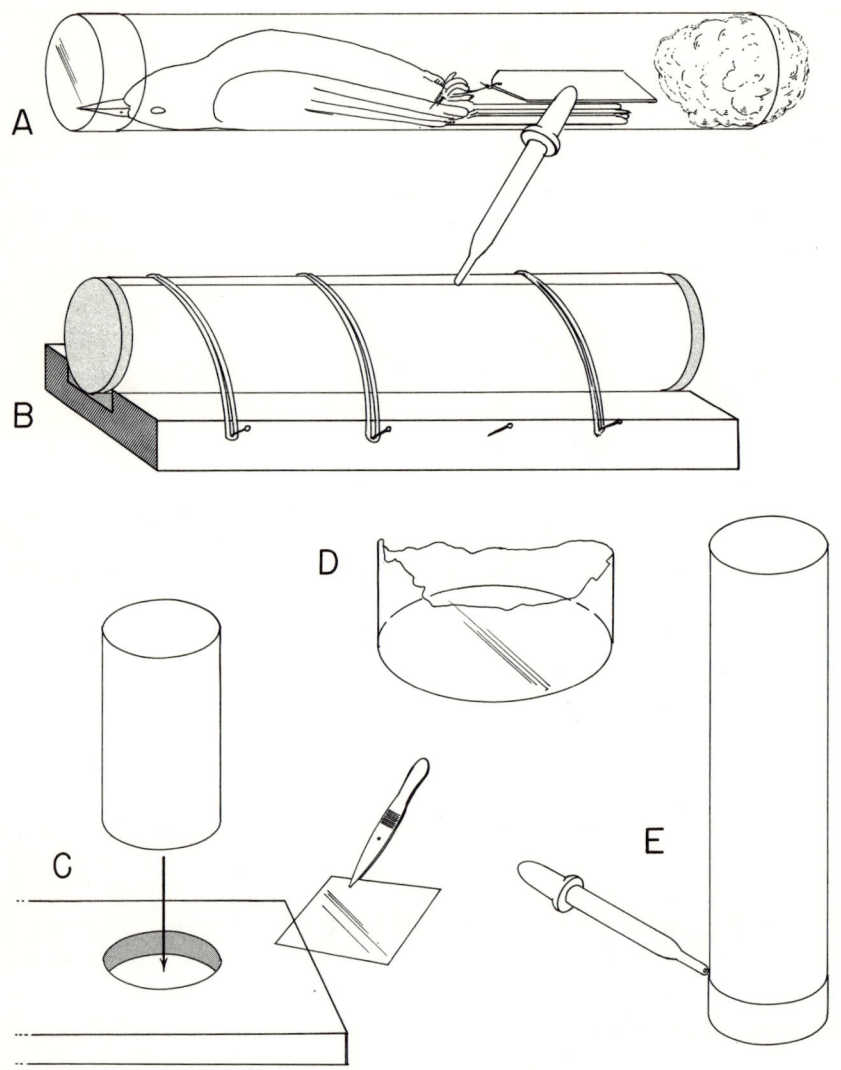

FIG. 21–7. Making plastic bird display tubes. A. Finished tube with specimen and a cotton plug. B. Forming the seam of the tube. C. Forming the end cap. D. The end cap prior to trimming. E. Attaching the end cap to the tube.

the cylinder and permit it to cool to room temperature. This heat process sets the plastic in a cylindrical rather than a flat form, which is essential if a good seam is to be made. Next, put one sheet of plastic around the cylinder, place the cylinder on a holding rack, and secure the plastic with rubber bands (Fig. 21–7B). The seam may be set by adding minute quantities of clear, model-airplane cement or glass cement (Duco Cement, or others) or a very small quantity of ethyl acetate. Be careful not to cement the plastic to the cylinder. Permit the plastic to dry thoroughly before removing.

Figure 21–7, C and D, demonstrates the procedure of making a plastic cap. Cut a hole in a piece of board slightly larger than the cylinder. Cut a square of plastic large enough to form the entire cap. Holding the square with a pair of forceps, dip the plastic in boiling water for 1 or 2 minutes. Then quickly transfer the plastic over the hole and immediately force the cylinder down into the hole to form the cap. Trim the plastic and remove it from the mold. Finally, put the plastic cap over one end of the bird tube and cement this in place with a very small quantity of ethyl acetate or any appropriate cement.

The opposite end of the tube is plugged with long-fiber, absorbent cotton. The cotton plug may be pushed in as far as necessary to keep the bird specimen from sliding back and forth in the tube. If the tubes are made long enough, both ends may be plugged with cotton, thus eliminating the process of making caps.

Storage of Skins

There are three general requirements for storing bird or mammal skins. Storage cabinets must be kept in a dry room; a damp, unheated room is not satisfactory. Second, the storage cabinet must be lightproof so that the colors of the specimens will not be altered. Finally, storage cabinets must be insect-proof and must be fumigated periodically with paradichlorobenzene. The design and construction of a suitable storage cabinet will be described in Chapter 22.

Measuring Bird Skins

Unlike mammals, which must be measured before mounting, birds' measurements may all be adequately taken from the prepared skin. With the exception of weight and the possible exception of total length, the measurements of prepared round mounts are very suitable for scientific work. Thus, many bird collectors omit measurement at the time of mounting. The standard measurements are shown in Fig. 21–8; they are the bill length, the total

396 Biological Techniques

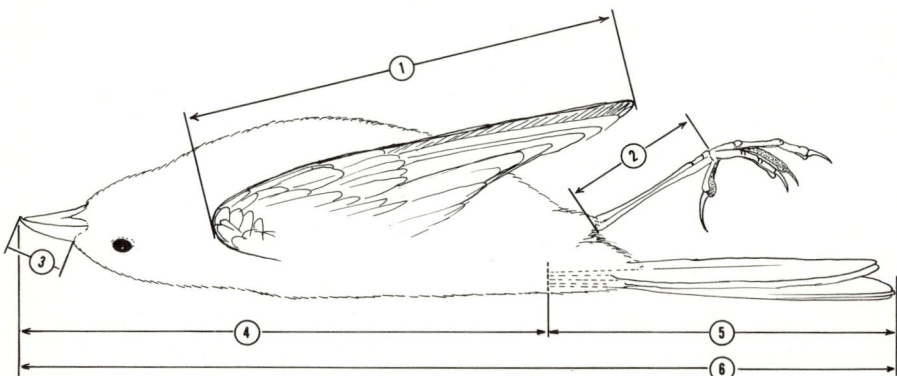

FIG. 21–8. Some standard bird measurements. (1) Wing length, (2) tarsal length, (3) bill length, (4) body length measured to base of tail feathers, (5) tail length, (6) total length.

length, the body length, the tail length, the wing length, and the "tarsal" length (tarso-metatarsus).

FIELD STUDY AND IDENTIFICATION

Perhaps one of the most enjoyable experiences of a field naturalist is the serious study of bird identification. It requires considerable time and a little equipment, but can become a hobby of unending interest. Possibly more is known about the natural history of birds than of any other animal group. Yet, there is still a lot to be learned concerning breeding, nesting, and distribution of birds. Many amateur bird-watchers have recorded and published excellent accounts of nesting behavior and the like. To gain some insight as to how really intriguing natural-history studies of birds can be—in procedure, experimentation, and results—every student should read Tinbergen's *The Herring Gull's World*. This is a truly excellent account and cannot be too highly recommended.

Equipment. Perhaps a field guide to bird identification is the first requisite. Many excellent field guides are now available, some of which are mentioned in the bibliography. The author finds the Peterson Field Guide Series quite adequate. Second, a pair of field glasses will greatly enhance bird identification. The 7 × 50 field glass is a popular size. This magnifies objects 7 times and has a lens opening of 50 millimeters. However, glasses with greater or lesser power are quite suitable. Consider the weight of the field glasses, as you will be required to observe birds for long periods of time. Just as important as the field guide and field glasses is the field notebook. Every trip into the field should have its own field number. The observer should note the type of terrain, the type of vegetation, the weather, and the date, and then list the birds observed. Such chronological listings eventually give a year-round picture of bird activity at the local level. Elsewhere in the field notebook, give species accounts. One to several pages may be devoted to a particular species of interest. Every new observation of the behavior of this species should be recorded in this section as well as in the general observation section. Finally, the field notebook should have a checklist of all birds observed, so that new species may be added, along with the date of observation and station number. The serious keeping of field notes cannot be overemphasized, as it is the first step toward truly scientific bird study.

Identification. Pettingill's work (1956), as well as many others, provides a sound basis for study of the external structures of birds, types of plumage, field identification, bird ecology, bird communities, mating, nest-building, egg-laying, and other activities; this work is not only descriptive, but also permits the student to perform actual laboratory studies. People pinpoint the basic problem of field identification when they ask you to identify some bird they saw by describing it as being "red" or "yellow and black." A number of things must be observed before the details of color, particular markings, and the like, are taken into account. Some of these factors are the size and general silhouette of the bird, the type of bill, the type of feet and legs, the markings of the bird, and its over-all color.

Get to know common birds well enough so that you can establish general size classes such as the crow, the pigeon, the robin, the sparrow, the nuthatch. When you become absolutely familiar with such sizes, you can first judge a bird as being slightly larger or slightly smaller than, for example, the sparrow. Next, its general silhouette will be of value. Whether the bird sits or stands erect, has a long or short neck, stands well above the ground or close to it, and so on, will help you to identify the specimen.

The type of bill is extremely important in separating groups of songbirds which may otherwise be similar in size and coloration. Bird bills are modified for the type of food they use. You will recognize the short cone-shaped bill as that of seed-eating birds, whereas the long cone-shaped bill belongs to insect eaters. Long cone-shaped bills which are flattened laterally and somewhat blunted on the end often belong to woodpeckers. Extremely long, tapered bills that are flattened laterally are used by "spear fishermen," birds which catch their food in aquatic situations. The hooked, flesh-tearing bills of hawks, owls, parrots (which tear plant flesh), and the like, separate those birds from others. Some birds such as crows, jays, and gulls have generalized bills which are usually long and somewhat cone-shaped like the insect eater's, may be partially hooked like the flesh eater's, and, in general, suggest that these birds may eat living or dead plant or animal material.

Bird feet and legs are also modified for their particular habitat and occupation. Songbirds have perching feet with three toes directed forward and one backward. Such birds generally do not run, but rather hop along the ground. Quail, pheasants, and other ground runners have similar-appearing feet, but have heavier pads beneath the toes. Birds of prey have hooked, flesh-tearing claws, whereas aquatic birds possess partially webbed or fully webbed feet. Leg length may denote the difference between aquatic birds and terrestrial birds in some instances.

Probably these few descriptions will suffice to introduce the procedure of bird identification. You will quickly classify different groups of birds, such as warblers, juncos, and so on, by their size class. By a process of elimination based on size, feet, bills, color, and markings you can locate and identify birds in your field guide.

NESTING STUDIES

The study of bird nesting offers a chance for first-class research, by students and amateurs as well as professional biologists. Nests located close to your place of residence or school are preferable, as they afford constant observation. Nest studies should begin as soon as possible during the breeding season. Notes should be taken, and dated, concerning all activities involved in reproduction. A typical reproductive pattern in a songbird might be as follows: A territory is established and defended by the male who usually arrives several days before the female of the species. By one means or another the male must attract a female to his territory and entice her to mate and nest in that territory. Thus, singing and courtship, nest-building, and copulation make up the early facets of reproduction. Notes on such activities should include diagrams and descriptions of courtship procedures, fighting and defense, nest-building techniques, discussions of materials used, and a diagram of the actual nesting site. By carefully watching the defense of the territory one may learn the actual size of the territory defended by a particular male. The time and duration of egg-laying and the number of eggs per nest are the next items to be recorded. The interest should then shift to the process of incubating the eggs, for this work may be shared by both parents or carried out only by the female. The amount of time spent by each individual on the nest or off the nest and the various activities of the individuals not occupying the nest should be recorded. Likewise, it should be determined if one or both parents feed the young, how many trips per day they make, the type of food they carry, the amount of food, and so on. The activities of the young are also of importance, in the realm of feeding, dominance, independence, and the activities leading toward flight. Finally, it is of interest when and how the family group disperses from the territory and whether the parents produce a second or even a third brood during a given season.

Obviously, the student must have an excellent vantage point from which to make such observations. Nesting birds often get accustomed to certain individuals near their nest and will behave quite normally in their presence. If branches obscure the observer's view he can remove these by tying a dark piece of string to the branch while the parents are away from the nest. The branch is then pulled back from the nest a little bit more each day, until perfect vision is obtained. Such studies are very time-consuming but are also very gratifying.

BIRD BANDING

Bird banding permits are issued in the United States to citizens, eighteen years of age or over, who can demonstrate that they can identify species and subspecies of birds. The U.S. Fish and Wildlife Service in Washington, D.C. should be contacted for banding permits. This agency issues a bulletin describing banding methods and equipment, and issues a supply of aluminum leg bands of various sizes and record forms for keeping data. As is true for bird collecting, the bird bander must also apply to his state for a banding permit. Once the federal permit is obtained the state permit is relatively easy to secure. In some states the bird bander must register his bird nets and other traps, if these are used. Birds are generally obtained from the nest, from traps, or from nets. There are several sources of mist nets, used for catching birds and bats, in the United States. Two sources are (1) the Bleitz Wildlife Foundation, 5334 Hollywood Boulevard, Hollywood, California, and (2) W. B. Davis, 712 Mary Lake Drive, Bryan, Texas.

The following information is taken from instructions issued by the Bleitz Wildlife Foundation. The mist net is made of exceedingly fine nylon and measures anywhere from 7 by 30 feet to 15 by 60 feet in over-all size. Such nets are strung between two or more poles in bird flyways, preferably against a background of mixed vegetation. Nets set along watercourses or

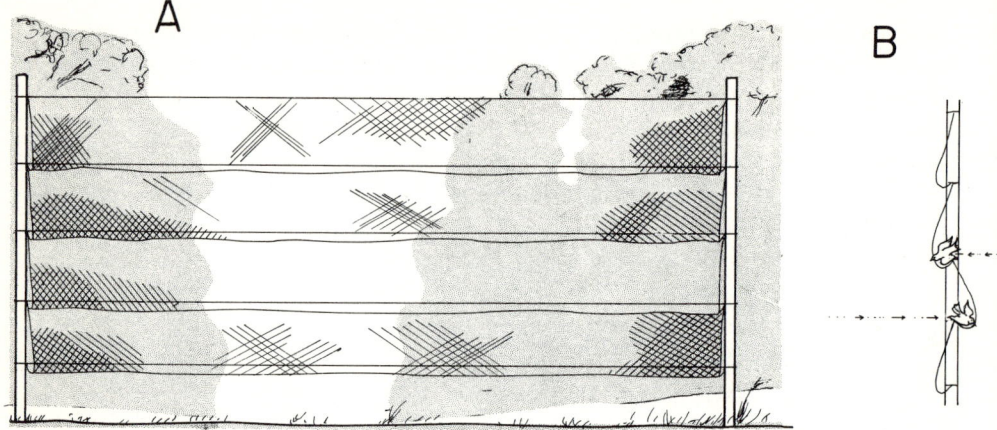

FIG. 21–9. Using the mist net. A. The mist net set in a bird flyway surrounded by trees. B. A side view showing birds entering the mist net. See text for an explanation.

near water holes are useful. One factor that is very important in locating the net is at least partial shade. The shade will not only obscure the net but will prevent entrapped birds from becoming overheated. Bird nets should never be left unattended and captured specimens should be removed promptly.

As shown in Fig. 21–9A, the fine mesh of the mist net is held between several supporting cords. Thus, each section of the net tends to "bag-out" to form a receptacle for a flying bird (Fig. 21–9B). Birds are removed from the net and placed into brown paper bags prior to banding. Once in the bag the bird remains quiet and will be unharmed, provided the bag is not left in the sun. Specimens are banded and released as quickly as possible. While this discussion is brief, it will give the individual an idea of what bird netting is like. Complete instructions are sent with each net purchased.

REFERENCES

Allen, A. A., 1930, *The Book of Bird Life*, Van Nostrand, Princeton, N.J.

Bendire, C., 1891, Instructions for Collecting, Preparing, and Preserving Bird Eggs and Nests, *U.S. Nat. Mus. Bull.*, 39:3–10.

Bent, A. C., 1919 and following, *Life Histories of North American Birds*, 19 vols. to date, U.S. National Museum, Washington.

Campbell, Bruce, 1953, *Finding Nests*, William Collins, London.

Chapman, Frank M., 1932, *Handbook of Birds of Eastern North America,* Appleton-Century-Crofts, N.Y.
Eliot, W., 1923, *Birds of the Pacific Coast,* Putnam, N.Y.
Fisher, J., and R. M. Lockley, 1954, *Sea Birds,* Houghton Mifflin, Boston.
Griscom, Ludlow, 1945, *Modern Bird Study,* Harvard Univ. Press, Cambridge, Mass.
Headstrom, R., 1949, *Birds' Nests: A Field Guide,* Ives Washburn, N.Y.
Murphy, R. C., 1936, *The Oceanic Birds of South America,* Macmillan, N.Y.
Peters, J. L., 1931–1948, *Birds of the World,* 6 vols., Harvard Univ. Press, Cambridge, Mass.
Peterson, R., 1947, *A Field Guide to the Birds,* Houghton Mifflin, Boston.
Peterson, R., 1948, *How to Know the Birds,* Houghton Mifflin, Boston.
Peterson, R., 1961, *A Field Guide to the Western Birds,* Houghton Mifflin, Boston.
Peterson, R., G. Mountfort, and P. Hollom, 1954, *A Field Guide to the Birds of Britain and Europe,* Houghton Mifflin, Boston.
Pettingill, O., 1951, *A Guide to Bird Finding East of the Mississippi,* Oxford Univ. Press, N.Y.
Pettingill, O., 1953, *A Guide to Bird Finding West of the Mississippi,* Oxford Univ. Press, N.Y.
Pettingill, O., 1956, *A Laboratory and Field Manual of Ornithology,* Burgess, Minneapolis.
Pough, R. H., 1946, *Audubon Bird Guide: Eastern Land Birds,* Doubleday, N.Y.
Pough, R. H., 1951, *Audubon Water Bird Guide,* Doubleday, Garden City, N.Y.
Pough, R. H., 1953, Audubon Guides, *All the Birds of Eastern and Central North America,* Doubleday, N.Y.
Ridgway, R., 1891, Directions for Collecting Birds, *U.S. Nat. Mus. Bull.,* 39(A): 5–27.
Scott, P., 1961, *A Coloured Key to the Wildfowl of the World,* Scribner's, N.Y.
Tinbergen, N., 1961, *The Herring Gull's World: A Study of The Social Behaviour of Birds,* The New Naturalist Series No. 9, Basic Books, N.Y.

CHAPTER 22

The Mammals

The mammals, which are so familiar to us, are tetrapod vertebrates which are characterized by having true hair, heterodont teeth, and mammary glands. In addition to general classroom studies in biology the mammals are currently being investigated in many fields. Current studies are being conducted on mammal taxonomy, ecology, population dynamics, food chains, natural history and behavior, physiology, and the like. Popular interest in the mammals has fostered a number of good textbooks, including field guides, and a number of good motion pictures on mammalian natural history. Many of the mammals, like the birds, can be studied and photographed in the field by those willing to devote the time and energy to this fascinating science.

COLLECTING TECHNIQUES

The small mammals can readily be collected once their natural history is understood. Like all other creatures, mammals are "found only where you find them." As poor as this statement seems, it points up the problem that most beginning collectors have in locating small or large mammals. It is surprising how numerous some of the small rodents are, on the one hand, and how few of us are really aware of their presence, on the other hand. Among the requisites for locating mammals is a knowledge of mammalian natural history, periods of behavior, habitat requirements and hiding-place requirements, food requirements and the distribution of food organisms, and, finally, knowledge of the telltale signs such as runways, droppings, and the like, which mammals leave as evidence of their presence. One of the excellent books dealing with this last-named topic is *A Field Guide to Animal Tracks* by Murie (1954). Among the many field guides dealing with mammal recognition, distribution, habitat, and signs are the following: Anthony (1942), Booth (1950), Burt and Grossenheider (1952), Cameron (1956), Palmer (1954).

Laws and Permits

A few mammals, such as the fur seal and the sea otter, are totally protected by law and the possession of any specimens of these species is illegal without a federal permit. A larger number of mammals is also classified by each state as fur-bearing mammals which may be taken in season with a trapper's permit. Write to your state fish and game department for full details on obtaining such a permit and on protected animals. Some of the predators are unprotected (though more of them should be protected) and most of the small rodents have no legal protection. Small rodents are generally so prolific that one may collect freely without endangering the species involved. Some species, such as the silver squirrel in western Washington, are disappearing, owing to natural causes, and thus should be protected and not collected. In western Washington the silver squirrel feeds primarily on acorns and the once abundant stands of oak trees are now slowly being crowded out by the fir tree. Thus the red squirrel, which is a cone-feeder, is slowly taking over and the silver squirrel is on the way out. Since collecting laws change from year to year, each collector should contact his state game department for the latest information before collecting.

Trapping

Snap Traps. The standard mouse and rat trap sold everywhere is excellent for collecting mice, rats, chipmunks, squirrels, weasels, shrews, and so on. Because of its smaller size, the mouse trap tends to damage many skulls, whereas the rat trap occasionally tears the skin of specimens. The Museum Special Rodent Trap (Animal Trap Company, Lititz, Pa.) is intermediate in size between the mouse and rat traps and has a "softer"

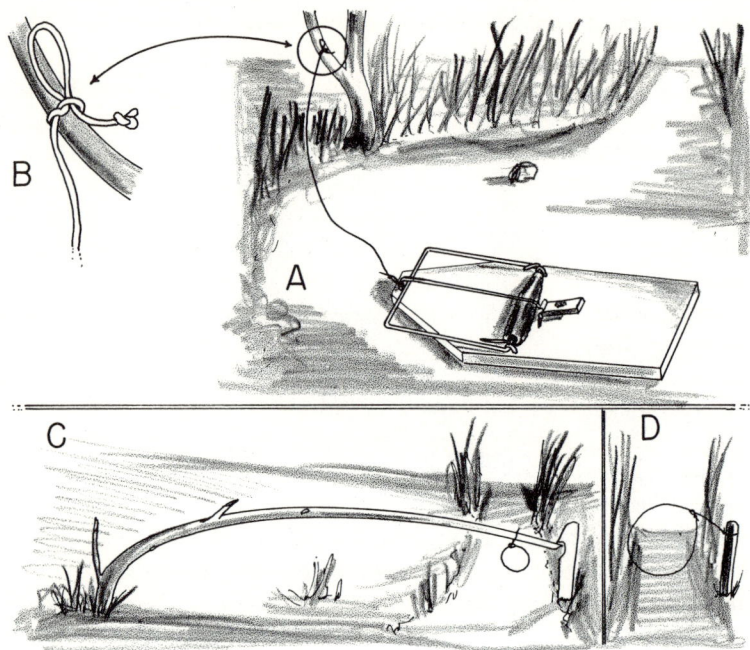

FIG. 22–1. Small mammal traps. A. Museum special set in runway. B. A handy knot for securing snap traps. C. A small snare set in a rodent runway. D. A wire snare set in a rodent runway.

spring. Thus, this trap (Fig. 22–1A) is a favorite of mammal collectors in that the bail seldom catches a specimen across the skull or crushes its body.

Many kinds of baits may be tried, but those containing peanut butter seem to work the best. A mixture of rolled oats, water, and peanut butter should be sticky enough to adhere to the trigger of the trap. Addition of chopped raisins and chopped nuts to the peanut butter-water-oat bait seldom improves the catch per unit of bait. At least, the author finds that he eats more of this bait than he uses on the traps. Plain cheese or plain bacon seldom work as well as the peanut-butter mixtures.

Each trap should be provided with a stout string about 20 inches long, so that it may be anchored to local vegetation. The trapper should also carry a piece of red or white cloth to make flag markers for relocating the traps. There are possibly as many ways of setting traps as there are trappers. Where there is a natural orientation of the vegetation in the area to be trapped, such as along a stream bed or path, the trapper may simply move along at random, setting one or two traps by each large clump of brush. The trap may be anchored quickly, using the combination overhand and slipknot (Fig. 22–1B). A strip of cloth should be tied to some brush in a conspicuous place near each set, in order to guide you back to your traps in the morning.

In open country, such as prairie grassland or desert and semidesert where the vegetation is sparse and lacks any natural formation, traps may be set every ten paces or at any likely looking site. When working in the open this way, relocate your traps by using cloth flags, by following a natural path and scratching a deep mark into the ground by each trap, or by dragging a heavy stick along the ground to make a furrow which will denote your trapping route the next morning.

Traps should be picked up very early in the morning, as specimens left in the sunshine soon begin to rot and may be attacked by ants. In the desert the author generally runs the trapline twice during the night and again the first thing in the morning, as both predators and the warm evening temperatures are effective in reducing the specimens. When it is impossible to mount the specimens immediately, place them in a plastic bag, along with their data, and put this in a camp freezer chest with dry ice. So preserved, the specimens may be handled at your leisure.

Weasels and some rats are attracted to fresh, bloody meat. For weasels, nail the trap onto a fence post or other object near a weasel runway, with the trigger between 6 and 8 inches off the ground. When the weasel stands up to investigate the bloody-meat bait, the bail of the trap will swing downward, catching the animal across the shoulders, and will kill it instantly.

When you are interested in ectoparasites the specimens must be isolated —at least, according to species and, preferably, by individuals. Fleas and lice are slow to leave their host animals and, thus, some of them will be present at the time of pickup. Therefore, if individual specimens are placed in small cloth or paper bags at the time of pickup the relationship between ectoparasite and host will be maintained. Locate the ectoparasites as described in Chapters 13 and 15.

Steel Traps. Steel foot traps consist of two heavy jaws which catch the mammal by the foot when it steps on the trigger. These traps are cruel, to say the least, and should not be used in random settings. It is desirable to select the largest-size foot trap designated for a particular kind of mammal, rather than a smaller size, in that the larger trap will grip the specimen higher up on the leg and thus prevent it from twisting or chewing its foot off. Needless to say, steel traps should be tended faithfully in the interest of the animal caught, or not used at all.

The kind of bait used with steel traps, if any, will depend on the type of animal to be caught. For example, wolves, coyotes, and foxes may be trapped at natural signposts (places where these animals habitually urinate) or baited with the urine from a female dog which is in heat, with the carcass of some animal, or with a stink bait. A stink bait is prepared by sealing fish or meat in a bottle half-filled with water. After two or three weeks' rotting this fluid makes an excellent lure. One of the many ways of using the stink bait is as follows: Dig a shallow hole large enough to accommodate the trap and its stake and chain, sprinkle the trap with the stink bait, set the trap and place it in the hole, with the pan about 2 inches from the surface of the ground. Next, carefully pack dirt around the outside of the jaws and over the springs, place some cotton between the jaws and under the pan, and put a piece of light tissue paper over the jaws to prevent dirt from getting underneath the trigger or pan. Finally, sprinkle dirt over the entire trap so as to conceal it. The set is complete when all traces of the trapper have been removed and an additional drop of the stink bait is placed over the trap site and on some of the surrounding vegetation. The addition of an unwashed tin can near the trap site may help the catch. The animal is caught when it attempts to dig up what it thinks is a buried portion of rotten meat.

Many traps are placed in such a fashion as to take advantage of the normal movements of the animals sought. For example, wild cats and wild dogs, as well as other large mammals, usually step over small limbs, logs, or stones that are present in their runways. Thus, the steel trap may be set in the place where the mammal is most likely to step when crossing such an object.

Commercial Live-Catch Traps. A number of commercially made live-catch traps, such as the Havahart trap and Sherman trap, are very humane and are thus recommended. The Sherman Company produces small, light-weight, all-metal mouse and rat traps which are collapsible and, therefore, take up very little room in a field pack. This is an important factor to consider when ordering live-catch traps.

Can and Bottle Traps. Many small rodents and shrews can readily be trapped by burying cans or bottles flush with the ground and then suspending an appropriate bait (peanut butter or meat, respectively) over the opening. These animals will simply wander into such traps and will find it impossible to escape. Since shrews are carnivorous and usually eat one another, most workers partially fill the can or bottle with water and thus drown the specimens as they are caught. Chipmunks readily enter milk bottles for peanuts. Place a milk bottle on its side and supply this with a handful of salted peanuts. After the chipmunk has filled its cheek pouches you can run up to the bottle and cap it before the chipmunk can empty out its cheek pouches and escape.

Trapping Bats. Bats are usually obtained by shooting, netting, or by mist nets similar to those used for birds. (There is a brief discussion on the use of mist nets and some addresses of dealers selling mist nets in Chapter 21, pp. 399). Mist nets are set for bats in places where these animals fly, such as between rows of trees, across pathways, and along or over small streams. Fruit-eating bats are more easily taken than insect-eating bats, since the latter use their "radar" more than the former. Likewise, on clear moonlit nights bats tend to navigate as much by vision as by radar, and thus fail to detect the net. Conversely, on very dark nights many bats will fly directly to the net then turn and go over it or under it since they have detected it by means of their radar.

Heavy gloves should be worn when removing bats from the net. Transfer live bats into canvas sacks and later kill them by putting the sacks into a cyanide bottle or chloroform bottle. Since the author was bitten by a bat in Costa Rica and had to undergo the rabies treatment, it seems expedient to mention that about 1-percent of the bats which have been tested prove to be rabid. Since bats, unlike all other mammals, show no symptoms of rabies, one must use extreme caution with these animals.

The mist net is also effective when a bat tree or bat cave is located. In Costa Rica the author helped put mist nets over openings into large hollow trees. Following this, a collector entered the hollow and fired .22-caliber bird-shot shells up into the hollow to evict the bats and cause them to fly into the net. Fire and smoke were also used, and all proved very successful.

Dalquest (1954), one of the pioneers at using the mist net for bats, has published an account of his collecting in tropical Mexico. His article should be consulted as it gives methods and procedures for collecting bats in great detail. Dalquest concludes that bat netting does not give a complete picture of the bat fauna in an area, as many species taken by shotgun are not taken in the nets. The reverse of this is also true, however. Bat nets should not be placed in the "dead air" of caves or buildings but rather in the open. He further suggests that nets should be placed close to the ground, as many species of bats fly between 1 and 2 feet above ground. Bat nets should be tended regularly and should not be left out overnight in the tropics. Dalquest recommends that notes be taken as to the time bats are caught in the nets, as there seems to be a definite correlation between the feeding habits of bats and the time of their activity.

Many species of bats may be collected by stretching a taut wire or dark-colored string about 4 inches above the surface of slow-moving streams or ponds. Bats fly low over such bodies of water at dusk in order to drink or to catch insects. Therefore, they will strike the wire and fall into the water. This renders them helpless and will permit you to retrieve them with a dip net.

Gopher and Mole Traps. Most burrowing rodents, such as ground squirrels, can be shot or taken in foot traps placed directly outside their burrows. Gophers and moles, however, present a special problem and must be taken with traps designed especially for them. The wire gopher trap should be set 5 or 6 inches inside the mouth of a gopher burrow after it has been opened and cleaned. The hole is then partially plugged with vegetation or soil. The gopher is trapped when it pushes soil down the runway in an attempt to shut out the small amount of daylight showing.

Mole traps, on the other hand, are set in the middle of mole runways which are filled with soft, rock-free dirt. The mole is trapped when it noses its way through the soft dirt while moving along its runway.

Snares. Snares are very effective for some mammals. One can buy large snares for mountain lions, coyotes, and similar animals. Small snares are easily made out of light copper wire for rabbits, squirrels, and so on. The rabbit snare can be attached to a small sapling which is held down by a stake (Fig. 22–1C), or attached directly to a stake next to a runway (Fig. 22–1D). Squirrels can be trapped by placing snares on downed logs or branches where these animals usually run. Place the snare in some vegetation so that the animal will automatically stick its head through the loop as it runs along.

Shooting Specimens

Little need be said about hunting techniques or equipment. It should be stated, however, that specimens should be shot with the least amount of damage possible. Head shots must be avoided, since they destroy the skull. When you are close to some specimen to be collected with a shotgun, aim beside the specimen rather than directly at it so that the animal will be killed with a minimum number of pellets.

As soon as a specimen has been shot, plug all of the shot holes with wads of absorbent cotton to prevent bleeding. If any blood occurs in the mouth of the animal, hang it up by its hind feet and permit this blood to drain out. Then, plug the throat with a large wad of cotton to prevent further bleeding. If it seems necessary, also plug the anus with cotton. Wash off all blood with fresh, cool water before it has a chance to set. Wrap the specimen in absorbent paper toweling or old newspaper or place the specimen in a small canvas sack. Specimens should be mounted as quickly as possible or frozen. However, it is best to wait at least one-half hour after the specimen has been shot, since there will be less bleeding in the mounting process.

Other Sources of Specimens

Alert mammalogists will notice a large number of mammal specimens while driving along the highways. The D.O.R. (dead on road) specimen is frequently very suitable for mounting, especially if it has been flipped to the side of the road and is fresh. Meadow mice (Genus *Microtus*), which are very difficult to trap in snap traps, can be obtained in huge numbers by following a farmer as he plows his field. A canvas bag and leather gloves are the only necessary equipment. Occasionally, gophers are evicted by the plow; these same animals can also be obtained when rivers periodically flood into cultivated fields.

Killing Injured Specimens

Trapped or wounded specimens must be dispatched immediately. Bear in mind that the skull cannot be injured, as this is needed for taxonomic work. Therefore, the best procedure for smaller mammals is to kill them by forcing the air out of the lungs. Place your foot on the chest of a foot-trapped specimen and hold it for a moment, or else place the specimen in a canvas bag and depress the lungs between your fingers or with your foot. Larger, dangerous mammals will have to be shot or drowned but should suffer the least amount of injury possible.

Salting Field Skins

Although smaller mammals may be prepared in the field, large specimens are usually salted, shipped to the laboratory, and then tanned commercially. Spread out the hide, fur side down, and cover it with a thick layer of ice cream salt or table salt. This salt will draw a lot of moisture from the atmosphere and from the skin itself, part of which may be drawn off. If necessary, scrape off the wet salt and dry it in an oven, then resalt the hide. When the hide is almost dry, shake off the excess salt, roll the hide fur side in, and pack it in a canvas or plastic bag. Be sure that a waterproof label has been attached to the hide giving all of the field data, which are also recorded in your field notebook.

MOUNTING AND PRESERVATION

Choice of Technique

The standard method of preservation for mammals is that of the round mount for small mammals and the tanned pelt for larger specimens. The taxidermy mount, which poses the animal in a lifelike way, is not used in museum collections (except in public displays), since each specimen requires much more space for storage. The round mount, on the other hand, takes up a very small amount of space and yet exhibits all parts of the pelt. In teaching collections, where storage space is not a major problem, mammals as large as the bobcat, badger, or fox are more realistic when prepared as round mounts rather than as pelts.

The case-skin mount, or flat-pelt technique, is very useful for small rodents when there is little time for standard mounting methods. Any small mammal intended for muscle study should be preserved in liquid. It is now preferable to preserve bats in liquid, although these may also be prepared as round mounts. Very small mammals such as shrews can be preserved in liquid and dried, using the "wet and dry mount." Therefore, the technique used will depend on the specimen and the circumstances. Regardless of the technique, however, all mammals must be measured in a standard way to ensure an accurate record.

Measurements and Other Data

Figure 22–2 demonstrates the procedure used for measuring mammals. These measurements are the total length, the tail length, the foot length, and the ear length (Fig. 22–2, A, B, C, and D, respectively). All measurements are recorded in millimeters. Take the total length measurement by placing the specimen on a ruler, as shown, and measuring from the tip of the snout to the end of the vertebral column (not to the tip of the hair on the tail). The tail length is again concerned with the length of the vertebral column (not the hair), and is measured from that point where the tail bends away from the body to the tip. The foot length is taken on the hind foot and may be somewhat variable depending on how worn the toenails are. A ruler (as shown) or a calipers (preferably) may be used for this measure-

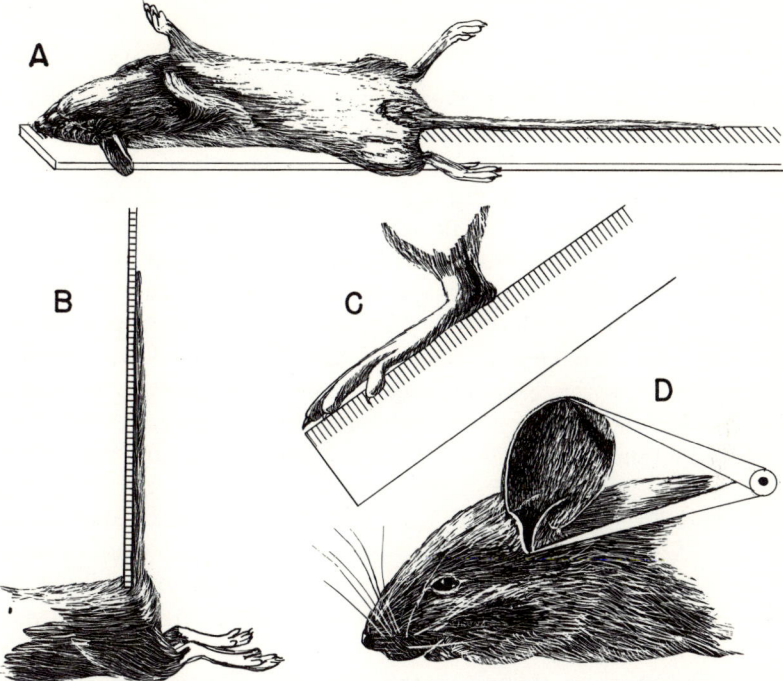

FIG. 22–2. Standard mammal measurements. A. Total length. B. Tail length. C. Foot length, measuring the hind foot. D. Measuring the ear: this may be done with a ruler, divider, or caliper.

ment. The ear length, which can also be measured with a ruler, calipers, or dividers, is taken from the notch of the ear to the tip of the ear. Sometimes a second measurement is made from the crown of the skull to the tip of the ear. The four major measurements are also recorded in the field notebook and on the specimen label.

The collector should be concerned with a number of other data. The weight should be recorded in grams, as accurately as possible. Note the size and position of the gonads, as well as other data associated with reproduction such as pregnancy, the size of the embryo, the size of the mammary glands, whether or not lactation is occurring, and so on. Pelage notes and color notes need not be taken, since these can be observed at any time on the preserved skin. For further information, see the section on labeling below.

The Mounting Kit

An old fishing-tackle box makes an excellent kit. The following things should be provided: a scalpel or sharp knife, a carborundum sharpening stone, a pair of fine-pointed scissors and a coarse pair of scissors or a bone shears, a fine-pointed pair of forceps and a 10- or 12-inch pair of forceps, a 30-centimeter ruler graduated in millimeters, a package of assorted sewing needles, a spool of No. 2 cotton or linen thread and a spool of heavy cotton or linen button thread, and a package of straight pins; a pair of dividers and a pair of calipers are desirable, but optional. In addition, provide a stout side-cutting pliers and an assortment of different sizes of wire to make up tail wires and leg wires. Stainless steel fish leader is preferable; galvanized wire is suitable; plain or ungalvanized wire is barely satisfactory. Galvanized wire can be obtained in small 1-pound coils in sizes of 16-gauge, 18-gauge, 20-gauge, and 22-gauge. These sizes are very useful for smaller rodents; 12-gauge and 14-gauge wire may be needed for rabbits and animals of similar size. A large (30 or 50 cubic centimeter) hypodermic syringe with a long needle is indispensable for removing the brains from skulls. Arsenic trioxide or Boraxo is used by most workers to poison the skin, and corn meal or hardwood sawdust must be on hand for removing grease and blood. Finally, a pound of long-fiber absorbent cotton is needed for making tail wires, leg wires, and artificial bodies for round mounts. Provide also skin labels, skull labels, waterproof ink, and dip pen.

Problems to be Anticipated

Fat Skins. During certain seasons of the year mammal skins contain large quantities of fat which must be removed. Fat left on the skin will eventually work through the pelt, discolor the fur, and may even cause the fur to slip. Once the skin has been removed from the carcass all small bits of meat and fat must be scraped away from it. Some workers wash fatty skins in warm water and detergent, paying particular attention to the concentrations of fat on the skin. When the fat has been extracted the skin is then squeezed, to remove most of the moisture, and dried. Copious amounts of corn meal or fine hardwood sawdust will quickly remove most of the moisture. Many museum men use a product known as "Blue Cloud" (the Don Company, Gardena, California) which is a drying agent developed for chinchilla baths. Any of these drying agents will shake out of the dried fur or can be removed by a small automobile-size vacuum cleaner or by blowing with an air hose.

Bloodstains. Blood should not be allowed to dry on the fur, but should be removed with clean cool water. Should some blood harden, however, mechanically dislodge it using your fingernail and a brush. Following this, clear water with a little detergent may be used with good results.

Determining Sex. The external genitalia usually are diagnostic for sex determination. Some mammals are difficult to sex if they are immature, if they are not in a breeding condition at the time of mounting, or if the openings for the penis or vagina are similar in appearance. Many beginning students label female deer mice (*Peromyscus*) as males because the clitoris in some species is as long as or longer than the penis, thus causing confusion. Ideally, a dissection should be made of the carcass to observe and measure the gonads.

Food Samples. Always examine the cheek pouches of small rodents and preserve or identify the contents. For a complete record, dissect the stomach and determine the type of food present. Record these data in the field notebook.

The Round Mount

Skinning. This account deals directly with the standard procedure used for small mammals. Some special problems encountered with certain species will also be dealt with; these include methods of dealing with the muscular tails of beavers, muskrats, and porcupines, and methods of dealing with fleshy feet of raccoons, beavers, and others.

Perform the skinning job on old newspaper. Replace these should they become soiled or greasy. Provide a large quantity of corn meal in a wide-mouth container so that it will always be handy. Finally, set up those tools that you anticipate using so that they will be within reach.

1. The incision. Weigh and measure the specimen, determine the sex, and record these data in your field notebook. Place the specimen on its back with its tail toward you, push the fur away from the midventral line of the body, grasp the belly skin between your fingers so as to pull it away from the body wall, and snip through the skin with a scissors. Now, slip the tip of the scissors underneath the skin and cut forward to the base of the sternum and then cut backward to the vent. If you have cut into the body cavity by mistake, plug the opening with cotton and add large quantities of corn meal to absorb any grease and blood. Following this, grasp the skin on one side of the incision and work it away from the body wall in the direction of the hind limb. On small rodents the skin is easily removed by simply pushing the body away from the skin with your fingers. In larger mammals, however, a great deal of connective tissue attaches the skin to the body and this must be broken with the butt end of a scalpel or very carefully cut. Two general dangers must be recognized here: (a) the skin may be torn or cut and (b) the skin may become stretched, making a "size 14" out of a "size 9" animal. Thus undue pulling should be avoided.

When the upper part of the leg is exposed, grasp the foot and push the leg forward, bending the knee as shown in Fig. 22–3B. Now work the skin down the lower part of the leg and sever the leg at the ankle. Pull the leg skin right side out. Repeat the process on the opposite leg.

At this point it should be noted that there is considerable difference in the technique for handling the lower limbs. Some workers leave the tibia and fibula of the lower leg in the finished skin by simply removing the tissue, adding arsenic, and wrapping the bones with cotton to replace the muscle tissue. Preferably, however, leg wires (discussed below) should be used in addition to these two bones or in place of them. In the former of these two methods, a wire is thrust into the foot all the way out to the end of the longest toe, and then is cut off even with the proximal end of the tibia bone. The bone is then poisoned with arsenic and the wire and bones are wrapped with cotton. The author prefers this last method, removing the tibia and fibula and replacing them with a leg wire and cotton. Nevertheless, the decision must be made as to which procedure will be used, for it involves removing the bones or leaving them intact.

2. The tail. After the knees have been worked out, cut the posterior end of the digestive tract away from the skin and plug the anal opening with cotton. Loosen the skin back around the body toward the base of the tail. In mice and other small mammals this is quickly accomplished and the tail is readily removed from the skin. Figure 22–3C shows the typical procedure for removing a mouse skin. Actually, the fingernails of the thumb and first finger are better substituted for forceps, but either method is satisfactory. Work the skin partly up the tail, not by pulling it, but by pushing it with your fingernails. As soon as there is room, place the blades of the forceps on either side of the tail (not the tail skin but the vertebral column) and by pulling the tail downward slip the tail out of the tail skin. The important thing here is that the forceps or your fingernails do not grip the skin and thus hold it tightly to the tail but rather are placed under the skin and thus push the skin up along the shank of the tail.

The tails of foxes, coyotes, and other large animals can be slipped in the same manner. The tail skin of any animal is most easily removed, how-

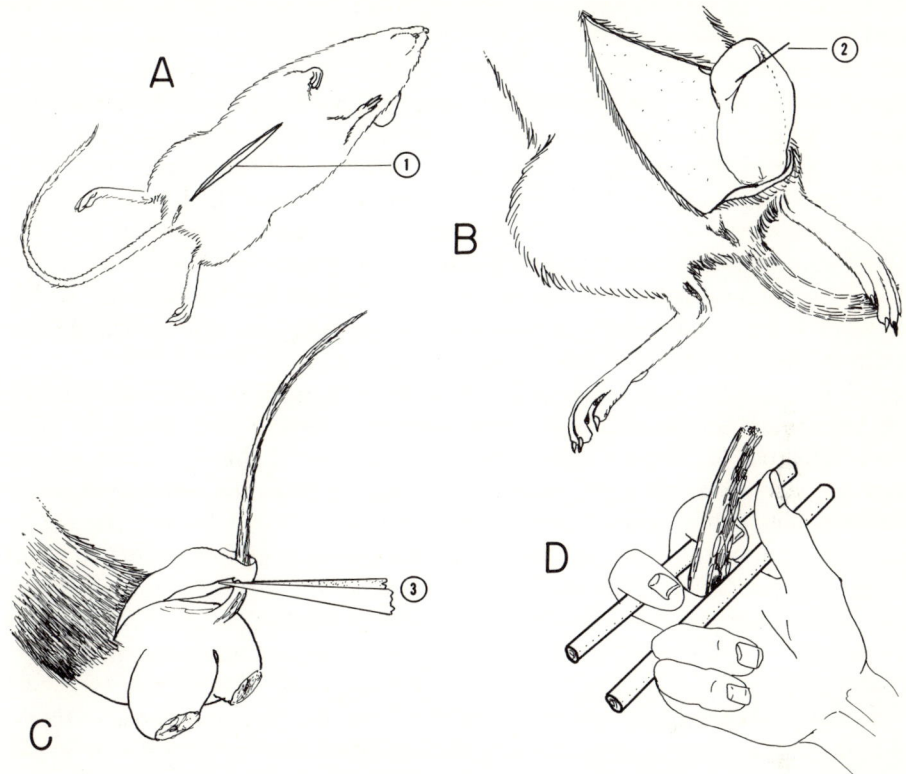

FIG. 22–3. Initial steps in skinning a small mammal. A. The primary incision (1). B. Severing the knee (2). C. Pulling the tail out of its skin with the aid of a pair of forceps (see text). D. A diagrammatic drawing showing how the tail of a large mammal is held for skinning (see text).

ever, soon after the death of the animal and may become very difficult with specimens that have been frozen for some time. To get a better grip on the tail of a large mammal cut two green sticks about ⅜ inch in diameter, place one on either side of the tail, and grip these with your fingers (Fig. 22–3D). Again, these are placed against the tail proper, not the tail skin, and are used to force the skin up the shank of the tail. If the skin refuses to slip, one must resort to cutting the tail skin open along the entire length of the ventral side to remove the vertebral column.

3. The body and neck. With the tail slipped from the tail skin, begin to turn the skin inside out and work it down over the length of the body. This can be done in a matter of seconds with small rodents, but care should be taken not to stretch the skin unduly. Sever the forelimbs at the elbows for ease in skinning, and then work the skin down each forelimb and sever the limb at the wrist. Continue to loosen the skin toward the head. About the head, neck, and shoulders connective tissues may attach the skin to the body. The skin of many rodents, rabbits, pikas, and other mammals will readily tear at these points and should be given additional care.

4. The head. Work the skin up over the base of the skull, loosening the connective tissue as you go. As the skin is worked toward the greatest diameter of the skull the funnel-like ear skin will be seen (Fig. 22–4A). Cut this with a fine-pointed scissors or scalpel where it enters the skull. If the ear skin is cut out beyond this point a large gaping hole will appear in the finished mount. After the ears have been freed, work the skin down over the skull until the eyes come into view. A great deal of care is needed in cutting the eyelids free from the eye itself. The author prefers to place the blade of a scalpel down on the eyeball immediately behind the skin and to free the skin by sawing the scalpel back and forth across the pupil while slowly moving the scalpel in an anterior direction (Fig. 22–4B). The eye is seldom ruptured by this procedure and the eyelid is almost never cut. The alternative procedure is to use a fine-pointed scissors and carefully separate the eyelid from the eye. Finally, work the skin forward, separating it along the margins of the mouth and across the cartilaginous end of the nose (Fig. 22–4C).

5. Final preparation. At this point all of the fat and muscle tissue clinging to the skin must be removed as described above. Next, the mouth is sewn closed while the skin is still turned inside out. The same procedure is used for small and large mammals. Pinch the skin of the lower jaw and pass a needle and thread through this, repeat this procedure with the skin of the upper jaw, and tie the mouth closed as shown in Fig. 22–4D. Now, if the skin is too dry moisten it slightly with water. Complete the preparation by dusting arsenic or Boraxo on the skin, roll it, and place it in a plastic bag to prevent drying during the preparation of the artificial body.

Problems with Tails. Those animals that use their tails for swimming or defense, such as the beaver, muskrat or porcupine, are provided with a great deal of connective tissue that attaches the tail skin to the vertebral column along its entire length. Treat these animals as follows: Split the tail skin along its ventral surface from base to tip and carefully separate the connective tissue by means of a scalpel. When the time arrives, make an artificial tail out of a piece of straight-grained wood. The artificial tail must be as long as the true tail plus one-third of the body length. Add additional

The Mammals 417

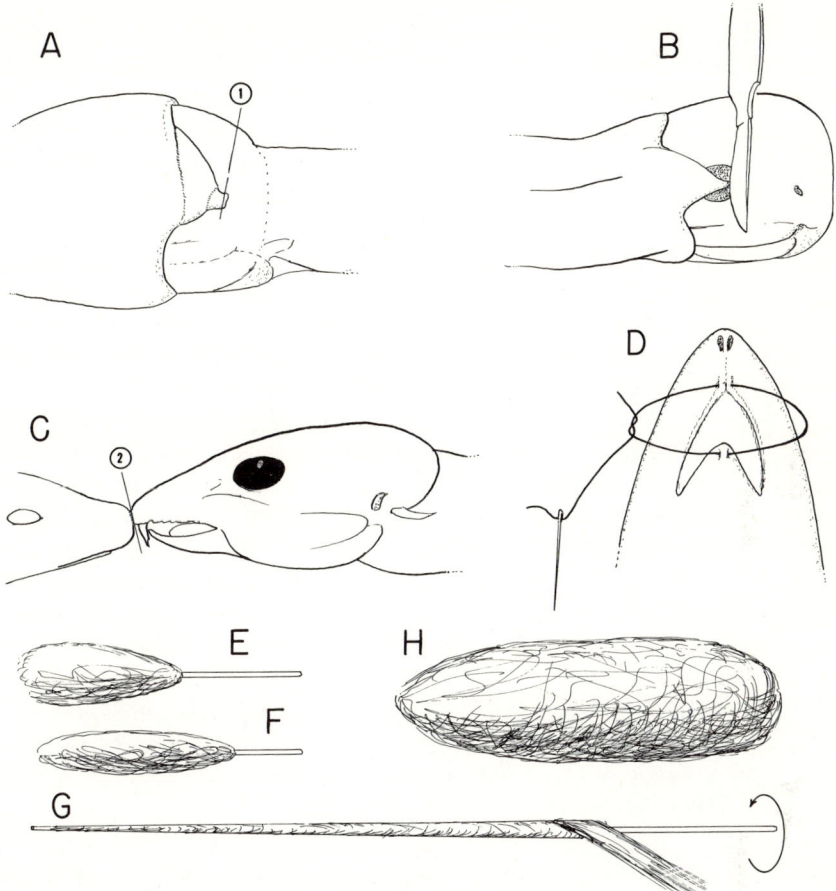

FIG. 22–4. Completing the skinning and preparing the stuffing. A. The ear technique. B. Working with the eye. C. Severing the nose. D. Stitching up the mouth. E–F. Leg wires with cotton applied. G. Twisting the tail wire and applying cotton. H. The artificial body made of cotton. See text for complete instructions. (1) Point of ear cut, (2) point of nose cut.

cotton if needed to fill out the space and sew the tail skin from tip to base with black thread.

Fleshy Feet. The hind feet of some mammals such as raccoons, bears, and others, should be treated in one of two ways. Work the skin down over the ankle and out toward the toes far enough to remove the metatarsals and the associated tissue. The second method, which is quite useful for any mammal that may be used in a round mount, is injecting the feet and toes with 10-percent formalin to preserve them during the drying process. This method is quite simple and very satisfactory.

Ears of Large Animals. This does not apply to the round mount technique, but will be discussed here with other problems. The ears of deer or other similarly large animals are generally skinned out so as to remove the cartilaginous tissue. This is not always essential for skins that are to be tanned, but it is the standard procedure, and is the sole procedure for taxidermy mounts.

Stuffing. The next task involves making an artificial body, leg wires, and a tail wire. Some practice may be needed in selecting the proper size of leg wire. This should be as large as possible and yet small enough to be worked part way up into the longest toe. The leg-wire length equals the length of the true leg, plus one third to one half of the body length. Cut all four leg wires to size. Now roll out a pound roll of long-fiber, absorbent cotton. Grasp a small quantity of this cotton from the surface of the roll and attempt to pull out as long a wisp as possible. The artificial leg is built up by wrapping a series of small thin wisps around the wire (Fig. 22–4, E and F, hind limb and forelimb, respectively). The bare part of the wire must be long enough to extend through the sole or palm of the foot out into the longest toe. Make the cotton portion only as large as the leg itself, or slightly smaller than the leg itself. When applying cotton to wires for tails or legs the job may be simplified by moistening the wire so that the cotton will stick to it when first applied.

Cut the tail wire as long as the tail (measure from the carcass) and one third of the body. Pull some fine cotton wisps and begin wrapping these on the tail (Fig. 22–4G), beginning near the tip of the wire and working down toward the base.

The artificial body for small mammals is made by separating a long thin layer of cotton from the main roll. The width of this cotton should be slightly longer than the animal's carcass. Begin rolling the cotton until a roll equal to or very slightly larger than the diameter of the carcass is obtained. To develop the tapered head and rump contours, do the following: Grasp the roll firmly near one end, grab the tip of the roll at that same end, and pull the tip off and away from the main roll. When the main body is smoothed up again a tapered "head" will be produced. Follow the same procedure for the rump end, grasping the butt end of the roll and breaking the roll off at the required length. The butt end should be blunt (Fig. 22–4H).

Make sure that the skin has been poisoned and that the mouth has been

The Mammals   419

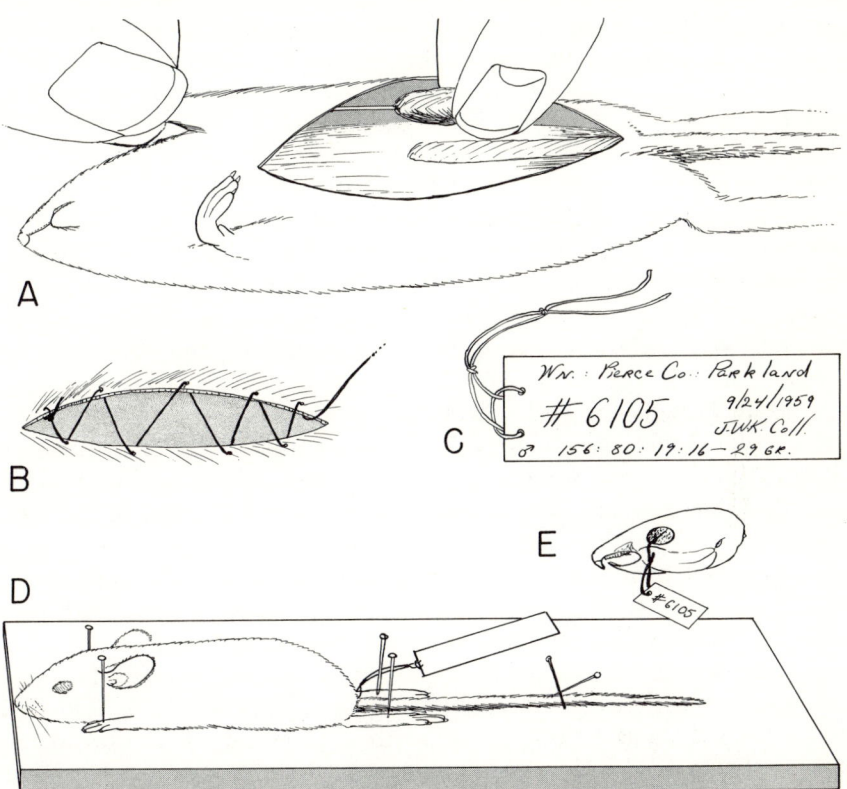

FIG. 22–5. The final mounting process. A. Inserting the foreleg wire. B. Stitching up the abdominal incision. C. A typical leg tag for a small mammal. Note the museum number and standard measurements given at the bottom. D. Pinning the specimen for drying. E. The partially clean skull with a label.

sewn up. Turn the skin right side out and insert the artificial body. In small mammals this is easily accomplished by grasping the artificial body along its entire length with a 12-inch pair of forceps. The plug is then introduced into the skin and the forceps are removed. Next, insert the tail wire and leg wires, as shown in Fig. 22–5A. With a little practice you will learn to place the leg wires in position with no difficulty, as follows: Hold the foot firmly between the fingers of one hand, stretch the leg away from the body so as to straighten the skin, and insert the leg wire. With just a little probing the wire will find a channel to pass down through the leg and then should be directed into the longest toe. Finally, sew up the ventral incision, as follows: Run a needle and thread through one side of the incision on the anterior end. Tie the thread to the skin rather than using a knot in the thread. Stitch back and forth from side to side as shown in Fig. 22–5B, but do not draw the incision closed until you have traversed the entire length. Beginning at the anterior end, close the incision and tie the thread off at the posterior end.

The last task is attaching a foot label to the right hind limb, smoothing the fur, reproportioning the body, and pinning out the specimen to dry. A skull label which bears the same number as that used on the leg tag must also be attached to the skull at this time. If the skin has been stretched it will be somewhat baggy and should be "worked together and pushed together" so as to fit the artificial body as closely as possible. If the skin is too badly stretched the artificial body should be made somewhat larger than that of the carcass. Usually the skin will shrink during the drying process and will fit the artificial body very closely. The specimen is placed in a position that will take up the least amount of storage room. The forelimbs extend up along the chin and are pinned close to the body (Fig. 22–5D). The tail is pinned straight back from the body in line with the median axis. The hind feet are turned over (in an unnatural position) so that the fur side is up and the pads are turned down. These are pinned close to the tail for the reason that they would be easily broken if spread far apart.

After the tail and legs have been pinned in position, smooth the fur with your finger or with a soft brush. The ears of rabbits should be folded back along the body and held together with a single stitch of thread. Some workers also fold back the ears of certain species of mice to protect the ears from damage. The author prefers to leave the ears sticking upward, since such specimens look more lifelike and are better as teaching aids for beginning classes. Finally, somewhere in this process, you should reach in through the eyelids and grasp the cotton of the artificial body and pull this out through the eyelid just far enough to keep the lids open. This is done to facilitate the study of colors around the eyelid. Place the specimen in a warm airy place to dry and protect it from direct sunlight or from flies which will deposit maggots on the skin.

Labeling. Figure 22–5C shows a typical foot label. This contains the geographic locality and the date of collection, the name of the collector, and the standard measurements, which are always given in the order of total length, tail length, foot length, and ear length. The sex and weight of the specimen are also noted. Finally, the acquisition number or specimen number is given. Mammal collectors usually keep an accession notebook along with their field notebook in which each specimen is given a serial number.

This number appears with the skin, skull, and skeleton. The same number is also entered in the museum records and thus represents the specimen in all of the records kept. The label is attached by means of a stout piece of cotton or linen button thread, tied as shown.

Posturing Bat Skins. While it is good to prepare an occasional bat skin with the wings outstretched, the standard procedure for round mounts is to pin the wings close to the body, as shown in Fig. 22–6A. This both conserves space and protects the wings from damage.

Skeleton or Skull. The skull should always be labeled, cleaned, and saved for future studies. Many important taxonomic characteristics are concerned with the bony structure of the skull and the dentition, thus making the skull a valuable specimen. On the other hand, the skeleton is generally discarded unless it is needed for particular studies. Remove as much flesh from the skull as possible and attach a skull label bearing the accession number (Fig. 22–5E). Next, remove the tissue around the opening of the foramen magnum to expose the posterior end of the brain. Insert a long hypodermic needle through the foramen magnum and pass this through to the anterior part of the brain. Now inject water into the brain cavity in order to force the brain back out through the foramen magnum. If the skull is cracked this procedure will usually fail and the brain must be picked out with a dissecting needle. If the eyes are left in while the worker is "blowing the brains" there will be little or no danger of rupturing the skull. Dry the skull without any arsenic and in a place where flies can not get at it. The skull is cleaned by dermestids at some later date. See Chapter 23 for details.

Flat Pelts and Case Pelts

Flat Pelts. Large mammals are skinned out, salted, and then tanned. Place the specimen on its back, make an incision down the midventral side of the body from the chin to the vent. Finally, make a cut across the body between each set of legs and extend each cut on up the median side of the leg, shifting over to the posterior side of the leg (Fig. 22–6B). Pull the skin free from the body, using the skinning knife where muscle or connective tissue tends to adhere to the skin. Skin out to the last digits in the feet, removing all bone and tissue, unless the intact paw is to be saved. Skin out the head, following the directions given under "Round Mount," above. When skinning the head cut the ears off initially and remove the cartilage after the pelt is taken from the carcass.

In removal of the entire pelt from a horned specimen the procedure will vary, depending on whether the horns or antlers are to be kept intact with

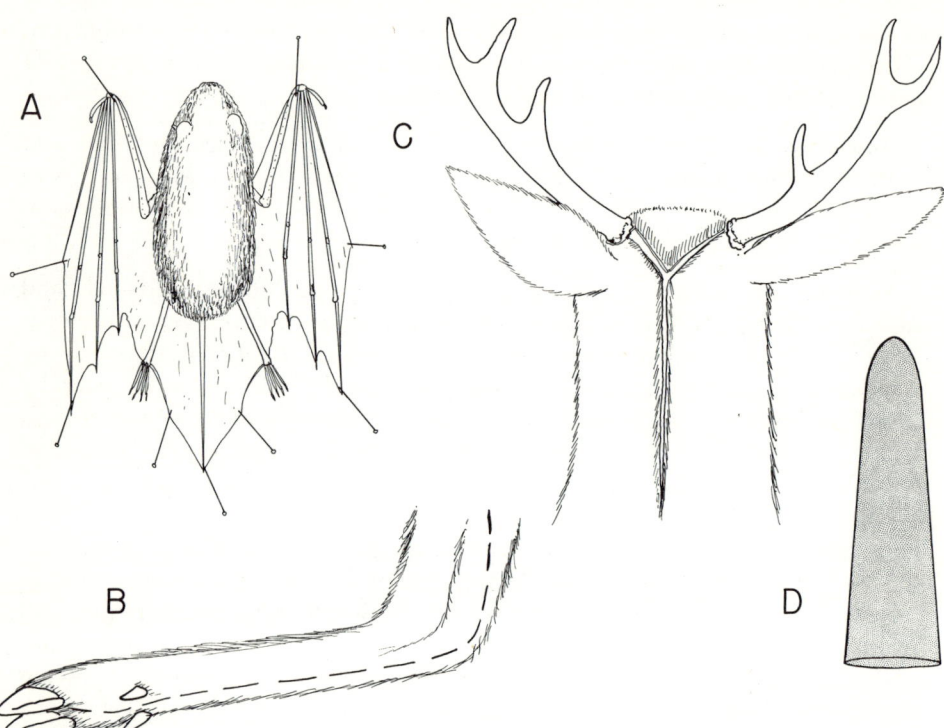

FIG. 22–6. Some techniques for special mammals. A. Pinning out a bat. B–C. Techniques for skinning large mammals. D. A stretching board for a mammal pelt. See the text for complete instructions.

the skull or whether they can be removed, and whether the entire pelt or just the head and neck pelt is to be saved. In the cases where only the head and neck skin are required or where the entire pelt is desired but where the antlers or horns must remain attached to the skull, the following method is used: skin out the specimen in the usual manner, but extend the ventral incision only as far as the base of the neck. Next, make a V-shaped incision extending from the base of the antlers or horns down to a point between the ears, and continue with a median-dorsal incision part way down the back of the neck (Fig. 22–6C). This incision permits the skin to be taken off the skull and horns after it has been loosened from the skull by the usual methods. If the horns may be removed from the skull, the specimen is skinned out in the usual manner, the horns are sawed off at the appropriate time, the skin is cut free from the base of the horns, and then the head is skinned down toward the snout in the usual fashion.

Remove all flesh, grease, and blood from the pelt, spread the pelt out in a shaded but airy place, and apply large quantities of salt to the undersurface as described under "Salting Field Skins."

Case Pelts. In some instances it is more convenient to make a case pelt of mammals such as the mustelids, rabbits, or other species. Such skins are generally prepared as round mounts or tanned skins at some later time. To skin the specimen simply make an incision between the two hind legs and extend this out to the hind feet as described above. Work the tail out of the tail skin as you would for a round mount. Next, work the skin down over the length of the body by turning it inside out, removing muscle and connective tissue as you go. This job is made a lot easier if the carcass is suspended by the hind limbs. Cut off the forelimbs at the elbows and finish skinning out the forelimbs once the pelt has been removed from the body. Finally, work the pelt down over the skull, following the same procedure as for the round mount. Leave the pelt inside out and put it over a stretcher board or stretcher wire (Fig. 22–6D). Remove all flesh, grease, and blood and place the pelt in a shaded, airy, and fly-free place to dry. Such pelts are generally not salted, although this procedure is optional. When the pelt is dry it should be removed from the stretcher to facilitate the free circulation of air inside. This is especially necessary to ensure the complete drying of the skin of the forelegs.

Case-Skin Mount. Anderson (1948) introduced a case-skin technique (in the first edition of his work, 1932) which is very useful when working with beginning students or when dealing with large numbers of specimens under field conditions. This technique may be used for mammals up to the size of small squirrels. In brief, the technique consists of skinning out the specimen as for a case pelt and attaching the skin to a permanent cardboard dryer, as shown in Fig. 22–7, A and B. The author uses a heavy grade of Bristol drawing board and carries three stock sizes in the field: $1\frac{1}{8}$ by 12 inches, 2 by 18 inches, and 3 by 18 inches. These stock cardboards can be cut down and cut off to fit most small mammals as needed.

This technique can be modified in many ways. The author uses the following method, which is patterned fairly closely after Anderson's original technique: Make an incision across the body which cuts through the anal opening and extends part way out on the inner side of the hind limbs. Skin out the limbs, severing both feet at the ankles. Skin out the tail in the manner described for the round mount. Work the skin down over the back, severing the forelimbs, first at the elbows, and then at the wrists, and

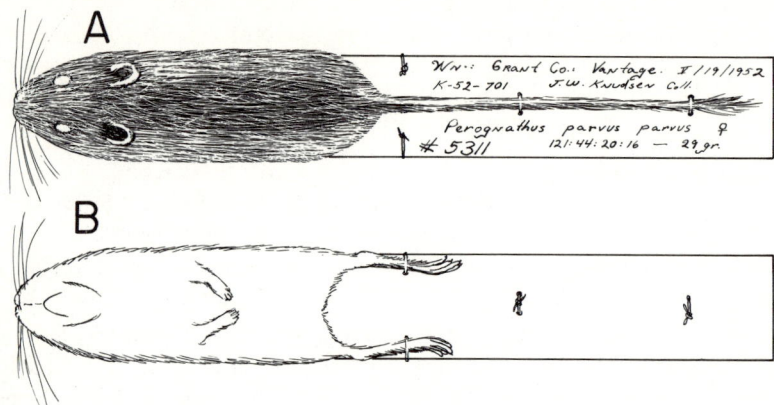

FIG. 22–7. The case-skin mount. A–B. Two sides of a rodent skin mounted on a cardboard stretcher. See text.

continue to remove the skin over the skull in the manner for the round mount. Next, sew up the mouth by the standard method, poison the skin with arsenic or Boraxo, and insert a tail wire which will extend halfway up into the body skin. Pull the skin over a cardboard stretcher which has been tapered to fit the head of the pelt. Attach the hind feet with a single loop of thread through the outside of the cardboard, and attach the tail with one or two loops of thread drawn through the cardboard stretcher as shown in Fig. 22–7. Some workers prefer leg wires, although these are not necessary if the skin is given decent treatment in the museum. The forelimbs and ears should be folded back along the pelt. If necessary, slip the pelt into an envelope to hold the limbs and ears in position while drying. Record the standard field data, measurements, collector's name and date, and other data on the white surface of the Bristol board.

Liquid Preservation

Any mammal specimens required for muscle studies should be carefully preserved in formalin. Since so many of the taxonomic characteristics of bats are lost in the round mount, it has also become standard procedure to preserve most bat specimens in liquid. Specimens should be measured, weighed, examined for ectoparasites, and logged in the field or accession notebook. Next, inject the body cavity and larger areas of muscle with 10-percent formalin. Finally, wash the specimen with water and a little detergent to remove the grease from the body and fur. Washing is essential to permit rapid penetration of the body by the formalin. Be sure to use labels that will not disintegrate in liquid. Store specimens in airtight museum jars (filled with 10-percent formalin) in a dark place to prevent color breakdown in the more highly pigmented species.

Wet and Dry Mounts

Although this is not a standard procedure, small mammals can be labeled, preserved in 10-percent formalin, and finally dried. Specimens treated in this manner should be injected in the body cavity, wrapped in cotton to keep the fur in place, and submerged for 24 hours in the preservative. When thoroughly dry such specimens will keep indefinitely in the standard museum drawer, but the tails and legs should be protected against breakage.

OTHER FIELD TECHNIQUES

Animal Tracks

Plaster casts of animal tracks are quite useful for natural history classes; thus, the technique for making casts of animal tracks will be given here. Animal tracks are best located on muddy banks near water sources. You will need the following things to make a plaster cast in the field: a series of tin cans of different sizes, a canteen of water, and a few pounds of plaster of Paris. Select a clear footprint that is free of extraneous debris. Place an appropriate-sized can over the print and build up a mud wall, as shown in Fig. 22–8B, remove the can. Mix a sufficient quantity of plaster to fill the wall and pour, as shown in Fig. 22–8C. When the plaster has set, carefully remove the cast, wash it in the water nearby, trim the excess plaster away from the cast, and set the cast aside to dry. Casts can be made with permanent metal rings, $5/8$ inch wide. Take $5/8$-inch wide strips of galvanized metal with you into the field. Cut off a strip so that it will form a suitable diameter, and punch two holes by means of a heavy nail through the ring where the two ends overlap (Fig. 22–8D). Insert a U-shaped piece of No. 14 galvanized wire through the holes to make a hanger, and twist the wire as shown. Place the metal ring over the animal track, support it with some mud on the outside, and pour in the plaster as above.

All of the prints obtained by the above method are negative prints and resemble the foot of the animal more than the footprint. Positive footprints

426 Biological Techniques

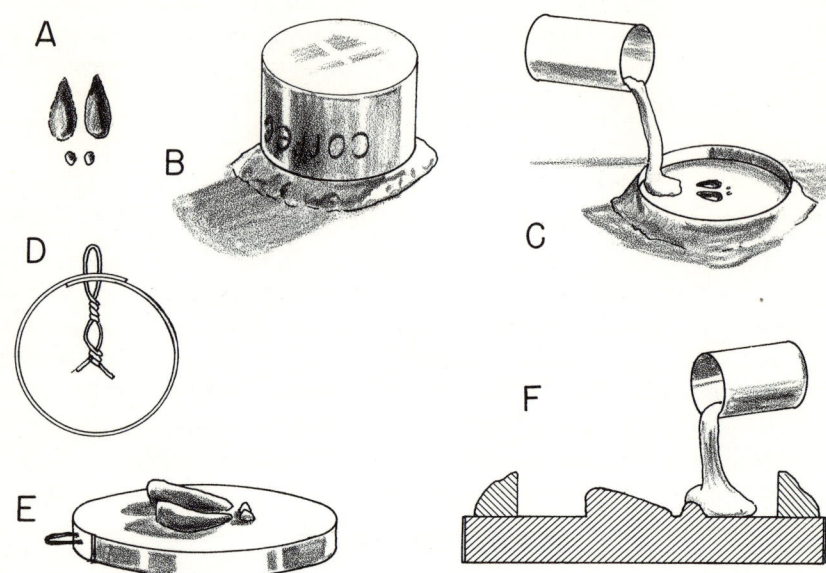

FIG. 22–8. Making plaster casts of animal tracks. A. Select clean, well-defined animal tracks. B. Forming a mud dam. C. Pouring the plaster of Paris. D. A preformed metal ring. E. A mammal track cast within a metal ring. F. Making a reverse casting of a mammal track. See text for full instructions.

can be made from the initial negative prints by coating the original casting with tincture of green soap, building a dam to receive the plaster, and pouring this full of plaster of Paris (Fig. 22–8F). The tincture of green soap serves as a parting compound to permit the positive and negative casts to be separated.

Scat

Scat, animal droppings, will yield considerable information about the feeding habits of certain mammals. This is especially true of predaceous mammals such as coyotes and foxes. Such droppings contain hair, bones, plant seeds, insect skeletons, and many other structures which can be identified. Dried scat is collected in paper bags, along with the field data (geographic location, date, general habitat, and the name of the species if this is known). Scat can be stored indefinitely if kept dry, but should be fumigated periodically with crystals of paradichlorobenzene to discourage dermestid beetles and clothes moths.

SKELETAL TECHNIQUES

The skulls and skeletons of mammals are best if fleshed, dried, and then cleaned by dermestid beetles. Other less convenient techniques such as boiling or rotting the flesh away from the bones may be used. Small mammals (either adult or embryonic) should be cleared, stained, and stored in glycerin to prevent bone damage that would occur from the normal cleaning procedure. Because the techniques for all vertebrates are so similar, these will be discussed in Chapter 23.

FUMIGATION TECHNIQUES

Newly arrived field skins or skins with known insect infestations should be fumigated before they are placed in the museum cabinets. Carbon bisulfide is a satisfactory fumigant, although it is very poisonous and should be used only by adults associated with academic institutions. Figure 22–9, A and B, shows the external and internal aspects of a fumigating box. This box measures 42 by 18 by 18 inches and is equipped with three sliding shelves. The door hinges on one end and is provided with a foam rubber seal and hardware (as described in Chapter 13 under "Insect Cases"). Note that two ordinary gas cocks are fused to the box, one to the top and the

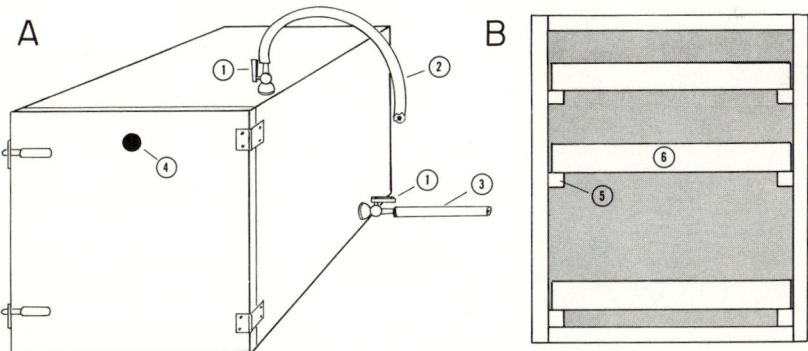

FIG. 22–9. A fumigating box for bird, mammal, or other animal specimens. A. An outside view of the box, showing the placement of hardware. B. A cross-section through the inside of the box showing the trays and tray runners. (1) Standard gas cock, (2) intake air hose (this should be provided with a water trap), (3) exhaust air hose, (4) a cork stopper, if it is desirable to introduce the fumigant through this small aperture, (5) tray runner, (6) tray.

other to the side. The entire inside of the box and the trays are painted with boiling hot paraffin to seal them and prevent them from absorbing the fumigant. Specimens are placed in the various trays, a dish of carbon bisulfide is placed on the top tray, the door is sealed, and the gas cocks are closed. After 24 hours both gas cocks are opened and compressed air is forced through the fumigating box. This takes the carbon bisulfide or other fumigant to the out-of-doors or to a chemistry hood. After the case is thoroughly aired it may be opened and the specimens transferred to the standard museum cabinets. Be sure to check your compressed air system; if necessary, construct a trap for condensed water that may have collected in the pipeline.

PERMANENT STORAGE CABINETS

The requirements for bird or mammal storage are as follows: The storage cabinet should be in a dry room with moderate heat to discourage mold. The cabinet itself must be lightproof and must be provided with a seal that will (1) prevent the entry of insect pests and (2) retard the loss of the paradichlorobenzene fumigant. Specimens are stored in paper-lined trays and arranged taxonomically. The mammal or bird case may be any size that is convenient, so long as the above requirements are met.

Figure 22–10 shows a single storage cabinet that is very suitable and can be made inexpensively. This cabinet measures 48 inches high, 38 inches wide, and 24 inches deep on the outside, not including the doors. The sides, top, bottom, and doors are made of $3/4$-inch plywood which is "sound on both sides." The back is made of $1/4$-inch plywood, sound on one side. This unit contains up to 19 trays, depending on the size of the specimens, which measure $36 1/4$ inches long by $23 1/2$ inches wide and $1 3/4$ inches deep on the outside dimension. The drawer runners (36 in all) are made of a fine-grained, medium-hard wood and measure $1/2$ inch by $1/2$ inch by $23 1/2$ inches.

To build this cabinet cut out two side boards 24 by 48 inches. Cut one top and one bottom board $36 1/2$ by 24 inches. Now, with a dado blade cut a groove $1/4$ inch deep and $3/8$ inch wide on the inside, back edge of the sides, top, and bottom. This recess is designed to receive the $1/4$-inch plywood back. Finally, cut a square 3 by 3 inches out of the lower front corner of the two side boards. This cutout permits the toe board to be recessed from the door surface. Finally, cut the $1/4$-inch plywood back $37 1/4$ by $44 1/4$ inches (or slightly under). Use finish nails and glue for all joints.

Assemble the main box by nailing the top and bottom boards between

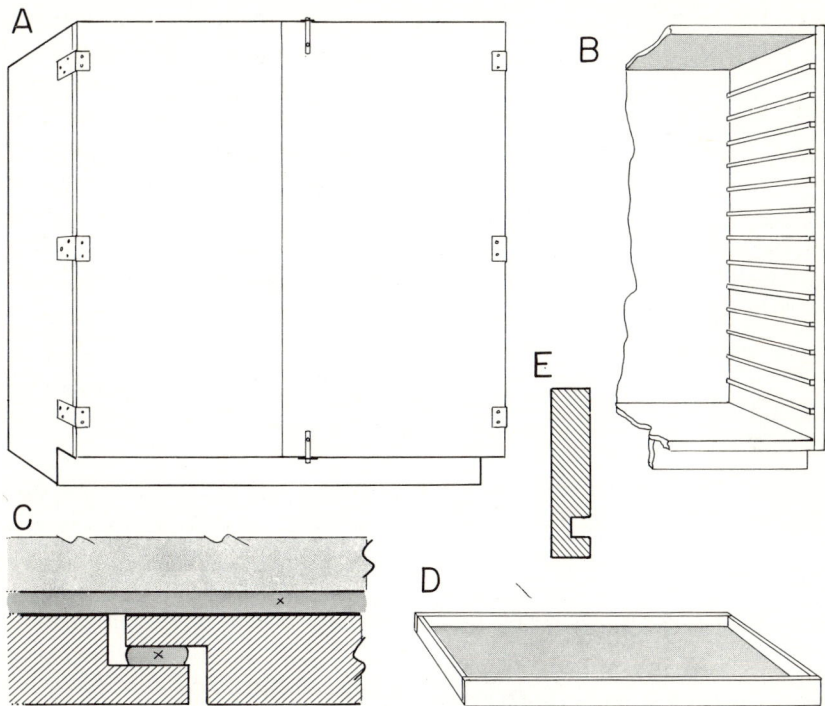

FIG. 22–10. A convenient mammal or bird storage cabinet. A. Front view, showing door hardware. B. A partial interior view showing the position of shelf runners. C. A diagrammatic view, showing the placement of rubber or plastic foam (X). D. The finished tray. E. Diagram of the tray side, showing the dado cut.

the two side boards, as shown in Fig. 22–10B. The undersurface of the bottom board, as you will note, is placed fully 3 inches above the floor level. Next, nail a toe board (3 by 38 inches) in place. Finally, insert the plywood back and use this to square the cabinet. When the glue has dried, turn the cabinet on its side and install the runners. Consult Fig. 13–17 (p. 258) for details of nailing the runner directly to the surface of the side board. The bottom drawer goes directly against the bottom of the cabinet. Each runner is separated by a distance of $1\frac{7}{8}$ inches. Therefore, cut a spacer $1\frac{7}{8}$ inches wide and use this for installing the runners. When the cabinet is on its side, put the spacer against the bottom board. Glue and nail the first runner in place. Then repeat the process, putting the spacer against the top of the first runner and working your way up the side of the cabinet until all of the 18 runners have been installed. Repeat this on the other side.

Cut the two doors 45 by 19¼ inches. Note in Fig. 21–10C that the two doors overlap ½ inch and have an additional ⅛-inch clearance on either side. Set the dado blade to cut $7/16$ inch deep and ⅝ inch wide. When the two doors are dadoed as shown, there will be room to compress a $3/16$-inch-thick piece of foam plastic or rubber between the doors.

Finish the outside of the cabinet with varnish. Install ⅛-inch-thick foam plastic around the entire outside lip of the cabinet to receive the doors, and a $3/16$- by ½-inch piece of plastic on one of the two doors where they overlap. Install the doors, using hardware as shown in Fig. 13–17.

The trays are made of fine-grained softwood measuring ½ inch wide and 1¾ inches high. With the dado blade cut a groove ¼ inch wide and ¼ inch deep, as shown in the cross-section (Fig. 22–10E). Cut the sides and ends of the trays to length. The tray bottoms are made of ¼-inch plywood, sound both sides and measuring 35¾ by 23 inches. Assemble the trays as follows: Glue and nail the two ends to one of the side pieces. Put glue in the dado cuts and slide in the ¼-inch bottom. Glue and add the remaining side to the tray, nail, and let dry. Finish with a coat of varnish or shellac.

REFERENCES

Anderson, R. M., 1948, Methods of Collecting and Preserving Vertebrate Animals, *Biol. Bull. No. 39*, pp. VII+164, Nat. Mus. Canada, Ottawa.

Anthony, H. E., 1942, *Mammals of America*, Garden City Books, N.Y.

Booth, E. S., 1950, *How to Know the Mammals*, William C. Brown, Dubuque, Iowa.

Bourlière, F., 1955, *Mammals of the World*, Knopf, N.Y.

Burt, W., and R. Grossenheider, 1952, *A Field Guide to the Mammals*, Houghton Mifflin, Boston.

Cahalane, V. H., 1947, *Mammals of North America*, Macmillan, N.Y.

Cameron, Austin W., 1956, *A Guide to Eastern Canadian Mammals*, Natl. Mus. of Canada, Ottawa.

Dalquest, W. W., 1954, Netting Bats in Tropical Mexico, *Trans. Kan. Acad. Sci.*, 57(1):1–10.

Glass, Bryan, 1951, *A Key to the Skulls of North American Mammals*, Burgess, Minneapolis.

Hall, E. R., and K. Kelson, 1959, *The Mammals of North America*, Ronald, N.Y.

Ingles, L. G., 1954, *Mammals of California and Coastal Waters*, Stanford Univ. Press, Stanford, Calif.

Murie, Olaus, 1954, *A Field Guide to Animal Tracks*, Houghton Mifflin, Boston.

Palmer, R., 1954, *The Mammal Guide*, Doubleday, Garden City, N.Y.

Sanderson, I., 1955, *Living Mammals of the World*, Hanover House (Doubleday), Garden City, N.Y.

Young, J. Z., 1957, *The Life of Mammals*, Oxford Univ. Press, London.

Zim, H., and D. Hoffmeister, *Mammals*, Simon & Schuster, N.Y.

CHAPTER 23

Vertebrate Skeletal Techniques

Osteological material has long served as the basis for studies of vertebrate relationships, classification, and evolution. Obviously, osteological material presents one of the few means of comparing present-day animals with fossil specimens. The study of bone structures is becoming increasingly important, especially in the field of taxonomy on the generic, specific, and subspecific levels.

Skeletal specimens are prepared in a number of ways, depending on the intended use. Small vertebrates, including embryological specimens, can be studied conveniently by clearing and staining, or if rare, by means of X-ray techniques. Larger vertebrate skeletons are usually cleaned and dried, and may be stored as disarticulated or partially articulated skeletons for museum and taxonomic purposes or as fully articulated and mounted skeletons for teaching and demonstration. Cartilaginous skeletons, such as those of sharks and rays, require different techniques of cleaning and preservation for classroom use.

CLEARING AND STAINING TECHNIQUES

Limitations and Uses

The clearing and staining technique, useful for small vertebrates, involves staining the osteological material and clearing all of the body tissues to make study possible (Fig. 23–1). As Davis and Gore (1947) point out, this method makes use of the Schultze (1897) technique of clearing tissue with KOH. They note that this technique is often only a supplement to other techniques. However, it has the advantage of preparing the entire skeleton with no loss or damage of small bones or distortion as a result of drying and shrinking of cartilaginous elements. Groups of bones, such as the girdles, are always intact and left in the proper spatial relationships to other bony ele-

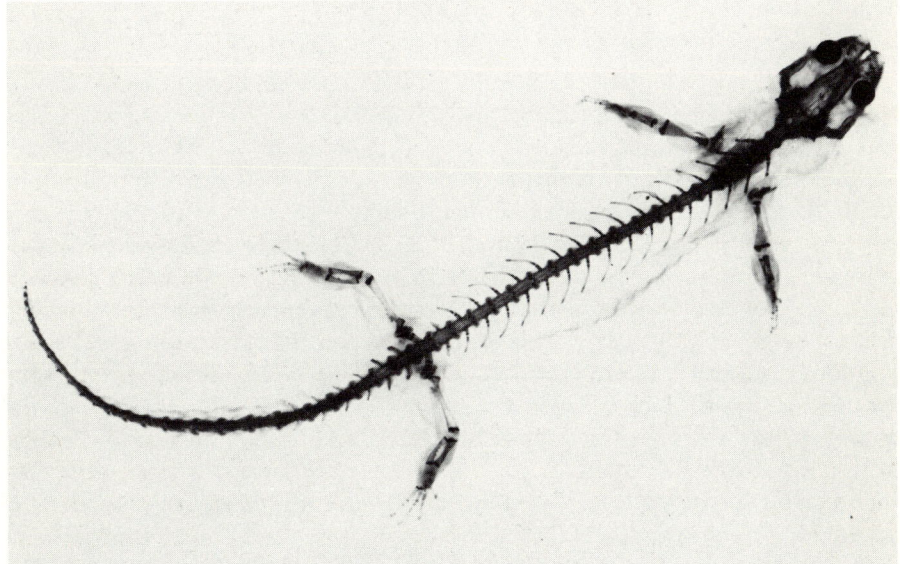

FIG. 23–1. A photograph of a specimen prepared by the clearing and staining technique, courtesy of Dr. David Wake. Note that the entire animal is present surrounding the skeletal structures, with the exception of the visceral mass. Cartilage, bone, and other structures stand out clearly.

ments. Therefore, whenever small vertebrates can be spared for "making skeletons," this technique should be one of the first used.

The author wishes to acknowledge that many of the discussions to be presented here have been obtained from members of the Vertebrate Laboratory at the University of Southern California. One point that everyone seems to agree on is that results of the clearing and staining method may at first be highly variable, and that the causes of such variation are somewhat difficult to understand. Eventually, each individual develops a system that "works" for him. The origin and age of the material may be significant in causing variable results. For example, fresh but unpreserved material, fresh material hardened in alcohol, fresh material hardened in formalin, specimens long preserved in alcohol, and specimens long preserved in formalin may all be successfully treated by one method or another, but probably not by the same method.

THE GENERAL METHOD

This technique follows Davis and Gore (1947) quite closely, and follows Hollister (1934) in part. The method is as follows: (1) If possible, use material originally fixed in 10-percent formalin for about 1 week. However, specimens fixed in 95-percent alcohol for the same amount of time may be used. Older alcoholic specimens are less satisfactory. Remove the viscera and skin from larger specimens, but never remove the skin from small delicate individuals. (2) Transfer specimens to distilled water for 24 hours, in order to soak out the formalin or alcohol. (3) Place specimens in a 2-percent solution of KOH for 12 to 24 hours. (4) Add a 3-percent hydrogen peroxide solution directly to the KOH to remove the pigment. If the specimens are not heavily pigmented, or if the pigment has been removed, go on to the next step. (5) Now, place the specimens in a fresh 2-percent KOH solution, after washing in distilled water for 12 to 24 hours; add 1 cubic centimeter of the stock alizarin red stain to the KOH, immediately. If the hydrogen peroxide treatment was omitted above, the stain may be added to the initial KOH solution.

(6) After 24 to 48 hours, or longer (observe the vertebrae to see how the stain is taking), place the specimens in a fresh solution of 2-percent KOH containing a few drops of glycerin. This will destain and clear the specimens simultaneously. The specimens should be left in this solution until most of the stain is removed and the flesh becomes quite clear. This may take a few days for some individuals and up to 1½ or 2 weeks for others.

If the specimens later prove to be unsatisfactorily cleared, they may be returned to the solution for additional clearing.

(7) Finally, the specimens should be transferred to pure glycerin by moving them through three solutions for 24 hours each, as follows: solution one, 20 cubic centimeters of U.S.P. white glycerin, 3 cubic centimeters of 2-percent KOH and 77 cubic centimeters of water; solution two, 50 cubic centimeters of glycerin, 3 cubic centimeters of 2-percent KOH, and 47 cubic centimeters of water; solution three, 75 cubic centimeters of glycerin and 25 cubic centimeters of water. The specimens are transferred from solution three into pure U.S.P. white glycerin, in which they are stored.

You may have to use the trial-and-error method to start with. One of the trouble points is leaving the specimens in the peroxide solution too long (usually they should be left no more than a couple of hours). Another point which may give trouble is that of not letting the specimens clear and destain long enough.

The stock solution of stain is made according to Hollister (1934) as follows: Make a saturated solution of alizarin in 5 cubic centimeters of glacial acetic acid, 10 cubic centimeters of white glycerin, and 60 cubic centimeters of a 1-percent solution of chloral hydrate.

Some Modifications

Fishes. The above technique works quite well for fishes, with some modifications. Hollister (1934) preferred fresh specimens that had not been preserved or hardened in either formalin or alcohol. Davis and Gore (working with reptiles and amphibians, 1947) found Hollister's method unsatisfactory. The formalin or alcohol method given above is satisfactory here. Fishes should be eviscerated if they are very small. Hollister and others recommend pricking through the skin and scales with a needle to permit the rapid entry of the various solutions and stain, and to prevent the build-up of bubbles under the skin during destaining. Hollister used KOH solutions ranging from 1 to 4 percent. She also found that it was essential to add additional stain for fishes with dense skeletons, since this is absorbed by the calcium salts. If the fish specimens are stout enough, the scales may be scraped off with a knife; if they are very delicate, scaling should wait until the clearing process has been completed. Finally, if difficulty develops in transferring the specimens into pure glycerin, precede the above-mentioned three solutions with the following: solution A, 5 cubic centimeters of glycerin, 90 cubic centimeters of 2-percent KOH and 5 cubic centimeters of water; solution B, 10 cubic centimeters of glycerin, 13 cubic centimeters of 2-percent KOH and 77 cubic centimeters of water.

Amphibians and Reptiles. The above technique works well for amphibians and reptiles. However, one should consult the paper by Davis and Gore (1947) which deals with these animals, for it gives a number of methods worth trying. For example, specimens were hardened in 95-percent alcohol, preferably. Davis and Gore used 4-percent KOH for preserved specimens and exposed these to sunlight for two days prior to bleaching, the latter being done in full-strength hydrogen peroxide. Other workers suggest pin-pricking of reptiles, as with fishes, in preference to skinning, unless ossified material is contained in the skin. Of course, relatively long snakes can readily be skinned and this procedure is recommended.

Birds and Mammals. Featherless birds and embryonic or newborn mammals may be treated as described above. Otherwise, bird feathers should be removed and mammals should be skinned as described in Chapter 22, with the exception that the feet are left intact. Remove the viscera, eyes, tongue, and other structures that may be too thick to clear well.

X-RAY TECHNIQUE

Uses and General Method

For some purposes X-ray equipment proves very useful for osteological studies. For example, vertebral counts may be made accurately and quickly without resorting to clearing and staining or other techniques. Frequently, specimens may be on loan or may be too rare to make into skeletons, as in the case of type material, and here again X-ray methods are most useful. Stereoscopic X-ray techniques give such a clear, three-dimensional view of the skeleton that they are very satisfactory for study or illustration. The use of X ray in the study of anatomy and systematic zoology is not new; consult Miller (1957) for an excellent and detailed discussion of this method.

The author is not qualified to discuss X-ray equipment as such. Those who have access to such equipment will probably also have access to a trained technician who can provide the needed information. There are two general types of X-ray units: the "hard-head," or high-voltage, unit and the "soft-head," or low-voltage, unit. The hard-head unit will get everything in focus, but the picture (for small vertebrates) is not sharp and tends to be rather grainy. Apparently the use of fine-grain film does not improve the picture, since the exposure time is too short (about 15 seconds) to give a clear image. Conversely, the soft-head machine may use exposure times of 10 to 15 minutes and will give sharp and clear detail. However, focusing is very critical, for it is done by controlling the kilovolt peak and the milliamper. Thus, considerable experimentation is required. Frequently, the vertebral column will be clearly defined while the feet will be burned out, or vice versa. The age of the specimens, the length of time they have been preserved, and even the preservative or the water used in making the preservative may have some influence on the success of soft-head X ray.

Hard-head X ray, therefore, is more useful for locating various structures (or even to determine if various bony structures exist), for vertebral counts, and the like. The hard-head X ray can be used to determine whether the time and trouble necessary for soft-head X ray will be worthwhile. Hard-head and soft-head X rays are taken with the head directed straight down on the specimen, which is placed above the X-ray plate.

Figure 23–2 shows diagrammatically the technique used by Dr. David Wake for making stereoscopic X rays at the U.S.C. Vertebrate Laboratory. In Fig. 23–2A the X-ray head is shown directly over the salamander specimen. The X-ray plate is arranged in a slot so that it can be moved from left to right. Note that half of this plate is covered with a lead shield. The first

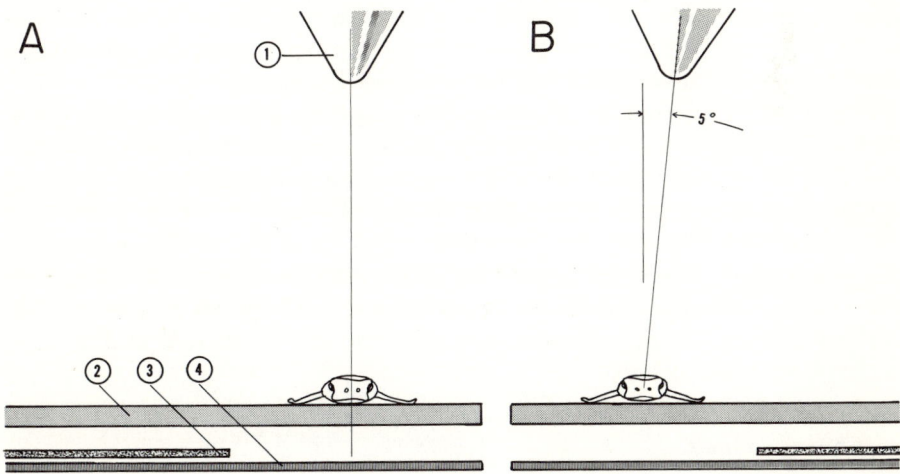

FIG. 23–2. The X-ray technique. A. The first exposure. B. The second exposure. See text. (1) X-ray head, (2) the X-ray table, (3) the X-ray shield, (4) the X-ray plate.

X ray is taken when the equipment is set up as shown. The second exposure (Fig. 23–2B) differs as follows: The X-ray plate, in its slot, has been moved so that the unexposed area is underneath the specimen. Second, the lead shield has been shifted to cover the previously exposed portion of the plate. Third, the X-ray head is moved over a short distance to assume a 5° angle from the original vertical position, as shown. After the second exposure the X-ray plate is processed. The X-ray plate is examined over a reading stand (an opaque glass plate with lighting from beneath) with an Army type, CF8 stereoscope such as that manufactured by the Abramas Instrument Corporation, Lansing, Michigan. This stereoscope is inexpensive and is designed for use by cartographers for reading twin aerial photos.

As mentioned above, stereographic X rays are amazingly clear and truly three-dimensional. When viewed from one side, the X-ray plate gives a convincing, three-dimensional illusion of the dorsal view of the skeleton. When the X-ray plate is turned over and read from the opposite side, the ventral view is obtained.

STANDARD SKELETAL TECHNIQUES

Before selecting some method at random, the reader should be advised that the techniques given below for fleshing skeletons do not all produce the same results. Although the objective of these methods is to remove the flesh from the skeleton the degree of efficiency or damage to the skeleton varies. For example, all could be used for preparing disarticulated skeletons, but only one is truly satisfactory for articulated skeletons.

If you are interested in mounting the skeleton on a base or in a glass-topped display box the technique is as follows: (1) Flesh out the specimen and disarticulate it as described below, (2) degrease, (3) bleach, and (4) mount.

Specimens of fish, amphibians, reptiles, birds, and mammals intended for study skeletons should be treated by dermestids if they are too large for the clear and stain technique. Only the largest fish, reptiles and most birds and mammals will require fleshing out. The rest may be treated as described under dermestids below.

Removing Flesh

Fleshing Out. When dealing with large fish, cut through the skin on the side of the body cavity and remove the viscera. Cut away skin and muscle back along the flanks of the body only in areas where no skeletal structures are involved. Dry and clean, as described under "Dermestid Technique" below. In working with the larger reptiles, birds, and mammals (especially mammals) first identify the specimen and then skin it and remove the viscera. Lucas (1899) recommends carefully detaching the legs from the body, taking special care not to lose the collarbones if these are present. They are small and not articulated in the cat family and very small in the weasel group, but totally lacking in deer, antelope, bears, and seals. Remove the muscle tissue at the point of origin and insertion, taking care not to destroy the ligamentous connections between the bones. Be careful also not to lose the patellae of knee bones of the hind limbs. Next, locate the hyoid apparatus and either remove this immediately or be conscious of it when disarticulating the skull from the spinal column. Disjoint the skull by carefully cutting the ligamentous connections between it and the atlas. Clean out the brain with a brain spoon or a hypodermic, as directed in Chapter 22. Remove the eyes and large bundles of cheek muscles, but take care not to cut soft bony projections from the skull. On larger specimens remove the organs from the rib cage and the flesh from between the ribs, taking care not to cut the cartilage joining the ribs to the breast bone. On very large mammals which must be shipped before cleaning can begin, it is well to separate the vertebral column immediately in front of and behind the rib cage. Unless you are ready to proceed immediately with a cleaning process other than that of the dermestid technique, dry the specimens thoroughly in some place where flies cannot deposit eggs on the flesh. When they are dry, wrap and ship the specimens or proceed with any of the cleaning techniques given below.

Dermestid Technique. Beetles of the family Dermestidae habitually feed on dead, preferably dry, animal matter, including muscle, hair, feathers, and the like. For this reason they can be extremely destructive if they get into insect collections or bird and mammal collections, but may also be used to clean skeletal material. Dermestids can be obtained from biological supply houses or, more simply, by stopping at the first flat, dry carcass of some bird or mammal that has been killed along the highway. Moist or juicy road kills will attract beetles other than the dermestids, whereas the dry, mummy-like specimens harbor dermestids. Tear open the road kill to locate the specimens. Dermestids may be maintained in a large 5-gallon can with a tight-fitting lid or any other suitable container. This container must provide free ventilation and yet be insect-proof. When not "working," dermestids should be supplied with dog biscuits, dried meat, and other kinds of food material to keep the colony going. It is the larval forms of the dermestids that actually do the cleaning; thus, maintaining a colony means rearing the young of dermestid beetles.

Prepare fish, amphibians, and small reptiles by removing the viscera and completely drying. Keep flies away from the specimens as they dry to keep maggots out of the flesh. Place small vertebrates in with the dermestids, but watch these daily. If the specimens are thoroughly dry the dermestids will first eat the flesh, then will begin to attack the cartilaginous joints between the bones, and finally will begin to attack softer bones, such as those of the fishes. If the flesh is wet inside, however, the dermestids will begin to eat the bones along with the flesh. When the bulk of the flesh has been removed take the skeleton away from the dermestids and finish cleaning with Clorox; this tends to reduce tissue, which can then be picked away with fine forceps.

Less care is required for larger reptiles, birds, and mammals with fully developed bones. When these specimens are thoroughly dried place them in with the dermestids until they are clean. Following cleaning, specimens may be degreased and even bleached if required.

If more than one skull or skeleton at a time is cleaned, each must be fully labeled to prevent confusion. For example, when large numbers of skulls are being cleaned each is placed in a small open cardboard box with its label, and put in with the dermestids. This prevents skulls, jaws, and other parts from becoming mixed up in the container.

Boil-and-Clean Technique. While many mammologists clean skulls or skeletons by this method, the boil-and-clean technique does more damage to the bones, bone sutures, and ligamentous connections than any other technique. Bone surfaces become very porous, sutures tend to disarticulate, and teeth tend to fall out of their sockets. To use this technique simply boil the specimens until the meat loosens from the bone. Pick the flesh loose, dry the skeleton thoroughly, and then spray the skeleton with a clear plastic spray to reduce the porosity of the bone surface.

Maceration in Water. If nonligamentous skeletons are needed, simply place the skeletal material, after it has been fleshed out, in a crock of cold water and let this stand at room temperature from 10 to 20 days until the meat is thoroughly rotted and loosened from the bones. Pour off the fluid contents, being careful not to lose any small bones. After drying, degrease and bleach if required. The following technique for preparing ligamentous skeletons for mounting is quoted from *Turtox Service Leaflet No. 9, 1958*, p. 2, by permission of the General Biological Supply House, Incorporated, Chicago.

Place the bones in a glass or earthen-ware container of suitable size and cover the bones with cold tap-water. (Note—never use acids or chemicals of any kind). Do not allow any foreign substance such as metal, wood, etc. to get into the maceration jar as these will discolor the bones. The maceration jar should be kept at room temperature and the water should be changed daily replacing it with fresh water each time. At first a great deal of blood will be evident at each change, but will gradually diminish as it is extracted from the bones. This will take two to three days depending on the temperature of the room. During this time bacterial action (rotting) is taking place on the flesh that was left adhering to the bones. When the bath becomes clear (third or fourth day), pour off and place the bones in a solution made up of one ounce of trisodium phosphate to each gallon of water. Stir well until the trisodium is dissolved, leave the bones in this solution twelve to twenty-four hours. This bath serves a dual purpose in that it halts the maceration and also swells and loosens the tissue remaining on the bones.

Cleaning

Remove the bones from the trisodium phosphate solution and let drain. Secure a small nail or hand brush with stiff bristles, an old toothbrush, chlorinated lime and lots of hot water. Proceed as follows: Dip the brush in hot water, then into the chlorinated lime; now brush the bones using short rapid strokes (use the toothbrush in small hard to get places and on small bones). The action of the hot water and lime plus the friction of the brush creates a burning action upon the tissues adhering to the bones and these virtually disappear as the liming progresses. (Caution—use of rubber gloves in this operation is recommended to protect the skin on the hands from becoming burned). Rinse in cold water frequently, watching the removal of the flesh carefully. Brush or lime until the flesh has been removed entirely but not the ligaments holding the bones in place. Rinse thoroughly in cold water and lay out to dry at room temperature.

Degreasing Methods

It is not customary to degrease smaller skeletons or skulls unless they are intended for display or student use. Larger bones should have several holes drilled into the shaft and head to permit the degreasing solution to penetrate. Carbon tetrachloride is preferred for degreasing, although white gas works well. As Martin (1964) notes, the solvents are extremely volatile and thus the container should have a small ratio of surface area to volume. He notes that "as both breathing of carbon tetrachloride fumes and absorption through the skin are extremely hazardous, the damaging effects being cumulative even for short, repeated exposures, special care should be taken to insure adequate ventilation through the entire degreasing process and rubber gloves should be worn." A layer of water on top of the carbon tetrachloride will limit evaporation. Specimens should remain in carbon tetrachloride from a few days up to a few weeks, depending on the size of the specimen and the amount of grease. When removing bone specimens carefully tip the container and pour off the grease accumulated on the surface. Martin further recommends covering large skulls with a piece of cloth while they are still under the carbon tetrachloride. When the skull and cloth are removed, only the cloth will pick up surface grease. Air-dry the cleaned bones.

Cleaning and Bleaching

Skeletal specimens that have become dirty from continuous classroom use should be washed with warm water and detergent. Following this, rinse with fresh water and dry. If the bones remain discolored they may be bleached with a 5-percent solution of carbon tetrachloride. If possible, submerge the bones in the solution. Freshly cleaned skeletal material will require up to 10 hours before bleaching is complete. Rinse the bones thoroughly with fresh water and dry.

Display and teaching skeletons are greatly improved if given a light coat of clear plastic spray. This seals the pores, prevents dirt from penetrating, and generally strengthens the bones.

Mounting Skeletons

Ligamentous skeletons described under "Maceration in Water," above, are most desirable for mounting, since disarticulated skeletons require gluing and wiring and are generally much weaker. Smaller skeletons are generally mounted in glass-top display boxes; larger specimens are mounted on a wooden stand.

Make a glass-top display box as described for the Riker mount (see Fig. 13–13, p. 244). Select a box which is deep enough to properly position the skeleton in a lifelike pose. Cotton is not used in displaying skeletons; rather, the specimen is mounted on a piece of dull, black cardboard cut to fit the bottom of the display box. Ligamentous skeletons of this size are seldom disarticulated at any of the joints and thus are naturally held together. If the specimen to be mounted is dry, soak it in water until the joints are pliable. Blot off the excess water, permit the bones to become superficially dry, and then position the specimen to fit the mounting box. While it is still pliable, place the skeleton on the black cardboard and attach it securely with very fine brass wire. For example, a bat would be placed on its back and wired at the following points: ankles, the vertebral column ahead of the pelvic girdle, the elbows, wrists, and the first joint of the longest finger. Lizards, salamanders, and small mammals are usually placed in a walking position and are wired, at least, at the following points: the ankles and the tail. When the specimen is wired to the cardboard, glue the cardboard in the bottom of the display box, add the glass, and seal the lid in place.

Only a few basic techniques are required to mount larger skeletons on a wooden base, but these must be modified for each skeleton to be mounted. The procedure for a mammal the size of a small dog or cat would be as follows. Cut, sand, and varnish a wooden base made from 1-inch-thick lumber. Insert a heavy brass or galvanized iron wire, which has been tapered at one end, down into the vertebral column, and cut the wire off so that it will project part way into the skull. Next, bend the vertebral column into a natural position while it is still wet. Make two metal supporting rods out of $3/16$-inch brass welding rod (larger for dogs). Thread one end of each so that it can be attached to the base by means of two nuts, one above and one below the baseboard. One rod should be attached to the vertebral column just ahead of the pelvic girdle, the other rod to the axis vertebra of the neck. To attach these rods to the vertebral column, simply split the rod in half for a length of 1 inch (by means of a hacksaw), bend the two halves out to form a Y, place the vertebral column in position, and finally bend the ends of the rod back around the vertebral column to hold it firmly. Note that the two rods are generally about the same height, although the anterior one usually projects upward just ahead of the forelimbs and then bends forward along the length of the neck. In this preparation the pelvic girdle and the ribs are attached by ligaments to the vertebral column. Mount the hind limbs by drilling through the head of the femur with a $1/16$-inch drill, and continue the hole right on through the hip socket (acetabulum). A single wire passes through both femur bones and both sockets and is bent to hold the legs in position. The forelimbs are attached by wiring each scapula to the second thoracic vertebra. Bend the knees and elbows, place the foot against the stand in a lifelike position, and secure the feet by pushing common pins through exposed ligaments and down into the base. Cut the pins close to the bone with an end-cutting pliers. The tail is attached to the base in the same fashion. When the skeleton is totally dry, spray it with a clear plastic spray and make a dust cover for it out of a polyethylene bag.

CARTILAGINOUS SKELETAL TECHNIQUES

The author has not had experience in preparing cartilaginous skeletons of sharks, rays, and other animals. Therefore, the following is quoted directly from the *Turtox Service Leaflet No. 9*, 1958, pp. 3–4, by permission of the General Biological Supply House, Incorporated, Chicago.

Preparing cartilaginous skeletons requires a great deal of skill and patience. As in any other tedious process one must expect to sacrifice some time and perhaps a few specimens for the sake of experience. The work is somewhat smelly and messy, but if the person is intent upon work of this sort, such matters are trivial.

The materials needed, including the specimens are not costly and can easily be obtained. An earthen-ware or glass container of suitable size for holding the specimens, a small amount of commercial hydrochloric acid, spirits of ammonia, specially denatured alcohol (formula 30) and plenty of running water are about the only supplies necessary.

The specimen most widely used for this type preparation is the salt-water Dogfish, *Squalus acanthias*. Essentially the same technique is required for preparing other cartilaginous skeletons. So in our discussion we will consider *Squalus* as our example. It is essential that the specimen be fresh. Formalin preserved specimens cannot be made into skeletons and alcohol preserved specimens are very difficult to handle and give very poor results.

The first step is to dissect out the pectoral girdle and its fins from the body walls, being careful not to cut into any of the cartilages of the gill branches which lie very close. Next, dissect out the pelvic girdle and eviscerate the body. Remove the head with the gills attached. Flesh the body, being careful not to cut too closely so as to cut away the rib cartilages. All fleshing should be done very roughly, leaving the major portions on the skeleton. The maceration process will remove the greater part of this flesh. Do not try to flesh the head and gills as this can be accomplished only after much experience. Instead, score the skin with a sharp scalpel, thus allowing the acid to work its way under the hard skin. The instructions having been followed thus far, the dogfish parts are ready to be placed in the acid solution.

An earthen-ware or glass container, long enough to take the specimen without the necessity of folding or bending, is best, but a container of sufficient size will answer the purpose. The acid solution is made up by adding one part of commercial hydrochloric acid to forty parts of cold water. Stir well with a wooden paddle and allow to stand for an hour before the specimen is placed in it. In placing the specimen in the solution be certain that it is completely covered. After remaining in the solution for from twelve to fifteen hours the maceration should be well enough along to permit some work. To test this rub your fingers along the soft parts and if the flesh falls away readily the specimen is ready to be removed. If not, leave for another hour, then test again. When ready, remove from the solution and gently scrape away the major portions of the loose flesh. Now place the parts in a spirits of ammonia bath (prepared by adding two ounces of liquid ammonia to each gallon of cold water). The ammonia neutralizes the acid and halts the maceration. Allow the parts to remain in the ammonia bath for about an hour. Do not clean any further at this time. It should now be ready for hardening.

Prepare alcohol baths of the following percentages: 25%, 50%, 75%, and 95% (denatured ethyl alcohol, formula 30 being used). Place the skeleton first in the 25% alcohol and allow it to remain four to five hours, whereupon it is placed in the 50% solution for 24 hours and then the 75% solution for a like amount of time. After placing in the 95% alcohol allow it to remain for six to seven days before proceeding with the final cleaning. The skeleton can also be left in the 95% alcohol indefinitely if necessary.

When ready to proceed with the final cleaning, remove the skeleton from the 95% alcohol, place it in cold water for about one hour to wash out the alcohol. Now go over the skeleton with a blunt scalpel, scraping away the flesh carefully. Forceps will be found quite useful at this stage to pick away any small bits of flesh adhering to the skeleton. This is a very tedious and painstaking task and must be performed with great care. Finally brush over the skeleton with a small tooth brush. During the latter operations the skeleton should have been worked upon in a shallow enameled pan filled with cold water and the water should be changed as often as necessary. The skeleton should now be very presentable, but may appear slightly stained or discolored. If discolored and the skeleton seems rather strong, place it in a 3% solution of hydrogen peroxide for one or two hours or better just long enough for the stains to disappear. Wash in cold water to remove all traces of the bleach. If the skeleton is to be mounted permanently as a display specimen it should now be tied on to a glass plate cut to fit the storage or museum jar. Tie in place wherever necessary to maintain a natural shape. Fill container with 5% formalin, let harden for 24 hours and seal.

If one prefers to have the skeleton available for handling it may be transferred from the bleach to a water wash and then placed back into 95% alcohol. The alcohol tends to distort and shrink the cartilages, but by placing the specimen in water for about 30 minutes before using it will be restored to normal proportions, and can be handled. Thus, the final storing solution depends upon the use to which the skeleton is to be put.

The foregoing is only a brief outline of this process, but quite practical results can be obtained if a little patience is exercised and more than one or two specimens are worked upon.

REFERENCES

Davis, D. D., and U. R. Gore, 1947, Clearing and Staining Skeletons of Small Vertebrates, *Fieldiana: Technique No. 4*:3–16.

Evans, H. E., 1948, Clearing and Staining Small Vertebrates, in toto, for Demonstrating Ossification, *Turtox News*, 26(2):42–47.

Hollister, G., 1934, Clearing and Dyeing Fish for Bone Study, *Zoologica*, 12(10): 89–101.

Lucas, F. A., 1899, Notes on the Preparation of Rough Skeletons, *Bull. U.S. Nat. Mus.*, 39:1–11.

Martin, R. L., 1964, Skull Degreasing Technique, *Turtox News*, 42(10):248–249.

Miller, R. R., 1957, X-Rays as a Tool in Systematic Zoology, *Syst. Zool.*, 6(1): 29–40.

Schultze, O., 1897, Uber Herstellung und Conservirung durchsichtiger Embryonen zum Studium der Skeletbildung, *Verhandl. der Anat. Gesellsch: Anat. Anz.*, 13:3–5.

Turtox Service Leaflet No. 9, 1958, General Biological Supply House, Inc., Chicago.

CHAPTER 24

Scientific Illustration

Perhaps my most compelling reason for writing this text, beyond realization of need for a compilation of techniques on collecting and preservation, was the fact that I would get to explore and experiment with illustrating techniques that I had never used before. I will use the first person in writing this chapter, for it will permit more lucid communication concerning a subject which is not so scientific in nature as the other chapters of this book. Scientific illustrating is really quite simple once a few principles and "tricks of the trade" are understood. The biggest problem for the average student is to get over the erroneous idea that he is not an artist and, thus, is not an illustrator. I have been through the process of getting a masters degree, a doctorate, and of publishing numerous articles containing illustrations, and thus I feel that I can see the problem of scientific illustrating from both the point of view of the scientist and that of the artist.

The scientific illustration should not depict the specimen as "the artist sees it," but rather, it should depict the specimen as it really is. Perhaps this is one of the most exciting aspects of scientific illustration; that is, it demands that you really look at the specimen and that you become a keen observer of minor details. When you become instinctively aware of color and shape as seen in living and preserved organisms your whole experience of field collecting will change with your new awareness of natural beauty.

In many ways a good scientific illustration, properly rendered, is more useful than a photograph. Photographs are often plagued with problems of lighting, shadow, depth of focus, and the like, and thus cannot emphasize important anatomical aspects—which are the reason for the illustration in the first place. The drawing has a complete depth of focus, whereas the photograph is often limited in thick-bodied specimens.

Unless a scientific illustrator is available, it is always better that each scientist make his own illustrations. The scientist can easily learn enough technique in illustrating to bring out and emphasize exactly what he desires in a good drawing. On the other hand, a lot of time and motion may be lost in attempting to communicate to an artist the scientific importance that the scientist attaches to a particular specimen. It is not essential that a student have courses in painting or drafting techniques. However, if you anticipate making scientific illustrations, begin collecting outstanding samples of illustrations in any biological field, especially in your own field of endeavor. These illustrations will become your teachers in art, and will show you how to overcome problems in illustration. The only thing necessary to becoming a competent illustrator is a desire to learn and a willingness to observe and practice. In this field, at least, there is not a thing that others can do that you cannot do.

THE STEPS INVOLVED IN ILLUSTRATING

The first problem solves itself. (1) You must know the intended use of the finished drawing; that is, whether it is for thesis work, for publication, or for some other use. The finished drawing will be reproduced regardless of the intended use, but the method and cost of reproduction differ greatly, depending on the way the drawing is rendered. (2) You must determine the size that the drawing will be in the thesis or publication. From this the size of the initial drawing is determined. (3) Following this, spacing and balance of the plate layout (if more than one drawing is included) must be considered. (4) Next, the most perfect specimen for illustrating must be selected and the initial drawing made. Because accuracy is imperative, some technique of making the initial drawing must be selected. (5) If the initial drawing was not made on a good quality of drawing paper it must be transferred at this time. (6) The means of rendering should now be decided upon (on the basis of the type of reproduction and the technique that will portray the specimen in the most satisfactory way) and the drawing rendered. (7) If individual drawings, intended for one plate, were made on separate pieces of paper, the plate paste-up is made. (8) Finally, some notation of specimen size, lettering, and numbering should be added before the drawing is completed.

ILLUSTRATION USES AND PROBLEMS INVOLVED

Publications

Almost any type of illustration can be reproduced by scientific journals. However, cost is a big factor which is often shared by the journal and the author of an article. Therefore, colored illustrations are almost never used, except in very special cases. This technically reduces the means of reproducing an illustration to the line cut and the black-and-white halftone engraving.

Line Cut. The line cut is not restricted to line drawings, but rather deals with any drawing using only solid black or white with no intermediate grays. Almost all of the illustrations in this text are of line cuts. The illustrator may use stippling, texture board, scratch board, Ben Day pattern, and so on, for line cut reproduction. The illustrator achieves the illusion of intermediate grays in stippling, for example, by placing dots closer together or farther apart. Thus, all of the values (shades from dark to light) can be achieved by this process. The line cut is the least expensive and, thus, the most desirable for commercial printing. The finished drawing is transferred photographically to a zinc or copper plate which is used in the final printing process.

Halftone Engraving. Ink washes, airbrush drawings, pencil or charcoal drawings, and photographs contain a wide range of values using true intermediary grays. Because these grays are not composed of small black dots placed at various distances, the line-cut process cannot be used. Rather, the illustration must be transferred to a copper plate through a complex series of screens which actually translate the pure grays into various-sized dots of pure black. Figure 24–1 shows the same photograph reproduced through a 50-line screen, 100-line screen, and 150-line screen. Figure 24–2 shows a portion of the 50-line screen enlarged to illustrate how initial gray areas are translated to areas of black and white.

FIG. 24–1. The same photograph reproduced by the halftone engraving process using (from left to right) a 50-line screen, a 100-line screen, and a 150-line screen, respectively.

Halftone engravings are very expensive as compared to line cuts. Therefore, consult the editor of the journal in which you plan to publish before rendering illustrations for halftone engravings.

Theses and Dissertations

Almost every technique of scientific illustration may be used in thesis or dissertation work, because of the many ways in which illustrations may be reproduced. Pen-and-ink drawings and photographs, however, are most frequently used. Every university has a rigid set of standards governing the acceptability of various styles of drawing. Usually, however, the student is free to select whatever means of illustration suits his particular topic best.

Students usually reproduce their illustrations by one of three methods, inasmuch as only three to five copies of the finished work are required. Simple, uncomplicated line drawings are often redrafted for each copy required. However, students have recently begun to use one of the photocopying processes which generally meet the illustration standards. For example, when only four or five copies are needed these may be obtained for 10 or 15 cents each by means of Xerox or other light-sensitive copying

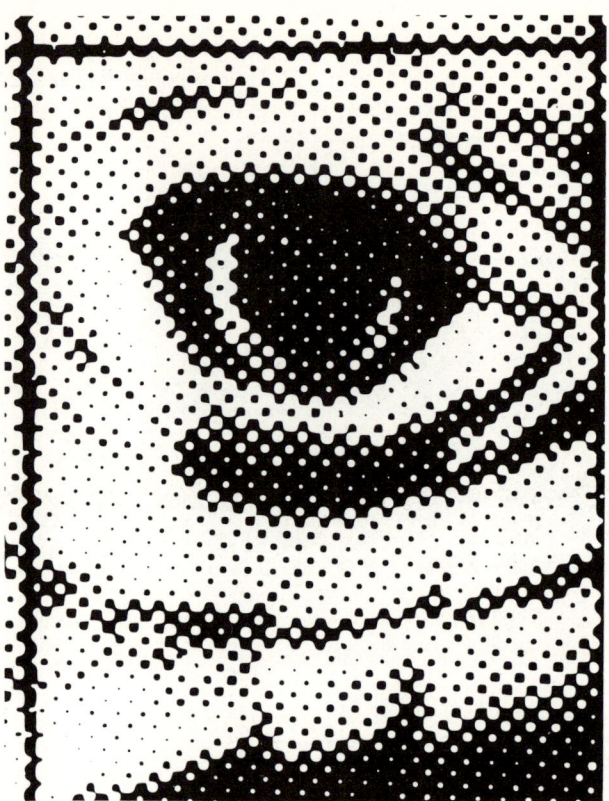

FIG. 24–2. A small portion of the 50-line screen reproduction of Fig. 24–1. This has been enlarged to show how shades of gray are translated into variously shaped black and white areas by the halftone and line process.

machines. If the equipment is in good working order the reproduced illustration will be sharp and black, while the paper will remain perfectly white. Thus, the copy often looks as good as or better than the original drawing. Many of the large blueprinting companies use a Xerox process from which several hundred copies can be made. The original is transferred to an electroplate which is used in making the copies. This process costs as little as 3 dollars for the first 50 copies and thus is excellent for reproducing maps, diagrams, plates, and graphs which are needed in large numbers.

A third method of reproducing all types of illustrations is the photographic method. Plates containing drawings, photographs, or a combination of the two may be reproduced in this way. The student will need access to a

35-millimeter single-lens reflex camera (or some other suitable camera) and a photographic darkroom equipped with an enlarger. Lightweight photographic papers (such as Kodabromide, A-3, or others) are available which have a smooth nonglossy finish, are approximately the same thickness as a sheet of 20-pound typing paper, and will stand up nicely in the bound thesis. Some students glue photographs to regular thesis paper. Although this practice may be acceptable, it is the least satisfactory. The most satisfactory method is that of printing directly on $8\frac{1}{2}$- by 11-inch paper. Any illustrations used in the thesis or dissertation must be contained within the limits of the blue line of the standard dissertation paper. Therefore, photographic paper may be held in a homemade holder with a shield made to the dimensions of the blue lines. This will permit printing only within the limits of the blue lines and will keep the rest of the sheet free of any printed material.

Prepare your drawings by any of the techniques described below, label them completely, include any caption that will accompany the drawing, and make a photographic negative (using a fine-grain film and developer). As soon as you have a suitable negative you can print as many copies of your drawing as required, using the full-size lightweight photographic paper.

I found that plates containing two or more photographs were tedious to produce by the standard method. This method consists of making shields to cover all the respective areas of the photographic paper so that each photograph can be printed sequentially on its respective area of the paper. This procedure requires two or more individual printings before the paper can be developed. You will find the following procedure much simpler: Print a sharp, clear photograph with good contrast of the proper size for each picture you wish to include on a plate. Cut the photographs to the finished size and mount them in their exact position on a piece of white drawing board or poster board. Number the photographs as they are to appear in the thesis or dissertation, and add any labels that are necessary. If the photograph is quite dark, so that the photograph number does not show up, place a $\frac{3}{8}$-inch round, white, gummed label in the lower left-hand corner of each photograph and place the number on this label. Finally, when the plate is prepared in the exact way that it is to appear in the thesis or dissertation, take a picture of the completed plate, using a fine-grain film. From the new negative the entire plate, along with its captions, labels, and numbers, is printed in one exposure, thus reducing the tedium of multiple printings. The finished product will show no appreciable loss of detail, may show slightly more contrast than the original, but will be absolutely satisfactory for the intended purpose.

PLATE SIZE AND LAYOUT

Plate Size

Scientific illustrations are usually made larger than the finished reproduction will be. Almost every illustration in this book was drawn twice as large as the finished engraving. The larger drawing permits you to work more freely and execute your lines with a greater degree of accuracy. Minor inaccuracies are reduced when the finished drawing is reduced. In addition, the finished drawings appear with finer lines and a greater degree of beauty than those that have not been reduced.

In most instances you can control the degree of enlargement of the initial drawing. However, when using the Camera Lucida for enlarging specimens seen through the microscope, you cannot always control the size of the finished enlargement, inasmuch as this is governed by the lens system of the microscope and other factors. Nevertheless, the initial drawing must always be of the same proportions as the finished reproduction. For thesis and dissertation work the finished reproduction must fall within the blue lines of the thesis paper. In scientific publication the drawing must be contained within the space occupied by the type on the printed page or on one column of the printed page. Figure 24–3 shows how the dimensions for an enlarged drawing may be determined. The gray area represents the size that the reproduced drawing is to be. First, extend the line from the lower left-hand corner up through the upper right-hand corner to establish a "corner line." Now, any two lines which parallel the top edge and the right edge of the initial drawing and cross at the corner line will be in the proper proportion. Lines showing enlargements of 2 times the original and $2\frac{1}{2}$ times the original drawing are shown. However, any dimension may be used, so long as the proportions remain the same.

Plate Layout

When many individual drawings occur on a single plate some planning should go into the layout to ensure proper balance and spacing. You must plan your drawings in such a way as to provide room for numbering and labeling, and you must bear in mind that the space between drawings undergoes just as much reduction as the drawing itself. Therefore, do not

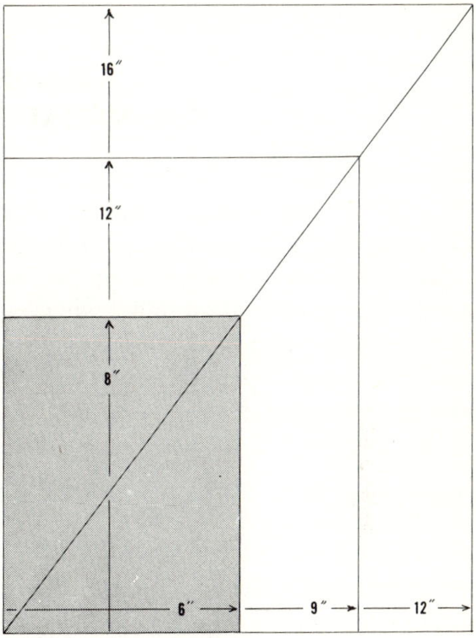

FIG. 24–3. A technique for determining the dimension of a drawing. See text for a complete discussion.

crowd every bit of available space on the enlarged plate with individual drawings, but provide ample space between them. I find it difficult to envision the required amount of space on enlarged drawings and, thus, have used the following procedure for every figure (plate) in this book. I mimeographed sheets of paper showing the maximum finished dimensions of the figures to be used in this book. As the book was written I made a rough pencil sketch of each individual drawing as it was to appear on the finished page. These drawings were extremely crude, but gave the correct spatial relationship and provided room for numbering, labeling, and so on. Using a pantograph (described below) I transferred a rough outline of each drawing to a large sheet of drawing paper; then I resketched each drawing within the dimensions assigned. It was during the resketching process that actual specimens were used, measured, and worked with in order to produce an accurate finished illustration. By this method proper spatial relationships and balance can readily be obtained.

When a series of Camera Lucida drawings or other drawings have been rendered individually, without prior thought to plate composition, you will have to follow some procedure like that described under "Plate Pasteup Techniques," below.

MAKING
THE INITIAL DRAWING

This heading refers to the first true drawing made from the specimen; it will be divided into three general categories: (1) microscopic subjects and small subjects, (2) medium and large subjects, and (3) equipment and mechanical drawings. In making the initial drawing you need not be concerned with problems of light, shadow, highlight, and reflected light, for these are all problems of rendering (to be discussed below). However, you should be concerned with perspective (especially in mechanical drawings) and foreshortening.

Perspective

A simple knowledge of perspective is essential to making good illustrations. Perspective deals with the effect of distance on the appearance of objects. Our mind has become so accustomed to accepting this effect that we judge the spatial relationship of things around us by perspective. For example, the artist knows that the railroad tracks are the same distance apart because they come together, and so on. Figure 24-4 is presented to show how the illusion of perspective affects our thinking and how we can take advantage of a few simple rudiments of perspective in scientific illustration. Figure 24-4A shows a rectangular box of uniform length, width, and thickness. However, if you compare the various dimensions with a ruler you will see that they are not the same. Sit in a chair about 4 or 5 feet away from the end of your desk or table. Note how the far end seems to be smaller than the near end. Hold up a ruler at arms length and "measure" the near width as compared to the far width, as shown in Fig. 24-4B. Now, in illustrating we often forget about reality when working on a blank piece of paper and may tend to draw the same desk with the near and far ends of the same dimensions, simply because we know that both ends are equal. Figure 24-4C shows how absurd this would appear. Notice how troublesome this particular figure is to the mind. The desk appears as if it were tilted toward you or as if the far end were much wider than the near end. Notice how the telephone poles in Fig. 24-4D seem to increase in length with distance, simply because they have all been drawn to the same scale, whereas they appear to be of the same height in Fig. 24-4E, simply because their length decreases with distance in the actual drawing. The pipe or tube in Fig.

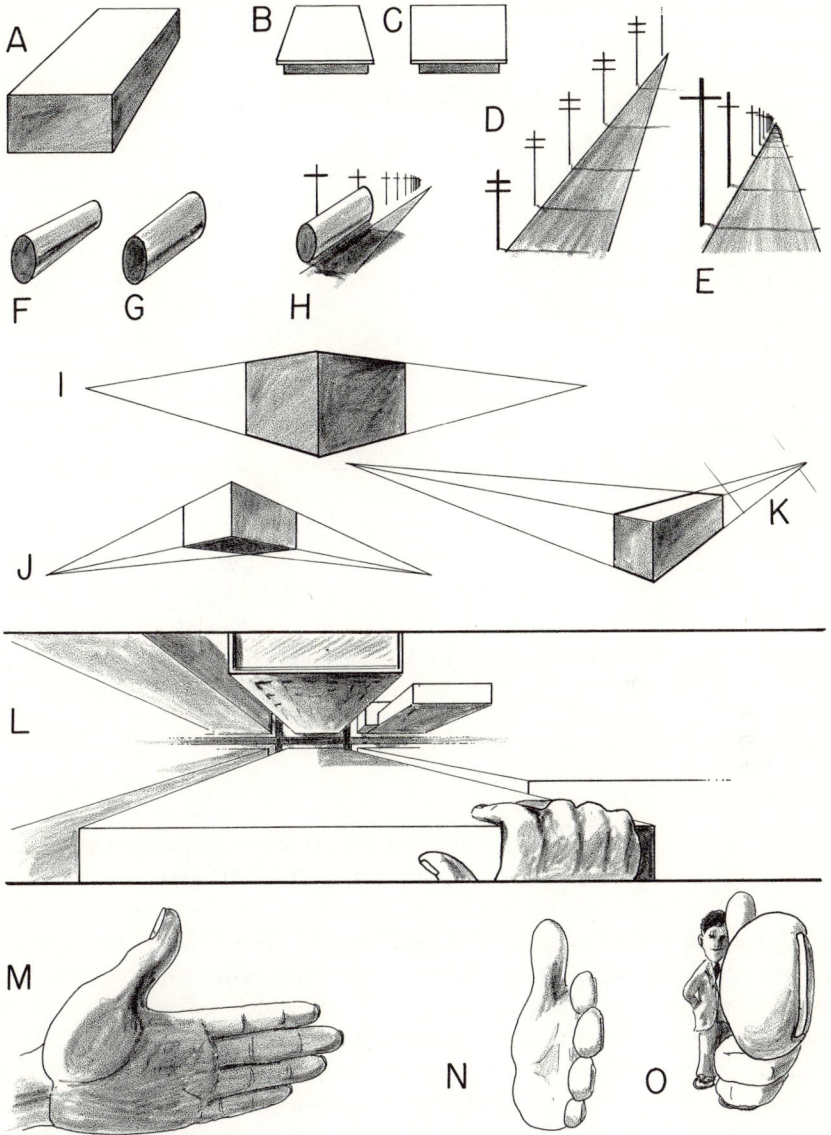

FIG. 24–4. Examples of the use and effect of perspective and foreshortening in illustrating. See the text for a complete description.

24–4F seems to be relatively long and of an equal diameter, whereas that in Fig. 24–4G is either very short or else becomes larger in diameter with distance. Notice how troublesome the same length of pipe is when placed in a scene where other things are in perspective. You will have this same difficulty when parts of a drawing are out of perspective, and until you develop an eye for perspective you will have trouble analyzing your own drawings. In other words, the illusion of the effect of distance on objects is more meaningful to the mind than the disputable fact that the true and actual size does not decrease with distance.

Figure 24–4, I, J, and K, shows the use of perspective on a box seen at eye level, from below, and from above. In each case the vertical sides of the box are parallel to the side margins of the printed page. In most of your drawing this will be true. There are many exceptions to this line of thinking, however; for example when you look down at the street from a very tall building, the vertical edges of the building no longer parallel the sides of a printed page (Fig. 24–4L). Each of the three drawings has two vanishing points from which all parallel lines radiate. You should make it a practice to use vanishing points when laying out cube-shaped objects, until you develop a feel for perspective in your illustration. In addition, you must recognize that plant and animal specimens are subject to the effect of perspective and that this must be accounted for to some degree in all drawings. Otherwise the drawing will be troublesome in appearance (see Fig. 24–4D), even though all of the relative measurements are correct as presented on the drawing paper.

Foreshortening

Figure 24–4, M, N, and O, deals with the illusion of perspective known as "foreshortening." Whereas Fig. 24–4M is two-dimensional (thus the fingers are all correct in their actual measurement), Fig. 24–4N is three-dimensional, with the forefingers foreshortened because they are directed toward you. In illustration of scientific specimens which are three-dimensional and complex in form (a crab, for example) certain parts of the anatomy will require foreshortening. Thus, you can use actual measurements from the specimen on all parts of the body in the two-dimensional plane, but must rely on perspective to add those structures in the third dimension (directed toward you or away from you). The only difference in foreshortening between N and O in Fig. 24–4 is that the hand in the first drawing appears at a normal distance away from you, whereas the second drawing (or cartoon) shows the pointing finger so close to you that it might well illustrate the way a biologist with bad manners appears to a worm sticking its head out of an apple. Foreshortening seems obvious to us when dealing with the human figure. Nevertheless, it becomes extremely troublesome if we forget to use it, since our drawings will be perfect according to measurement and still "seem" poorly proportioned.

Microscopic and Small Subjects

There are several means of transferring the image of a microscopic or small object to the drawing paper. These include the Camera Lucida, the microscope-slide projector, the ocular micrometer, and the ocular grid.

Camera Lucida. The Camera Lucida may be obtained at most scientific supply companies. It consists of an attachment which fits on top of the ocular of the microscope. By means of a prism and a mirror you actually see the specimen under the microscope and, simultaneously, your drawing paper at one side of the microscope. After the Camera Lucida has been properly adjusted you simply trace the outline of the specimen as it seems to be projected on the drawing paper next to the microscope. If the drawing paper is held in the same plane as the specimen under the microscope, and if the image is sharply in focus at the time of the tracing, you will achieve an accurate enlargement of the specimen. By placing a stage micrometer under the microscope after tracing the specimen, you may project a millometric rule to be used either to measure the specimen drawn on your paper or to indicate size in the drawing.

Microscope-Slide Projectors. There are a number of good projectors to be used with microscope slides. Such projectors contain the low-, medium-, and high-power objectives of the standard microscope and project the enlarged image of the specimen down to a drawing surface. Thus, relatively large specimens can be projected safely and traced accurately with this device. The microscope-slide projector is better for transparent specimens, but may also be used for opaque specimens. By projecting a scale of a stage micrometer you can establish a rule, either for measuring the projected specimen or for indicating the size of the enlarged drawing.

Ocular Micrometers and Grids. The ocular micrometer and the ocular grid are glass discs housed in the ocular of the microscope. The micrometer has a small scale ruled on its surface; the grid is made up of a series of lines which form a grid pattern. By calibrating the ocular micrometer or grid with a stage micrometer of known metric units, you can easily measure a specimen as seen under the microscope and recreate its measurements on a piece of drawing paper. Obviously, this can be a slow and tedious process,

458 Biological Techniques

since the ocular must be rotated to place the micrometer or grid in the proper position for measuring the given structure. Nevertheless, for extremely small specimens this system is quite suitable.

Medium and Large Subjects

A number of methods are available for transcribing the image of specimens of medium and large size to the drawing board. These include direct measurement, the opaque projector, the Camera Lucida, the grid technique, the photograph, and the pantograph.

Direct Measurement. Since almost all scientific illustrations are larger than the actual specimens, some system of enlarging direct measurements must be achieved. Basically, measurements may be made with a pair of calipers or with a pair of dividers and a ruler. These measurements are generally multiplied by some factor and then transcribed to the enlarged drawing. If a pair of dividers is used to take the measurements, you may simply step off a given unit four, five, or more times to make the enlargement. Figure 24–5A shows a scale used for enlarging basic measurements. Note at the bottom of the scale that the first distance (1) is equal to the original length of the specimen, whereas the combined distance of (1) and (2) is equal to the length of the enlarged drawing. To use this scale make the initial measurement with the dividers (3) and locate this on the scale, then open the dividers at that point (4) and obtain the enlarged measurement.

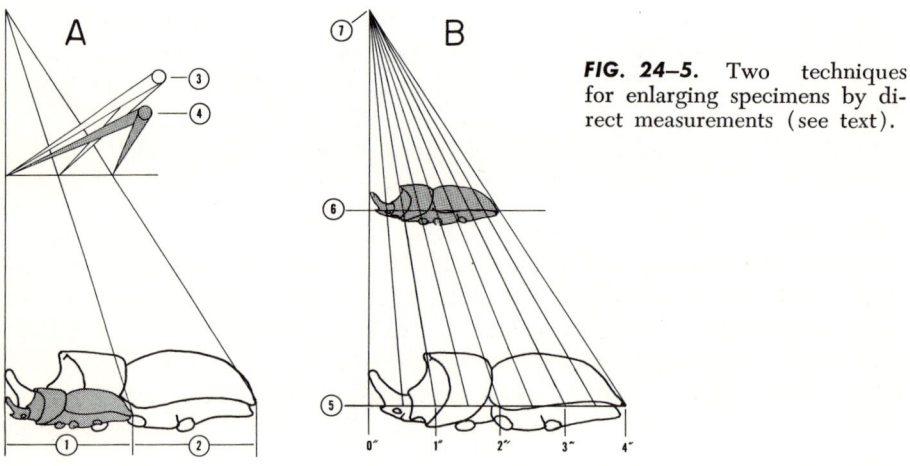

FIG. 24–5. Two techniques for enlarging specimens by direct measurements (see text).

Figure 24–5B shows another scale for making enlargements. In this case the length of the enlarged drawing is established at some known unit at the bottom of the scale (5)—say, 4 inches. Next, the true length of the actual specimen is established on the scale between the 0 line and the 4-inch line (6). Now, place a piece of white paper across the line on Fig. 24–5B (6). Mark the place where the paper crosses each of the lines as they ascend toward the vanishing point (7). The marked sheet of paper now becomes a ruler for measuring the specimen, since each unit corresponds to a unit (in inches) at the bottom of the scale.

The Opaque Projector. This projector is extremely useful for enlarging all kinds of biological specimens, including soft-bodied organisms such as jellyfishes. The particular projector that I use has a 1000-watt bulb which I find preferable to the 500-watt bulb. The bad feature about the projector is the extreme heat that is generated from the light source. However, this is quickly overcome by placing specimens in a water bath which will reduce the effect of heat on the actual specimen. The drawing of the crab *Zozymus*, used several times in this chapter, was originally projected in this fashion. You should carefully arrange the specimen in a suitable dish, propping it with pieces of clay if necessary. Attempt to get the body as level as possible, with all of the various structures well displayed. Two-dimensional objects, such as plant structures, may be covered with a glass plate. Finally, add water to the water bath and place the specimen directly in the center of the stage in the opaque projector. Project straight ahead (avoid directing the light up or down or to either side), focus, place a piece of drawing paper in the proper position, and trace the specimen. The opaque projector has an amazing depth of focus and thus requires few periods of focusing during the tracing process. You should note that the main lens housing is controlled by a focusing knob, but may also be moved in and out manually after release of a screw clamp. If you have trouble projecting large objects to a size small enough for your purposes, simply release the lens housing and manually slide the lens forward.

The Camera Lucida. A Camera Lucida for enlarging, reducing, and reproducing drawings (e.g., the French-made Universal Camera Lucida No. 111) is made by at least two firms. Unlike the Camera Lucida familiar to scientists, this unit works with macroscopic objects and may be used to transfer the image of large opaque specimens to drawing paper in the same way that the microscope attachment is used. This is a tool of the commercial artist, but would be well worth the cost (about 100 dollars) for anyone doing large quantities of scientific illustrations.

Grid Techniques. The grid has long been used as an aid in transferring the image of a specimen to the drawing board. For example, carefully draw a grid of identical size on two sheets of glass. Mount these two sheets so that one grid is directly above the other grid, then place the specimen beneath the two grids. Now by keeping the two grids perfectly lined up (in other words, by always looking in a vertical direction) you can accurately observe and transfer the outline of a specimen to a sheet of paper which has also been marked with a grid.

A single grid would not be so suitable for use as described above, even if the point of observation remained constant. However, if a specimen is mounted between 8 and 10 feet from the artist, with a single grid directly in front of it, and if the artist looks at the specimen only through a small, short-ranged monocular or telescope, an accurate drawing free of local distortion (parallax) may be achieved.

Photographic Technique. The Polaroid camera may be useful for making instant recordings of such things as dissections which are in a constant state of change. The pantograph may be used to transfer the image from the photograph to the drawing board. Slides (35-millimeter) can also be made of specimens, where the projected image is simply traced on the drawing paper.

The Pantograph. Two-dimensional objects such as floral structures, leaves, and so on, may be placed under a sheet of glass and traced with the pantograph directly to the drawing board. Drawings may also be enlarged or reduced with this device.

Mechanical Drawing

You may use considerable freedom in working up drawings of instruments and equipment. You should, however, carefully read the section on perspective and foreshortening, as these are extremely important. Establish vanishing points, keep vertical lines parallel to the margins of the drawing paper, and maintain approximate proportions between various structures. So long as measurements are given on the drawing or in the narrative the function of the drawing is that of advising the reader how the measurements apply and how the equipment is constructed.

Rulers and Ruling Devices. A number of inexpensive mechanical aids will greatly improve the quality of the finished drawing. Obtain a good-quality, wooden ruler with a metal edge that is set above the paper level. If you have the problem of ink working under the ruler and spoiling the

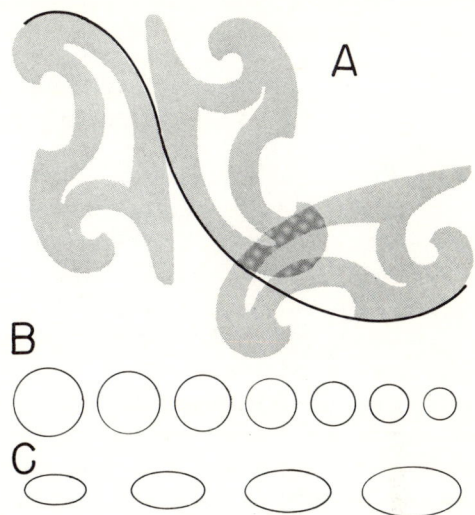

FIG. 24–6. Ruling devices. A. Using the French curve. B. A circle template. C. An oval template or ellipse.

drawing, do one of two things: (1) turn the ruler over if it is thicker above the metal edge than below, or (2) tape a strip of cardboard under the ruler to elevate the metal edge. The T-square is also ideal for the initial layout and the final inking, and may be used with all types of drafting pens except the old-fashioned stick pen or crow-quill pen. The drawing triangles (30°/60° and 45°) are extremely useful, but are less suitable for inking the final drawing. The French curve is another indispensable ruling guide. Figure 24–6 shows how various parts of the curve may be used to lay out a continuous, free-flowing curved line. You may use the French curve for inking the final drawing by elevating the curve with a piece of cardboard. Whenever it is necessary to ink two parallel curved lines, always ink the greater curvature first and the lesser curvature second. This will permit you to observe the greater curvature while inking the lesser curvature, and thus to keep the two lines an equal distance apart. I have found two other ruling devices indispensable. One is the circle template which gives 22 or more perfect circles, suitable for pencil or ink work, ranging from $\frac{1}{16}$ inch up to $1\frac{1}{2}$ inches (Fig. 24–6B). The other template is used for making ovals, such as the top oval of a test tube, jar, tin can, and so on, as seen from one side and slightly above. This template is called the ellipse and gives ovals ranging from $\frac{3}{8}$ inch up to 7 inches in the long dimension. The 30° ellipse is the most useful and the most commonly used. The ellipse is satisfactory for both pencil and ink work.

TRANSFERRING THE ORIGINAL DRAWING

Unless the original drawing was made on drawing paper it will have to be transferred by one of several simple techniques. An inexpensive pantograph is very satisfactory. If you are not familiar with this device, it is a unit which fastens to a drawing table, and has a stylus which is used to trace around the original drawing while a pencil recreates the same drawing on the drawing paper. The pantograph may be used for enlarging or reducing, as required.

Professional illustrators use the light box, which follows the principle of tracing a drawing against a windowpane. The light box consists of a sheet of white glass with a source of light underneath. The original drawing is placed on the light box, the drawing paper is placed over this, and the tracing is completed. I use this technique frequently, but simply put a sheet of plate glass between two chairs and place a gooseneck lamp on the floor underneath the glass. In a darkened room the results are excellent.

An older technique for tracing original drawings is that of smudging the back side of the drawing with soft pencil lead. The drawing is then transferred to drawing paper when you retrace the outline of the original with a pencil or ballpoint pen. You should never use typewriter carbon paper for tracing, in that it is very difficult to erase. However, Saral papers, used like carbon paper, erase cleanly and take ink perfectly. Saral paper comes in red, blue, yellow, and graphite. When using Saral papers or the smudge technique, lightly tape one edge of the original drawing down to the drawing paper. You may then periodically lift the original to check your progress on the finished paper. By this method the original can be returned to its exact position repeatedly.

METHODS OF RENDERING

Choice of Technique

The method of rendering your drawing must be based on two factors: First, the technique must be suitable for the method of reproduction in-

FIG. 24–7. Creating artificial textures by selecting the wrong means of rendering. A. Rough texture board. B. Pen-and-ink technique. Both drawings were reduced to approximately one half of their original size. See text.

tended; Second, the method must be compatible with the specimen that is being drawn. For example, the simple line drawing may be appropriate for all types of biological specimens, whereas other, more complex methods may make the specimen look either real or quite artificial. Figure 24–7 shows two mammal drawings; one was rendered with stipple board, whereas the other was handled in the style of a pen-and-ink drawing. Note how texture created by the stippling develops a nicely proportioned body, but seems artificial in that it does not resemble hair. Things that are glassy-smooth as another example, should not have rough texturing (rather, a very fine stippling) and so on.

Some Common Considerations Figure 24–8 shows four drawings which illustrate a problem common to all types of rendering. These problems deal

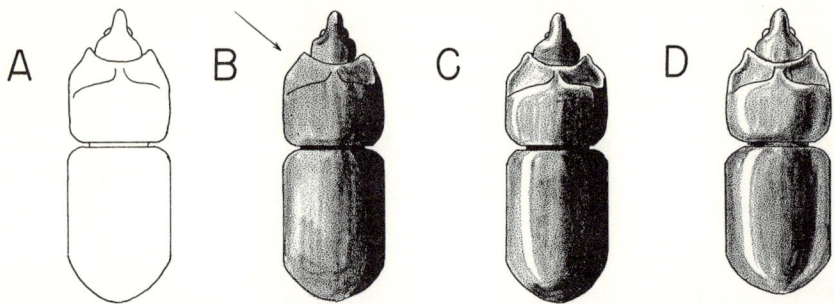

FIG. 24–8. Determining the direction of light, the use of highlight, and reflected light in rendering a drawing. These drawings were made on fine Coquille board and are reduced to one half of the original size.

with the direction of light, shadow, highlighting, and reflected lighting or backlighting, respectively. Figure 24–8A is a simple outline drawing. Now if we assume that light comes toward the specimen from the direction of the arrow (Fig. 24–8B) and from slightly above the level of the body, a strong shadow should be developed on all of the negative sides, as shown. However, certain surfaces of the body will reflect an unusual quantity of light and thus produce highlights. Look around you and you will notice highlights reflected from all objects receiving direct light. Since you cannot illuminate the drawing paper at the point of the highlight, you must render the rest of the structure in tones of gray in order to emphasize the highlighted areas. Highlights are achieved by leaving that surface unrendered, or by painting the surface white after other rendering has been completed. Figure 24–8D overcomes the harshness of the shadow shown in B and C, respectively, by showing reflected light. In nature almost every shaded surface is subject to reflected light. We are accustomed to seeing reflected light on objects; thus, illustrations portraying reflected light seem more real and lifelike to us than those with complete shading.

Correcting Mistakes

Although every artist attempts to render his drawings with no mistakes at all, accidents or miscalculations may always occur. This is not disastrous in most scientific illustrations, since the mistakes are never seen in the finished reproduction, especially those made by the line-cut process. Any drawing rendered with black India ink may be corrected by either chipping away the India ink with a scalpel if it covers only a minute area, or painting it over with an opaque white made exclusively for retouching and correction. This material is available at all art-supply stores. Clean erasures are usually satisfactory for correcting mistakes made in pencil or charcoal drawings. However, opaque retouch white may be required to remove unwanted lines or smudges.

The Line Drawing

Uses and Reproduction. The line drawing is one of the most useful means of rendering scientific drawings where the outline of a specimen is critical. It is applicable to all kinds of biological specimens, especially transparent individuals which are difficult to render in any other way. The line drawing also serves as the basis for other forms of rendering, including stippling, texture board, Craftint board, Ben Day patterns, and the like. The line drawing can be reproduced by the line-cut method, photographic techniques, and the Xerox or other photocopy methods.

Drawing Paper. There are hundreds of drawing papers intended for many different uses. One must be concerned with both the texture and the thickness of the paper. Drawing paper comes in various thicknesses from 1-ply, 2-ply, 3-ply, 4-ply, and so on, to mounted drawing board which consists of a thin sheet of drawing paper glued to a heavy cardboard backing. The heavier boards are often referred to as illustration board or Bristol board (lighter papers may also go by the name of Bristol). Plate or kid finish is satisfactory for line drawing and other pen-and-ink work. I have used a 2-ply, plate-finish paper (extremely smooth) throughout this entire text, with the exception of some mounted kid-finish illustration board and some 2-ply kid-finish drawing paper. The plate-finish paper has the slight advantage of creating less friction between the pen point and the paper. It permits the point of a pen to spread and produce a thicker line, which is highly important in some forms of rendering. The illustration board has the advantage of being suitable for pen-and-ink work and for water-color washes, airbrush, and other techniques. It is opaque, however, and thus drawings cannot be transferred to illustration board by means of a light box (see above).

Pens. The crow-quill pen is ideal for rendering line drawings of the anatomy of some organisms because of the very fine point, but is not very satisfactory for ruled drawings. An assortment of pen points to be used with an ordinary stick pen will also prove useful for various aspects of illustration. The Rapidograph fountain pen has a professional, mechanical drawing tip and is extremely useful for both ruled drawings and specimen drawings. I have used sizes 00, 0, 1, 2, and 2½ for rendering this text. The line is always of a uniform width and the pen is more than satisfactory for working with templates or rulers. The adjustable draftsmen's pen is also suitable for ruled drawings, for it produces clean, uniform lines of any practical width desired. You will simply have to practice with various pens and pen points to become accustomed to their uses and limitations. Finally, a good grade of black India ink should always be used. It is essential that the lines be uniformly dark in all parts of your drawing.

Procedures and Problems. Lay out the drawing, using a 3H or 4H pencil and being careful to keep the working surface as clean as possible. When you begin rendering, place a sheet of scrap paper over the working surface which is not being rendered to avoid getting fingerprints or dirt on the drawing. Fingerprints will usually cause the ink to "feather out," especially if your hands are sweaty. Keep your pen points clean and free of debris, wiping them frequently on a cotton cloth. Prepare a pen cleaner by mixing ten drops of household ammonia with an ounce of water. If mechanical drafting pens become clogged with ink, use this solution for cleaning them. Permit the drawing to dry 10 minutes or more before erasing the pencil lines. The Kneaded Rubber eraser is very satisfactory for cleaning pencil lines away from ink drawings. The art gum eraser is also satisfactory, as are various cleaning pads which contain powdered art gum.

Figure 24–9 shows various line drawings without any other form of rendering, such as stippling. These drawings are reproduced in the original size, with the exception of Fig. 24–9D which was reduced to one third the original size. Note how the thickness of the line (compare Fig. 24–9B with Fig. 24–9A) can make the drawing seem realistic or very crude, even though the initial layout, in both cases, was correct. It is quite easy to see how a slight bit of stippling could improve any of the individual drawings. This figure was drawn on mounted Bristol board with a kid finish. The crow-quill pen was used for all drawings except Fig. 24–9C which was done with the 00 Rapidograph pen. For other examples of line drawing using the pen-and-ink technique, consult Fig. 24–7 of this chapter and Fig. 5–1 (p. 73). Almost all other figures in this text contain some examples of the line-drawing technique.

Stippling

Uses and Reproduction. The stippling technique is one of the most useful and versatile means of rendering line drawings, for several reasons: (1) a complete range of values (shades of gray) may be obtained; (2)

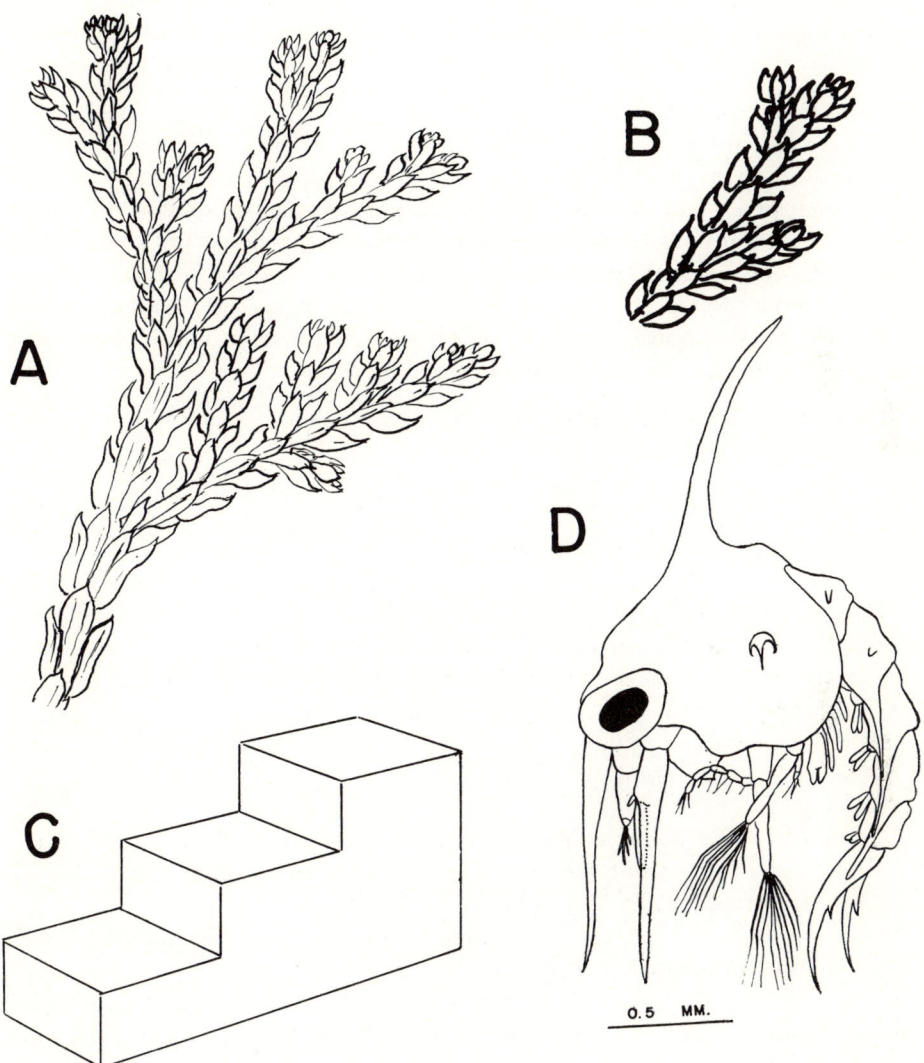

FIG. 24–9. Various line drawings without rendering. A–C. No reduction. D. Reduced to one third of the original drawing. See text.

stippling may be used to indicate shadow on perfectly smooth objects without producing a surface texture (by using fine, round dots); (3) stippling may be used to indicate various textures by using clusters of dots or by elongating individual dots. Drawings stippled with India ink may be reproduced by the inexpensive line-cut method or any other technique desired.

Materials Needed. Any paper well suited for line drawings (see above) will also be suitable for stippling. The crow-quill pen, stick pen, and the size 0 Rapidograph pen are all useful in stippling. You should select a particular pen point on the basis of its performance on a given type of paper and according to the amount of reduction expected for a particular drawing. When drawings are to be reduced more than half of the original size, slightly coarser stippling may be essential. It should be emphasized, however, that stippling should be as fine as possible while permitting complete reproduction at the intended reduced size.

Procedures and Problems. The drawing should be laid out on some suitable paper with a 3H or 4H pencil. With the layout completed, determine the angle of light which will display the specimen to its greatest advantage. For example, in Figs. 24–10 and 24–11 the light angle is from the upper left-hand side of the specimen, coming in from a fairly low angle. Highlights are left white or may be painted white with an opaque, white retouch paint (see "Correcting Mistakes," above). If you are unsure of how to shade a particular specimen, first direct a desk lamp from the proper angle toward the specimen being drawn. Second, place a sheet of tracing paper over your drawing and make a trial shading with a soft lead pencil (2B or 3B) or by stippling with pen and ink. Furthermore, study samples of effective stippling, if possible for specimens similar to those you are rendering.

Some of the problems which can quickly be overcome with a little practice are as follows: Any fingerprints left on the drawing will cause India ink to feather out and make large blotches rather than fine dots. Overcome this by placing a sheet of scrap paper under your fingers as you hold the drawing. Another problem is that a pen may drip ink simply because too much ink was carried on the point at a given time. Shake off the excess ink and wipe the pen point on a piece of cotton cloth each time you dip the pen. Another point to be cautious about is that of picking up lint or other debris on the pen point, so that coarse irregular dots are formed. Finally, you must guard against producing unwanted textures by lazy or hurried stippling. Figure 24–10A shows uniform stippling made with a firm, round dot in contrast to Fig. 24–10B which creates a texture because of elongation of the dot. Figure 24–10C shows stippling which is too coarse for specimen

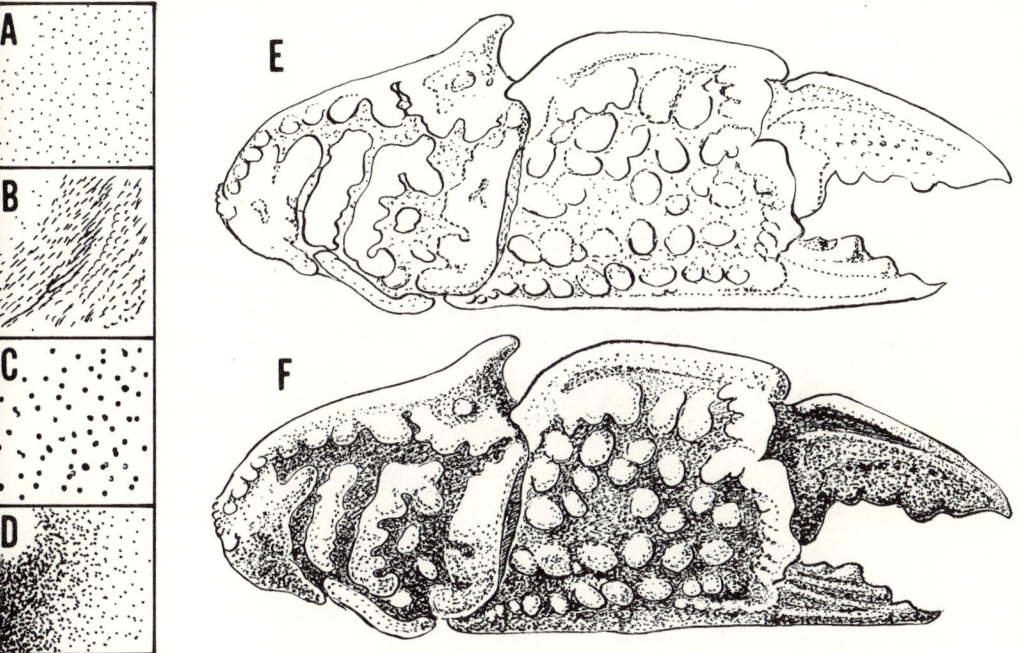

FIG. 24–10. Techniques in stippling. See text for details. This figure was reproduced at the original size.

drawings; Fig. 24–10D shows a gradation of stippling from dark to light. If mistakes are made in stippling, these may be corrected with opaque white, and restippled.

Figure 24–10, E and F, shows the large hand of the crab *Zozymus* reproduced at the actual size of the drawing. The initial drawing was made using the opaque projector, with the specimen in a water bath. Next, the specimen was rendered as a line drawing and then the stippling was developed in a uniform way over the entire specimen (Fig. 24–10E), with only the crow-quill pen. The point here is that you should tone the entire specimen simultaneously, rather than rendering any single area to completion. By the latter procedure some areas will get out of control and become much too dark. Thus, it is desirable to darken and shade all areas simultaneously to avoid this problem. Figure 24–11 has been reduced to half the size of the original drawing to show the effect reduction has on the final reproduction. Compare Fig. 24–11 with Fig. 24–10 which was reproduced at its full size.

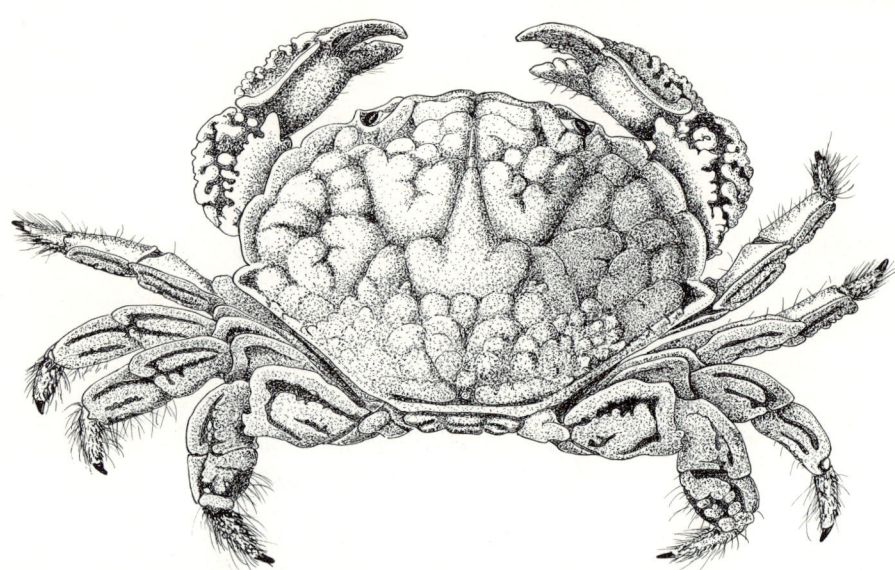

FIG. 24–11. The crab *Zozymodes* rendered by line drawing and stippling. This drawing was reduced to one half the original size.

The layout for Fig. 24–11 was made by using the opaque-projector and water-bath technique. The back, pinching hands, and opposing groups of legs were rendered as individual units. The entire back was done at one time, the entire surface being rendered, but special attention given to the shaded areas in the beginning. The most difficult problem was controlling the texture. The stippling was rendered with the fine point of a crow-quill pen. Yet, unless the pen point was wiped clean every few minutes the dots tended to become much too coarse. There also seems to be a natural tendency to group dots unless a conscious effort is made to distribute them equally. When all of the component parts of this drawing were rendered, the entire drawing was studied, some parts were darkened, and one or two areas were given highlights with opaque white. Note how one part of the anatomy is separated from another by working lighted areas against shaded areas. Note also how rows of dots may indicate a surface line in a lighted area, while solid ink lines are used to emphasize a similar line in the shaded area. This drawing required 10 hours of stippling after the initial layout was made.

Textured-Board Techniques

Uses and Reproduction. The textured board is one of the most fascinating materials used by professional artists that is completely suitable for fine-quality scientific illustrations. Textured boards are really heavy papers

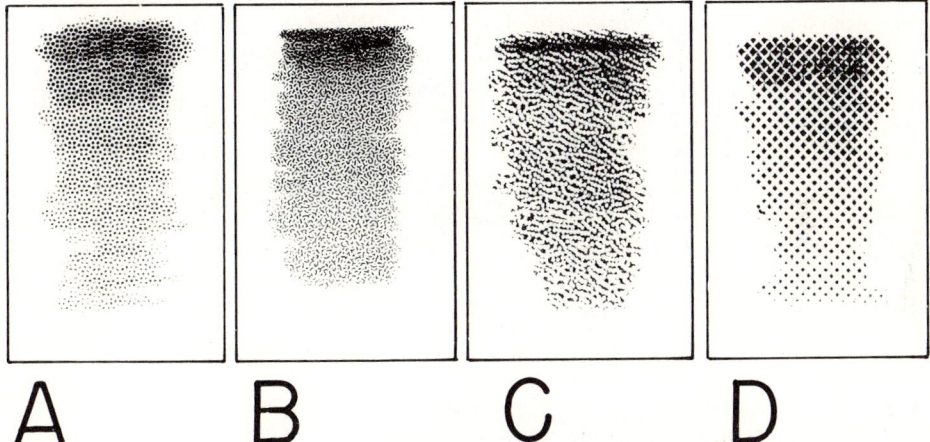

FIG. 24–12. Four types of texture board reproduced at the actual size. A. Dot board. B. Fine stipple board. C. Coarse stipple board. D. Halftone texture.

which have a raised surface in various textures. The author is indebted to the Grumbacher Company, Incorporated (460 West 34 Street, New York) for sending generous supplies of various textured boards. Figure 24–12, A through D, shows four samples of these boards which have been rendered by simply rubbing a black pencil over the surface. A fairly complete range of values (shades of gray) is obtained by pressing the pencil firmly or by reducing the pressure. Furthermore, drawings made in this medium can be rendered in minutes, as compared to hours of laborious stippling that would otherwise be required. In addition to their contribution to speed, textured boards are very economical and can be reproduced by all techniques, including the inexpensive line-cut process. Many of the figures in this book have been rendered on textured boards (see Figs. 11–1, 17–1, 18–1, 22–8, 24–4, and others) as have those in other textbooks, for example Brown (1961) and Barnes (1963).

Materials Needed. Textured boards are commonly known as Coquille board, Repro board, or Glerco board. These are available at any large professional artists' supply store or may be ordered by your own supply store. Figure 24–12A shows the Grumbacher dot pattern which, when reduced, is very effective in biological work for producing either values of shading or natural pigmentation, as in fish illustrations (see Fig. 16–1 for the effective way in which this pattern was used upside-down). Figure 24–12B shows the fine stipple board which is very similar to Coquille board and can be used in almost any type of biological illustration. Figure 24–12, C and D, shows coarse stipple board and coarse halftone board, respectively. Because of their textures they are less suitable for most types of scientific illustration, except for mechanical drawings in which various degrees of shading are required. For rendering, I have used the black Litho crayon (black wax pencil), the L. and C. Hardtmuth "Negro" 350, No. 2 black pencil, and the Eagle Prismacolor black 935 pencil. Others are also suitable. Solid black areas are achieved with India ink; poster white or retouch white may be used for creating highlights in shaded areas.

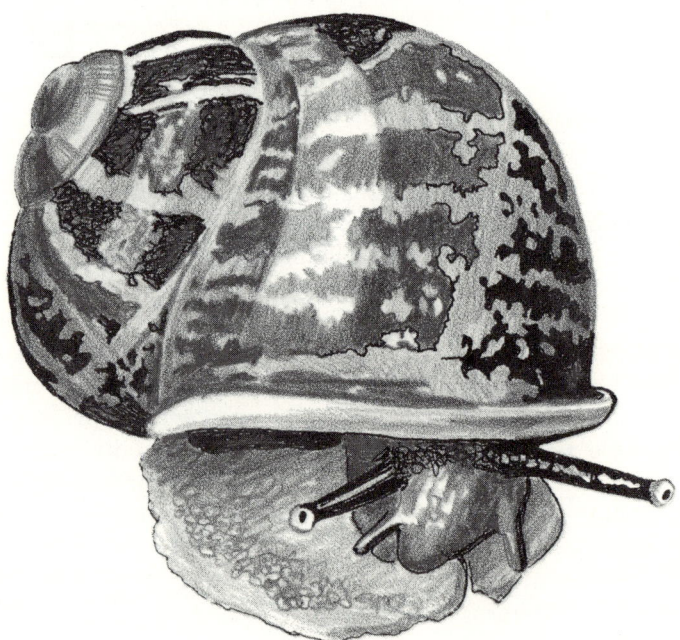

FIG. 24–13. The land snail Helix drawn on fine Coquille or stipple board, and reduced to 50 percent of the original size.

Procedure and Problems. For Fig. 24–13 I selected the Coquille board and used India ink and pencil for rendering. This drawing was made quite large (1½ times the reproduced drawing) so that the Coquille texture would not make the shell appear too rough. The initial layout of the snail *Helix* was made from a Kodachrome slide directly on Coquille board with a light blue pencil (which will not photograph in the line-cut process). Next, the lines were inked in and the solid black areas were painted in with India ink. Finally, the entire drawing was rendered with a pencil, working from the dark areas to the light areas. Areas of highlight were circled in blue and were purposely avoided in the initial rendering. Because of the great speed possible in Coquille-board drawings, it was very easy to harmonize the entire drawing and to pull it into a value balance. Note how the dark pigments and light pigments of *Helix* are treated in the shaded areas and the highlighted areas; it is all a matter of relative value. This entire drawing was rendered in 45 minutes, but would have required 12 to 15 hours of stippling by the conventional method.

Textured Scratch Board

Uses and Reproduction. Textured scratch board consists of a heavy drawing paper which is coated with pure white clay embossed with various textures on its surface. The author is indebted to the Charles J. Ross Company (1525 Fairmount Avenue, Philadelphia) for providing samples of textured scratch board. The Ross Company makes fifteen different patterns of Ross drawing board, some of which are shown in Fig. 24–14, A through D. This medium is used a great deal for cartoon and newspaper reproduction and, to a lesser extent, for biological illustration. However, there are so many advantages to using textured drawing boards or textured scratch boards that one should seriously consider it for his own uses. The advantages of textured scratch board are: (1) speed of rendering, (2) ease of creating highlights or removing mistakes, (3) the low cost of line-cut reproduction.

The four samples of Ross drawing board shown in Fig. 24–14 have three blotches: the upper blotch was made with a black pencil and shows the range of values (shades of gray) possible. The middle blotch was also made with a black pencil and shows how highlights may be scratched in the clay surface. The lower blotch on each sample was made with India ink and shows how lines or the embossed pattern can be brought out by scraping away the painted surface.

Materials. My experience has been that textured scratch board is not

474 Biological Techniques

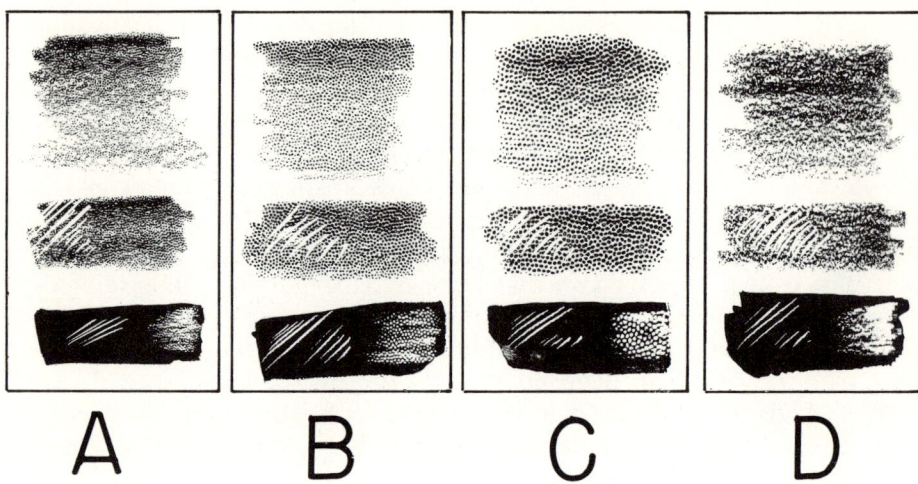

FIG. 24–14. Four patterns of Ross board, which is a clay-coated, textured paper, reproduced at the actual size. A–D. Represent textures 1, 1½, 2 and 5 respectively.

readily available, except at dealers who supply commercial artists. If this is your problem, have your local art supply store write directly to the manufacturer for a brochure and price list. Any of the pencils or crayons mentioned under textured board (above) are suitable for Ross drawing board. Scrapers and knives used for scratching the surface are illustrated and discussed under "Smooth Scratch Board," below.

Procedure and Problems. The crab *Zozymus* was again illustrated (Fig. 24–15), in order to compare the time required by the texture board technique with that of the standard stippling technique used in Figure 24–11; about 1 hour was required for Fig. 24–15, whereas 10 were required for Fig. 24–11.

The first step in making this illustration was mounting a piece of textured scratch board on a sheet of heavy, white illustration board with rubber cement. This was done to avoid cracking the clay-coated surface. The initial drawing was made using the opaque-projector and water-bath technique; the outline was made with a light blue pencil. Next, most of the major lines were inked in with a crow-quill pen. Some of the darkest areas on the legs were painted black with India ink. When painting scratch board use an artist's water-color brush; avoid leaving pools of ink to dry on the surface of the scratch board. Finally, the entire drawing was rendered with

Scientific Illustration   475

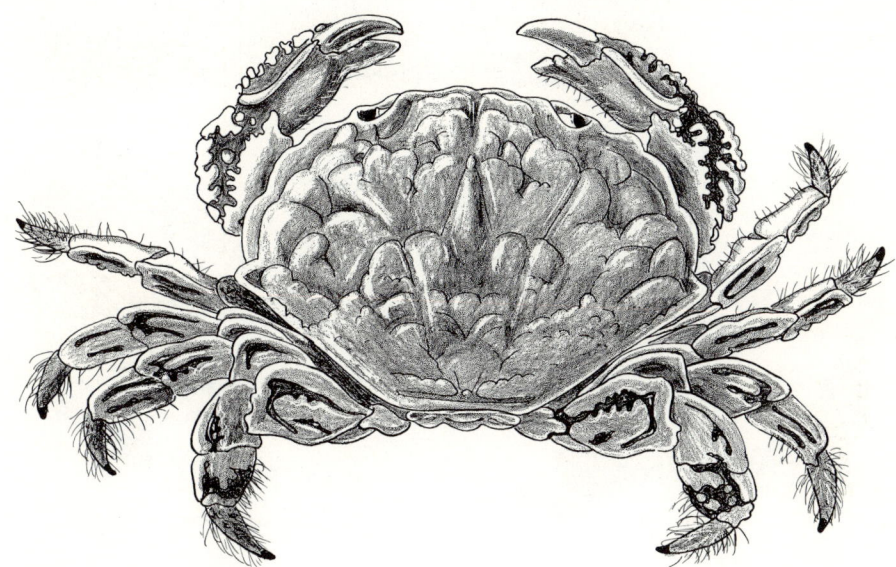

FIG. 24–15. The crab *Zozymodes* drawn and rendered on 1½ Ross board and reduced to one half of the original size.

a black pencil: first the back; then the pinching hands; finally, the legs. The effect of highlighting and reflected lighting were then achieved by scraping away the darkened areas. If you scrape lightly enough it is possible to go back and retexture an area. Finally, the black hairs on the legs were created with a crow-quill pen and the white hairs were made by scratching away the surface. For other examples of textured Ross board see Figs. 11–4, C and D, 11–5, 11–6, 14–1, 15–1, and others.

Smooth Scratch Board

Uses and Reproduction. Plain scratch board consists of heavy drawing paper coated with a smooth layer of pure white clay. Scratch-board drawings are made by painting certain areas black and then scratching away white highlights. This medium is used for all kinds of book, magazine, and newspaper art, as well as in some forms of biological illustration. The brilliant contrast between black and white achieved by this medium may be used for dramatic effect.

Materials. Scratch board is available in most large art-supply stores, or

FIG. 24–16. Scratchboard techniques. A. Various textures possible with scratchboard. B–C. Scratchboard cutters. D. A stencil knife useful in cutting scratchboard. E–G. Various steps in scratchboard drawing. See text.

can be ordered from the manufacturers. Figure 24–16A shows some of the many textures that can be developed on a black surface. Figure 24–16, B and C, shows the two common scraping knives (drawn to actual scale) which are fitted into stick-pen holders. A sharp scalpel, stencil cutter (Fig. 24–16D), or model-airplane knife may also be of value.

Procedure and Problems. Scratch board is very suitable for illustrating plants, birds, or hairy animals such as mammals and insects. The procedure is illustrated in Fig. 24–16, E, F, and G. I began by mounting a piece of scratch board on a sheet of heavy white illustration board with rubber cement. The initial drawing was made with a medium lead pencil. One problem that develops is that after the entire drawing is painted black, it is difficult to locate some of the more important lines. This can be overcome by redrawing such lines with a medium-hard drawing pencil. By using just the right amount of pressure, you will imprint these lines on the surface. These can be seen, after the illustration has been painted black, by holding the scratch board so that light is reflected from the indented lines. An ink pen and paintbrush were used to render the surface black (Fig. 24–16E). Next, rendering was begun by scratching away the black surface (Fig. 24–16E). Note that the scratched lines must always be in the direction in which the hair lies on the body, in order to give the drawing a natural appearance. Any excess ink can be scraped away as you proceed, should parts of the initial drawing be out of proportion. Likewise, you may paint out scratch lines and render them a second time, if need be. The entire drawing was scratched at one time, the different areas gradually being rendered until the drawing reached a state of completion (Fig. 24–16, F and G). Finally, the example drawings were cleaned with a Kneaded Rubber eraser and wrapped in a sheet of tissue paper for protection. Fig. 24–17 shows how scratch board may be used to render feathers or "fur."

The only problem not discussed in this example is that of moisture. If at any time you are unable to render a clean, neat line it is probably because moisture has accumulated in the clay surface. Should this occur, simply dry the scratch board in an oven or a warm, dry room.

Ben Day Patterns

Ben Day screens consist of different arrangements of lines, dots, or other patterns printed on transparent sheets which are fastened directly to pen-and-ink drawings. These are probably better known today by two popular tradenames, Craftint and Zip-a-Tone. Of the literally hundreds of Zip-a-Tone and Craftint patterns available, Fig. 24–18 lists ten which were commonly

478   Biological Techniques

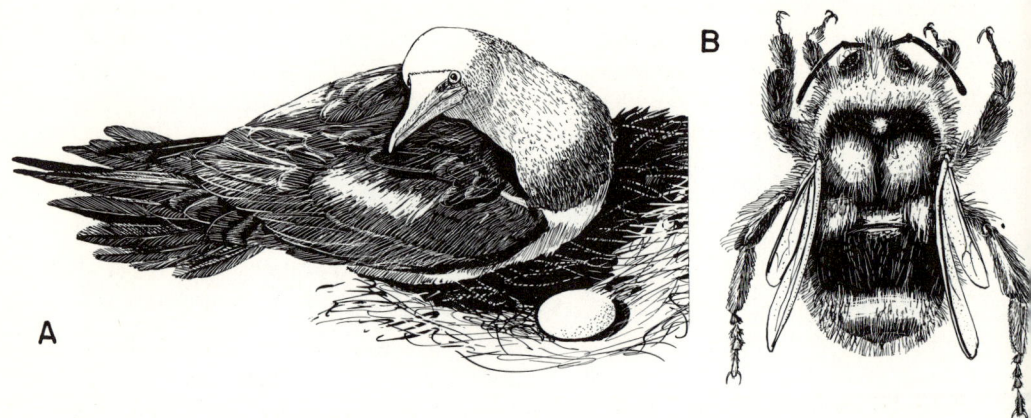

FIG. 24–17. Two additional scratchboard drawings showing the application for feathered and hairy animals. Reduced to one half of the original size.

used in this book. To use, place Ben Day patterns directly over the finished ink drawing and "tack" them lightly to the drawing by rubbing the pattern with a fingernail or the back of a spoon. Following this, cut away the excess material with a sharp knife. Finally, affix the pattern to the drawing by placing a sheet of paper over the pattern and vigorously rubbing with the back of a spoon, as shown in Fig. 24–18, B, C, and D, respectively.

Figure 24–19 shows the same ten patterns reduced to half the original size. Ben Day patterns are very useful for making graphs and charts, toning line drawings, or making ecology maps or distribution maps (Fig. 24–19, B, C, and D, respectively). In Fig. 24–19C the dragonfly head was first laid out by pencil, was inked and partially stippled, and was then provided with Ben Day patterns to both tone the face and create the illusion of the compound eye. All of these techniques are suitable for any form of reproduction, including the inexpensive line-cut process. Many illustrations in this text are toned with these patterns.

Pencil Techniques

Uses and Reproduction.   Pencil drawings are very effective and lifelike, are quickly and easily rendered, and require only simple materials. However, the pencil medium produces true shades of gray and must be reproduced (like photographs) by the more expensive halftone engraving process.

Scientific Illustration 479

Nevertheless, the effectiveness of pencil drawings will often compensate for the higher cost of reproduction.

Materials. Almost any paper is suitable for pencil drawing. The only thing to consider is the "tooth" or degree of roughness of the paper. Papers range from extremely rough to a smooth plate finish. Avoid the rougher papers, since these produce an unwanted texture. If you use a lightweight paper, such as 20-pound bond, mount this on a piece of illustration board with rubber cement before rendering. A hard (3H or 4H) pencil should be used for the initial drawing. The pencil, or pencils, used for rendering will depend entirely on the paper, for a given pencil will perform differently

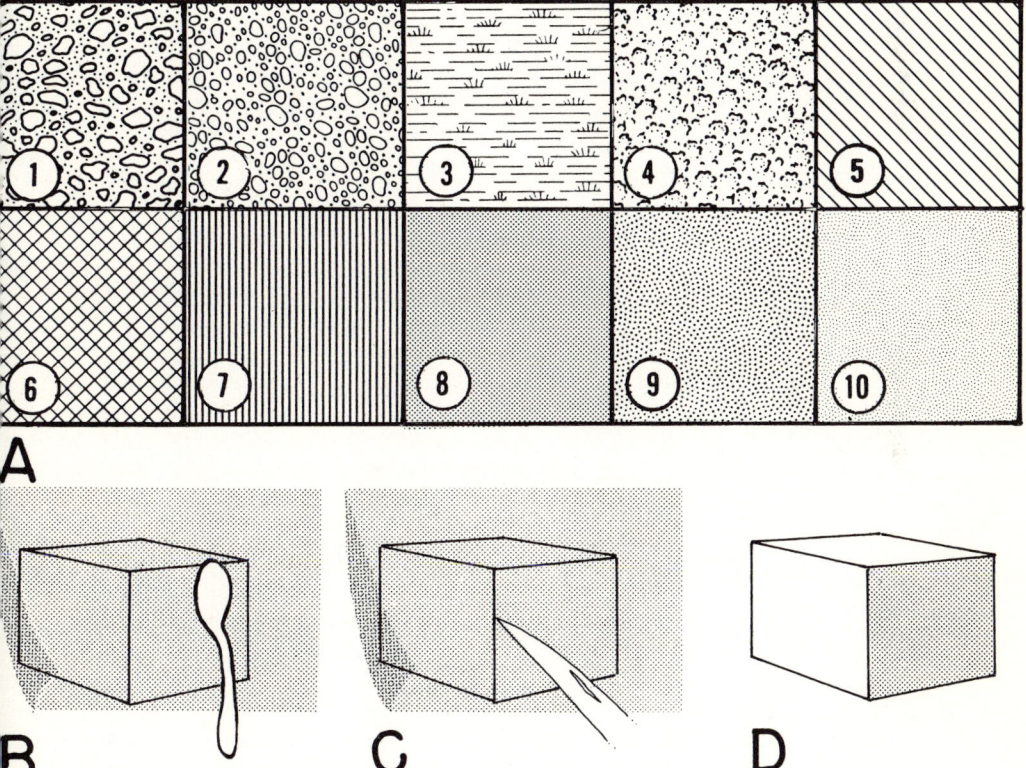

FIG. 24–18. Ben Day patterns. A. Ten different Ben Day patterns made by Craftint and Zip-a-Tone which were commonly used throughout this text. None of these have been reduced in size. B–D. Using Ben Day patterns for shading drawings (see text).

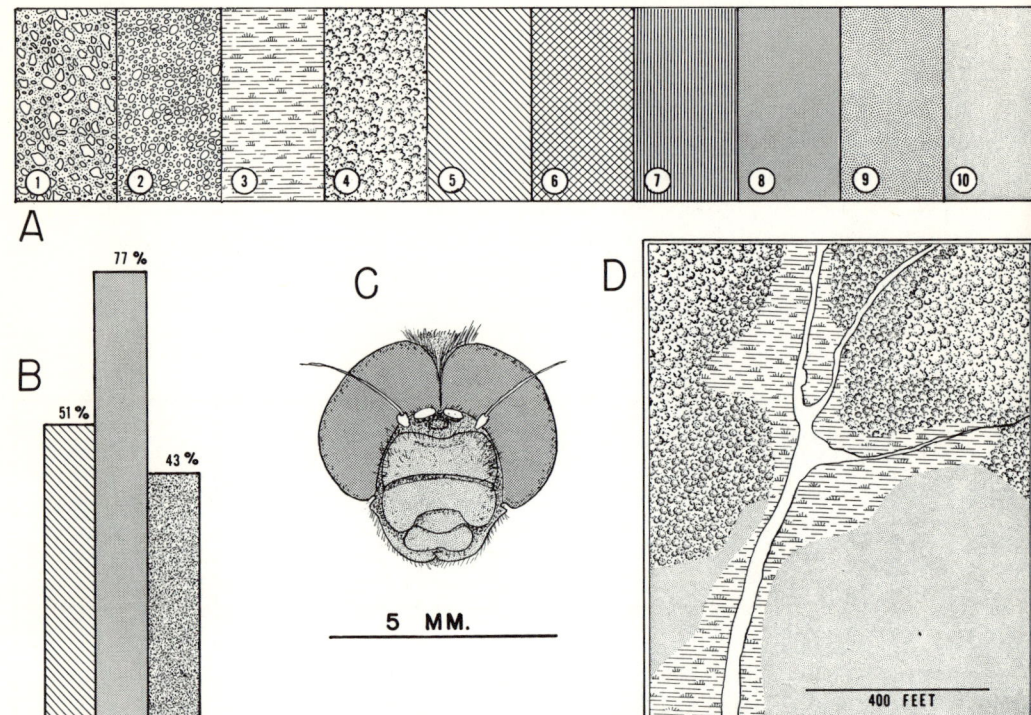

FIG. 24–19. Uses of Ben Day patterns. This figure has been reduced to 50 percent of the original size. A. The same ten patterns shown in Fig. 24–18. B. Ben Day patterns used in tables or graphs. C. Ben Day patterns used for shading. D. Ben Day patterns used for map-making.

on different grades of roughness. The standard No. 2 pencil or the 2B and 4B drawing pencils will probably suffice. You should provide also a Kneaded Rubber eraser and a Pink Pearl eraser, the former for cleaning smudges and the latter for creating highlights. Other useful items include a metal erasing template and a paper smudge (Fig. 24–20B). Finally, a can of spray fixative is essential to protect the finished drawing.

Procedure and Problems. Figure 24–20, D and E, shows two examples of how effective pencil drawings can be in biological illustration. The former drawing, which will be our example, was drawn on plate-finish drawing paper with 2B, 4B, and 6B pencils. Figure 24–20E, on the other hand, was drawn on paper with a great deal of tooth, and was therefore rendered with H, 2H, 3H, and 4H drafting pencils. A preliminary sketch was made of the

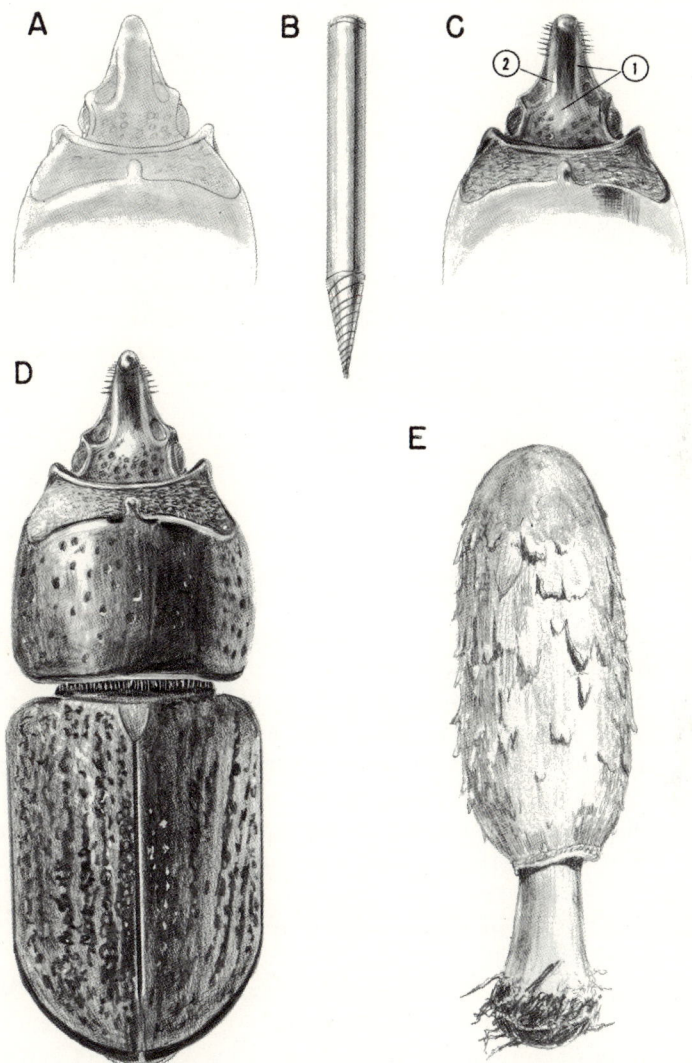

FIG. 24–20. Rendering drawings by pencil. These drawings have been reproduced by the half-tone engraving process and are reduced to about 70 percent of the original drawings. A and C. Steps in pencil rendering. B. A paper smudge used for blending. D. Finished drawing of a rhino beetle. E. An inky cap mushroom, final drawing. See text.

rhino beetle with a 4H pencil. Following this, a general light tone was applied to all areas of the drawing, except highlights, with a 2B pencil. I smoothed out the pencil marks, using both my fingers and the paper smudge (Fig. 24-20A). Next, the darker shadows and sculpturing of the body were deepened in tone with the 2B and 4B pencils. The 6B pencil was used occasionally where the darkest tones were required. Highlights were achieved by using the Pink Pearl eraser for moderate tones and opaque white, applied with a brush, for the bright tones (Fig. 24-20C). Following this, the entire drawing was given three light coats of spray fixative, used as directed on the can.

By using various pencils, erasers, and smudges you will quickly learn to render pencil drawings. Since these can be reproduced by any of the photographic processes, they are suitable for thesis and dissertation work, but are more expensive for publication, as mentioned above.

The Airbrush

The airbrush consists of a miniature spray-paint gun powered by compressed air or carbon dioxide from a cylinder. It is used commercially for the advertising arts, for technical illustration, and for photo retouching. Its application in biological work is primarily that of rendering drawings of equipment, in making wall charts, or in museum work where specimens, models, or display backgrounds require painting or toning. I have found the airbrush an almost indispensable tool in making biological displays such as that shown in Fig. A-2 (p. 496). The airbrush produces true tones of gray, and thus requires the more expensive halftone-engraving technique for reproduction in publications. In addition, airbrush rendering is not fully suitable for illustrating plant or animal specimens in that an abnormal amount of effort would be required for rendering the anatomy of a complex organism, such as the crab *Zozymus* illustrated in this chapter. For these reasons, only the simplest aspects of airbrush rendering will be given.

Several well-constructed airbrushes on the market are suitable for technical illustrations. Figure 24-21 shows an airbrush drawing of the Thayer and Chandler airbrush. Free airbrush drawings produce very soft tones which lack sharp definition. For this reason, a masking sheet is used so that sharp, clean lines of separation can be produced. The masking sheet, or frisket, is made by applying rubber cement to one surface of a sheet of tracing paper. When the rubber cement is dry, but still slightly tacky, the frisket is placed over the drawing, cement-side down. Next, the frisket is cut away with a scalpel or stencil cutter from that area of the drawing which is to receive the darkest tone. This area is then painted with the airbrush so

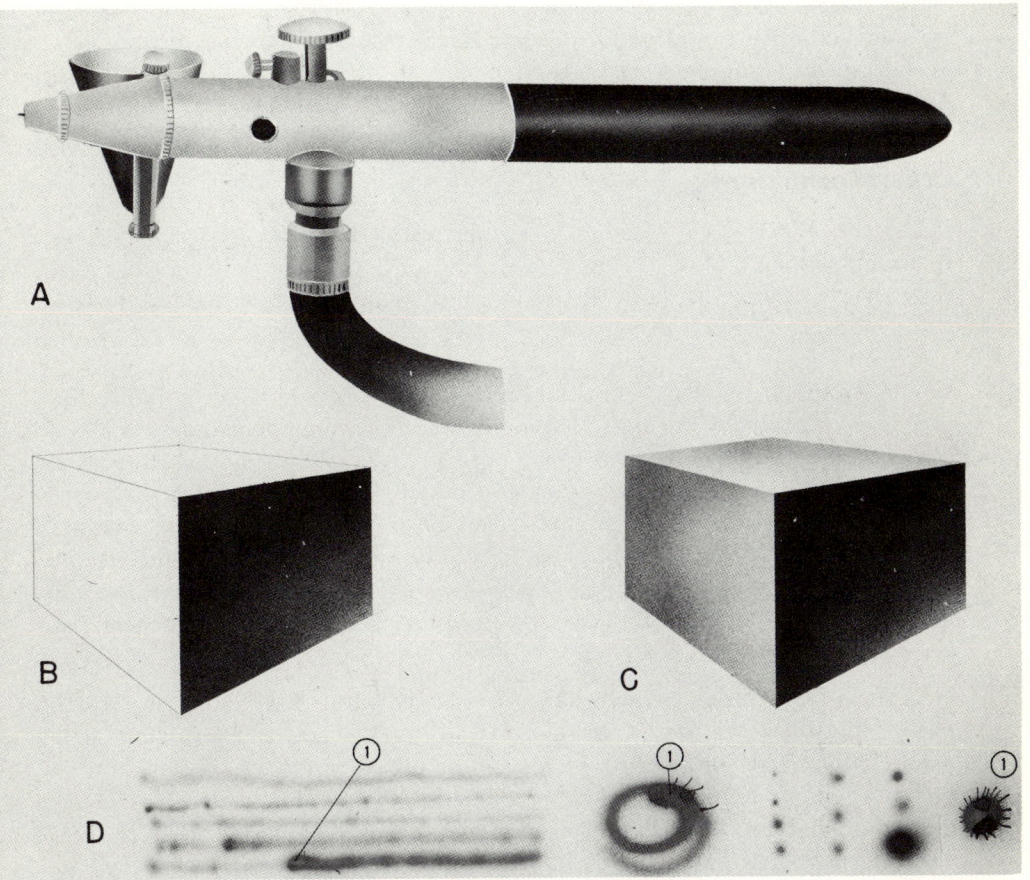

FIG. 24–21. Airbrush drawing technique. A. An airbrush drawing of an airbrush. This is the author's first attempt at a technical drawing. B–C. Shading a block with the airbrush (see text). D. Fine lines and dots obtained by the airbrush. (1) These errors were made by holding the airbrush too close to the paper and using too much air.

as to produce the tone and texture desired (Fig. 24–21B). Following this, new areas of the frisket are removed and painted; always work from the darkest toward the lightest tones, until the drawing is completed (Fig. 24–21C).

The airbrush carries a small paint cup in which the desired pigment is mixed, usually with water. The flow of air and the amount of paint released are controlled by a finger lever. The area sprayed may be controlled and adjusted, from a broad area down to relatively narrow lines, by adjustment of the nozzle (Fig. 24–21D). Thus, the amazing degree of control that can be achieved with the airbrush permits the rendering of technical drawings of biological equipment and also lends itself to painting and toning of natural history displays.

Other Techniques

The Ink Wash. Ink washes resemble airbrush drawings somewhat in their soft values (shades of gray), but are quite exacting in detail because they are rendered with both the crow-quill pen and the fine-pointed, watercolor paintbrush. Such drawings are extremely useful for illustrating protozoans and other microscopic organisms, tissues, and the like. Ink washes should be done on heavy illustration board or on any paper suitable for pen and ink which has been mounted on illustration board with rubber cement. The 4H pencil is used to lay out the preliminary drawing. India ink is diluted in small containers so as to produce a wide range of gray tones. These gray tones may be applied as washes to large areas with a paintbrush or they may be applied as stippling, fine lines, or other textures with the crow-quill pen or paintbrush. Such drawings may be reproduced photographically in dissertation work, but require the halftone engraving for publications.

Craftint Doubletone Drawing Board. The Craftint Manufacturing Company (18501 Euclid Ave., Cleveland, Ohio) makes both a Doubletone and a Singletone drawing board which is widely used in advertising art and in technical illustrations of machinery and equipment. The reader is referred to Fig. 13–5H (p. 220) for an example of Craftint Doubletone rendering. You will note that black, white, and two shades of gray are achieved by this process (in Singletone drawing board only one shade of gray may be developed). Chemical developers are painted over the ink drawing to bring out the two shades of gray where they are desired. Thus, with amazing speed, the entire drawing can be rendered and ready for engraving in a matter of minutes. A generous Student's Packet of Doubletone drawing board containing a number of Doubletone sheets, Doubletone developer, illustrations showing uses of Doubletone, a pattern chart of Singletone and Doubletone patterns, and other information may be obtained for 1 dollar from art supply stores or by writing directly to the company.

The Doubletone procedure is as follows: The initial drawing is laid out with pencil and is then inked with India ink. Next, the darkest shades of gray are brought out with the chemical developer; the excess developer is blotted away. Following this, the lighter tones are developed and blotted. Finally, any additional black areas are painted in with India ink and the drawing is then cleaned with a soft eraser.

PLATE PASTEUP TECHNIQUES

It is necessary to paste up the finished figure or plate whenever the individual drawings were executed on separate pieces of paper. If the plate size and layout were predetermined (see above) the procedure is as follows: (1) Cut a piece of heavy illustration board large enough to receive all of the individual drawings; (2) arrange the drawings in their predetermined positions; (3) trim the excess paper around the individual drawings, leaving as much of the individual paper intact as possible. After this, (4) reposition the individual drawings on the illustration board and mark their exact position with a faint blue pencil line. Next, (5) paint the entire surface of the illustration board with a coat of rubber cement and paint the backside of each individual illustration with a similar coat of rubber cement. When this has dried, (6) carefully position each individual drawing over the illustration board and, when it is in place, lower the drawing so that it contacts the illustration board. The two seemingly dry coats of rubber cement will immediately fuse and weld the picture to the illustration board. After this has been completed, (7) clean off the excess rubber cement with an eraser or, preferably, a rubber pick-up. You can make a rubber pick-up by painting rubber cement on a sheet of glass or the surface of a bottle and permitting it to dry. When it is dry, roll the cement up into a ball and repeat the process until a fairly large ball of rubber cement is attained. When this rubber pick-up is applied to the illustration board it will immediately remove any excess rubber cement from the drawing surface.

If the plate or figure size and the layout were not determined ahead of time it is essential to make a plate enlargement diagram (Fig. 24-3) on a large sheet of paper, as directed above. With this completed, arrange the individual drawings in various positions, with two objectives in mind: (1) to use as small an area as possible for the over-all figure, keeping within the proper proportions, and (2) to leave sufficient space between the drawings to prevent a crowded appearance after the plate has been reduced to the finished size. If necessary, make a light box (see p. 462) out of a large sheet of glass and arrange the drawings on this. With this technique you can see one drawing through another, when they are lighted from beneath, and can thus arrange the entire layout without trimming any of the drawings ahead of time. A second trick that will save considerable time, when the light box is used, is to place a clear polyethylene sheet of plastic over the arranged drawings, and record the area to be occupied by each drawing on the plastic with a fountain pen. The plastic sheet then serves as a guide for trimming the individual drawings. When the plate layout and dimensions are finally determined, proceed as directed in the paragraph above.

METHODS OF LETTERING AND NUMBERING

Two types of lettering and numbering procedures are applicable to biological illustrating (1) the lettering set, and (2) preprinted letters and numbers which are transferred to the drawing. A mechanical lettering set (Fig. 24–22A, courtesy of the K & E Co., Inc.) consists of a template with numbers and letters, a pen holder or scriber, and a pen with India ink. I have found the K & E Doric Lettering Set, which has 3 sizes of capital letters and numbers (Fig. 24–22B) to be suitable for all of the requirements of graduate work and publication. The K & E Leroy Lettering Set has a larger number of templates of capitals, lower-case letters and numbers, and pen-point sizes. This set permits a greater degree of flexibility in lettering.

Figure 24–22C shows a number of Zip-a-Tone and Craftint letters and numbers of the Ben Day pattern type. These are printed on clear plastic sheets and are simply cut out and transferred directly to the finished drawing. Figure 24–22D shows the Cello-Tak Transfer Type letters and numbers. These are preprinted on sheets of wax-impregnated paper and are transferred to the drawing as follows: The sheet of Transfer Type is placed over the drawing so that the desired number or letter is in its proper location. The letter or number is then transferred to the drawing by rubbing the Transfer Type paper with a lead pencil. Transfer Type differs from the Ben Day pattern in that the letter or number is transferred, in the form of a pigment, directly to the drawing without a plastic sheet. Artists' supply stores will stock or order all forms and Ben Day patterns and Transfer Type.

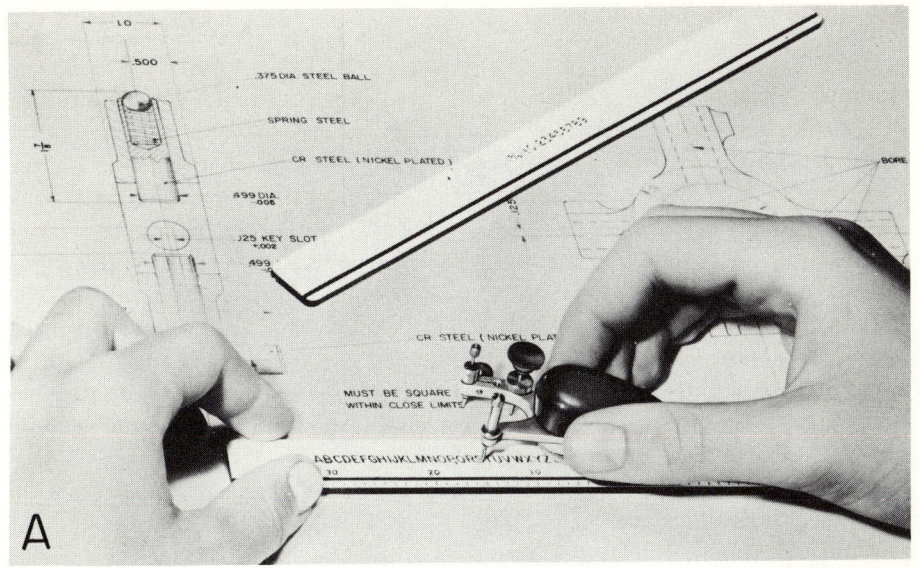

FIG. 24–22. Various lettering devices and techniques available. A. Using the K & E Leroy (trademark) lettering set, courtesy of the K & E Company. B. Some of the letters and symbols available with the K & E Doric (trademark) lettering set, actual size. C. Some of the many patterns of letters and numbers available in the Ben Day pattern series. D. Some of the many letters and numbers available in the Transfer Type (trademark) series. See text.

DENOTING SPECIMEN SIZE

It is often essential that some clear indication be given of specimen size, so that your illustrations may be put to practical use by other workers. The older method of indicating size as "× 5," "× 19½," and so on, or, conversely, "1/25" or some other fraction, is as difficult for the reader to visualize as it is for the artist to determine. For example, a number of magnifications must first be determined for the initial enlarged drawing and then redetermined for the finished reduction before the final figure can accurately be recorded in the accompanying text. Furthermore, with a "× 17¼" approximation of size it is difficult to make a "ruler" to measure the various parts of the pictured specimen. By far the simplest method of indicating size is that of placing a line beneath the drawing and indicating the exact size of that line as it applies to the drawing. Consult Fig. 24–9D and Fig. 24–19C for examples of this technique. With this method it is possible to make a "ruler" and remeasure the entire specimen if desired. Of more importance is the fact that such an indication is immediately intelligible to any reader. Furthermore, once the size notation is placed on the initial drawing no additional computations are required. Thus, when working with a large specimen simply measure some unit of the anatomy and record this unit as a line below the drawing in the exact size that it appears in the drawing. When using the Camera Lucida, the microscope-slide projector, or the opaque-projector method of layout (see discussion on pp. 457), project a stage micrometer (or a millimeter ruler for the last method) and record some unit of measurement of the drawing.

PROTECTING THE FINISHED DRAWING

Drawings which have been rendered with Ben Day patterns, stippling, texture board, scratch board, pencil, and the like should be covered with a sheet of tissue paper or plastic film to reduce surface abrasion and soiling. Second, never roll drawings; always ship them flat between several pieces of heavy cardboard or plywood. Take care to double wrap the drawings so that they cannot move around within the package or suffer from mishandling in the postal service. Write the words "Photograph—Do Not Bend" on mailing packages. Finally, register and insure packages of illustrations in order to gain a better degree of handling in transit.

REFERENCES

Allen, J. Zellers, 1948, *Airbrush Techniques,* Graphicraft Publications, Ferndale, Mich.
Bacon, C. W., 1951, *Scraperboard Drawing,* Studio Publications, N.Y.
Barnes, Robert D., 1963, *Invertebrate Zoology,* W. B. Saunders, Philadelphia.
Brown, Relis B., 1961, *Biology,* Heath, Boston.
Kautzky, Theodore, 1954, *Pencil Broadsides,* Reinhold, N.Y.
Maurello, S. Ralph, 1963, *Commercial Art Techniques,* Tudor, N.Y.
Techniques, Higgins Ink Co., Inc., Brooklyn, N.Y.

APPENDIXES

APPENDIX A

Some Display Methods

While the number of biological display methods is practically endless, only a few simple techniques which are applicable to classroom display will be given here.

SPECIMENS PRESERVED IN LIQUID

Specimens preserved in liquid are displayed in glass jars. Some specimens need not be mounted inside the jar, since they are large enough and firm enough to display themselves. Most specimens, however, are attached to a piece of glass (a glass blank) in order to show the morphology to its greatest advantage. The most desirable display jars are tall, straight-sided museum jars equipped with airtight plastic lids. The 16-ounce and 32-ounce sizes prove most satisfactory. Nothing is more distracting than an entire array of jars of different sizes and shapes on a display shelf. If museum jars are too expensive for classroom display, adopt some common jar which is used generally. For example, many brands of instant coffee come in tall, straight-

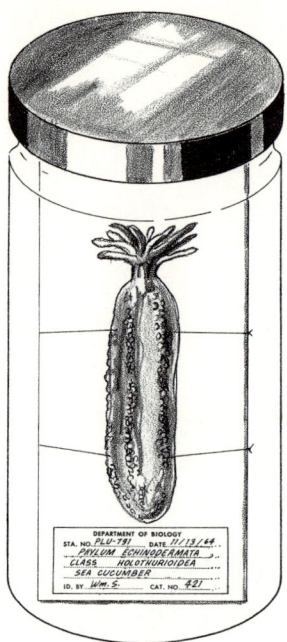

FIG. A–1.
Mounting a specimen in liquid, on a sheet of glass.

sided, plain jars. If the lids are painted black and a thin layer of soft wax (see Appendix C) is poured on the sealer, these jars prove very satisfactory.

Cut the glass blank (Fig. A–1) as large as possible, to prevent it from moving around in the jar. The blank is roughly $\frac{4}{5}$ the inside diameter of the mouth of the jar in width, and is equal to the inside height of the jar. Measure your jar, cut a piece of cardboard to the anticipated size, and test this in the jar to make sure it will lean from the lower front of the jar to the upper back of the jar. Hardware stores with modern glass-cutting tables will cut strips of glass to any desired width. Clear window glass is the most convenient and useful type. Occasionally, opaque black or white glass may enhance the appearance of some specimens. Cut all of the glass blanks to a uniform size so that display bottles will have a uniform appearance.

The specimen label may be glued to the outer surface of the specimen jar or to the lower end of the glass blank. To achieve the latter, roughen the end of the glass blank with sandpaper and glue a standard label to the glass by means of waterproof and alcohol-proof cement, such as Murrayite, Duco Cement, Bond Cement, or others. The label should be printed in India ink, and both the ink and the glue should be permitted to dry for 15 or 20 minutes. Attach the display specimen to the glass blank with black or white thread. Pass a piece of thread through the body of the specimen with a long needle, place the specimen on the glass blank, wrap the thread around the blank (see Fig. A–1) and tie the thread so that it will hold the specimen in the desired position.

Fill the bottle with 70-percent alcohol or 5-percent formalin and lower the glass blank into position. Wipe all preservative away from the mouth and lip of the jar. Finally, put on a lid which has a plastic liner, a soft wax liner, or a piece of polyethylene bag. If the last is to be used, place a piece of polyethylene over the top of the jar, screw on the lid, and cut away the surplus plastic.

DRY SPECIMENS

Modified Insect Cases

Chapter 13 discusses materials and techniques for making insect cases (with or without pinning bottoms). As noted, these can be made to any size, may be equipped with a glass top, and can be made to any depth desired. Many large dried specimens (lobsters, crabs, corals, and so on) are handsomely displayed by attaching these to a piece of heavy, white Bristol drawing board cut to fit inside of the case. Use very fine brass wire to attach the specimen, passing the wire around the specimen and through small holes, perforated in the Bristol board. When the specimen is firmly mounted and labeled, glue the Bristol board into the bottom of the display box.

Insect boxes with pinning bottoms are also excellent for making teaching displays, since labels and specimens may be attached directly to the pinning base. If specimens are properly selected they may be pinned or glued into such a display. Labels may be typed or printed on white paper, and glued to the display. Figure A–2 shows a teaching display constructed by the author.

Riker Mounts

Riker mounts consist of glass-topped boxes filled with pure white cotton upon which specimens and labels are mounted. They are very suitable for insects, plant material, and other medium-sized biological specimens. Chapter 13 gives full instructions for the construction and use of the Riker mount.

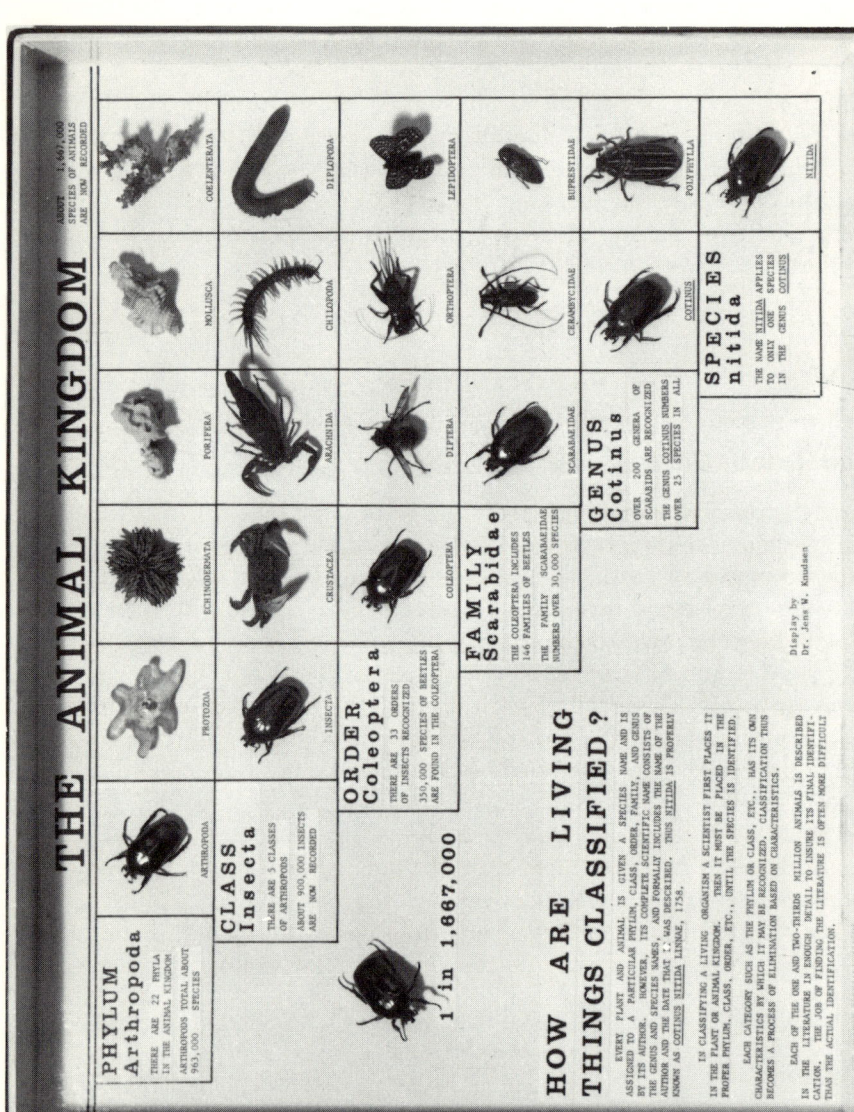

FIG. A-2. A photograph of a display made in a converted insect-mounting box.

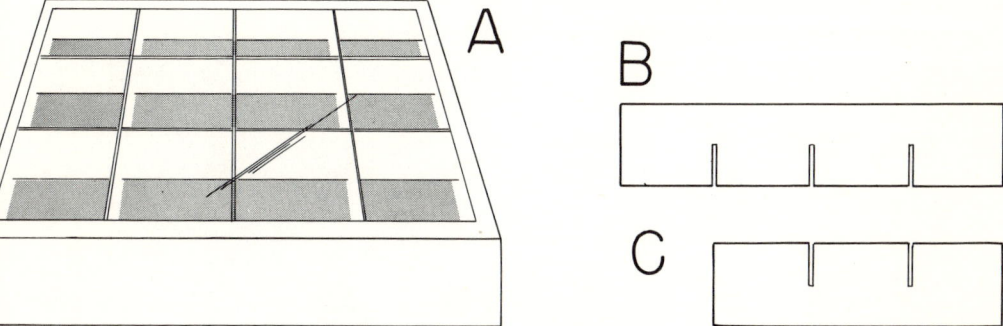

FIG. A–3. Making partitions for glass-topped display boxes.

Partitioned Boxes

Frequently, small to medium-sized specimens are best displayed in a partitioned, glass-topped box such as that shown in Fig. A–3. The box should be constructed like the insect pinning box, as described in Chapter 13. An alternative method is to use the technique for the Riker mount, also given in Chapter 13. For example, if a large supply of uniform boxes is available (such as stationery boxes which are 2 by 9 by 11½ inches), prepare these with a glass top as you would for the Riker mount; spray the outside of the box black and the inside white. Make the partitions out of heavy white cardboard, as shown in Fig. A–3.

APPENDIX B

Slide-Making

Histological techniques and some of the more elaborate whole-mount techniques are beyond the scope of this book. The main intent has been to present simple slide-making techniques for permanent preservation. There are many excellent texts dealing with histological, histochemical, histopathological and/or microbiological techniques to which you should refer. Some of these are listed in the references at the end of this section.

A DEFINITION OF TERMS AND TECHNIQUES

Fixation

"Fixation" refers to the preservation of an intact organism, or some portion of that organism, in such a way as to render the cells, tissues, or gross structures most suitable for slide preparations. The more common fixatives, listed in Appendix C, include Bouin's fixative, chrom-acidic fixative, corrosive sublimate, FAA (formol-acidic-alcohol), and Kleinenberg's solution.

When fixatives are required, these are discussed in each chapter under methods of killing and preserving specimens.

Staining

Specimens are stained with various dyes in order to make whole structures, certain tissues, certain cells, or certain cellular structures stand out in such a fashion that they may be studied when the specimen is mounted on a microscope slide. Vital stains are those which are mixed with water and can be used in moderate concentrations to stain living organisms or portions of living organisms. Obviously, such stains are very useful in natural history studies. Most stains, however, are used after fixation, while the specimen is held in water or in some solution of alcohol. The more important stains cited in this text are listed in Appendix C.

Dehydration

Dehydration in slide-making refers to the removal of water from specimens to be mounted. This is achieved by replacing the water with alcohol. Dehydration, therefore, generally implies that the specimen be placed in 35-, 50-, 75-, 85-, 95-percent, and absolute alcohol, in order to remove all traces of water. Specimens may be placed in vials or small stacking dishes for dehydration. Small specimens should be left in such vials while the alcohols are subsequently changed. If specimens are large enough, however, they may be lifted from one alcohol solution to the next by means of a wire loop. The amount of time that a specimen must remain in each solution of alcohol depends entirely on the size of the organism and the nature of its tissue. Very delicate specimens should be moved slowly through a long series of alcohol solutions, gradually increasing in strength, to prevent shrinking or contracting the tissues. It is a good practice to use at least two washes of absolute alcohol to ensure that all water is removed. A number of the mounting media listed below require little or no dehydration.

Rehydration

Rehydration simply refers to the return of a specimen to water from some concentration of alcohol. This is the reverse of dehydration and is achieved in the same way. That is to say, the specimen should be moved through increasingly diluted concentrations of alcohol to the final water level.

Clearing

The term "clearing" may be used in more ways than one in connection with slide-making. However, it generally refers to that process which follows complete dehydration and precedes the actual mounting of the specimen. The function of clearing is simply to change opaque or semiopaque material to a glassy-clear nature, so that stained structures become readily visible. Xylene, xylol, oil of cloves, oil of cedar, Terpineol, or even lactophenol may be used with different techniques.

When you add one of the clearing agents to fully dehydrated specimens note carefully whether the remaining alcohol turns slightly milky in color. This signifies that some water still remains in the specimen and in the alcohol, and that the specimen will ultimately turn black and opaque. Should water be detected, return the specimens to fresh absolute alcohol until every trace of water is removed.

Mounting Media

Some of the more common mounting media will be listed here. Others with a limited scope of use are listed separately in various chapters.

Balsam. Canadian balsam has long been a standard mounting medium. It is a true gum derived from certain trees, and is sold in a dry or in a prepared state. Specimens must be fully dehydrated and cleared before mounting in balsam. Balsam will continue the clearing process after the specimen has been mounted. The refractive index of balsam is quite suitable for most specimens, but such things as diatoms and mites become "lost" in balsam because of the similarity of the refractive indices. Balsam has a tendency to darken with age and, therefore, is sometimes replaced by many synthetic mounting media.

Balsam Substitutes. Permount, Piccolite, Kleermount, and many other synthetic mounting media are excellent as substitutes for balsam. These media may be thickened or diluted to the desired consistency, dry quickly, and do not have a tendency to darken with age. Like balsam, they are used for specimens which have been completely dehydrated and cleared prior to mounting.

Euparal. Euparal is an excellent mounting medium for most invertebrate types. It comes both in a clear and a green form (Euparal Vert, used for contrast with hematoxylin stains). Specimens are dehydrated to 90-percent alcohol, transferred to Euparal Essence, and then to Euparal itself, or directly from 90-percent alcohol to Euparal. Euparal clears and mounts simultaneously.

Glycerin Jelly. Many soft-bodied specimens are mounted in glycerin jelly by first transferring them to glycerin and then placing them in glycerin jelly. The glycerin jelly is heated upon a slide in order to receive the specimen and the coverslip. Glycerin jelly mounts are considered temporary unless the coverslip is sealed with Murrayite or some other waterproof cement.

Turtox CMC Mounting Medium. There are two forms of the Turtox mounting medium (General Biological Supply House, Chicago), CMC-10 and CMC-S. These media will kill, mount, and clear specimens, and one (CMC-S) will stain specimens after they are mounted. Thus, specimens may be transferred directly from water or alcohol into these mounting media. The author has had good success in adding various dry stains to the Turtox CMC-10 to obtain other colors. For example, fast green, orange G, eosin, methylene blue, and others work quite well.

Techniques for Balsam or Permount. In many places throughout this book references have been made to Appendix B and the methods of mounting in balsam or Permount. Thus, they will be briefly reviewed herein. Whether or not maceration or staining is required must be determined independently. Dehydrate specimens fully through absolute alcohol, clear, and mount. Use coverslip supports (broken glass slide, etc.) or plastic rings, if necessary. Add the coverslip in the manner shown in Fig. 6–3 (p. 105).

Slides and Coverslips

Although the standard 1- by 3-inch microscope slide is used for the majority of slide mounts, special microscope slides are available. Slides with concavities may be obtained from any supply house for mounting delicate or thick-bodied specimens. As an alternative (concave slides are expensive), inexpensive plastic rings or pieces of broken coverslip (or slide) may be used to support the coverslip over soft-bodied specimens. Plastic rings, which are made especially for microscope slide use, are first cemented to the slide and the medium and specimen are then added. Still another alternative to using depression slides is that of spinning a cell directly on the slide. For this, a commercial turntable is used. A glass slide is placed on the turntable, and a ring of gold size is spun on the slide to the desired

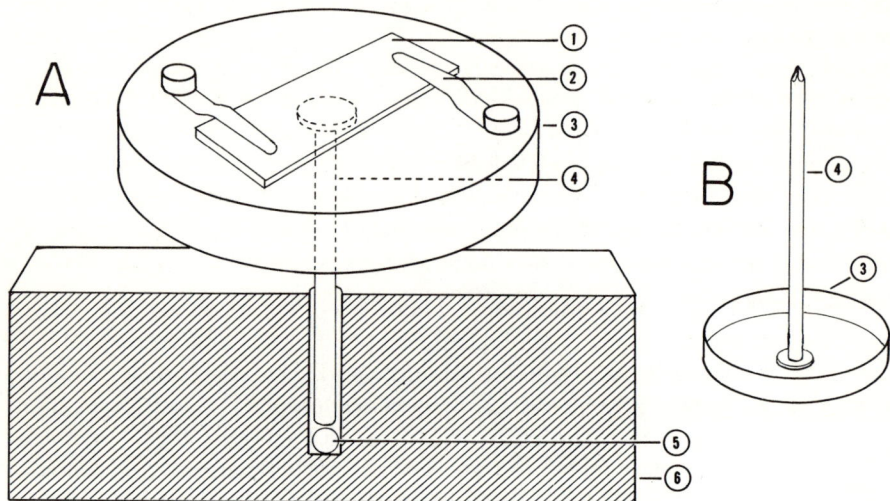

FIG. B–1. A homemade turntable for ringing slides. A. A view of the turntable with the base cut away. B. Forming the turntable. See text. (1) Glass slide, (2) microscope clip, (3) jar lid, (4) large nail, (5) ball bearing, (6) wooden block.

thickness. When this is dry, the medium and specimen are placed in the center of this ring and are covered by a glass coverslip.

You can make your own turntable with a block of wood, a ¼-inch ball bearing, a large spike, the lid of a gallon bottle, and two microscope clips, as shown in Fig. B–1. Remove the sealer from the lid, place a large spike directly in the center of the lid, and pour the lid full of lead. When the lead has hardened, cut the spike off 1½ inches below the lid. Drill a ¼-inch hole in a block of wood, insert a ball bearing, and rest the turntable upon this (Fig. B–1). Finally, drill two small holes to receive the microscope clips. This turntable will work very satisfactorily for making spun cells.

Coverslips or coverglasses come in various sizes, shapes, and thicknesses. Always select glass in preference to plastic coverslips for permanent mounts. The No. 2 thickness (or heavier) is desirable for whole mounts, whereas the No. 1 thickness is essential for thin tissue sections.

Maceration of Tissue—Arthropods

When insects and some other arthropods are mounted for the purpose of studying the exoskeletal structures, it is essential to remove all muscle and other tissues from the specimen. Tissues are macerated either by placing the specimens in a 10-percent solution of KOH for 12 or more hours or by gently boiling the specimen in KOH. In the former technique, you should check specimens periodically under the dissecting microscope to determine when all of the tissues have been removed. In the latter technique, boil the specimens in a small beaker over a hot plate or in a test tube over a Bunsen burner. Use extreme caution, as KOH may be violently expelled from the container. Therefore, always direct the test tube away from you and never hold your face near the mouth of a beaker. When specimens are sufficiently macerated, wash them in several changes of clean water and one change of 30-percent acid alcohol (see Appendix C).

REFERENCES

Baker, J. R., 1958, *Principles of Biological Microtechniques,* Methuen and Company, London.

Conn, H. J., 1953, *Biological Stains,* Biotech Publications, Geneva, N.Y.

Cowdry, E. V., 1952, *Laboratory Technique in Biology and Medicine,* Williams and Wilkins, Baltimore.

Davenport, H. A., 1960, *Histological and Histochemical Techniques,* Saunders, Philadelphia.

Galigher, A. E., 1934, *The Essentials of Practical Microtechnique,* published privately.

Gray, P., 1954, *The Microtomist Formulary and Guide,* Blakiston, Philadelphia.

Gray, P., 1958, *Handbook of Basic Microtechnique,* McGraw-Hill, N.Y.

Gridley, M. F., 1957, *Manual of Histologic and Special Staining Technics,* Armed Forces Institute of Pathology, Washington, D.C.

Gurr, E., 1956, *A Practical Manual of Medical and Biological Staining Techniques,* Interscience, N.Y.

Humason, G. L., 1962, *Animal Tissue Techniques,* Freeman, San Francisco.

Lillie, R. D., 1954, *Histopathologic Technique and Practical Histochemistry,* Blakiston, Philadelphia.

Mallory, F. B., 1944, *Pathological Technique,* Saunders, Philadelphia.

APPENDIX C

Reagents and Solutions

Only the general and broadly used reagents and solutions are listed herein. Those with limited application are treated in the text: see the Index. For a complete treatment of stains and reagents consult any of the references given in Appendix B.

ALBUMEN (MAYER'S)
    Eggwhite 50 cc.
    Glycerin 50 cc.
    Sodium salicylate 1 gm.
    Mix together and shake in a clean bottle until emulsified; filter (this may require several days).

ALCOHOL
    The word "alcohol" always refers to ethyl or grain alcohol.

ALCOHOL, ABSOLUTE
    This has *no* water; it is 100% or 200 proof. So-called "100% alcohol" (a trade term) may have up to 0.5% water.

ALCOHOL, ACID
    Alcohol of the proper percentage 100 cc.
    Hydrochloric acid, conc. 6 drops

ALCOHOL, ALKALINE
   Add a few drops of 0.1% sodium bicarbonate to the 70% alcohol wash.

AMMONIA WATER
   Ammonia (NH$_4$OH)   2 drops
   Water   500 cc.

BERLESE'S MEDIUM (GRAY, 1952)
   Water   10 cc.
   Acetic acid, glacial   3 cc.
   Dextrose syrup   5 cc.
   Gum acacia   8 gm.
   Chloral hydrate   75 gm.
   Mix water, acid, and dextrose. Dissolve the gum in this mixture (requires over a week, with occasional stirring; avoid air bubbles). When in solution, stir in the chloral hydrate.

BOUIN'S FIXATIVE
   Picric acid, saturated aqueous sol.   75 parts
   Formalin, commercial   25 parts
   Acetic acid, glacial   5 parts

CARMINE, ACETO-
   Water   25 cc.
   Acetic acid (glacial)   25 cc.
   Dry carmine stain
   Mix acid and water slowly. Add dry stain to mixture in excess of that which initially dissolves, heat to 95° C. for 10 minutes, filter. Use with an equal amount of 70% alcohol.

CARMINE, BORAX- STAIN
   Carmine   1.5 gm.
   Borax   2 gm.
   Water   50 cc.
   Alcohol, 70%   50 cc.
   Mix carmine, borax, and water, boil for 30 minutes. Add alcohol, age for 2 days, filter. Very useful for small invertebrates.

CHROM-ACETIC FIXATIVE
   Chromic acid, 1%   100 cc.
   Acetic acid, glacial   5 cc.

CHROM-OSMIC MIXTURE
   Water   99 cc.
   Chromic acid   1 cc.
   Osmic acid   1 cc. of 1%
   Acetic acid, glacial   10 cc.

CORROSIVE SUBLIMATE

Saturate water and filter (5 gm. mercuric chloride per 100 cc. water). This is used either hot (50° to 60° C.) or cold. Do not inhale fumes; wash vigorously if it contacts the skin. Do not use metal forceps or containers. For wet specimens, wash after fixing and transfer to alcohol, 50%. Add an iodine solution, drop by drop, until iodine does not lose its color. This will remove remaining corrosive sublimate and prevent specimens from turning black. For dried specimens, do not soak out before drying.

CORROSIVE SUBLIMATE, ACETIC

Saturated sol. corrosive sublimate   100 cc.
Acetic acid, glacial   5 cc.

CRYSTAL VIOLET

Crystal violet   3 gm.
Distilled water   80 cc.
Ethyl alcohol   20 cc.
Ammonium oxalate   0.8 gm.

Dissolve crystal violet in alcohol, add ammonium oxalate and water.

FAA   FORMAL-ACETIC-ALCOHOL: For plants or animals

Formaldehyde, commercial   10 parts for animals, 2 parts for plants
Alcohol, 95%   50 parts
Acetic acid   2 parts
Water   40 parts

FORMALIN

This term always refers to a solution of formaldehyde. Formaldehyde comes in a saturated solution of about 39% or 40%, and thus may be referred to as commercial formaldehyde. *Always* treat commercial formaldehyde as 100% formalin when making a formalin solution. In other words, 10 parts water and 1 part commercial formaldehyde make a 10% formalin solution. Obtainable at drugstores or biological supply houses. Ask for a commercial grade.

FORMALIN, BUFFERED NEUTRAL

For long-term storage where neutral or slightly basic (pH 7.5) formalin is required, add 6 oz. of hexamine to each quart of formaldehyde.

FORMALIN, NEUTRAL

For general use, add borax or Boraxo, check with litmus paper to be sure that a neutral or basic pH is obtained. Coral may be crushed and added to formalin if nothing else is available.

FUCHSIN, ACID
    0.5 gm. dissolved in 100 cc. of distilled water. Acidify with a drop or two of HCl before using. Excellent for Crustacea.
FUCHSIN, CARBOL-
    Solution A
        Basic fuchsin   0.3 gm.
        Ethyl alcohol (95%)   10 cc.
    Solution B
        Phenol   5 gm.
    Distilled water   95 cc.
    Mix solutions A and B.
GILSON'S FIXATIVE
    Mercuric chloride   5 gm.
    Nitric acid, 80% sol.   5 cc.
    Acetic acid, glacial   1 cc.
    Alcohol, 70%   25 cc.
    Water   220 cc.
        Filter after 3 days.
GRAM'S IODINE
    Iodine, c.p.   1 gm.
    Potassium iodide   2 gm.
    Water, distilled   300 cc.
        Grind iodine and potassium iodide in a mortar, add water, transfer to a graduated cylinder, add the remainder of the water, and mix.
GRAY AND WESS' MEDIUM (GRAY, 1952)
    Polyvinyl alcohol   2 gm.
    Acetone, 70%   7 cc.
    Glycerin   5 cc.
    Lactic acid   5 cc.
    Water   10 cc.
        This has a low refractive index.
HAUG'S SOLUTION (from Gray, 1952)
    Alcohol, 95%   70 cc.
    Water   30 cc.
    Phloroglucinol   1 gm.
    Nitric acid   5 cc.
        Heat the acid slowly in a warm water bath (never have a flame nearby) and dissolve phloroglucinol into acid. Cool and add to water. Add alcohol and mix.
HEMATOXYLIN, EHRLICH'S ACID
    Hematoxylin   2 gm.

Absolute alcohol   100 cc.
Distilled water   100 cc.
Glycerin   100 cc.
Acetic acid, glacial   25 cc.
Potassium alum   10 gm.

Dissolve hematoxylin in the alcohol and acid. Dissolve alum in heated water. Mix together. Place in stoppered bottle and age until it turns a dark red (up to several weeks). Ready for use. Keeps for years.

KLEINENBERG'S SOLUTION

Picric acid, saturated sol.   100 cc.
Sulfuric acid, concentrated   2 cc.
Distilled water   300 cc.

Mix acids, filter, add water.

METHOCELLULOSE SLIDE MOUNTING MEDIUM (after Baker and Wharton, 1952)

Methocellulose   5 gm.
Carbowax, 4,000   2 gm.
Diethylene glycol   1 cc.
Alcohol, 95%   25 cc.
Lactic acid   100 cc.
Distilled water   25 cc.

METHYL CELLULOSE

Dissolve 10 gm. of methyl cellulose in 90 cc. of distilled water.

METHYLENE BLUE

Methylene blue   0.3 gm.
Ethyl alcohol   30 cc.
Distilled water   100 cc.
Potassium hydroxide   0.01 gm.

Dissolve stain in alcohol, then mix with water and KOH.

MONK'S MEDIUM

White Karo syrup   5 cc.
Certo (fruit pectin)   5 cc.
Water   3 cc.

Make fresh or add thymol to preserve.

NOLAND'S FIXATIVE

Saturated aqueous solution   80 cc.
Formaldehyde (100% formalin)   20 cc.
Glycerin   4 cc.
Gentian violet (pre-moistened in 1 cc. of water after weighing)   20 mg.

Used with protozoans.

SAFRANIN
>   Safranin   0.25 gm.
>   Alcohol, 95%   10 cc.
>   Water   100 cc.

SALINE FOR COLD-BLOODED VERTEBRATES
>   Sodium chloride   7 gm.
>   Distilled water   1000 cc.

SALINE FOR WARM-BLOODED VERTEBRATES
>   Sodium chloride   8.5 gm.
>   Distilled water   1000 cc.

SALINE, RINGER'S SOLUTION
>   Sodium chloride   8 gm.
>   Sodium bicarbonate   0.2 gm.
>   Potassium chloride   0.2 gm.
>   Distilled water   1000 cc.
>   Calcium chloride   0.2 gm.
>   Mix calcium chloride with a little of the water. Mix the remainder together. Mix both solutions together.

SOFT WAX
>   1 part Paraffin
>   1 part Vaseline
>   For sealing coverslips—liquefy with heat and apply to edges of coverslip with a cotton swab.

VITAL STAINS
>   Methylene blue
>   Methylene green
>   Janus green B
>   Aniline yellow
>   Neutral red
>   Crystal violet
>   Bismarck brown
>   Mix with water as directed on the container.

WRIGHTS' STAIN
>   Dry Wrights' stain   0.1 gm.
>   Methyl alcohol, absolute (acetone-free)   60 cc.
>   Grind in mortar, mix, filter. Cover blood smear 3–5 minutes; next, add distilled water until a metallic film appears. Let stand 4 minutes, wash in distilled water, and dry.

APPENDIX D

Narcotizing Agents

ALCOHOL (ETHYL). Add, drop by drop, to the culture water over a period of an hour or more until a 5- to 10-percent solution is obtained. Let stand until organisms are insensitive. Otherwise, use as directed.

BENZOCAIN. A new local anesthetic obtainable at drugstores. No permit needed. Add to water as directed.

CHLORAL HYDRATE. Used like Chloretone. Obtainable at drugstores or biological supply houses.

CHLORETONE. Made by Parke-Davis. Obtainable at drugstores or at General Biological Supply House. Chicago. Add to water as directed.

CHLOROFORM. Add to culture water by placing tip of pipette under water so that the drops sink. Cover container. Obtainable at drugstores.

CLOVE OIL. Add to culture dish by placing tip of pipette under water, so that drops sink to bottom. Useful for quieting shrimp and narcotizing vertebrates and invertebrates. Obtain at drugstores or biological supply houses.

EPSOM SALTS. This crude form works as well as, or better (echinoderms) than, the pure form, magnesium sulfate. Obtainable at markets or drugstores. Add crystals to water as directed for specific animals.

MENTHOL. Add crystals to culture water as directed for specific animals;

cover container. Obtainable at drugstores and biological supply houses.

NICOTINE. May be useful for many invertebrates. For microscopic organisms, fill a bottle with cigarette smoke and cover opening with an inverted slide containing a drop culture. For large organisms, place tobacco in culture water.

STALE OR PUTRID WATER. Habitat water previously boiled, or provided with some organic material, and allowed to stand at room temperature, will kill invertebrates, often in an expanded condition, as the oxygen is used up.

INDEX

Acanthocephala, 130, 152
  collecting, 152
  preserving, 153
  slide mounts, 153
Acid fuchsin, 274
Aeration, 16
Agar, 40
Airbrush, 482
Albumen, 504
Alcohol, 504, 510
  absolute, 504
  acid, 504
  alkaline, 505
Algae, 26
  collecting fresh-water, 33
  collecting marine, 29
  coralline, 32
  culturing, 34
  encrusting, 32
  fresh-water, 33
  herbarium mounts, 30
  liquid preservation of, 30
  marine, 27
  morphology of, 27
  preserving, 35
  rolled specimens, 31
  storage methods, 36
Ammonia water, 505
Amoeba, 103
Amphibians, 346
  characteristics, 346

Amphibians (*Continued*)
  collecting, 349
  conservation, 348
  dry skins and color, 356
  field data, 348
  labeling and records, 355
  narcotizing, 354
  preserving, 353, 355
  shipping, 357
  skeletal techniques, 358
  storage, 356
  transporting, 353
Amphineura, 183
Amphioxus, 315
  collecting, 315
  preserving, 316
Amphipods, 272
Anemones, 125
Angiospermae, 72
Annelida, 156
Anoplura, 234
Anostraca, 271
Araneae, 286
Arrow worms, 165
Aschelminthes, 130, 141
Aspirators, 212
Asteroidea, 301
Aurelia, 130

Bacteria, 39
  cleaning glassware, 44

516   Index

Bacteria (*Continued*)
  collecting, 45
  culture destruction, 50
  culture storage, 50
  culturing, 39
  equipment, 42
  incubator, 43
  identification, 40
  media, 40
  pour plate, 48
  staining, 49
  sterilization, 42, 44
  streak plate, 47
Baermann funnel, 149
Balsam, 500
  substitutes, 500
Bark, 75, 87
  mature, 88
Barnacles, 278
Bats, posturing skins, 421
  trapping, 407
Beetles, 233
Ben Day patterns, 477
Benzocain, 510
Berlese funnel, 213
Berlese's medium, 296, 505
Beroe, 125
Bird mounting, blood, 380
  determining sex, 380
  fat skins and grease, 380
  mounting kit, 379
  mounting methods, 379
  problems to anticipate, 380
  skinning techniques, 381
Birds, 373
  banding, 399
  care of specimens, 375
  collecting techniques, 374
  display tubes, 393
  ectoparasites, 381
  eggs, 392
  field data, 378
  field equipment, 396
  field identification, 397
  field specimens, 378
  field study, 396
  flat skins, 390
  killing injured, 376
  labeling, 390

Birds (*Continued*)
  large-headed, 386
  laws and permits, 374
  measuring skins, 395
  nesting studies, 398
  nests, 391
  parts for teaching, 391
  shooting, 375
  skeletal techniques, 392
  storage of skins, 395
  stuffing, 387
  trapping and netting, 377
  wet, dry mounts, 390
  wings and feet, 387
Blood smear, 101
Bony fishes, 318
Borax-carmine, 134
Bouin's fixative, 505
Brachiopoda, 187
  collecting, 187
  preserving, 188
Brachiopods, 167
Bryozoa, 193
  burrowing, 199
  characteristics, 193, 200
  collecting, 194, 201
  cultures, 202
  dry slide, 198
  fresh-water, 200
  intact-colony slides, 202
  narcotizing, 197, 201
  preserving, 197, 201
  slide mounts, 198
  trapping, 195
Butterflies, 235, 240
Butterfly boards, 228

Cabinets, bird and mammal, 428
  insect, 251, 257
  shell, 189
Cactus, 81
Caecilians, 350
Camera lucida, 457, 459
Carmine, aceto, 505
  borax-stain, 505
Cartilaginous skeletons, 443
Case pelts, 423
Case-skin mounts, mammals, 423
Centipeds, 297

Cephalochordata, 315
Cephalopoda, 185
  characteristics, 185
Cercaria, 136
Cestodaria, 138
Cestodarians, collecting, 138
  preserving, 138
Cestoidea, 138
Chaetognatha, 165
Chitons, 183
  characteristics, 183
  collecting, 183
  preserving, 184
  radula, 184
Chloral hydrate, 510
Chloretone, 354, 510
Chlorocresol crystals, 237
Chlorocresol method, 225
Chloroform, 510
Chrom-acetic fixative, 505
Chrom-osmic mixture, 505
Cirripedia, 278
  narcotizing, 278
  preserving, 278
Cladocera, 271
Clam shrimps, 271
Clams, 179
  anatomy, 179
  fresh-water, 181
  marine, 180
  preserving, 182
  shells, 182
Clearing, 500
Clove oil, 159, 510
Club mosses, 71, 76
Coelenterata, 119
Coleoptera, 234
Collecting, bacteria, 45
  higher fungi, 56
  lichens, 52
  molds, 45
Collecting methods, aquatic, 20
  exposed intertidal zones, 7
  night light, 14
  pelagic, 13
  subintertidal, 8
Collecting sites and techniques, 1
Collembola, 234
Color notes, 23

Conchostraca, 271
Cones, 87
Conifers, 72
Conservation, need for, 1
Copepods, 271
Copying processes, 449
Corals, 127
Corrodentia, 234
Corrosive sublimate, 506
  acetic, 506
Crab traps, 281
Crabs, 280
Crayfish, 280
Crinoidea, 308
Crocodilians, 370
  killing, 370
Crustaceans, 266
  collecting, 280
  collecting methods, 267
  dissection, 276
  dried specimens, 284
  fluid mounts, 275
  killing, 273
  large, 280
  larval, 272
  narcotizing, 273
  preserving, 283
  preservatives, 266
  rearing, 277
  skeletal studies, 285
  slide mounts, 273
  small, 267
  traps, 281
Crystal violet, 506
Ctenophora, 119, 128
Culture destruction, 50
Culture methods, protozoa, 101
Culture tubes, 42
Culturing, algae, 34
  bacteria, 39
  mold, 39
Cyclostomes, 318

Damsel flies, 237
Dehydration, 499
Dermal ossicles, 309
  sea cucumber, 309
Dermaptera, 234
Diatoms, 35

Difflugia, 103
Diplura, 234
Dipper, 212
Diptera, 234
Display methods, 493
    dry specimens, 495
    partitioned boxes, 497
    specimens in liquid, 493
Display tubes, birds, 393
Dissection instruments, 250
Dragonflies, 235
Drawing paper, illustration, 465
Dredges, 9
    biological, 10
Dredging, 9
Droppings, mammal, 426
Dryers, botanical, 80
Drying oven, echinoderm, 303

Earthworms, 160
    collecting, 161
    narcotizing, 162
Echinoderms, 299
    collecting, 299
    drying oven for, 303
    larvae, 306
    mop drag, 300
Echinoidea, 306
Echinuroidea, 164
Eggs, trematoda, 136
Electric shockers, fish, 333
Embioptera, 234
Entoprocta, 130
Envelopes, insect, 224
Environment, marine, 2
Ephemeroptera, 234
Epsom salts, 510
Eucestoda, 138
Euglena, 98, 104
Euparal, 500

FAA, 506
Fairy shrimps, 271
Feather stars, 308
Ferns, 72, 76
Field notebooks, 22
Field notes, 22
Fish, 318
    clearing and staining, 435

Fish (*Continued*)
    collecting, 321
    color preservation of, 340
    drying, 342
    electric shockers, 333
    measurements, 320
    morphology, 319
    nets, 323
    otoliths, 320, 341
    poisons, 335
    preserving, 339
    puffers, 341
    scales, 320, 340
    skeletal techniques, 343
    skeleton, 435
    taxonomic features, 318
    transportation of, 338
    trap net, 332
    traps, 331
Flower, aquatic, 81
    color, 91
Fluid mounts, crustaceans, 275
Foraminifera, 110
    collecting, 100
    dry-mounting, 111
Foreshortening, in illustrations, 456
Formal-acetic-alcohol, 506
Formalin, 506
    buffered, 506
    neutral, 506
Free-living, 271
Freeze-drying technique, 250
French curve, 461
Fresh-water environment, 18
    currents, 19
    food, 19
    light, 18
    oxygen and carbon dioxide, 19
    substrate, 19
    temperature, 18
Frog calls, 352
Frogs, 250
Fruit, 75, 85
Fuchsin, acid, 507
    carbol-, 507
Fumigation, mammal, 427
Fungi, 38
Funnel trap, 215, 281

Gastropoda, 168
Gastrotricha, 141, 145
  collecting, 145
  culturing, 145
  narcotizing, 146
  preserving, 146
Gilson's fixative, 507
Glochidia, 182
Glycerin jelly, 151, 501
  mounts, 144, 276
Gophers, 408
Gorgonians, 126
Gram's iodine, 507
Gram's stain, 50
Grasshoppers, 233
Gray and Wess' medium, 296, 507
Gregarina, 104
Gymnospermae, 72

Haemogergarines, 101
Hagfish, 318
Halftone engraving, 448
Hard-head X ray, 437
Haug's solution, 507
Heavy liquids, 111
Helix, 170, 177
Hematein, 140
Hematoxylin, Ehrlich's acid, 507
Hemichordata, 311
Hemichordata, collecting, 311
  preserving, 312
Hemiptera, 233, 234
Herbarium, 74
  insect pests, 93
  paper, 83
  plastic mounting, 85
  storage, 94
Hermit crabs, 284
Heteropod, 170
Higher fungi, 54
  collecting, 56
  preserving, 57
Hirudinea, 163
  collecting, 163
Holothuroidea, 308
Homoptera, 233, 234
Hook and line fishing, 337
Hornworts, 67
Horny corals, 126

Horsehair worms, 151
Horsetails, 72, 76
Hydra, 119
  collecting, 122
Hydroids, colonial, 122
  narcotizing, 123
  preserving, 123
Hydroid polyps, 123
Hydrozoa, 124
Hymenoptera, 234

Immature insects, 218
Incubator, 43
Inkwash, 484
Insects, 204
  adults, 222
  boards, 230, 243
  cabinets, 251, 257
  carding, 239
  case layout, 254
  collecting, 222
  collecting sites, 217
  coverslip mount, 239
  dissection of, 249–251
  eggs, 249
  envelopes, 224
  evisceration and stuffing, 235
  field data, 227
  field drying, 226
  field pinning, 223
  field storage, 219, 223
  genitalia, 248
  juveniles, 222
  killing, 219
  labels, 231
  larvae, 249
  larval inflation, 245
  liquid storage, 255
  mounting, 227, 233
  mounting equipment, 227
  mouth parts, 248
  papering, 224
  pin removal, 245
  pinning, 233
  pinning blocks, 230
  pinning dowel, 230
  pinning forceps, 231
  pins, 201
  pointing, 238

Insects (*Continued*)
   preserving, 219
   rearing, 261
   rearing cages, 262
   relaxing chamber, 237
   relaxing specimens, 237
   Riker mounts, 243
   shipment, 261
   slide mounts, 247
   storage, chlorocresol method, 225; liquid, 224; problems in, 259; techniques, 251
   techniques for various orders, 234
   trapping, 214; tin can, 215
   unit trays, 255
   where to pin, 233
   whole mounts, 248
   winged, 240
Intertidal zones, 4
Isoptera, 234

Jellyfish, 120, 124

Killing bottle, disposal, 220
Kinorhyncha, 141
Kleermount, 500
Kleinenberg's solution, 508

Labels, 92
Lamp shells, 187
Lampreys, 318
Larvae, trematoda, 136
Leaf characteristics, 89
Leaf venation, 89
Leech, 163
   collecting, 163
Lepidoptera, 234
Lettering illustrations, 486
Lichens, 51
   collecting, 52
   preserving, 53
Light trap, 216
Limax, 170, 177
Limpets, 170
Line cut, 448
Line drawing, 465
Liverworts, 62, 67
   collecting, 68

Liverworts (*Continued*)
   liquid preservation, 69
   preserving, 68
Lizards, 362
   arboreal, 363
   fixing, 367
   labeling, 367
   positioning, 367
   preserving, 366, 368
   snare, 363
Lobsters, 280
Lower chordates, 311
Lycopsida, 71

Mallophaga, 234
Mammals, 402
   bloodstains, 413
   can traps, 406
   case pelts, 423
   case-skin mount, 423
   collecting techniques, 402
   commercial live-catch traps, 406
   determining sex, 413
   droppings, 426
   ears, 418
   fat skins, 412
   field skins, 409
   flat pelts, 421
   fleshy feet, 417
   fumigation, 427
   head, 416
   killing injured specimens, 409
   labeling, 420
   laws and permits, 403
   liquid preservation, 424
   measurements, 410
   mounting, 410
   mounting kit, 412
   round mount, 413
   shooting specimens, 408
   skeletal technique, 427
   skeleton or skull, 421
   skinning, 413
   snap traps, 403
   snares, 408
   steel traps, 405
   stuffing, 418
   tails, 414, 416
   tracks, 425

Mammals (*Continued*)
  trapping, 403
  wet, dry mounts, 425
Marine environment, 2
  bays and inlets, 6
  currents, 3
  exposed sand, 6
  light, 4
  other factors, 4
  other zones or habitats, 6
  rocky coasts, protected, 5; unprotected, 4
  substrate, 3
  temperature, 4
  tides, 3
  waves, 3
Mechanical drawing, 460
Mecoptera, 234
Medusae, 124
  collecting, 124
  narcotizing, 124
  preserving, 124
Menthol, 510
Merostomata, 297
Methocellulose, medium, 508
Methyl cellulose, 508
Methylene blue, 508
Microscope-slide projectors, 457
Millipedes, 297
Minnow trap, 331
Minuten nadeln, 238
Miracidium, 136
Mites, 295
  collecting, 295
  preserving, 296
  slide mounts, 296
Molds, 39
  collecting, 45
  culturing, 39
  microscopic study of, 51
Moles, 408
Mollusks, 167
  storage, 188
Monk's mounting medium, 276, 508
Mosses, 62
  bulk-dry, 66
  collecting, 64
  herbarium mounts, 66
  liquid preservation, 67

Mosses (*Continued*)
  preserving, 64
  storage, 64
Moths, 235, 240
Mounting insects, 227
Mounting media, 500
Mounting medium, Berlese's, 296
  Gray and Wess', 296
Museum special trap, 403
Mushrooms, 54
  drying techniques, 58
  liquid preservation, 58
  morphology, 54
  preserving, 57
  spore print, 57
  storage, 59

Nematocysts, 122
Nematoda, 141, 147
  characteristics, 147
  fixing, 150
  parasitic, 148
  plant, 150
  preserving, 150
  soil, 149
  study of, 148
Nematomorpha, 151
  collecting, 151
  preserving, 152
Nemertea, 130, 140
  collecting, 140
  narcotizing, 141
  preserving, 141
Nereis, 160
Net, aerial, 208
  aquatic-dip, 211
  beat, 208
  Birge cone, 268
  construction, 205
  dredge, 211
  fabrics, 205
  fish, 323
  fyke, 331
  ground, 210
  handle, 207
  mesh size, 324
  otter trawl, 325
  plankton, 13
  seines, 324

522 Index

Net (*Continued*)
  stream, 212
  surface trawl, 330
  sweep, 208
  umbrella, 208
Neuroptera, 234
Nicotine, 511
Noland's fixative, 508
Notostraca, 271
Nudibranches, 170
Nudibranchia, 177

Obelia, 120, 124
Octopi, 185
  narcotizing, 186
Ocular micrometers, 457
Odonata, 234
Oligochaeta, 160
  collecting, 160
  narcotizing, 161
  preserving, 163
Opaque projector, 459
Operculum, 174
Ophiuroidea, 301
Orange-peel buckets, 11
Orthoptera, 234
Otter trawl, 325
Oysters, 179

Pantograph, illustration, 460
Paramecium, 103
Parasitic insects, 219
Pelecypoda, 179
Pedicellaria, sea stars, 305
Pelts, flat, 421
Pencil techniques, 478
Periostracum, 174
Permount, 500
Perspective, illustration, 454
Philodina, 142
Phoronida, 164
Photocopying processes, 449
Physalia, 120, 125
Piccolite, 500
Pinning blocks, insects, 230, 235
Pinning dowel, insect, 230
Pipette, fine, 102
Pistols, .22-caliber, 363
Planaria trap, 132

Plankton nets, 13
  slide preparation, 15
  treatment, 15
Plants, chlorophyll, 91
  collecting, 75
  color of, 91
  display methods, 92
  field drying, 81
  field notes, 82
  flowering, 72
  higher, 71
  identification of, 83
  labels for, 92
  leaves, 89
  liquid preservation of, 91
  mounting, 83
  press, 76, 80
  storage, 93, 94
  succulent, 81
Plaster cages, 262
Plate layout, illustration, 452
Plate size, illustration, 452
Platyhelminthes, 130
Plecoptera, 234
Pleurobrachia, 128
Poison bottles, 219
  cyanide, 219
  cyan-o-gas, 221, 237
  liquid, 220
  use of, 221
Poisons, fish, 335
Pollen, 86
Polychaeta, 156
  collecting, 157
  narcotizing, 158
  preserving, 160
  slide mounts, 160
Porifera, 114
Preservation, bulk field, 14
  lichens, 53
Priapuloidea, 141, 165
Protected rocky coasts, 5
Protozoa, 97
  amoeboid, 103
  ciliate, 103
  collecting, 98
  culture methods, 101
  enzyme activity, 107
  euglenoid, 104

Protozoa (*Continued*)
  fecal smears, 105, 110
  fresh-water, 98
  of insects, 104
  isolating specimens, 110
  marine, 99
  parasitic, 100
  permanent slides of, 108
  slowing movement, 106
  stain-kill methods, 108
  study of, 104
  termites, 104
  wet mount, 104
Protura, 234
Pseudoscorpions, 292, 293
Pteropsida, 72
Publications, 448
Puffers, fish, 341
Pure culture techniques, 46

Radiolaria, 110
Radula, 174, 175, 184
Rays, 318
Rearing cages, insect, 262
Rearing insects, 261
Redial stage, 136
Rehydration, 499
  dried specimens, 267
Rendering, 463
Reptiles, 359
  collecting, 360
  collecting equipment for, 361
  field data, 360
  shipping, 371
  skeletal techniques, 371
Rhizocephala, 279
Ribbonworms, 140
Riker mounts, 243, 495
Ringer's solution, 509
Rose bengal stain, 99
Rotifera, 141, 142
  anatomy, 143
  collecting, 142
  culturing, 142
  fixing, 143
  glycerin jelly mount, 144
  narcotizing, 143
  slide mounts, 144
  trophi, 145

Roundworms, 147
Ruler, wooden, 460

Sacculina, 279
Safranin, 509
Salamanders, 350
Saline, 509
Samplers, bottom, 11
  scoop, 11
Sand dollars, 306
Sand screen, 157
Sandworms, 156
Scaphopoda, 184
Scat, 426
Scientific illustration, 446
  airbrush, 482
  Ben Day patterns, 477
  correcting mistakes, 464
  denoting size, 488
  direct measurement, 458
  drawing paper, 465
  foreshortening, 456
  grid techniques, 460
  ink wash, 484
  lettering, 486
  pantograph, 460
  pencil techniques, 478
  pens, 465
  perspective, 454
  plate layout, 452
  plate pasteup, 485
  plate size, 452
  protecting drawings, 488
  publication, 448
  smooth scratch board, 475
  steps of, 447
  stippling, 466
  textured-board, 471
  textured scratch board, 473
  theses and dissertations, 449
  transferring drawings, 462
Scorpions, 286, 292
Screen, sand, 157
Scyphozoa, 124
Sea-corn, 170
Sea cucumbers, 299, 308
  dermal ossicles, 309
  preserving, 300
Sea lilies, 308

Sea pens, 126
Sea stars, 299
   drying methods, 302
   histological study, 305
   narcotization of, 302
   pedicellaria, 305
   skeletal techniques, 306
   storage, 306
   tube-feet, 305
Sea urchins, 306
   drying methods, 307
   positioning spines, 307
   preservation of, 306
Sea walnuts, 128
Sea whips, 126
Seeds, 85
Sharks, 318
Single-lens reflex camera, 451
Siphonaptera, 234
Siphonophora, 125
Sipunculoidea, 164
Skeletal techniques, 432
   amphibians, 436
   birds, 436
   boil and clean, 440
   cartilaginous, 443
   cleaning, 442
   clearing, 433
   degreasing, 441
   dermestid, 439
   fleshing out, 438
   maceration in water, 440
   mounting, 442
   staining, 433
   standard method, 438
Skeletal X ray, 436
Skeleton, amphibians, 358
   birds, 392
   crustacea, 285
   echinoderms, 306
   mammals, 427
   reptiles, 371
   sponge, 116
Skin diving, 337
Slide, glycerin jelly, 144, 151
   hanging-drop, 105
   making, 498; fixation, 498; staining, 499; terms, 498

Slide (*Continued*)
   smears, 109
   technique, 105
   wet mounts, 103
Slugs, 168, 177
Smooth scratch board, 475
Snails, 168
   collecting, 169
   dry shells, 172
   egg laying, 178
   fresh-water, 171
   narcotizing, 175, 177
   preserving, 178
   radula, 174
   terrestrial, 171
   traps, 171
Snake bags, 361
Snakes, 364
   arboreal, 363
   fixing, 367
   forceps, 364
   labeling, 367
   positioning, 367
   preserving, 366, 368
Soft-head X ray, 437
Soft wax, 509
Specimens, shipping dry, 21
   shipping preserved in liquid, 21
Sphenopsida, 72
Spicules, 117
Spiders, 286
   collecting, 286
   collecting materials, 287
   collecting webs, 289
   egg sacs, 289
   photographing webs, 290
   preserving, 289
   tarantulas, 288
   trap-door, 288
   true, 286
   tuning-fork technique of capturing, 288
Spiny-headed worms, 152
Sponge, 114
   collecting, 115
   culture, 117
   dry, 116
   fixation, 116
   preserving, 115

Sponge (*Continued*)
  skeleton, 114
  transport, 117
Spongin, 116
Sporocyst, 136
Squids, 185, 186
  narcotizing, 186
Stain, borax-carmine, 134
  hematein, 140
  rose bengal, 99, 109
Stale or putrid water, 511
Stippling, illustrations, 466
Storage cabinet, plants, 94
Storage problems, insects, 259
Sun spiders, 292, 293
Surface trawl, 330

Tadpole shrimps, 271
Tapeworms, 138
  preserving, 139
  primitive, 138
Techniques, collecting, 1
Textured-board, 471
Textured scratch board, 473
Thermometers, 13
Thysanoptera, 234
Thysanura, 234
Ticks, 294
  collecting, 294
  preserving, 294
  slide mounts, 294
Tissue maceration, 502
Tracheophyta, 71
Tracks, mammal, 425
Transporting live material, 15
Traps, fish, 331
  museum special, 403
Trawl, otter, 325
  surface, 330
Trematoda, 135
  collecting, 135
  eggs, 136
  larvae, 136
  monogenetic, 136
  narcotizing, 136
Triangles, drawing, 461

Trichonympha, 104
Trichoptera, 234
Trophi, 145
Tube worms, 156
Tubularia, 120
Tunicates, 313
  collecting, 313
  preserving, 314
  sessile, 313
Turbellaria, 131
  culturing, 132
  fresh-water, 131
  marine, 132
  preserving, 133
  slide preparation, 134
  study of, 133
Turntable, 502
Turtles, 369
  collecting, 365
  dried specimens, 370
  killing, 369
  preserving, 369
  traps, 365
  wet-dry specimens, 370
Turtox, CMC, 501
Tusk shells, 184

Unit trays, 191, 255
Unprotected rocky coasts, 4
Urochordata, 313

Vasculum, 74
Velella, 125
Ventilators, 80
Vital stains, 107, 509

Wasps, 235
Water fleas, 271
Webs, spider, 289
Wire loop, 42
Wire needle, 42
Wood, 87
Wrights' stain, 509

X-ray technique, 436

Zonation, 17